Color Quality of Semiconductor and Conventional Light Sources

Color Quality of Semiconductor and Conventional Light Sources

Tran Quoc Khanh, Peter Bodrogi, and Trinh Quang Vinh

Authors

Prof. Dr. Tran Quoc Khanh
Technische Universitaet Darmstadt
Lab of Lighting Technology
Hochschulstraße 4A
64289 Darmstadt
Germany

Dr.-Ing. Peter Bodrogi
Technische Universitaet Darmstadt
Lab of Lighting Technology
Hochschulstraße 4A
64289 Darmstadt
Germany

Dr.-Ing. Trinh Quang Vinh
Technische Universitaet Darmstadt
Lab of Lighting Technology
Hochschulstraße 4A
64289 Darmstadt
Germany

Cover credit
Eye Image from DigitalVision Ltd.

All books published by **Wiley-VCH** are carefully produced. Nevertheless, authors, editors, and publisher do not warrant the information contained in these books, including this book, to be free of errors. Readers are advised to keep in mind that statements, data, illustrations, procedural details or other items may inadvertently be inaccurate.

Library of Congress Card No.: applied for

British Library Cataloguing-in-Publication Data
A catalogue record for this book is available from the British Library.

Bibliographic information published by the Deutsche Nationalbibliothek
The Deutsche Nationalbibliothek lists this publication in the Deutsche Nationalbibliografie; detailed bibliographic data are available on the Internet at <http://dnb.d-nb.de>.

© 2017 Wiley-VCH Verlag GmbH & Co. KGaA, Boschstr. 12, 69469 Weinheim, Germany

All rights reserved (including those of translation into other languages). No part of this book may be reproduced in any form – by photoprinting, microfilm, or any other means – nor transmitted or translated into a machine language without written permission from the publishers. Registered names, trademarks, etc. used in this book, even when not specifically marked as such, are not to be considered unprotected by law.

Print ISBN: 978-3-527-34166-5
ePDF ISBN: 978-3-527-80344-6
ePub ISBN: 978-3-527-80346-0
Mobi ISBN: 978-3-527-80347-7
oBook ISBN: 978-3-527-80345-3

Cover Design Schulz Grafik-Design, Fußgönheim, Germany
Typesetting SPi Global, Chennai, India
Printing and Binding Markono Print Media Pte Ltd, Singapore

Printed on acid-free paper

Contents

Preface *xi*

1 **Introduction** *1*
References *9*

2 **Color Appearance and Color Quality: Phenomena and Metrics** *11*
2.1 Color Vision *11*
2.2 Colorimetry *16*
2.2.1 Color-Matching Functions and Tristimulus Values *17*
2.2.2 Chromaticity Diagram *19*
2.2.3 Interobserver Variability of Color Vision *20*
2.2.4 Important Concepts Related to the Chromaticity Diagram *21*
2.2.5 MacAdam Ellipses and the $u' - v'$ Chromaticity Diagram *24*
2.3 Color Appearance, Color Cognition *26*
2.3.1 Perceived Color Attributes *26*
2.3.2 Viewing Conditions, Chromatic Adaptation, and Other Phenomena *28*
2.3.3 Perceived Color Differences *29*
2.3.4 Cognitive Color, Memory Color, and Semantic Interpretations *29*
2.4 The Subjective Impression of Color Quality and Its Different Aspects *31*
2.5 Modeling of Color Appearance and Perceived Color Differences *35*
2.5.1 CIELAB Color Space *36*
2.5.2 The CIECAM02 Color Appearance Model *37*
2.5.3 Brightness Models *41*
2.5.3.1 The CIE Brightness Model *43*
2.5.3.2 The Ware and Cowan Conversion Factor Formula (WCCF) *44*
2.5.3.3 The Berman *et al*. Model *44*
2.5.3.4 Fotios and Levermore's Brightness Model *45*
2.5.3.5 Fairchild and Pirrotta's L^{**} Model of Chromatic Lightness *45*
2.5.4 Modeling of Color Difference Perception in Color Spaces *45*
2.5.4.1 CIELAB Color Difference *45*
2.5.4.2 CAM02-UCS Uniform Color Space and Color Difference *46*
2.6 Modeling of Color Quality *48*

2.6.1	Color Fidelity Indices 49
2.6.1.1	The CIE Color-Rendering Index 49
2.6.1.2	The Color Fidelity Index of the CQS Method 52
2.6.1.3	The Color Fidelity Index CRI2012 (nCRI) 53
2.6.1.4	The Color Fidelity Index R_f of the IES Method (2015) 56
2.6.1.5	RCRI 57
2.6.1.6	Summary of the Deficiencies of Color Fidelity Metrics 57
2.6.2	Color Preference Indices 57
2.6.2.1	Judd's Flattery Index 57
2.6.2.2	Gamut Area Index (GAI) in Combination with CIE R_a 58
2.6.2.3	Thornton's Color Preference Index (CPI) 58
2.6.2.4	Memory Color Rendition Index R_m or MCRI 58
2.6.2.5	The Color Preference Indices of the CQS Method (Q_a, Q_p) 60
2.6.3	Color Gamut Indices 61
2.6.3.1	The Color Gamut Index of the CQS Method (Q_g) 62
2.6.3.2	The Feeling of Contrast Index (FCI) 62
2.6.3.3	Xu's Color-Rendering Capacity (CRC) 62
2.6.3.4	Gamut Area Index (GAI) 62
2.6.3.5	Fotios' Cone Surface Area (CSA) Index 62
2.6.3.6	The Color Gamut Index R_g of the IES Method (2015) 62
2.6.3.7	Deficiencies of Color Gamut Metrics 63
2.6.4	Color Discrimination Indices 63
2.7	Summary 64
	References 65
3	**The White Point of the Light Source** *71*
3.1	The Location of Unique White in the Chromaticity Diagram 74
3.2	Modeling *Unique White* in Terms of L − M and L + M − S Signals 77
3.3	Interobserver Variability of White Tone Perception 78
3.4	White Tone Preference 83
3.5	The White Tone's Perceived Brightness 85
3.6	Summary and Outlook 87
	References 89
4	**Object Colors – Spectral Reflectance, Grouping of Colored Objects, and Color Gamut Aspects** *91*
4.1	Introduction: Aims and Research Questions 91
4.2	Spectral Reflectance of Flowers 94
4.3	Spectral Reflectance of Skin Tones 96
4.4	Spectral Reflectance of Art Paintings 97
4.5	The Leeds Database of Object Colors 98
4.6	State-of-the-Art Sets of Test Color Samples and Their Ability to Evaluate the Color Quality of Light Sources 100
4.7	Principles of Color Grouping with Two Examples for Applications 114
4.7.1	Method 1 – Application of the Theory of Signal Processing in the Classical Approach 120

4.7.2	Method 2 – the Application of a Visual Color Model in the Classical Approach *121*
4.7.3	Method 3 – the Application of Visual Color Models in the Modern Approach *121*
4.7.4	First Example of Color Grouping with a Specific Lighting System Applying Two Methods *122*
4.7.5	Second Example of Applying Method 3 by Using Modern Color Metrics *123*
4.8	Summary and Lessons Learnt for Lighting Practice *125*
	References *126*

5 State of the Art of Color Quality Research and Light Source Technology: A Literature Review *129*

5.1	General Aspects *129*
5.2	Review of the State of the Art of Light Source Technology Regarding Color Quality *132*
5.3	Review of the State of the Art of Colored Object Aspects *141*
5.4	Viewing Conditions in Color Research *142*
5.5	Review of the State-of-the-Art Color Spaces and Color Difference Formulae *145*
5.6	General Review of the State of the Art of Color Quality Metrics *154*
5.7	Review of the Visual Experiments *160*
5.8	Review of the State-of-the-Art Analyses about the Correlation of Color Quality Metrics of Light Sources *161*
5.9	Review of the State-of-the-Art Analysis of the Prediction Potential and Correctness of Color Quality Metrics Verified by Visual Experiments *166*
	References *171*

6 Correlations of Color Quality Metrics and a Two-Metrics Analysis *175*

6.1	Introduction: Research Questions *175*
6.2	Correlation of Color Quality Metrics *177*
6.2.1	Correlation of Color Metrics for the Warm White Light Sources *178*
6.2.2	Correlation of Color Quality Metrics for Cold White Light Sources *184*
6.3	Color Preference and Naturalness Metrics as a Function of Two-Metrics Combinations *189*
6.3.1	Color Preference with the Constrained Linear Formula (Eq. (6.2)) *192*
6.3.2	Color Preference with the Unconstrained Linear Formula (Eq. (6.3)) *194*
6.3.3	Color Preference with the Quadratic Saturation and Linear Fidelity Formula (Eq. (6.4)) *195*
6.4	Conclusions and Lessons Learnt for Lighting Practice *196*
	References *198*

7	**Visual Color Quality Experiments at the Technische Universität Darmstadt** *201*
7.1	Motivation and Aim of the Visual Color Quality Experiments *201*
7.2	Experiment on Chromatic and Achromatic Visual Clarity *204*
7.2.1	Experimental Method *205*
7.2.2	Analysis and Modeling of the Visual Clarity Dataset *208*
7.3	Brightness Matching of Strongly Metameric White Light Sources *212*
7.3.1	Experimental Method *213*
7.3.2	Results of the Brightness-Matching Experiment *216*
7.4	Correlated Color Temperature Preference for White Objects *218*
7.4.1	Experimental Method *218*
7.4.2	Results and Discussion *223*
7.4.3	Modeling in Terms of LMS Cone Signals and Their Combinations *223*
7.4.4	Summary *225*
7.5	Color Temperature Preference of Illumination with Red, Blue, and Colorful Object Combinations *225*
7.5.1	Experimental Method *226*
7.5.2	Results and Discussion *230*
7.5.3	Modeling in Terms of LMS Cone Signals and Their Combinations *230*
7.5.4	Summary *233*
7.6	Experiments on Color Preference, Naturalness, and Vividness in a Real Room *234*
7.6.1	Experimental Method *234*
7.6.2	Relationship among the Visual Interval Scale Variables Color Naturalness, Vividness, and Preference *238*
7.6.3	Correlation of the Visual Assessments with Color Quality Indices *239*
7.6.4	Combinations of Color Quality Indices and Their Semantic Interpretation for the Set of Five Light Sources *240*
7.6.4.1	Prediction of Vividness *240*
7.6.4.2	Prediction of Naturalness *241*
7.6.4.3	Prediction of Color Preference *241*
7.6.5	Cause Analysis in Terms of Chroma Shifts and Color Gamut Differences *243*
7.6.6	Lessons Learnt from Section 7.6 *246*
7.7	Experiments on Color Preference, Naturalness, and Vividness in a One-Chamber Viewing Booth with Makeup Products *246*
7.7.1	Experimental Method *247*
7.7.2	Color Preference, Naturalness, and Vividness and Their Modeling *251*
7.8	Food and Makeup Products: Comparison of Color Preference, Naturalness, and Vividness Results *256*
7.8.1	Method of the Experiment with Food Products *257*
7.8.2	Color Preference, Naturalness, and Vividness Assessments: Merging the Results of the Two Experiments (for Multicolored Food and Reddish and Skin-Tone Type Makeup Products) *258*

7.8.3	Analysis and Modeling of the Merged Results of the Two Experiments *261*	
7.8.4	Effect of Object Oversaturation on Color Discrimination: a Computational Approach *265*	
7.9	Semantic Interpretation and Criterion Values of Color Quality Metrics *268*	
7.9.1	Semantic Interpretation and Criterion Values of Color Differences *268*	
7.9.1.1	Semantic Interpretation of Color Fidelity Indices *270*	
7.9.1.2	Color Discrimination *272*	
7.9.1.3	Criterion Values for White Tone Chromaticity for the Binning of White LEDs *273*	
7.9.2	Semantic Interpretation and Criterion Values for the Visual Attributes of Color Appearance *276*	
7.10	Lessons Learnt for Lighting Practice *277*	
	References *280*	

8 Optimization of LED Light Engines for High Color Quality *283*
8.1 Overview of the Development Process of LED Luminaires *283*
8.2 Thermal and Electric Behavior of Typical LEDs *295*
8.2.1 Temperature and Current Dependence of Warm White LED Spectra *295*
8.2.1.1 Temperature Dependence of Warm White pc-LED Spectra *295*
8.2.1.2 Current Dependence of Warm White pc-LED Spectra *297*
8.2.1.3 Current Dependence of the Color Difference of Warm White pc-LEDs *297*
8.2.2 Temperature and Current Dependence of Color LED Spectra *299*
8.3 Colorimetric Behavior of LEDs under PWM and CCD Dimming *300*
8.4 Spectral Models of Color LEDs and White pc-LEDs *302*
8.5 General Aspects of Color Quality Optimization *305*
8.6 Appropriate Wavelengths of the LEDs to Apply and a System of Color Quality Optimization for LED Luminaires *311*
8.6.1 Appropriate Wavelengths of the LEDs to Apply *311*
8.6.2 Systematization for the Color Quality Optimization of LED Luminaires *315*
8.6.2.1 Conventional Structures of LED Luminaries in Real Applications *315*
8.6.2.2 Schematic Description of the Color Quality Optimization of LED Luminaries *315*
8.6.2.3 Algorithmic Description of Color Quality Optimization in the Development of LED Luminaires *318*
8.6.2.4 Optimization Solutions *319*
8.7 Optimization of LED Light Engines on Color Fidelity and Chroma Enhancement in the Case of Skin Tones *320*
8.8 Optimization of LED Light Engines on Color Quality with the Workflow *323*
8.8.1 Optimization of the LED Light Engine on Color Quality Using the RGB-W-LED Configuration *323*

	8.8.2	Optimization of the LED Light Engine on Color Quality with the R_1 - R_2 - G - B_1 - B_2 - W - LED - configuration *327*
	8.9	Conclusions: Lessons Learnt for Lighting Practice *333*
		References *334*
9		**Human Centric Lighting and Color Quality** *335*
	9.1	Principles of Color Quality Optimization for Human Centric Lighting *335*
	9.2	The Circadian Stimulus in the Rea *et al.* Model *338*
	9.3	Spectral Design for HCL: Co-optimizing Circadian Aspects and Color Quality *344*
	9.4	Spectral Design for HCL: Change of Spectral Transmittance of the Eye Lens with Age *348*
	9.5	Conclusions *354*
		References *355*
10		**Conclusions: Lessons Learnt for Lighting Engineering** *357*
		Index *365*

Preface

In the history of the development of electrical light sources from 1882 to date, relevant theoretical and technological progress has been achieved making the development of thermal radiators, discharge lamps, LED/OLED, and laser modules possible. Parallel to this dynamic process, lighting science and engineering and color science have made a big effort to find out new methods, parameters, metrics, and measurement methodologies to describe the color perception and color quality impression of light source users of different subjective characteristics in order to accurately model and evaluate lighting systems and lighting situations.

The development of semiconductor light sources on the basis of LED, OLED, and lasers from 1994 has offered lighting engineers and the users of illuminating systems novel possibilities for varying the spectral, temporal, and spatial light distributions of light sources and lighting systems. However, the color-rendering index (CRI) has remained to date the only official metric for the description of the color quality of lighting despite its deficits due to the outdated color space, chromatic adaptation formula, test color samples used in its definition, and the lack of its semantic interpretations in terms of easy-to-understand categories (e.g., "good" or "moderate" color rendering).

CRI is not able to be used for the evaluation of lighting installations according to visually relevant color quality attributes such as color preference, color naturalness, and color vividness. New methods for the modeling of color perception and new experiments for the determination of the relationship between the visual assessments of the subjects about color quality and usable color quality metrics (or combinations of color quality metrics) are therefore necessary.

Accordingly, the aim of this book is to analyze the state of the art of color quality research concerning the concepts of color space, color difference perception, white point appearance, the interaction between scene brightness, color temperature, and color preference, the relationship between colored objects, light sources, and the physiological and cognitive processing of the corresponding human photoreceptor signals. The correlation among currently discussed color quality metrics and their deficits as well as new models of color quality constitute the subject of several chapters of the present book. Basic and advanced knowledge will be presented on color quality to optimize light source spectral power distributions according to color quality (incorporating visual effects) and also the nonvisual effects of light for human well-being in the context of Human Centric Lighting.

The content of this book is based on the research of the present authors in the last few years. The authors would like to thank the students Yuda Li, Dragana Stojanovic, Taras Yuzkiv, and Eva Zhang who conducted the visual experiments and all students and scientific members of the Laboratory of Lighting Technology of the Technische Universität Darmstadt (Germany) for their participation in several sessions during the color quality experiments.

The authors would like to acknowledge the Federal Ministry of Education and Research (Germany) for the financial support of their LED projects.

Darmstadt, Germany, July 2016

Tran Quoc Khanh
Peter Bodrogi
Trinh Quang Vinh

1

Introduction

The technology of electrical light sources started in 1880 with the development of the tungsten lamp based on the thermal radiation principle by Thomas Edison. From the beginning of the twentieth century until 1990, a variety of light source types for indoor and outdoor lighting such as fluorescent lamps, compact fluorescent lamps, sodium and mercury high pressure lamps, and halogen metal lamps have been manufactured. Parallel to light source development, the disciplines of lighting science and illuminating engineering have been established by the use of conventional photometry and conventional illuminating design methods, which are based on the spectral luminous efficiency function $V(\lambda)$.

The $V(\lambda)$-function is valid for photopic vision in order to evaluate light sources from the point of view of illumination quality. In this process, quantities such as glare metrics, detection and identification probabilities, contrasts, luminance level and illuminance level, and illuminance uniformity have been developed in order to evaluate the illumination quality of light sources. These parameters have found their way to lighting and technological standards for the development of lighting products for international trade and lighting design. The use of $V(\lambda)$ is, however, misleading because it "represents only one of the possible responses of the retina of the eye to stimulation by electromagnetic radiation" [1], the response of the luminance channel with a spectral sensitivity limited to the central band of the visible spectrum around 555 nm, thus ignoring or underweighting rays of reddish and bluish wavelengths that are important to render or emphasize the colors of the colored objects in the lit environment.

With the appearance of modern triband fluorescent lamps, technological development has led to the necessity of defining a more color-related figure of merit: the color rendering index (CRI R_a) was introduced in 1965. It was based on the numeric evaluation of color differences and used test color samples to describe the color appearance change under a test light source in comparison to a reference illuminant of similar correlated color temperature (CCT). (More recently, the color rendering property has been called *color fidelity*.) The general CRI uses eight desaturated Munsell colors and the visually nonuniform $U^*V^*W^*$ color space and the aim was to optimize light source spectra so as to cause small color differences compared to a reference light source of the same CCT. But this color difference has to be computed independent of the direction of the

Color Quality of Semiconductor and Conventional Light Sources, First Edition.
Tran Quoc Khanh, Peter Bodrogi, and Trinh Quang Vinh.
© 2017 Wiley-VCH Verlag GmbH & Co. KGaA. Published 2017 by Wiley-VCH Verlag GmbH & Co. KGaA.

color shift between the test light source and the reference light source mixing lightness, chroma, and hue shifts between them.

Although the definition and the calculation method of the CRI have been improved during the last three decades, color rendering has remained the only internationally recognized metric to evaluate the color quality of light sources and its deficiencies turned out to be relevant as more and more flexible light source spectra became available for the advanced illuminating systems of lighting practice.

Since 1994, with the development of white LED light sources based on very bright and efficient blue LED emission and novel phosphor-converting materials, lighting industry and users of lighting systems have obtained a new technological platform for varying the spectral power distribution of a certain light source with previously unknown flexibility. This new era of spectral design is characterized by previously unknown possibilities to optimize the wavelength of the blue LED chips between 405 and 465 nm and a huge variety of broadband phosphor systems between 510 nm (cyan-green) and 660 nm (deep red).

Since about 2000, several types of optimal LED light sources have been combined in modern, so-called multi-LED light engines with separately addressable and optimizable LED channels including white phosphor-converted LED channels, colored partially phosphor-converted LED channels as well as colored pure semiconductor LEDs in the most recent so-called hybrid multi-LED illuminating systems.

The result of the above-described LED revolution has been a plethora of different spectral power distributions with very different color quality properties (including not only the variation of color fidelity but also of color harmony, color vividness, color naturalness for a wide range of white tones between warm white and cool white) from the same, flexible LED light engine. But a system of unambiguous numeric metrics starting from the spectral power distribution of the light source and providing indices for the different color quality properties available for lighting practitioners (lighting engineers, developers of LED light sources and LED lighting systems, lighting architects, lighting designers, scientists and students, facility managers, and system planners) to systematically evaluate and consciously design the color quality properties of these illuminating spectra by considering different applications (e.g., shop lighting or museum lighting) and different users (e.g., according to region of origin or age) is currently missing.

But from recent literature (from the twenty first century) it is well known that it is the color quality of high-quality illumination that mainly determines its user acceptance. Accordingly, from about 2000 on, new models to describe color perception have arisen (including the CIECAM02 color appearance model and the CAM02-UCS color space) and this knowledge has been used to develop more advanced metrics and indices to describe the different aspects of color quality, including new ways to describe color fidelity and color preference.

With the above new possibilities of LED spectral design in mind, industry and research need a *system* of practically relevant color quality metrics in order to rank, optimize, and select the color semiconductor LED light sources and phosphor-converted LED light sources to be used in the future's LED light engines and LED illuminating systems (as well as other, conventional light sources such as fluorescent lamps or other discharge lamps). Light source spectra should be adapted to the actual lighting application such as shop lighting, high-quality lighting of colored museum objects in galleries, or hotel, conference hall, restaurant, and retail lighting.

In the last few years, new lighting applications of increasing importance have emerged and have given rise to human centric lighting (HCL) considerations concentrating on the various properties of the human users and their interpersonal variations – for example, for hospitals, schools, and homes for elderly light source users. In such applications, the comprehensive modeling of color quality needs to be co-optimized with dynamic lighting control and the modeling quantities of nonvisual effects (including the so-called circadian effect).

Lighting engineers need a new, single criterion parameter or spectral optimization target parameter (similar to the concept of Perceived Adequacy of Illumination, PAI [2]) constructed from multiple measures for general lighting practice. This parameter should combine the following items: illuminance level, a measure of chromatic visual performance, CCT dependence, nonvisual factors, and a generally usable color quality measure possibly consisting of a color fidelity index and another, usable measure that describes how *saturated* object colors appear under the given light source.

The task of color quality research is to develop color quality (color rendition) metrics that correlate well with the visual assessments of the users of lighting systems and the designers of light sources and luminaires regarding the criteria "color preference," "color naturalness," and "color vividness." If high color vividness, high color preference, or high color naturalness should be ensured by the light source illuminating the colored objects, then a saturation term must be added according to the current development of lighting technology (at the time of writing at 2016). Concerning CCT, this parameter (as an important parameter for HCL design) has a strong influence on the comfort, well-being, and visual performance of the light source users. Combined with different types of colored objects (e.g., bluish objects or reddish objects), the CCT of the light source also influences subjective color preference assessments.

If a usable model of lighting quality is at the lighting engineers' disposal, this will allow a more efficient, human centric optimization of the illuminating system fostering user acceptability and the turnover from the sales of such modern and acceptable lighting systems. This gain, in turn, leads to the possibility of further technological improvements to be described by more developed models and lighting quality metrics. This cyclic nature of technological development and the scientific development of modeling lighting and color quality aspects is illustrated in Figure 1.1, together with the factors of the optimization of illuminating systems.

As suggested in Figure 1.1, the technological development of light sources (especially the occurrence of high spectral flexibility, e.g., instead of fluorescent lamps with fixed spectra, the emergence of spectrally tunable multi-LED engines) results in the development of new concepts and models of lighting engineering and color science, enabling a better description of visual perception, color quality, and visual performance. These advanced models, in turn, allow for new possibilities of spectral design for better user acceptance, for example, by

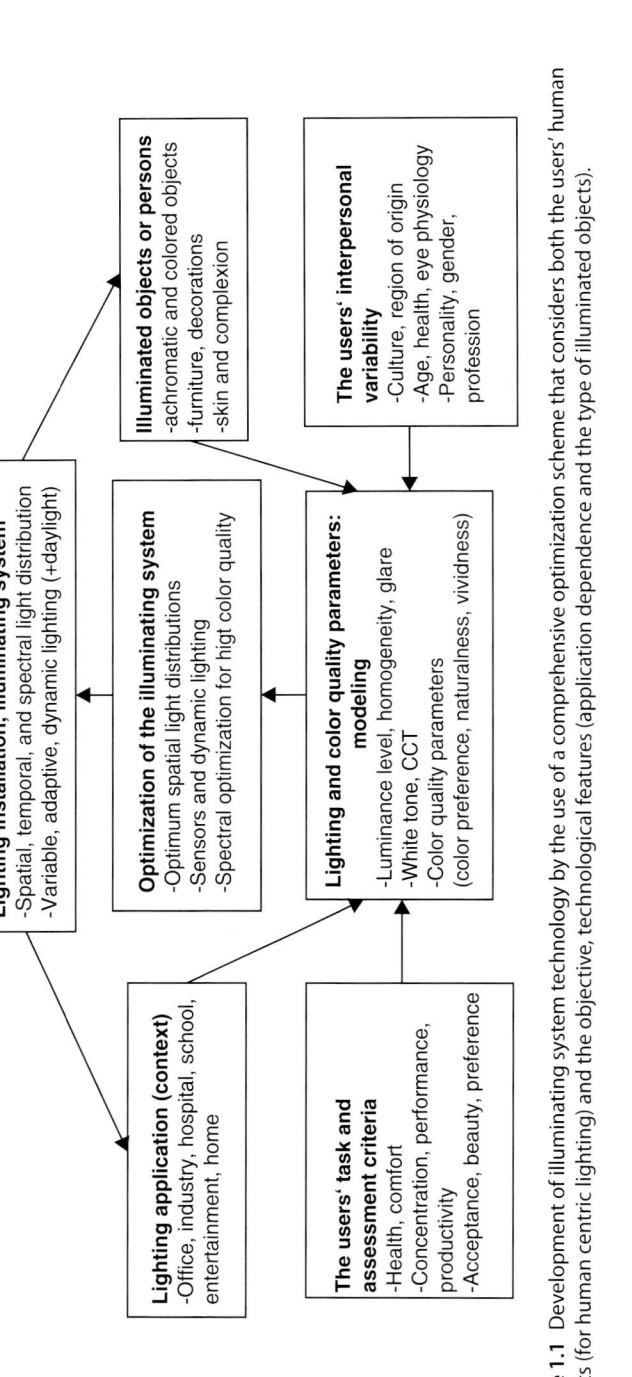

Figure 1.1 Development of illuminating system technology by the use of a comprehensive optimization scheme that considers both the users' human aspects (for human centric lighting) and the objective, technological features (application dependence and the type of illuminated objects).

ensuring an appropriate level of object oversaturation by the use of appropriate color quality metrics.

To help understand the concepts necessary to accomplish the comprehensive lighting quality optimization scheme of Figure 1.1, the present book will describe in the systematic view of its chapters, the following *visual* aspects of lighting and color quality evaluation and design:

1) Appreciation and preference of the *white tone* of the light source including metameric white light for different observers and field sizes.
2) *Illuminance level and visual clarity:* in order to ensure good visual performance at a workplace, illuminance levels of > 500 lx are needed but visual clarity depends on CCT. Brilliantly lit scenes represent an important color quality effect, for example, in high-quality shop lighting.
3) *Color temperature:* the preference of white light in warm, neutral, or cool (daylight) tones between 2500 and 7000 K depends on the individual user's characteristics and also the type of colored objects being illuminated (e.g., bluish objects only).
4) *The so-called "color quality" (or "color rendition") properties:* besides the color fidelity (earlier it was called *color rendering*) aspect of color quality that describes the realness and truth of a lit arrangement of colored objects, other aspects of color quality like vividness (colorfulness), naturalness, attractiveness, preference, and object color discrimination ability come into play and need to be taken into account (depending on the lighting application) if the aim of the lighting engineer is a modern, highly qualitative lighting design.

In the last 10 years, the above aspects have been the subject of intensive discussions and several psychophysical color quality experiments in many lighting laboratories all over the world. The resulting new descriptor quantities (color quality metrics) have given rise, according to the above-mentioned cycle, to significant technological developments driven by the dynamic development of LED light source technology. Many of these psychophysical experiments indicated that the general color rendering index, CIE CRI R_a, in its currently standardized and recommended form, cannot account for the visual scaling results of users in terms of color preference, color naturalness, or color discrimination. Therefore, a systematic analysis of all color quality aspects is needed together with a set of specific and usable principles to *optimize* the spectral power distributions of LED lighting systems, correspondingly. According to the above intentions, the present book deals with the issues to be described below.

The human visual system's relevant properties will be summarized in Chapter 2 together with all important visual color phenomena to understand how color appearance comes into existence and behaves, including human eye physiology, chromatic (opponent) channels and the color attributes lightness, brightness, colorfulness, chroma, and hue. Subjective aspects of color quality and their objective descriptor quantities (naturalness, preference, color gamut, color vividness) will also be described in Chapter 2 together with cognitive color effects such as color memory and color semantics. Cognitive color aspects are relevant to developing decision criteria in terms of limits of continuous lighting quality variables or color quality indices for lighting practice (e.g., limits of an index to distinguish between "acceptable"/"not acceptable"; or among "excellent," "moderate," or "bad" color quality).

The issues of white lighting, whiteness perception, white tone acceptance, and white tone preference will be dealt with in detail in Chapter 3. The reason is that the white point (white tone) of the light source is an important visual prerequisite to ensure good color quality in a lit environment with arrangements of colored objects. The interaction between light source spectral power distributions and the spectral reflectance of colored objects causes color appearance distortions in color space in comparison to the most natural light source for human beings, that is, daylight. To quantify these distortions affecting color discrimination and color gamut properties, different sets of colored objects from the real world (skin tones, flowers, paint colors, etc.) have been analyzed and grouped after their spectral reflectance in a systematic manner in Chapter 4.

A detailed study of scientific literature will summarize the main tendencies of visual experiments on color quality including the observers' scaling results and their correlation with existing color quality metrics. The aim of this part of the book (Chapter 5: literature review) is to demonstrate the nature and the importance of color quality issues for lighting practitioners by the brief presentation of these experiments (research questions, hypotheses, questionnaires, light sources, colored objects, and results) in a usable way for lighting practitioners.

Research questions include, for example, which figure of merit should be the second color quality metric besides the color fidelity metric if the perceived color quality of a lit scene could be described by the aid of two-color quality metrics. A systematic characterization of all color quality metrics discussed in recent literature (metrics of color fidelity, color gamut, color preference, color naturalness, and color memory) and the correlations among the different color quality metrics (including color preference and naturalness metrics as functions of two-metrics combinations) will be elaborated in Chapter 6.

In the analyzed literature, several pieces of missing information have been identified by the authors of the present book. Therefore, a series of own visual color quality experiments were carried out (at the Laboratory of Lighting Technology of the authors and other coworkers at the same laboratory) with different LEDs and other light sources at different CCTs and object saturation levels. These results will be presented in a systematic way in Chapter 7 with the final aim of establishing a novel color quality metric and finding a relevant semantic interpretation of the acceptable value ranges for this metric. This is essential to supply lighting practitioners with a concrete, numeric guidance to optimize their light sources, possibly the multi-LED light engines of the future.

The visual experiments on color quality presented in Chapter 7 include studies on the role of the illuminance level above which good color quality comes into play, brightness matching in a comparative study between different, metameric light sources, the effect of the chromaticity (white tone) of the light source and the effect of the changing relative spectral power distribution of the light source that causes the interaction between the emission spectra and the spectral reflectance of the illuminate objects, and the visual effects of changing color quality in the user's human visual system. Results from color preference experiments at different color temperatures (2700–6500 K) including white points at different chromaticities on the Planckian locus (and also below and above it, with different spectral power distributions) will be described in Chapter 7.

In our color quality experiments described in Chapter 7, different contexts such as beauty product lighting (skin tones and cosmetics) and food (grocery product) lighting have been investigated by selecting a set of representative colored objects for the above-mentioned applications (e.g., makeups or fruits). Some of the experiments were designed to find out the critical amount of chroma enhancement (ΔC^*) at which color preference or naturalness are rated to be maximal and above which these two subjective attributes are rated to be "no more acceptable." Experiments have been performed partially in a viewing booth but also in a real experimental room with three-dimensional immersion, observation, and visual color quality assessment.

Using the current color quality metrics obtained from literature and also the novel color quality metrics derived from the own experiments, a systematic and detailed guidance including workflow diagrams will be provided for the lighting practitioner about how to design and optimize the LED light engine's spectra by selecting different blue chips, colored semiconductor LEDs, and phosphor systems in Chapter 8. The aim of Chapter 8 is to help design such LED spectra (as a result of a comprehensive spectral optimization) that exhibit the best compromise between color fidelity, colorfulness, and color preference including the effects of long-term color memory, color discrimination, and the avoidance of unacceptable color gamut distortions.

By the use of all-inclusive, advanced LED modeling, color quality for different lighting applications can be co-optimized with HCL aspects including circadian optimization. This will be shown in Chapter 9 of the book. Finally, Chapter 10 recapitulates the lessons learnt from the book for the practice of lighting engineering including Figure 10.1 summarizing all color quality aspects and their optimization. As these spectral design and color quality optimization principles are not LED specific, they can also be applied to any light source including conventional (e.g., fluorescent lamps) or the future's OLED-based light engines.

According to the above-described aims and content, Figure 1.2 summarizes the chapter structure of the book.

As can be seen from Figure 1.2, the introduction, the foundations, and the conclusions are located at the first (upper) level of the structure. White point issues, the prerequisite of successful color quality design, constitute an individual level, the next intermediate one. After ensuring a correct white point, the block of four chapters (within the dashed line frame) dealing with color quality experiments and color quality metrics (including the most important issue of the colored objects to be illuminated by the light source) follow. Based on this knowledge, it is possible to carry out the spectral optimization of the light source, the

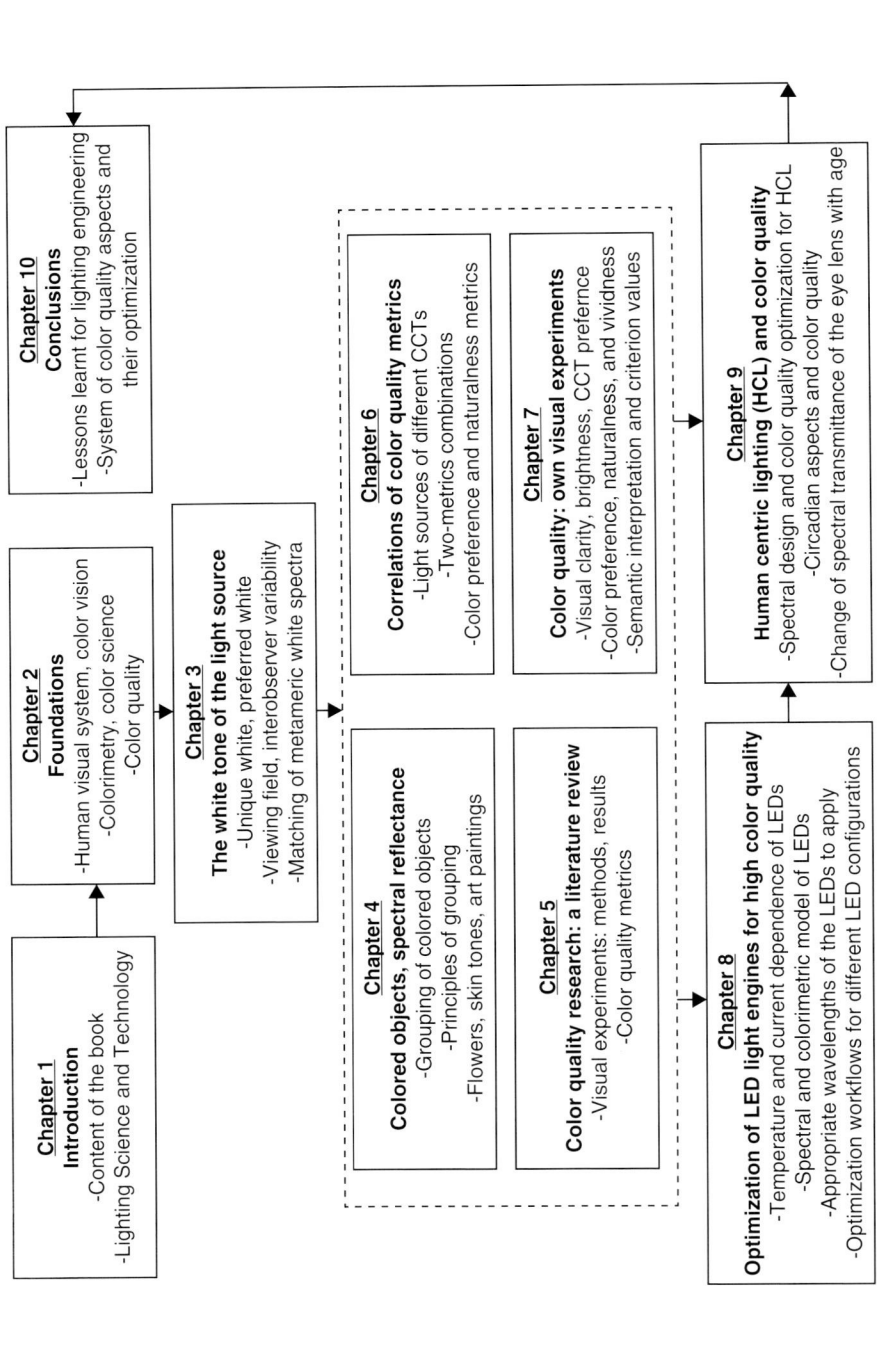

Figure 1.2 Chapter structure of the book.

final aim of the present book. The corresponding two chapters are situated at the lowest level of Figure 1.2, with an arrow pointing toward the conclusions, located, in turn, at the upper level of the book's framework.

References

1 Boyce, P. (2015) Editorial: the problem with light. *Light. Res. Technol.*, **47**, 387.
2 Cuttle, K. (2010) Opinion: lighting criteria for the future. *Light. Res. Technol.*, **42**, 270.

2

Color Appearance and Color Quality: Phenomena and Metrics

This chapter introduces the most important properties of the human visual system and their mathematical description for the reader to be able to understand the modeling of different color appearance phenomena. Color appearance models, in turn, constitute the basis for the definition of all types of color quality metrics used to model the perceived color quality of light sources, the subject of the present book. The most important visual color quality aspects (e.g., color fidelity, color preference, color naturalness, color vividness, resemblance to long-term memory colors) and the corresponding color quality metrics are described. Semantic interpretations of color quality metrics (e.g., "excellent," "good," "moderate," or "bad") will also be introduced as they provide numeric values to define user acceptance limits in terms of suitable color quality metrics for LED lighting system design. The most important concepts of basic and advanced colorimetry and color science are recapitulated based on existing knowledge from the literature cited in this chapter including several text modules and figures that are taken from a previous work of the present authors (and other authors) from the same publisher [1].

2.1 Color Vision

Visual perception (including the perception of color) is the result of a psychophysical process: the human visual system converts electromagnetic radiation reaching the human eye (the stimulus) into neural signals and interprets these neural signals as different psychological dimensions of visual perception (e.g., shape, spatial structure, motion, depth, and color) at later processing stages in the visual cortex [1]; see Figure 2.1. The understanding of human color perception and color vision mechanisms is essential for lighting engineering in order to model and optimize the color quality of light sources by applying the methodology of colorimetry and color science.

As can be seen from Figure 2.1, the electromagnetic radiation of the light source is reflected from the color sample (here: lilac color) which has a certain spectral reflectance. After the reflection, the spectrally selective sample (the lilac rectangle in Figure 2.1) changes the spectral radiance distribution of the light source and this reflected light provides the color stimulus for the human observer who

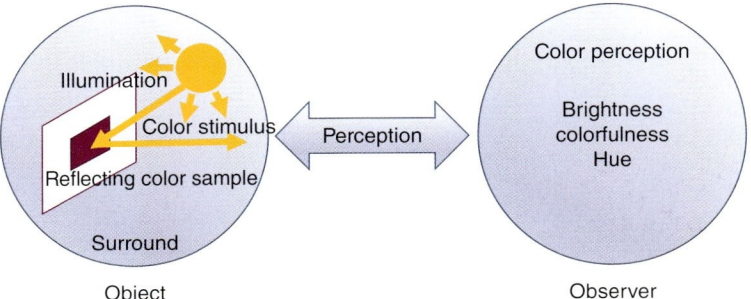

Figure 2.1 Illustration of how color perception (right) comes to existence from a color stimulus (left). (Khanh *et al*. 2014 [1]. Reproduced with permission of Wiley-VCH.)

processes the color stimulus in the visual brain. The result of this processing is the color perception of the color sample with the perceived color appearance attributes brightness, colorfulness, and hue (latter concepts will be defined later) [1]. In the example of Figure 2.1, the lilac color sample is a homogeneous colored plane surface (so-called surface color) characterized by its spectral reflectance $\rho(\lambda)$, which is the ratio of the reflected radiant power to the incident radiant power at every wavelength or by its spectral radiance factor $\beta(\lambda)$, the ratio of the radiance of the material to the radiance of the perfectly reflecting material that is irradiated in the same way. Such homogeneous surface colors are used as test color samples (TCSs) in the calculation methods of the different color quality indices; see Section 2.6 and Chapter 4.

To understand color vision mechanisms, first the structure and functioning of the human eye including the retina (the biological image-capturing device of the human observer) shall be understood. Figure 2.2 illustrates the structure of the human eye.

As can be seen from Figure 2.2, the human eye is an ellipsoid with an average length of about 26 mm and a diameter of about 24 mm. The eye is rotated in all directions by the aid of eye muscles. The outer layer is called *sclera*. The sclera is continued as the transparent *cornea* at the front. The *choroidea* supplies the *retina* with oxygen and nutrition. The retina is the photoreceptive (interior) layer of the eye, also containing the visual preprocessing cells (see below). The *vitreous body* is responsible for maintaining the ellipsoid form of the eye. It consists of a suspension of water (98%) and hyaluronic acid (2%). The optical system of the human eye is a complex, slightly decentered lens system projecting an inverted and downsized image of the environment onto the retina. The cornea, the anterior chamber, and the iris constitute the front part of this optical system and then, the posterior chamber and the biconvex eye lens follow [1]. The lens is held by the zonule fibers. By contracting the ciliary muscles, the focal length of the lens can be changed. The visual angle intersects the retina at the *fovea* (centralis), the location of the sharpest vision. The most important optical parameters of the components of the eye media include refractive indices (ranging typically between 1.33 and 1.43) and spectral transmission factors. All parameters vary among different persons considerably and are subject to significant changes with

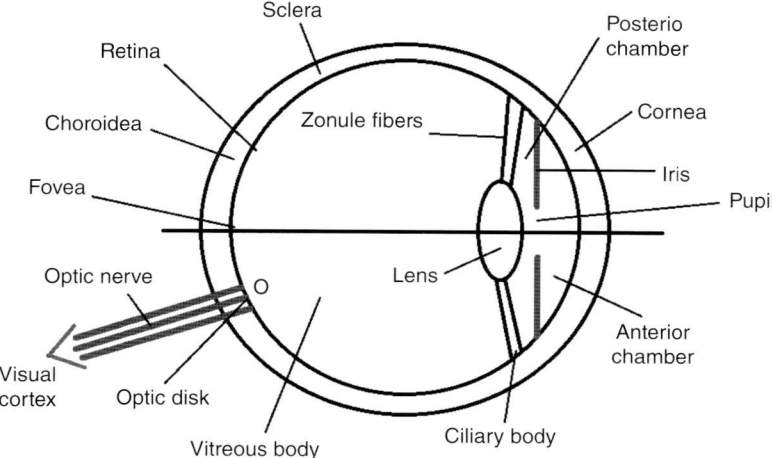

Figure 2.2 Structure of the human eye (the optic nerve is also called *visual nerve*). O, optic disk – the point at which the optic nerve passes through the eye and transmits the preprocessed neural signals of the retina toward the visual cortex. (Khanh *et al.* 2014 [1]. Reproduced with permission of Wiley-VCH.)

aging. Especially accommodation, visual acuity, and pupil reactions are impaired with increasing age. The spectral transmission of the eye media decreases with age significantly, especially for short wavelengths [1]. After having reached the retina, light rays have to travel through the retinal layers and in the central retina also, the so-called macula lutea (a yellow pigment layer that protects the central retina), *before* reaching the photoreceptors placed at the *rear* side of the retina. The *optic disk* (also called *optic nerve head* or *blind spot*) is the point (designated by O in Figures 2.2 and 2.3) at which the optic nerve passes through the eye. The retina is blind at the location O (as the density of rods and cones equals zero there) [1].

The retina (a layer of a typical thickness of 250 μm in average) is part of the optical system of the eye and, with its photoreceptor structure, also part of the visual brain. The retina contains a complex cell layer with two types of photoreceptors, rods, and cones. Both the rod receptors and the cone receptors are connected to the nerve fibers of the optic nerve via a complex network that computes neural signals from receptor signals. The retina contains about 6.5 millions of cones and 110–125 millions of rods while the number of nerve fibers is about 1 million. The density of rods and cones is different and depends on retinal location. Recently, a third type of photosensitive cell has been discovered, the so-called ipRGC (intrinsically photosensitive retinal ganglion cell containing the pigment melanopsin) responsible for regulating the circadian rhythm; see Chapter 9. Figure 2.3 shows rod density and cone density as a function of retinal location [1] while the inset diagram of Figure 2.3 shows the LMS cone mosaic (resembling an electronic image capturing device) of the rod-free inner fovea.

As can be seen from Figure 2.3, there are no receptors at the position of the optic disk or blind spot as the optic nerve exits the eye at this place (designated

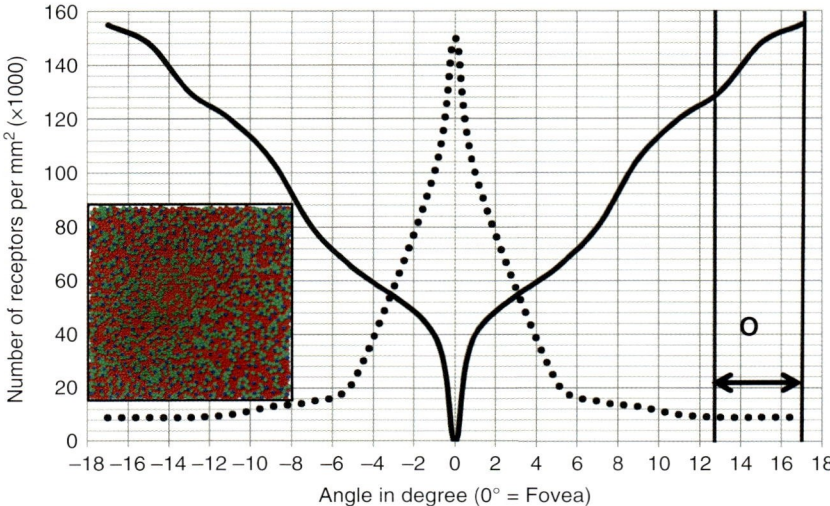

Figure 2.3 Rod density (continuous curve) and cone density (dots) as a function of retinal location (abscissa: α in degrees) drawn after Østerberg's data [2]. O, optic disk (blind spot); inset diagram, cone mosaic of the rod-free inner fovea subtending about 1.25°, that is, about 350 μm; red dots, long-wavelength-sensitive cone photoreceptors (L-cones); green dots, middle-wavelength-sensitive cones (M-cones); blue dots, short-wavelength-sensitive cones (S-cones). (Source of the inset diagram: Figure 1.1 from Gegenfurtner and Sharpe 1999 [3]. Reproduced with permission of Cambridge University Press.)

by O). The fovea is located in the center of the macula lutea region. A characteristic value to represent the diameter of the fovea is 1.5 mm corresponding to about 5° of visual angle. The fovea is responsible for best visual acuity according to the high cone receptor density; see the cone density maximum in Figure 2.3. Outside the fovea, cone diameter increases up to about 4.5 μm, cone density decreases, and rod (diameter of rods: 2 μm) density increases to reach a rod maximum at about 20° (temporally rather at 18°) [1].

Rods are responsible for night-time vision, also called *scotopic vision* at luminance levels lower than 0.001 cd/m². Rods are more sensitive than cones but they become completely inactive above about 100 cd/m². Cones are responsible for daytime or photopic vision (at luminance levels of about 10 cd/m² or higher). The transition range between scotopic (rod) vision and photopic (cone) vision is called the *twilight* or *mesopic range* in which both the rods and the cones are active. Acceptable color quality can only be expected in the photopic range (see Chapter 7). Besides pupil contraction, the transition between rod vision and cone vision constitutes a second important adaptation mechanism of the human visual system to changing light levels. There is a third adaptation mechanism, the *gain control* of the receptor signals. Photoreceptors contain pigments (opsins, certain types of proteins) that change their structure when they absorb photons and generate neural signals that are preprocessed by the horizontal, amacrine, bipolar, and ganglion cells of the retina to yield neural signals for later processing via the different visual (and nonvisual, e.g., circadian) pathways [1].

There are three types of cone, with pigments of different spectral sensitivities, the so-called L (long wavelength sensitive), M (middle wavelength sensitive), and S (short wavelength sensitive) cones (see Figure 2.3) that yield the so-called L, M, and S signals for the perception of homogeneous color patches and for colored spatial structures (e.g., a red-purplish rose with fine color shadings). The number of different cone signals (three) has an important psychophysical implication for color science that the designer of an LED light source should be aware of: color vision is trichromatic, color spaces have three dimensions (Section 2.5), and there are three *independent* psychological attributes of color perception (hue, saturation, brightness). The relative spectral sensitivities of the L, M, and S cones are depicted in Figure 2.4 together with some other important functions to be discussed below [1].

The spectral sensitivities in Figure 2.4 were measured at the cornea of the eye. They incorporate the average spectral transmission of the ocular media and the macular pigment at a retinal eccentricity of 2°; they are the so-called Stockman and Sharpe 2000 2° cone fundamentals [4–6]. As can be seen from Figure 2.4, the spectral bands of the L, M, and S cones yield three receptor signals for further processing in the human visual brain. From these signals, the retina derives two so-called opponent or chromatic signals: (i) L − M (the red–green opponent signal or its mediating neural channel yielding a signal for the visual

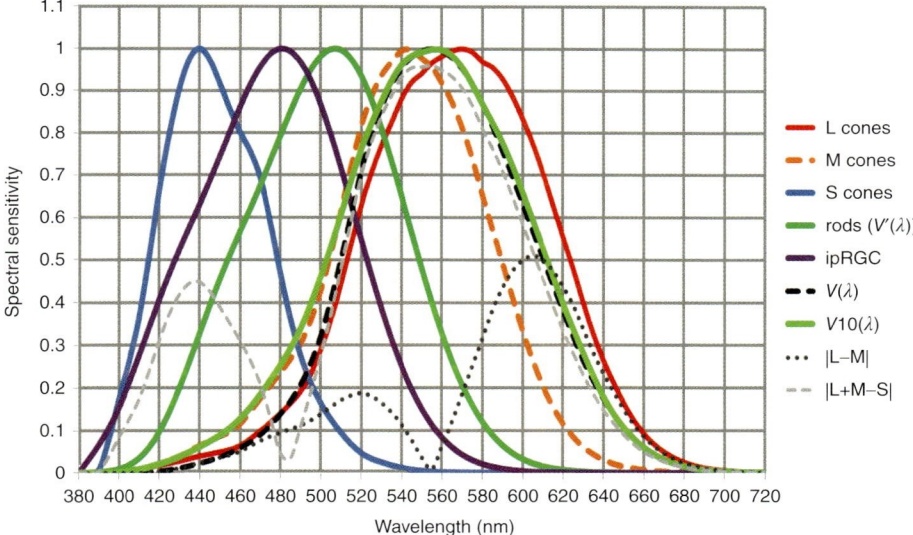

Figure 2.4 Relative spectral sensitivities of the L, M, and S cones (the so-called Stockman and Sharpe 2000 2° cone fundamentals; for a 2° viewing field) [4–6] as well as other visual mechanisms that use the LMS cone signals as input. The sensitivity of the ipRGC mechanism is also shown (see Chapter 9). The spectral sensitivity of the rods (dark green curve) is approximated by the so-called $V'(\lambda)$ function. $V(\lambda)$, luminous efficiency function (the basis of photometry, for stimuli of standard viewing angle, about 1–4°); $V_{10}(\lambda)$, its alternative version for stimuli of greater viewing angle (about 10°). (Khanh *et al.* 2014 [1]. Reproduced with permission of Wiley-VCH.)

brain to compute the perceived red or green content of the stimulus) and (ii) $L + M − S$ (the yellow–blue opponent signal or channel yielding a signal for the visual brain to compute the perceived yellow or blue content of the stimulus) and one so-called achromatic, $L + M$. The $L + M$ channel is usually considered as the luminance channel. The most important role of the luminance channel is that it enables the vision of fine image details and the temporal resolution of fast events. In the above signals, the "+" and "−" characters are only symbolic. In color vision models, the L, M, and S signals are usually weighted (see also Section 7.4), for example, with the weighting factors a and b to yield $(L + aM − bS)$ [1]; the modeling of color appearance (the so-called color appearance models) will be described in Section 2.4.

As can be seen from Figure 2.4, the L, M, and S cone sensitivity curves have their maxima at 566, 541, and 441 nm, respectively [3]. In photometry, for stimuli subtending 1–4° of visual angle, the spectral sensitivity of the $L + M$ channel is approximated by a standardized function, the so-called luminous efficiency function ($V(\lambda)$), which constitutes the basis of photometry while for spatially more extended (e.g., 10°) stimuli, the so-called $V_{10}(\lambda)$ function (the CIE 10° photopic photometric observer [7]) is used. Note that for more extended stimuli that cover, for example, a 10° viewing field, different LMS spectral sensitivities shall be used – for example, the so-called Stockman and Sharpe 2000 10° cone fundamentals [4].

Standard radiometric and photometric quantities (e.g., radiance, luminance, irradiance, illuminance) will not be defined here; they are introduced, for example, in [1]. For practical applications (see, e.g., Figure 9.10 in which the relative spectral radiance of the channels of a multi-LED engine are shown together with the spectral sensitivity of the mechanisms), it is important to compare in Figure 2.4 the spectral sensitivity of the rod (R) mechanism approximated by the so-called $V'(\lambda)$ function with the spectral sensitivities of the L, M, and S cones, the spectral sensitivity of the two chromatic mechanisms ($L − M$ and $S − (L + M)$), with $V(\lambda)$ and $V_{10}(\lambda)$ (that roughly represent the $L + M$ signal as mentioned above) and also with the already mentioned ipRGC mechanism [1]. In lighting practice, receptor signals shall be computed by multiplying their relative spectral sensitivity by the relative spectral radiance of the stimulus at every wavelength and integrating; see Eqs. (9.1)–(9.3).

2.2 Colorimetry

The understanding of the basics of colorimetry is very important because it contains mathematical models of human color vision suitable for straightforward application in lighting practice if the spectral radiance distribution or some other physically measurable quantities of the stimulus are at the lighting engineer's disposal. Colorimetry is the science and metrology that quantifies human color perception and recommends methods to derive colorimetric quantities from instrumentally measured spectral radiance distribution of the stimulus. By the use of these colorimetric quantities it is possible to predict the psychological magnitude of color perceptual attributes and use them for the characterization, design, and optimization of a lit environment.

The science of a more advanced modeling of sophisticated color perceptual phenomena is called *color science* (including, e.g., the so-called color appearance models; see Section 2.5). A further aim of colorimetry is to define such quantities that are simple to measure instrumentally (without measuring spectral power distributions (SPDs)) yet enable lighting engineers to characterize a stimulus in such a way that its color perception or a property of its color perception can be easily described and understood [1]. Such quantities include tristimulus values, chromaticity coordinates, and correlated color temperature (CCT); see below.

2.2.1 Color-Matching Functions and Tristimulus Values

Although it would be plausible to start from a quantity that is proportional to the signals of the three cone types (L, M, and S) according to Section 2.1, in current standard colorimetric practice, LMS spectral sensitivities (see Figure 2.4) are *not* used to characterize electromagnetic radiation [1], to describe the color stimulus, and, in turn, to quantify color perception. Instead of LMS, for color stimuli subtending 1–4° of visual angle, the so-called *color-matching functions* of the CIE 1931 standard colorimetric observer [8] are applied as the basis of standard colorimetry, denoted by $\bar{x}(\lambda), \bar{y}(\lambda), \bar{z}(\lambda)$. For visual angles greater than 4° (e.g., 10°), the so-called CIE 1964 standard colorimetric observer is recommended [8] with the $\bar{x}_{10}(\lambda), \bar{y}_{10}(\lambda), \bar{z}_{10}(\lambda)$ color matching functions; see Figure 2.5. It should be noted that, from historical reasons, the $\bar{y}(\lambda)$ color-matching function equals the $V(\lambda)$ function, the basis of (classic) photometry [1].

The aim of the use of the color matching functions in Figure 2.5 is to predict which spectral radiance distributions result in the same color appearance (in other words, to predict *matching color stimuli*) provided that the viewing condition is the same, that is, both stimuli are imaged to the central retina for an average observer (the standard colorimetric observer). From the spectral radiance distribution of the color stimulus measured by a spectroradiometer, the so-called *XYZ tristimulus values* are computed by the aid of the color-matching functions via Eq. (2.1). If the *XYZ* tristimulus values of two color stimuli are the same then they will result in the same color perception (in other words, there will be no *perceived color difference* between them). As can be seen from Eq. (2.1), as tristimulus values result from integration, two color stimuli with different SPDs can have the same tristimulus values and visually match despite this difference. Two such matching color stimuli are called *metamers* and the effect is called *metamerism*. Colors that match this way are called *metamers*. An example of seven (nearly) metameric warm white multi-LED SPDs can be seen in Figure 7.35. Although the seven white tones they provide match visually (almost perfectly), the visual

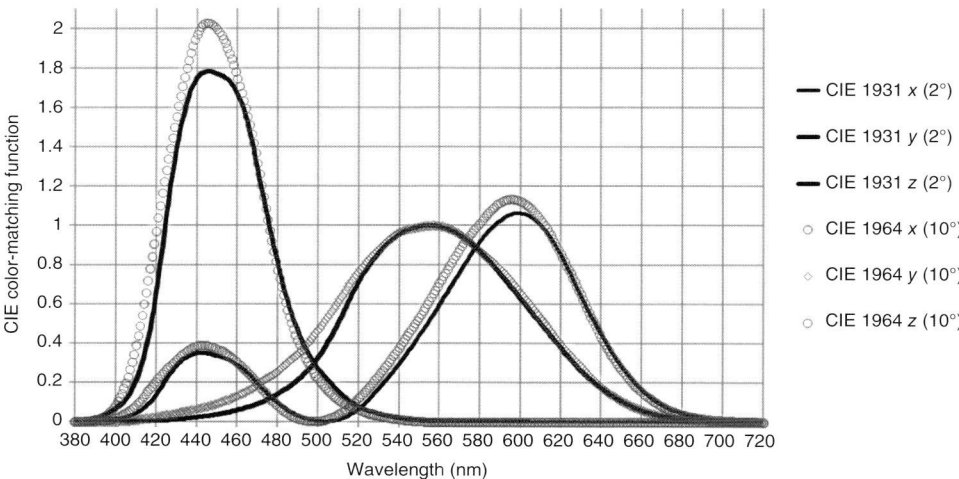

Figure 2.5 Black curves, color-matching functions of the CIE 1931 standard colorimetric observer [8] denoted by $\bar{x}(\lambda), \bar{y}(\lambda), \bar{z}(\lambda)$ for color stimuli subtending 1–4° of visual angle; open gray circles, color matching functions of the CIE 1964 standard colorimetric observer [8] denoted by $\bar{x}_{10}(\lambda), \bar{y}_{10}(\lambda), \bar{z}_{10}(\lambda)$ for color stimuli subtending greater than 4°. The $\bar{y}(\lambda)$ color matching function was chosen to equal $V(\lambda)$. (Bodrogi and Khanh 2012 [9]. Reproduced with permission of Wiley-VCH.)

appearance of the colored objects they illuminate differs (see Section 7.7).

$$X = k \int_{360\text{nm}}^{830\text{nm}} L(\lambda)\bar{x}(\lambda)d\lambda$$
$$Y = k \int_{360\text{nm}}^{830\text{nm}} L(\lambda)\bar{y}(\lambda)d\lambda \qquad (2.1)$$
$$Z = k \int_{360\text{nm}}^{830\text{nm}} L(\lambda)\bar{z}(\lambda)d\lambda$$

In Eq. (2.1), $L(\lambda)$ denotes the spectral radiance distribution of the color stimulus $L(\lambda)$ and k is a constant. According to the scheme of Figure 2.1, in case of reflecting color samples, the spectral radiance of the stimulus ($L(\lambda)$) equals the spectral reflectance ($R(\lambda)$) of the sample multiplied by the spectral irradiance from the light source illuminating the reflecting sample ($E(\lambda)$); see Eq. (2.2) (for diffusely reflecting materials) [1].

$$L(\lambda) = \frac{R(\lambda)E(\lambda)}{\pi} \qquad (2.2)$$

In Eq. (2.2), value of k is computed according to Eq. (2.3) [8].

$$k = 100 / \int_{360\text{ nm}}^{830\text{ nm}} L(\lambda)\bar{y}(\lambda)d\lambda \qquad (2.3)$$

As can be seen from Eq. (2.3), for reflecting color samples, the constant k is chosen so that the tristimulus value $Y = 100$ for ideal white objects with $R(\lambda) \equiv 1$. For self-luminous stimuli (if the subject observes, e.g., a white LED light source directly), the value of k can be chosen to equal 683 lm/W [8] and then, as $\bar{y}(\lambda) \equiv V(\lambda)$, the value of Y will be equal to the *luminance* of the self-luminous stimulus. For color stimuli with visual angles greater than 4°, the tristimulus values X_{10}, Y_{10}, and Z_{10} can be computed substituting $\bar{x}(\lambda), \bar{y}(\lambda), \bar{z}(\lambda)$ by $\bar{x}_{10}(\lambda), \bar{y}_{10}(\lambda), \bar{z}_{10}(\lambda)$ in Eq. (2.1). As can be seen from Figure 2.5, the two sets of color-matching functions, that is, $\bar{x}(\lambda), \bar{y}(\lambda), \bar{z}(\lambda)$ and $\bar{x}_{10}(\lambda), \bar{y}_{10}(\lambda), \bar{z}_{10}(\lambda)$ differ significantly. The consequence is that two matching (homogeneous) color stimuli subtending, for example, 1° of visual angle will not match in a general case if their size is increased to, for example, 10° [1].

2.2.2 Chromaticity Diagram

The so-called chromaticity coordinates (x, y, z) are defined by Eq. (2.4).

$$x = \frac{X}{X+Y+Z} \quad y = \frac{Y}{X+Y+Z} \quad z = \frac{Z}{X+Y+Z} \quad (2.4)$$

The diagram of the chromaticity coordinates x, y is called the *CIE 1931 chromaticity diagram* or the *CIE (x, y) chromaticity diagram* [8]. Figure 2.6 illustrates how color perception changes across the x, y diagram.

As can be seen from Figure 2.6, valid chromaticities of real color stimuli are located inside the so-called *spectral locus,* which is the boundary of quasi-monochromatic radiations of different wavelengths and the so-called

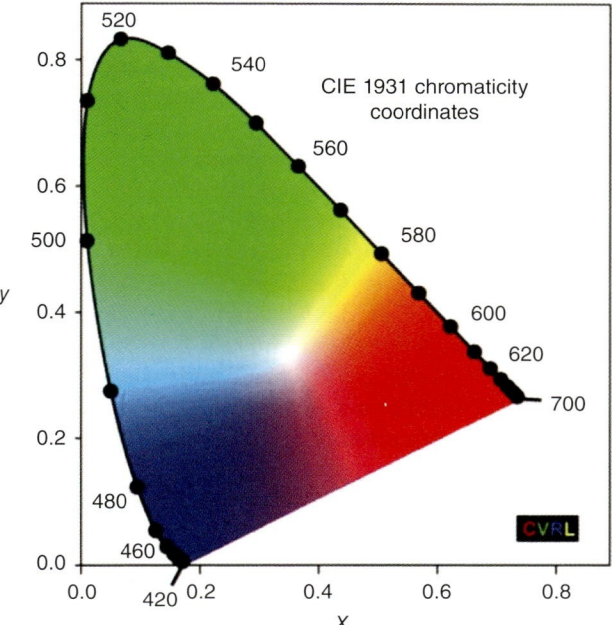

Figure 2.6 Illustration of how color perception changes across the CIE (x, y) chromaticity diagram [8]. The curved boundary of colors with three-digit numbers (wavelengths in nanometer units) represents the locus of monochromatic (i.e., most saturated) radiation. (Stockman *et al.* 2004 [10]. Reproduced with permission of Wiley-VCH.)

purple line in the bottom of the diagram. White tones can be found in the middle with increasing *saturation* (to be defined exactly below) toward the spectral locus. The perceived *hue* changes (purple, red, yellow, green, cyan, blue) when going around the region of white tones in the middle of the diagram [1].

2.2.3 Interobserver Variability of Color Vision

In the previous considerations, interindividual differences of color perception were neglected and a hypothetic average observer, the CIE 1931 standard colorimetric observer [8], was considered; see Figure 2.5. In reality, however, the LMS cone sensitivities (see Figure 2.4, which shows typical curves of an observer of normal color vision), and, consequently, the color-matching functions of the individual observers (still considered to have normal color vision) vary in a range [9] that was obtained in past experiments [11, 12]. There are also observers with anomalous or deficient color vision. Some observers have less L-, M-, or S-cones and exhibit protanomalous, deuteranomalous, or tritanomalous color vision, respectively. If one of the cone types is completely missing then they are called *protanope, deuteranope,* or *tritanope observers.* The issue of anomalous color vision is out of the scope of the present book.

It should be noted that, even within the range of normal trichromatic observers, according to the observer's genotype, spectral sensitivity maxima of the L-, M-, and S-cones might be shifted by up to 4 nm with important consequences to the white point of the light source; see Chapter 3 of this book. Even within the limits of normal trichromatic color vision, there is a large variability of retinal structures, especially concerning the ratios of the number of L- and M-cones varying between 0.4 and 13 [13]. The post-receptoral mechanisms of color vision are very adaptable and – at least in principle – able to counterbalance the variability of retinal LMS structures [9]. There are, however, large variations among the subjects at the later stages of neural color signal processing including the perception of color differences their color cognition, color preference, vividness, naturalness, and color harmony assessments and their long-term memory colors (latter concepts will be explained in Sections 2.3 and 2.4).

The interobserver variability of 10° color-matching functions is depicted in Figure 2.7 based on 49 individual observers of the Stiles–Burch (1959) dataset [11]. In Figure 2.7, there are 53 individual 10° color-matching functions altogether because 4 of the 49 observers repeated the observation. The CIE 1964 standard 10° colorimetric observer $\bar{x}_{10}(\lambda), \bar{y}_{10}(\lambda), \bar{z}_{10}(\lambda)$ (see Figure 2.5) is primarily based on this dataset, partly also on Speranskaya's (1959) 10° data [14]. Figure 2.7 was created by the present authors by using the so-called corrected individual RGB color-matching data downloaded from the web [4] and applying the same fixed values of the 3×3 RGB–XYZ conversion matrix of the Appendix B.3 of CIE Publ. 15:2004 (Colorimetry) [8] to each one of these individual RGB color-matching data.

As can be seen from Figure 2.7, there is considerable interobserver variability even among these 49 observers of normal color vision. The effect of this variability on white tone perception will be investigated in Chapter 3.

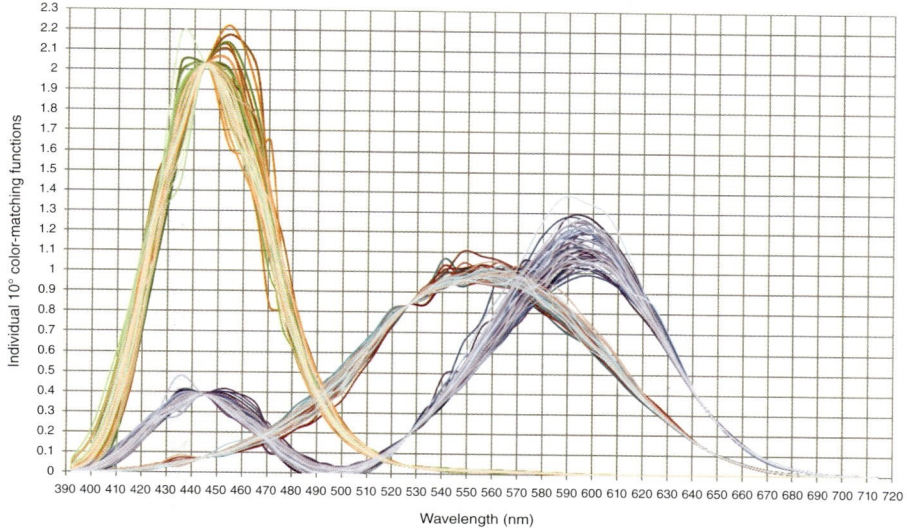

Figure 2.7 Individual 10° color-matching functions of 49 observers (+4 repetitions) [11] on which the CIE 1964 standard 10° colorimetric observer (Figure 2.5) is primarily based. (Stiles and Burch 1959 [11].)

2.2.4 Important Concepts Related to the Chromaticity Diagram

For practical applications, it is necessary to further analyze the x, y chromaticity diagram and introduce some further essential concepts. As can be seen from Figure 2.6, white tones (e.g., warm white or cool white) are located in the middle region of the x, y chromaticity diagram. The region of white tones deserves further attention; see Figure 2.8.

As can be seen from Figure 2.8 (compare with Figure 2.6), spectral color stimuli (of quasi-monochromatic electromagnetic radiation) constitute the left, upper, and right boundary of the region of the possible chromaticities of real color stimuli. In Figure 2.5, some spectral color stimuli (such as 680, 640, 600 nm, etc.) between 380 and 680 nm are marked. The purple line connects the chromaticities between the two extremes on the left (at the 380 nm point) and on the right (at the 680 nm point) [1].

One of the important concepts that shall be introduced is the curve of *blackbody radiators* (blackbody locus or Planckian curve); see the black line in the middle of Figure 2.8. This corresponds to the chromaticities (x, y) of blackbody radiators at different temperatures between 2000 K (yellowish tones) and 20 000 K (bluish tones) in the figure. The figure contains two marked data points, 2700 and 4000 K while 20 000 K is located at the left end of the black line. An important blackbody radiator is the CIE illuminant "A" because it represents typical tungsten filament lighting (its chromaticity is well known among light source users). Its relative SPD is that of a Planckian radiator at a temperature of 2856 K [1].

Associated with the blackbody locus, the concept of CCT shall be defined here according to its relevance to assess the color rendering of light sources. The CCT

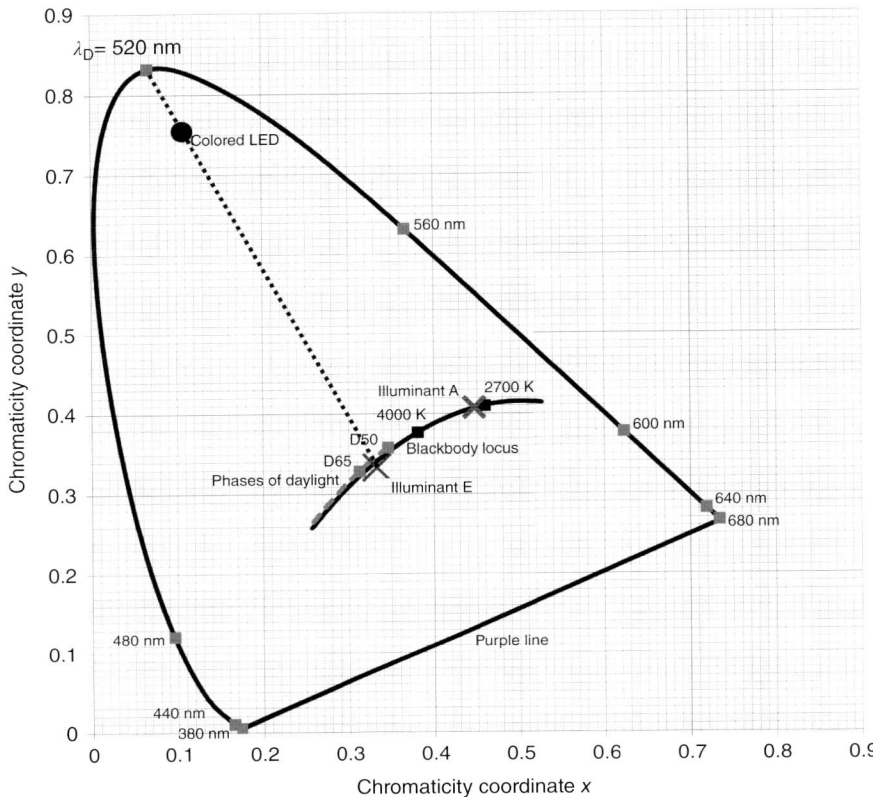

Figure 2.8 Important chromaticity points and curves in the x, y chromaticity diagram. The gray cross represents the chromaticity of CIE illuminant "A" (T_{cp} = 2856 K). Illuminant E: equienergetic stimulus (its spectral radiance is constant at every wavelength, $x = 0.333$; $y = 0.333$); filled black circle, a colored LED as an example to explain the concept of dominant wavelength ($\lambda_D = 520$ nm in this example) and colorimetric purity. (Khanh et al. 2014 [1]. Reproduced with permission of Wiley-VCH.)

of a test light source (T_{cp}; unit: kelvin) of a given chromaticity is equal to the temperature of the blackbody radiator of the nearest chromaticity. To compute the value of the CCT, the chromaticity point of the test light source shall be projected onto the blackbody locus and, by the use of Planck's law, the temperature shall be computed. The projection to the blackbody locus shall occur in a uniform chromaticity diagram [8] and not in the x, y diagram as the latter diagram is perceptually nonuniform; see Figure 2.10. Standard colorimetry uses the so-called u, v diagram to this purpose, as defined by Eq. (2.5).

$$u = 4x/(-2x + 12y + 3)$$
$$v = 6y/(-2x + 12y + 3) \qquad (2.5)$$

The chromaticities of average *phases of daylight* of different CCTs represent a further remarkable set, which is located slightly above the blackbody locus (see the gray curve in Figure 2.8). Figure 2.8 shows them between $T_{cp} = 5000$ and

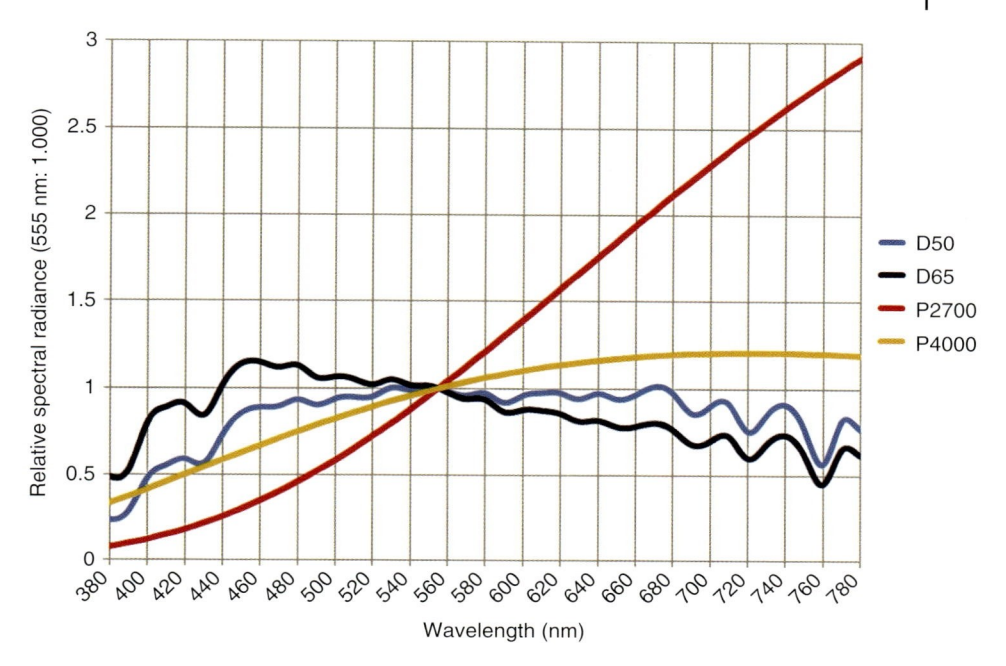

Figure 2.9 Relative spectral radiance distributions of two selected blackbody radiators (at 2700 and 4000 K) and the daylight illuminants D50 and D65; rescaled to ≡1 at 555 nm; see also Figure 9.3. (Khanh et al. 2014 [1]. Reproduced with permission of Wiley-VCH.)

$T_{cp} = 20\,000$ K. In lighting engineering, daylight illuminants D65 ($T_{cp} = 6504$ K) and D50 ($T_{cp} = 5003$ K) are especially important as they are often used as target white points when optimizing the spectral radiance distribution of a light source. They are marked in Figure 2.8. Figure 2.9 shows the relative spectral radiance distributions of two selected blackbody radiators (at 2700 and 4000 K) and the daylight illuminants D50 and D65 [1].

As can be seen from Figure 2.9, the blackbody radiator at 2700 K (P2700) has a high amount of emission in the yellow and red wavelength range (this is why its white tone is yellowish, the so-called warm white). This red–yellow amount decreases for P4000, D50 and D65, respectively. But the blue content of these light sources increases in this order. This is the reason why the daylight illuminant D65 has a slightly bluish white tone (the so-called cool white) [1].

There are two colorimetric quantities that are sometimes useful to characterize the color stimulus of colored (i.e., non-phosphor-converted) chip-LEDs concerning their hue and saturation (see Section 2.3). The concept of *dominant wavelength* can be used to characterize *hue* while *colorimetric purity* helps quantify *saturation* [15]. These two quantities can be calculated from the x, y chromaticity coordinates of the emission of the colored chip-LED. They are useless for white LEDs. For white LEDs, CCT (for hue) and the distance of their chromaticity point from the blackbody or daylight loci (for saturation) are the useful measures; see below.

The dominant wavelength (e.g., of a colored LED; denoted by λ_D) is equal to the "wavelength of the monochromatic stimulus that, when additively mixed in suitable proportions with the specified achromatic stimulus, matches the color stimulus considered" [8]. The achromatic stimulus is usually chosen to be the equienergetic stimulus E (its spectral radiance is constant at every wavelength and $x = 0.333$; $y = 0.333$). An example can be seen in Figure 2.8: the filled black circle represents the chromaticity point of a colored LED. It is connected with the chromaticity point of illuminant E ($x = 0.333$; $y = 0.333$) and this line is extended toward the spectral locus. The wavelength on the spectral locus at this intersection point is the dominant wavelength ($\lambda_D = 520$ nm in this example).

The colorimetric purity [8] (of a colored LED; denoted by p_C) is defined by Eq. (2.6).

$$p_C = \frac{L_d}{L_d + L_n} \tag{2.6}$$

In Eq. (2.6), L_d is the luminance of the monochromatic stimulus ($\lambda_D = 520$ nm in the example of Figure 2.8) and L_n is the luminance of the achromatic stimulus (illuminant E) that are necessary to match the color stimulus, that is, the colored LED in the example of Figure 2.8; see the filled black circle. The narrower the bandwidth of the LED, the closer its chromaticity point gets to the spectral locus and the less achromatic stimulus is needed to match its chromaticity. Consequently, narrowband LEDs have a high colorimetric purity.

2.2.5 MacAdam Ellipses and the $u' - v'$ Chromaticity Diagram

As mentioned above, the disadvantage of the CIE x, y chromaticity diagram (Figure 2.8) is that, concerning color difference perceptions (see Section 2.3), it is perceptually not uniform in the following sense. In Figure 2.6, observe that a distance in the green region of the diagram represents less change of perceived chromaticness (i.e., hue *and* saturation) than the same distance in the blue–purple region [1]. The so-called MacAdam ellipses [16] quantify this effect around the different chromaticity centers, that is, the centers of the ellipses in Figure 2.10. Roughly speaking, perceived chromaticity differences are hardly noticeable between any two chromaticity points inside the ellipse (for a more precise definition of the MacAdam ellipses, see [16]). Note that the ellipses depicted in Figure 2.10 represent 10 times magnified versions of the experimental data, for better visibility.

As can be seen from Figure 2.10, MacAdam ellipses are large in the green region of the CIE x, y chromaticity diagram while they are small in the blue–purple region and the orientation of the ellipses also changes. To overcome these difficulties, the x and y axes were distorted so as to make identical circles from the MacAdam ellipses and this resulted in the so-called CIE 1976 uniform chromaticity scale diagram (or simply u', v' diagram) defined by Eq. (2.7).

$$u' = 4x/(-2x + 12y + 3)$$
$$v' = 9y/(-2x + 12y + 3) \tag{2.7}$$

The concept of u', v' chromaticity difference between two chromaticity points (e.g., two slightly different white tones) can be defined as the Euclidean distance

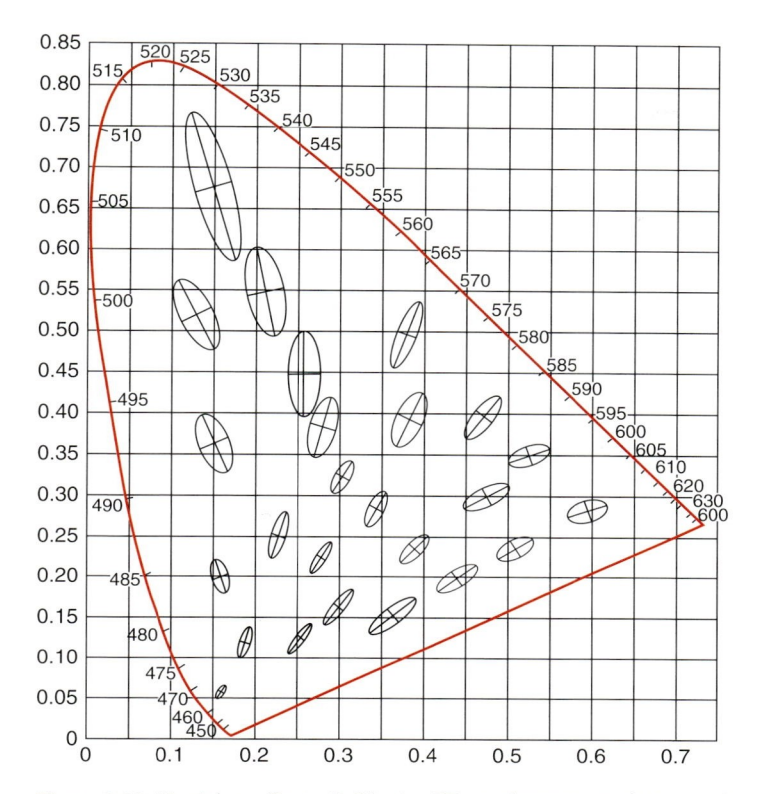

Figure 2.10 MacAdam ellipses [16] in the CIE x, y chromaticity diagram. Abscissa: chromaticity coordinate x, ordinate: chromaticity coordinate y. Roughly speaking, perceived chromaticity differences are not noticeable inside the ellipses. For a more precise definition of the MacAdam ellipses, see [16]. Ellipses are magnified 10 times. (Redrawn after [16] and reproduced from [17] with permission from the *Deutsche Lichttechnische Gesellschaft e.V. and Huss Verlag, Berlin*.)

on the plane of the u', v' diagram: $\Delta u'v' = \sqrt{(\Delta u')^2 + (\Delta v')^2}$. The u', v' diagram is perceptually *more* uniform than the x, y diagram [1]; compare with Figure 5.10. This means that equal $\Delta u'v'$ values *should* represent equal *perceived* chromaticity differences in *any* part of the u', v' diagram and in *any* direction at every chromaticity center provided that the relative luminance difference of the two color stimuli is small, for example, $\Delta Y < 0.5$. Therefore, the shape of the MacAdam ellipses in Figure 2.10 *comes closer* to a circular shape when they are depicted in the u', v' diagram. This means that the u', v' diagram is more useful (although far from perfect) than the x, y diagram to evaluate differences of perceived hue and saturation without lightness difference (see Section 2.3). In lighting practice, $\Delta u'v'$ values are often used to define an optimization criterion for the white tone of the light source depicted in the $u' - v'$ (or sometimes in the $u - v$) chromaticity diagram. During the spectral optimization of a light source, its white point should often be located at a target chromaticity on the blackbody (or, in other words, Planckian) locus or on a locus of the phases of daylight (see Figure 2.8) at a certain CCT. This criterion can be mathematically expressed, for example, by

$\Delta u'v' < 0.001$, the distance between the current white tone chromaticity during the optimization and the target chromaticity. This issue will be further elucidated in the next chapters.

2.3 Color Appearance, Color Cognition

Although color stimuli can be fully described in the system of tristimulus values (X, Y, and Z), this description yields a nonuniform and nonsystematic representation of the color perceptions corresponding to these color stimuli. The psychological attributes of the perceived colors (lightness, brightness, redness–greenness, yellowness–blueness, hue, chroma, saturation, and colorfulness; see below) cannot be modeled in terms of XYZ values directly. To derive a model that predicts the magnitude of these perceptual attributes (a so-called color space or color appearance model), mathematical descriptors shall be defined for each one of these attributes (the so-called numeric correlates; see Section 2.5). Numeric correlates can be computed from the XYZ tristimulus values of the stimulus and the XYZ values of their background and surround parameters (see Section 2.5). The above-mentioned perceived (psychological) color attributes are defined below.

2.3.1 Perceived Color Attributes

Hue is the attribute of a visual sensation according to which a color stimulus appears to be similar to the perceived colors red, yellow, green, and blue, or a combination of two of them [18]. *Brightness* is the attribute of a color stimulus according to which it appears to emit more or less light [18]. *Lightness* is the brightness of a color stimulus judged relative to the brightness of a similarly illuminated reference white (appearing white or highly transmitting). Lightness is an attribute of *related colors* (see above).

Colorfulness is the attribute of a color stimulus according to which the stimulus appears to exhibit more or less chromatic perceived color. For a given chromaticity, colorfulness generally increases with luminance. In a lit interior (a built environment that is important for the application of white LED light sources), observers tend to assess the *chroma* of (related) surface colors or colored objects. The perceived attribute *chroma* refers to the colorfulness of the color stimulus judged in proportion to the brightness of the reference white.

Saturation is the colorfulness of a stimulus judged in proportion of its own brightness [18]. A perceived color can be very saturated without exhibiting a high level of chroma, for example, the color of a deep red sour cherry is saturated but it exhibits less chroma because the sour cherry is colorful compared to its (low) own brightness but it is not so colorful in comparison to the brightness of the reference white. Figure 2.11 illustrates the three perceived attributes, hue, chroma, and lightness.

In a so-called *color space* (which has three dimensions, e.g., the Munsell color system), the three perpendicular axes and certain angles and distances carry psychologically relevant meanings related to the above-defined perceived color attributes. A schematic illustration of the structure of a color space can be

Figure 2.11 Illustration of three attributes of perceived color: (a) changing hue, (b) changing lightness, and (c) changing chroma. (Derefeldt *et al.* 2004 [19]. Reproduced from Color Res. Appl. with permission of Wiley.)

seen in Figure 2.12. A color space (or color solid) can be a collection of painted samples ordered according to an artists' concept (e.g., the Munsell color system) or a mathematical model that derives numeric scales correlating with the magnitude of the perceived color attributes. Such mathematical models include the interaction of the SPD of the light source with the spectral reflectance of the objects (see Figure 2.1, left) and the modeling of the processing of the colored image in the human visual system.

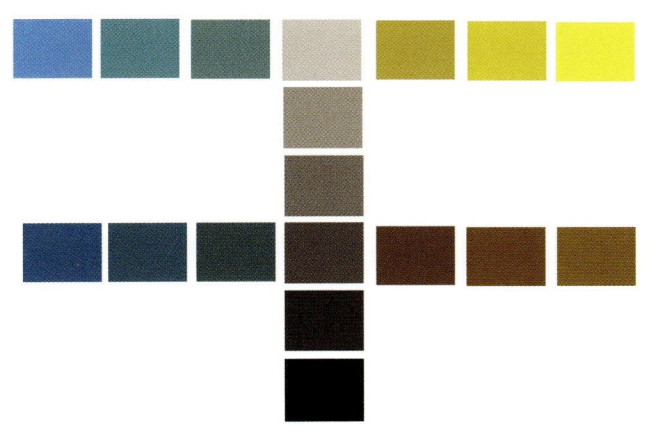

Figure 2.12 Schematic illustration of the general structure of a color space. Lightness increases from black to white from the bottom to the top along the gray lightness scale in the middle. Chroma increases from the gray scale toward the outer colors of high chroma. The perceptual attribute of hue varies when rotating the image plane around the gray axis in space (blue and yellow being opponent color perceptions). (Bodrogi and Khanh 2012 [9]. Reproduced with permission of Wiley-VCH.)

2.3.2 Viewing Conditions, Chromatic Adaptation, and Other Phenomena

The perceived attributes of the color appearance of a color stimulus depend not only on the color stimulus itself but also on the so-called *viewing conditions* in its visual environment. Viewing condition parameters influencing color appearance will be described in Section 2.5. Here, the most relevant viewing condition parameter, the so-called *adapted white point* or *adapted white tone* (e.g., the large white wall surfaces in a room illuminated by an LED light source) will be elucidated. This white point can be, for example, warm white (with a minor but perceptible yellowish shade in the white tone of the luminaire) or cool white (with a slight bluish shade). The chromaticity of the white tone that predominates in the visual environment influences the color perception of the object colors significantly. The reason is not only the spectral change of the stimulus triggered by the different light source emission spectra that illuminate the objects (see Figure 2.1, left) [1]. There is another, visual effect (that *partially* compensates for these *spectrally* induced changes): *chromatic adaptation*.

For example, if a cool white LED light source is switched on in the room instead of a warm white LED light source then the state of chromatic adaptation (the adapted white point in the scene) changes from "warm white adapted" to "cool white adapted." The mathematical description of chromatic adaptation is an important component of color spaces and color appearance models. The result of chromatic adaptation is that the perceived colors of the colored stimuli (colored objects) do *not* change *so much* (this is the tendency of the so-called *color constancy*) – despite the significant spectral changes caused by the change of the emission spectra (e.g., cool white vs. warm white) that illuminate the objects [1].

This is true as long as the light source that illuminates the objects has a shade of white (a white tone that can exhibit slight yellowish or bluish shades). If the objects are illuminated by nonwhite light sources (e.g., a poor LED lamp with a greenish shade in its "white" tone) then color constancy breaks down and a "strange" greenish shade is perceived on every object [1]. To experience a feeling about chromatic adaptation, look at the still-life in Figure 2.13a illuminated by a warm white light source about 1.5 min (cover the lower image) and then look at the same still-life illuminated by a cool white light source (cover the upper image). Fixation shall be maintained for both images at the white fixation cross in the middle.

If the still-life of Figure 2.13 is first illuminated by the warm white light source (Figure 2.13a) and the illuminant is changed to cool white (Figure 2.13b) then the process of chromatic adaptation starts immediately after the illuminant change: the bluish shade on the still-life under the cool white light source diminishes continuously with elapsing time. At the end of the chromatic adaptation process, the state of adaptation of the observer stabilizes (after about 90 s [20, 21] at the adapted white point of the cool white light source.

In addition to the above-described effect of chromatic adaptation, there are some other color appearance phenomena. Some of them are briefly mentioned here:

1) *Bezold–Brücke effect:* The hue impression changes with luminance level.
2) *Stevens effect:* Perceived contrast increases with luminance level.

Figure 2.13 Illustration of chromatic adaptation. Example for two different states of chromatic adaptation: the same colorful still life of the same colored objects under a warm white (a) and a cool white (b) light source. (Khanh *et al.* 2014 [1]. Reproduced with permission of Wiley-VCH.)

3) *Hunt effect:* Perceived colorfulness increases with luminance level.
4) *Helmholtz–Kohlrausch effect:* Perceived brightness increases with the saturation of the stimulus strongly depending on the hue of the stimulus.
5) *Simultaneous color contrast:* The perceived color of a (small, e.g., 2°) color stimulus is influenced by its (colorful) background.

2.3.3 Perceived Color Differences

Perceived color differences have three visual attributes, lightness difference, chroma difference, and hue difference. Any adjacent two rectangles in Figure 2.11a–c illustrate perceived hue, lightness, and chroma differences, respectively. In the most general case, two color stimuli evoke all three visual color difference components that result in more or less total perceived color difference. Compare any two of the colored rectangles in Figure 2.12 and try to decide which one of the two rectangles has more perceived lightness, chroma, and hue. Then select another pair and try to decide visually whether this pair exhibits more total perceived color difference than the previous one. The reader can see that there are small, medium, and large color differences. The magnitude of perceived color differences can be described as Euclidean distances in a color space or by the so-called color difference formulae; see Section 2.5 (there are different color difference formulae for small, medium, and large color differences).

2.3.4 Cognitive Color, Memory Color, and Semantic Interpretations

The distinction between perception and cognition is that while perception refers to immediate mapping of objects or events of the real world into the brain, cognition refers to subsequent higher order processes of semantic and verbal classification of the perceptions [22, 23] or to the mental imagery of the same objects or events. The term *cognitive color* may be defined as follows [19]. The result of the color module of the early stages of visual processing is *perceived color* with its three continuous perceptual attributes, hue, colorfulness, and brightness (or hue, chroma, and lightness). After this early visual processing stage, color perception is often classified by the subject into a (so-called "semantic") category.

Cognitive color means one from the discrete set of such categories [19]. This set of categories depends on the task of the observer, for example, the set of simple color names in a hue naming experiment, for example, "red," "orange," "yellow," "green," "blue," or "magenta"; or the set of so-called *long-term memory colors* of familiar objects (e.g., blue sky, skin, or green grass) often seen by the subject in the past and stored in the subject's long-term color memory in a color memory experiment. The set of categories can also express a subjective quality judgment about a colored object or a combination of colored objects under a certain light source: the light source can evoke, for example, a "very good," "good," "moderate," or "bad" color quality judgment about the colored objects it illuminates (compare with Figure 2.1, left and see Section 2.4) including whether the arrangement of the colored objects illuminated by the current light source evokes a harmonic or a disharmonic impression.

In a visual task in which the color memory of the observer is involved, the concept of cognitive color plays an important role [19]. In a typical color-matching task that includes color memory, observers memorize their perceived color of an "original" color stimulus, after which the original color disappears. After a given time interval, their task is to reproduce their short-term memory color by selecting one color from the perceived colors of several "actual" color stimuli [24]. In the visual experiments, the "original" color stimulus and the selected "actual" color stimulus are significantly and systematically different (so-called *memory color shift*). The memory color shift is often directed toward a color prototype [25] or, in other words, a *long-term memory color* of a familiar object that was similar to the "original color." The reason is that the observer tends to categorize the original color and to remember only that category, that is, to remember a cognitive color. For example, it seems more economical for the observer to remember the expression "green grass" instead of the particular shade of green grass of the original color. The selected "actual" color stimulus tends to represent the cognitive color remembered; it comes from the region of color space that corresponds to the cognitive color that was remembered [19].

The texture (spatial color structure) of the "green grass" or the shape of a face (as parts of the whole visual impression) influences the process of recalling long-term memory colors. The shape and/or the texture of the colored object can be considered as the "viewing context" of color perception. Viewing context helps recall long-term memory colors because if the shape/texture is visible then a category (e.g., "grass") can be identified easily, which is used to recall a color perception [26]. The "actual" color perception of a given colored object under a given light source might be different from the long-term memory color recalled on the basis of the shape/texture of this object (e.g., a banana exhibits a disturbing red–orange tone under the given light source instead of the typical yellow long-term memory color of the banana) and this contradiction results in the deterioration of the perceived color quality of the lit environment (see Section 2.4), and, consequently, in a "bad" color quality judgment under the given light source. The *preferred color appearance* of a given object (i.e., its color appearance preferred by most observers, although there is a huge interpersonal variability of color preference judgments depending on culture, region of origin, gender, profession, and age) is usually similar but not necessarily equal to the long-term memory color of that object [27–29].

Categorical judgments about the *similarity of color appearance* of two perceived colors (e.g., "very good," "good," "moderate," "low," or "bad" similarity) are very important for the so-called *semantic interpretation* of color differences and color fidelity indices (see Section 2.6). Any perceived color difference can be predicted by the use of a color difference formula (see Section 2.5) and, based on experimental categorical judgment results [30], every value of this color difference metric can be associated with a category. For example, if the value of the color difference metric is in a range between zero and a very small value, then there will be a "very good" similarity between the two color appearances. Similarly, every category is associated with a range on the scale of the color difference metric. This is the so-called semantic interpretation of color differences, which provides an important tool to define acceptance criteria for a light source. Visual attributes of color quality (see Section 2.4) can also be associated with semantic interpretations; see Figure 7.28.

2.4 The Subjective Impression of Color Quality and Its Different Aspects

The concept of color quality (in other words, color rendition) has already been mentioned in Section 2.3 as the reader is expected to have a feeling about what this concept means. But the authors believe that it is worth elucidating the concept in more detail, from the point of view of a lighting engineer, as follows. Color quality is a general subjective impression about the color appearance of all colored objects in a lit environment (mostly in interior lighting, e.g., home, shop, or office lighting but sometimes also in exterior lighting, e.g., park and festival lighting or garden lighting). The color quality judgment of the subject about this subjective impression may concern (consciously or unconsciously) different visual criteria (aspects or attributes) of color quality, for example, the naturalness or vividness of the colored objects. These aspects will be defined below.

Besides the type of color quality aspect (e.g., naturalness or vividness) considered by the subject, the following factors influence color quality impression and color quality judgments:

1) the combination of the colored objects in the lit environment;
2) the purpose and context of illumination, for example, office, home, shop, or school lighting;
3) the SPD of the light source (light sources with gaps or missing components in their visible spectrum cause low color quality judgments as such light sources distort the color appearance of the objects);

4) the white tone (or white point) of the light source (interacting with the predominating color of the actual colored objects, for example, reddish or bluish objects; see Section 7.5);
5) the brightness level (high color quality can be expected only at high enough luminance levels; see Section 7.2); and
6) the cultural background, region of origin, education level, gender, profession, and age of the subject; also fashion, advertisements, and actual social conditions.

Obviously, it is the relative SPD of the light source that has the most important influence on color quality. Therefore, lighting engineers often speak about the color quality of a given light source. This means the color quality impression of a representative subject about a certain representative set of colored objects in the environment illuminated by the given light source. In the following, the subjective aspects or visual attributes of color quality will be defined.

The first aspect is color fidelity (also called *color realness*), which means the similarity of the color appearance of the colored objects illuminated by the given (so-called "test") light source compared to the color appearance of the colored objects illuminated by a "reference" light source associated by the test light source. The *reference light source* is a Planckian (in other words thermal or blackbody) radiator at lower CCTs of the test light source and a phase of daylight at higher CCTs (see Figure 2.9). The reference light source has the same CCT as the test light source but the course of their relative SPD functions can be very different. As the reference light source is not visible in most situations of lighting practice, this aspect is not easy to assess visually. Naïve subjects in real situations never assess the color fidelity aspect and they are usually unable to do so (even if they are instructed) in a color quality experiment that comes close to a real application (e.g., experiment in a real room). Instead of color fidelity, they usually assess the aspects color naturalness, preference, and vividness defined as follows:

- *Color naturalness*: "Subjective extent of the similarity of color appearance of an identified object (e.g., a red rose) under the current light source compared to its long-term memory color."
- *Color preference* (*also called attractiveness or pleasantness*): "Subjective extent of how the subject likes the color appearance of the colored objects under the current (test) light source taking all colored objects in the scene into consideration."
- *Color vividness*: "Saturated brilliant color appearance of all colored objects in the scene."

Figure 2.14 illustrates the changing color appearance of an object scene illuminated by different light source spectra for high color fidelity, high naturalness, high preference, and high vividness. The first image corresponds to a light source spectrum that lowers the chroma of the objects (so-called *desaturating* light source spectrum), resulting in very low general perceived color quality.

As can be seen from Figure 2.14, the most significant change occurs in terms of the chroma of the objects. This can be expressed as the perceived overall chroma difference of the colored objects between the current lighting situation and the

2.4 The Subjective Impression of Color Quality and Its Different Aspects

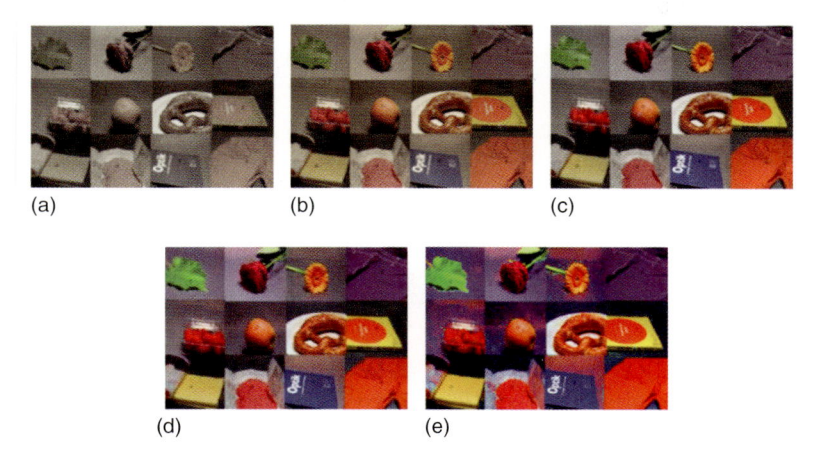

(a) (b) (c)

(d) (e)

Figure 2.14 Changing color appearance of an object scene illuminated by different light source spectra for high color fidelity (b), naturalness (c), preference (d) and vividness (e). (a) corresponds to a desaturating light source spectrum evoking a very low general color quality assessment.

reference situation. The latter is depicted in Figure 2.14b. The reference situation has high color fidelity and, by definition, zero overall chroma difference. This perceived chroma difference is negative in the case of Figure 2.14a (desaturating light source with lower chroma than the reference situation) and it increases toward the right of Figure 2.14.

Most long-term memory colors result from repeated observations during the life of the observer at higher illuminance levels (usually between 10 000 and 100 000 lx) under the viewing conditions of natural daylight (see Figure 2.15). In typical interior lighting, however, illuminance levels typically range between 20 and 2000 lx (partially from power-saving reasons and also depending on the application). At low illuminance levels, good color naturalness (and acceptable visual clarity, see below and in Table 7.2) cannot be expected. This shall be compensated for by increasing the saturation of the object colors (and possibly increasing the contrast among the object colors) by the conscious spectral design of the light source according to the scheme of Figure 2.14 in order to compensate for the loss of perceived colorfulness (Hunt effect) and the loss of perceived contrast (Stevens effect) with decreasing luminance level.

As an example, Figure 2.15 shows the result of a horizontal illuminance and CCT measurement during a typical summer afternoon (with sunshine and changing partially overcast sky) in Germany.

As can be seen from Figure 2.15, horizontal illuminance varied between about 10 000 and 80 000 lx while CCT was between 5000 and 7500 K. At 18:14 hours, there is a local minimum of illuminance (about 10 000 lx) caused by a cloud. At the same time, there is a sharp local CCT maximum indicating the spectral change caused by the cloud.

Besides color fidelity, naturalness, preference, and vividness, there are also some further color quality attributes including white tone quality, visual clarity, color discrimination ability, and also, color harmony:

Figure 2.15 Horizontal illuminance (lx) and CCT (K) during a typical summer afternoon (with sunshine and changing partially overcast sky) in Germany. (Image source: Technische Universität Darmstadt with acknowledgement to Mr. Sebastian Schüler (PhD student at the time of writing) who carried out the measurement.)

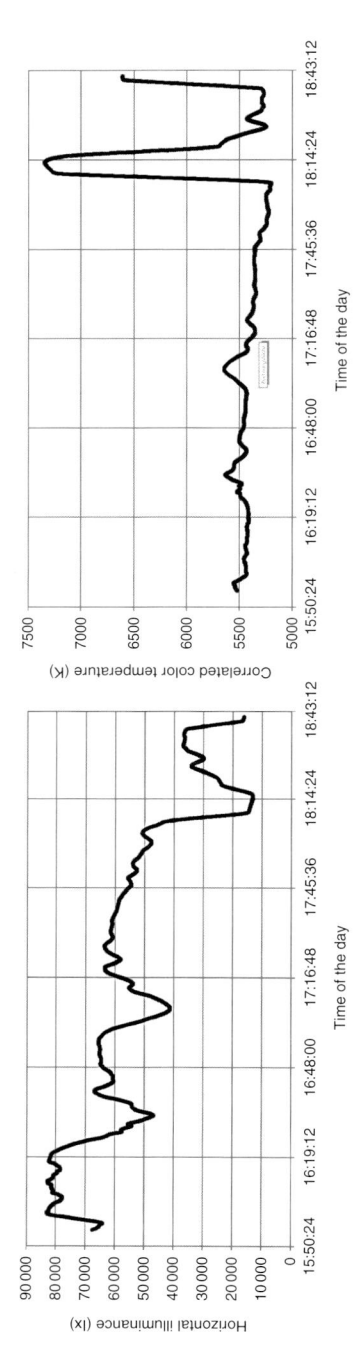

- White tone quality (see Chapter 3) means that the white tone predominating in the lit environment (i.e., the white point of the light source itself mixed with reflections from different white or colored objects in the room; see Figures 7.14 and 7.21) should be preferred by the observers, depending on the combination of the white point and the type of the colored objects (red, blue, colorful, or white; see Figures 7.14 and 7.21) being illuminated. This white tone preference has two components: (i) the CCT of the light source (e.g., warm white, neutral white, or cold white; e.g., warm white for home lighting and the illumination of reddish objects) and (ii) the (slight) amount of a certain hue perceived in the white point (e.g., a preferred warm white contains a certain amount of yellow); see Chapter 3. Therefore, white tone preference depends on the spectral composition of the light source in the different wavelength ranges (important ranges are 420–480; 510–560; 580–610; and 620–680 nm).
- Visual clarity is the clear visibility of continuous color transitions, fine color shadings on the object surfaces, and clearly visible contrasts (large color differences and/or large luminance differences) between the different colored objects; see Section 7.2.
- Color discrimination ability means that colored objects of similar color, that is, with only small color differences between them (e.g., wires, paintings, or yarn) should be easily differentiated visually under the given light source. This means that these small color differences among the colored objects should remain visible; see Section 7.8.4. This is a visual performance-related aspect of color quality.
- Color harmony means the preference of color combinations, an aesthetic judgment about the relationship among the color appearance of some selected (visually important) objects in the scene [31].
- Further aspects of color quality include the general impression of brightness and the feeling of space (three-dimensionality).

It will be pointed out in the subsequent chapters that the use of $V(\lambda)$-based quantities (e.g., photopic luminance or luminous efficacy) should be avoided during the optimization of light source color quality. Owing to the limited spectral band of the $V(\lambda)$ function (Figure 2.4), the use of LER (luminous efficacy of radiation) as one of the optimization criteria results in a loss of many important rays (i.e., rays that stimulate the $L - M$ or $L + M - S$ chromatic mechanisms) outside the spectral band of the $V(\lambda)$ function. These wavelengths are important for good color quality because they stimulate the chromatic mechanisms. Finally, it should be noted that all aspects of color quality constitute an important component of human centric lighting (HCL) design; see Chapter 9.

2.5 Modeling of Color Appearance and Perceived Color Differences

After the introduction of the basics of colorimetry and the visual attributes of color appearance, the present section focuses on the prediction of the magnitude of visual color appearance attributes in a *color space* or by a *color appearance model* in terms of numeric scales (or numeric correlates) to be computed from the tristimulus values of the color stimulus and a set of numeric parameters that characterize the viewing condition including the adapted white point and other viewing conditions characteristics. Perceived color appearance attributes (e.g., lightness) have been scaled by observers subjectively and mathematical models have been constructed to correlate with the answer of the observers. Imagine that the observer sees several color stimuli (e.g., the ones depicted in Figures 2.1 and 2.12) after each other and in any case, he or she should tell a number between 0 (black) and 100 (white) for lightness. A numeric (metric) predicting the value of lightness can also be calculated from the color appearance model and this value correlates well with the mean answer of the subjects. Similarly, the percentage of unique red (a red color that does not contain neither yellow nor blue) can also be assessed visually in terms of a number and then compared with the hue prediction of the color appearance model.

2.5.1 CIELAB Color Space

As can be seen from Figure 2.12, lightness increases from black to white from the bottom to the top along the gray lightness scale in the middle of color space. At every lightness level, chroma increases from the gray scale toward the most saturated outer colors [1]. The perceptual attribute of hue varies when rotating the image plane around the gray axis of the color space. To predict these visual properties, CIE colorimetry recommends two color spaces, CIELAB and CIELUV, the so-called CIE 1976 ("uniform") color spaces [8]. Here, only the more widely used CIELAB color space will be described. The XYZ values of the color stimulus and the XYZ values of a specified reference white color stimulus (X_n, Y_n, Z_n) constitute its input and the CIELAB L^*, a^*, b^* value its output; see Eq. (2.8).

$$L^* = 116 f(Y/Y_n) - 16$$
$$a^* = 500[f(X/X_n) - f(Y/Y_n)] \quad (2.8)$$
$$b^* = 200[f(Y/Y_n) - f(Z/Z_n)]$$

In Eq. (2.8), the function f is defined by Eq. (2.9).

$$f(u) = u^{1/3} \quad \text{if } u > (24/116)^3$$
$$f(u) = (841/108)u + (16/116) \quad \text{if } u \leq (24/116)^3 \quad (2.9)$$

The CIELAB formulae of Eq. (2.8) try to account for chromatic adaptation by simply dividing the tristimulus value of the color stimulus by the corresponding value of the adopted reference white. In reality, chromatic adaptation is based on the gain control of the L, M, and S cones. The modeling of photoreceptor gain control by the reference white (white point) control of the XYZ tristimulus values is a very rough approximation. This is one of the reasons why CIELAB is only of limited applicability for lighting practice [1].

The output quantities of the CIELAB system represent approximate correlates of the perceived attributes of color: L^* (CIE 1976 lightness) is intended to describe the perceived lightness of a color stimulus. Similarly, CIELAB chroma (C^*_{ab}) stands for perceived chroma and CIELAB hue angle (h_{ab}) for perceived hue; see Eq. (2.10). The quantities a^* and b^* in Eq. (2.8) can be considered as

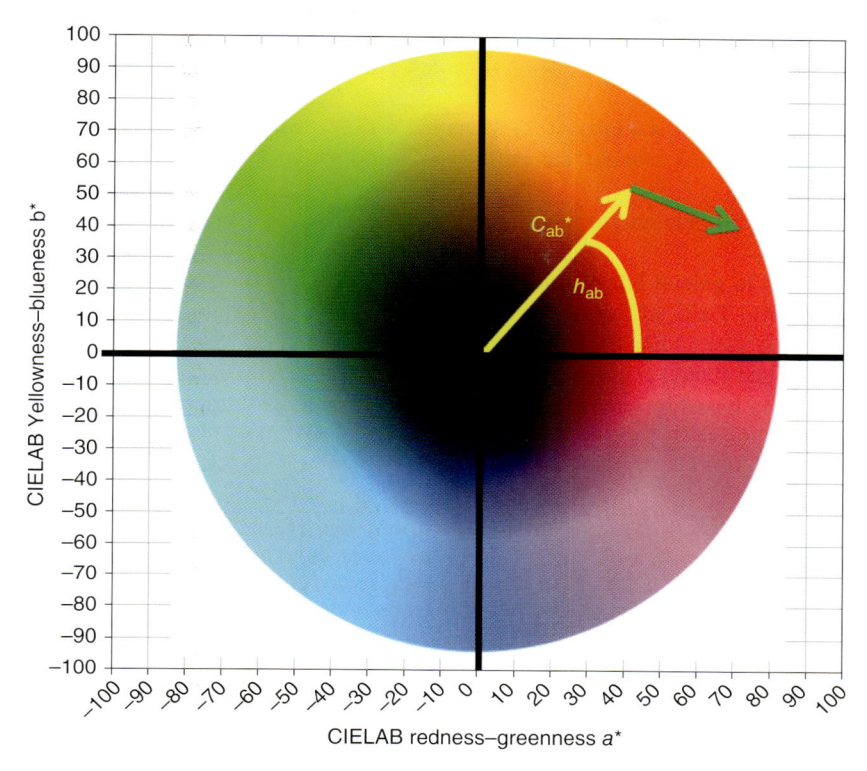

Figure 2.16 Illustration of a CIELAB $a^* - b^*$ diagram with the correlates of chroma (C^*_{ab}) and hue angle (h_{ab}). The L^* axis is perpendicular to the middle of the $a^* - b^*$ plane in which L^* is constant. At the end of the yellow arrow, there is a color stimulus with CIELAB chroma C^*_{ab} and CIELAB hue angle h_{ab}. The green arrow shows a color difference (between a red and an orange stimulus), here in the $a^* - b^*$ plane only, that is, without a lightness difference (ΔL^*). (Khanh et al. 2014 [1]. Reproduced with permission of Wiley-VCH.)

rough correlates of perceived redness–greenness (red for positive values of a^*) and perceived yellowness–blueness (yellow for positive values of b^*). L^*, a^*, and b^* constitute the three orthogonal axes of CIELAB color space. Equation (2.10) shows how to calculate C^*_{ab} and h_{ab} from a^* and b^* [1].

$$C^*_{ab} = \sqrt{a^{*2} + b^{*2}}$$
$$h_{ab} = \arctan(b^*/a^*) \quad (2.10)$$

Figure 2.16 shows the CIELAB $a^* - b^*$ diagram with the numeric correlates of chroma (C^*_{ab}) and hue angle (h_{ab}). In Figure 2.16, the CIELAB hue angle (h_{ab}) changes between 0° (at $b^* = 0, a^* > 0$) and 360° around the L^* axis (the achromatic axis that stands perpendicular to the middle of the $a^* - b^*$ plane).

2.5.2 The CIECAM02 Color Appearance Model

To apply the CIELAB color space, it is important to read the notes of the CIE publication [8] carefully: CIELAB is "intended to apply to… object colors of the

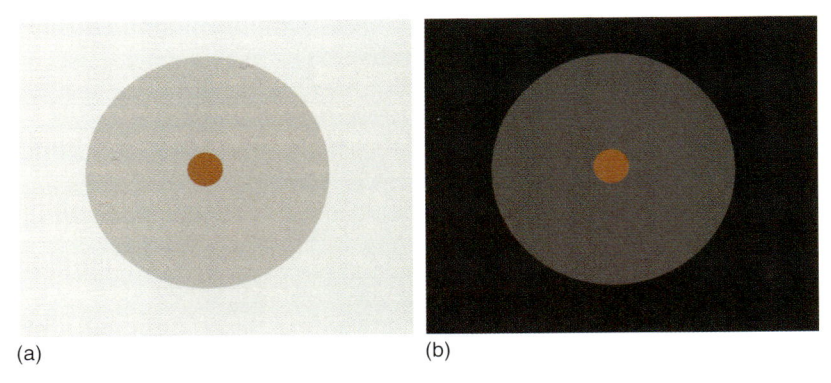

Figure 2.17 Two viewing conditions with the same color stimulus. Color stimulus (brownish color element considered): small filled circle in the middle, typically of a diameter of 2°; background: large gray filled circle, typically of a diameter of 10°; surround: the remainder of the viewing field outside the background; adapting field = background + surround. (a) Color stimulus on a bright adapting field and (b) the same stimulus on a dark adapting field (just an illustration, not intended to be colorimetrically correct). (Khanh et al. 2014 [1]. Reproduced with permission of Wiley-VCH.)

same size and shape, viewed in identical white to middle-gray surroundings by an observer photopically adapted to a field of chromaticity not too different from that of average daylight." To describe the color appearance of color stimuli viewed under different viewing conditions from incandescent (tungsten) light to daylight, from dark or dim to average or bright surround luminance levels or color stimuli on different backgrounds, the so-called color appearance models shall be used. Figure 2.17 illustrates the effect of different viewing conditions on the color appearance of a color element considered.

As can be seen from Figure 2.17, the color perception of the color stimulus (color element whose color appearance is being assessed by the human visual system) depends not only on the tristimulus values of the stimulus and on the adapted white point but also on other characteristics of the viewing conditions (the same stimulus on the right appears to be lighter). Below, the most widely used so-called CIECAM02 color appearance model [32] will be described, which is able to account for the change of viewing conditions. The CIECAM02 model computes numeric correlates (mathematical descriptor quantities) for all of the above-defined perceived attributes of color (e.g., chroma, saturation, brightness, etc.) for the color element considered. These CIECAM02 quantities correlate better with the perceived magnitude of all color attributes than the CIELAB quantities (L^*, C^*_{ab}, h_{ab}) [1].

The CIECAM02 model uses six parameters to describe the viewing condition of the color stimulus: L_A (adapting field luminance), Y_b (relative background luminance between 0 and 100), F (degree of adaptation), D (degree of chromatic adaptation to reference white, e.g., the white walls of an illuminated room), N_c (chromatic induction factor), and c (impact of the surround). In the absence of a measured value, the value of L_A can be estimated by dividing the value of the adopted white luminance by 5. For typical arrangements of color objects illuminated by a light source in a room, the value of $Y_b = 20$ can be used [1].

The values of F, c, and N_c depend on the so-called surround ratio computed by dividing the average surround luminance by the luminance of the reference white (e.g., the white walls in an illuminated office) in the scene. If the surround ratio equals 0, then the surround is called *dark*; if it is less than 0.2, then the surround is *dim*; otherwise, it can be considered an *average* surround. The values of F, c, and N_c equal 0.8, 0.525, and 0.8 for dark, 0.9, 0.59, and 0.95 for dim, and 1.0, 0.69, and 1.0 for average surrounds, respectively. For intermediate surround ratios, these values can be interpolated. In the CIECAM02 model, the value of D (degree of chromatic adaptation to reference white) is usually computed from the values of F and L_A by a dedicated equation. But the value of D can also be forced to a specific value instead of using that equation. For example, it can be forced to be equal to 1 to ensure complete chromatic adaptation to reference white for high illuminance levels (e.g., $E_v = 700$ lx) in a well-illuminated room [1]. Figure 2.18 shows the block diagram of the computational method of the CIECAM02 color appearance model. Owing to their complexity, its defining equations [32, 34] are not repeated here. A free worksheet can be found in the web [33].

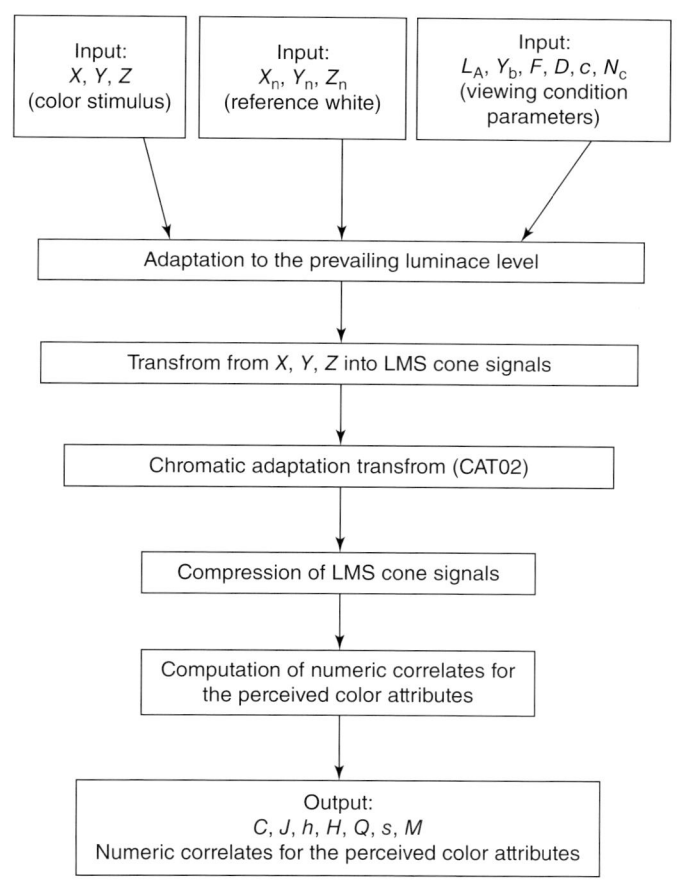

Figure 2.18 Block diagram of the computational method of the CIECAM02 color appearance model [32–34]. (Khanh *et al.* 2014 [1]. Reproduced with permission of Wiley-VCH.)

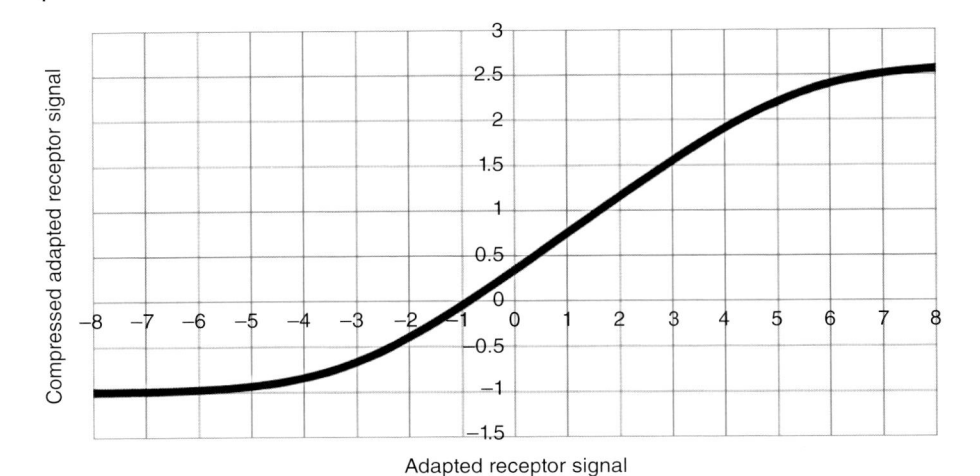

Figure 2.19 Illustration of the compression of the adapted receptor signal (L, M, or S) in the CIECAM02 color appearance model ($\log_{10}$–$\log_{10}$ diagram). (Khanh et al. 2014 [1]. Reproduced with permission of Wiley-VCH.)

As can be seen from Figure 2.18, the CIECAM02 model computes an adaptation factor to the prevailing luminance level, that is, an average luminance value of the adapting field. Then, the tristimulus values X, Y, Z (that can be measured and computed promptly by the most widely used standard instruments and their software including spectroradiometers and colorimeters) are transformed into LMS cone signals. Then, in turn, chromatic adaptation is modeled by an advanced chromatic adaptation transform, the so-called CAT02. CAT02 represents a significant enhancement compared to the neurophysiologically incorrect "chromatic adaptation" transform of CIELAB that simply divides a tristimulus value by the corresponding value of the reference white. Adapted receptor signals are then, in turn, compressed [1]. Figure 2.19 illustrates this important signal compression step.

As can be seen from the log–log diagram of Figure 2.19, the cone signal compression phase is nonlinear: this is an important feature of the human visual system. Note that there is noise for small adapted receptor signals while the compressed signal converges to an upper limit for high input signal values. After this compression phase, the numeric correlates of the perceived color attributes are computed as the output of the model (see Figure 2.18): C (CIECAM02 chroma), J (CIECAM02 lightness), h (CIECAM02 hue angle), H (CIECAM02 hue composition), Q (CIECAM02 brightness), s (CIECAM02 saturation), M (CIECAM02 colorfulness), a_M (CIECAM02 redness–greenness), and b_M (CIECAM02 yellowness–blueness) [1].

To visualize the quantities C, J, h, a_M, and b_M, refer to Figure 2.16: they are analogous to CIELAB C^*_{ab}, L^*, h_{ab}, a^*, and b^*, respectively. But the correlation between them and the magnitudes of the corresponding perceptual color attributes is significantly enhanced compared to CIELAB, as mentioned earlier. CIECAM02 H (hue composition) varies between 0 and 400 and not between 0° and 360°: 0

corresponds to the so-called *unique red* (without any yellow and blue perception), 100 to *unique yellow* (without red and green), 200 to *unique green* (without yellow and blue), and 300 to *unique blue* (without red and green), while 400 is the same as 0 (red) [1].

As a summary, it can be stated that CIECAM02 has the following advantages compared to CIELAB: (i) viewing condition parameters can be set to represent predominating viewing conditions in the environment illuminated by the light source; (ii) reference white can be changed in a wide range (from warm white to cool white) reliably; (iii) CIECAM02 provides more numeric correlates for all perceived attributes of color (colorfulness, chroma, saturation, hue, brightness, and lightness); (iv) numeric scales of these correlates correspond better to color perception than CIELAB correlates; and (v) a CIECAM02 based advanced color difference formula and a uniform color space (CAM02-UCS) were established (see below) [1, 35].

2.5.3 Brightness Models

In this section, alternative ways to CIECAM02 of modeling perceived brightness and perceived lightness will be described. As mentioned earlier, the brightness impression of the room illuminated by the light source is an important aspect of lighting quality. The brightness perception of the light source itself and the lightness perception of the colored objects illuminated by the light source depend not only on luminance (a $V(\lambda)$ based quantity) but also on chromaticity. The brightness impression of a white wall illuminated by a warm white and a cool white LED light source is different even if their luminance is the same. Also, the lightness impression of a saturated red object and an unsaturated yellow object of the same relative luminance is different. This is the Helmholtz–Kohlrausch effect mentioned in Section 2.3. This effect is also called *brightness–luminance discrepancy* [1, 36]. Unfortunately, this effect cannot be described by the CIECAM02 color appearance model accurately. An example for the Helmholtz–Kohlrausch effect is shown in Figure 2.20.

The reason of the brightness–luminance discrepancy shown in Figure 2.20 (in the photopic luminance range) is that, in the human visual system, it is not only the luminance channel that contributes to brightness perception but also the two chromatic channels, (L − M) and (L + M − S); see Figure 2.4. The signal of the luminance channel is approximately equivalent to the sum of the weighted L-cone and M-cone signals that can be assigned to the $V(\lambda)$ function, at least approximately. Figure 2.21 shows a scheme about how these mechanisms merge to provide a brightness signal in the human visual system [1].

As seen above, brightness perception results from the sum of the luminance channel and two chromatic channels in the brain and hence perceived brightness cannot be modeled correctly by the luminance channel represented by the $V(\lambda)$ or $V_{10}(\lambda)$ functions alone [1, 37]. The attempt to use alternative spectral luminous efficiency functions – instead of $V(\lambda)$ or $V_{10}(\lambda)$ – which are based upon brightness matching for quasi-monochromatic (i.e., very saturated) 2° and 10° stimuli, fails for the general case of spectrally broadband stimuli (i.e., most color stimuli in the everyday life of a lighting engineer apart from lasers). The reason is

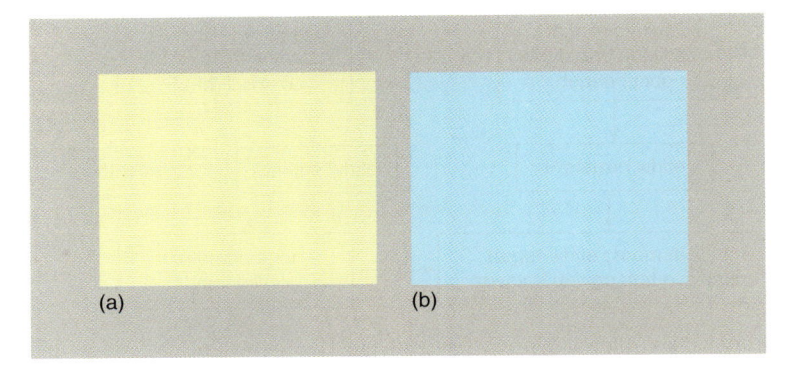

Figure 2.20 Heterochromatic brightness matching experiment. (b) The bluish color stimulus is being changed until the perceived brightness of the two color stimuli becomes the same. (a) The color stimulus (yellowish) remains constant. When the perceived brightness of the two rectangles matches visually, the luminance of the two color stimuli is still different: the bluish stimulus has a lower luminance than the yellowish stimulus [36]. (Khanh *et al.* 2014 [1]. Reproduced with permission of Wiley-VCH.)

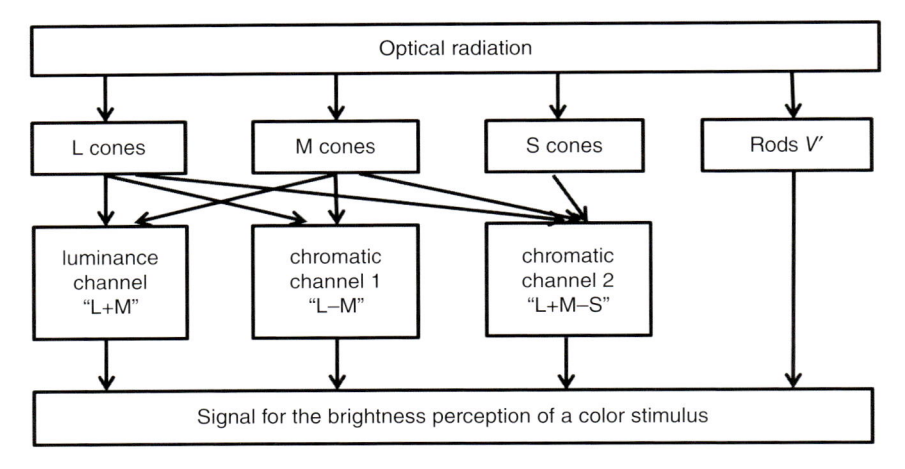

Figure 2.21 Scheme of human vision mechanisms that contribute to brightness perception. The sum of the L- and M-cone signals constitutes the luminance channel. The difference of the L- and M-cone signals represents chromatic channel 1 and the difference of the (L + M) and S-signals chromatic channel 2; compare with Figure 2.4. In the mesopic (twilight) range of vision, there is also a rod contribution with the spectral sensitivity $V'(\lambda)$. In the photopic range (generally speaking for $L > 10\,\text{cd/m}^2$, but the latter value depends on several factors), rods become inactive. (Khanh *et al.* 2014 [1]. Reproduced with permission of Wiley-VCH.)

the so-called spectral non-additivity error: as the *differences* of cone signals contribute to the human visual brain's brightness signal (see Figure 2.21), spectral integration of the SPD of the stimulus with a weighting function (*any* spectral luminous efficiency function) leads – in a mathematically obvious manner – to an error.

2.5 Modeling of Color Appearance and Perceived Color Differences

There are alternative brightness perception models without incorporating the spectral non-additivity error in order to solve the problem of brightness–luminance discrepancy. In this section, four selected models will be described:

1) the CIE brightness model [38, 39];
2) the Ware and Cowan conversion factor (WCCF) formula [40];
3) the Berman et al. model [41, 42]; and
4) the Fotios and Levermore [43] model.

The common feature of these models is that they provide the quantity L_{eq} (*equivalent luminance*; also designated by B, the abbreviation of brightness) from the SPD of the stimulus (e.g., a white wall illuminated by a warm white LED or a cold white LED). Equal values of L_{eq} (or B) of stimuli of different SPDs should result in equal brightness impressions and a set of different stimuli should be ordered by increasing perceived brightness according to their increasing L_{eq} values correctly (but a series of color stimuli with uniformly ascending L_{eq} values is not necessarily perceived to be a uniform brightness scale [1]).

2.5.3.1 The CIE Brightness Model

The structure of the CIE brightness model [38, 39] is shown in Figure 2.22. This model is intended to be valid for all luminance levels, even for mesopic (twilight) luminance levels that include scotopic (rod, $V'(\lambda)$) contribution. Mesopic levels, where both the rods and the cones are active, range typically between 0.001 and 3.0 cd/m² but there are many parameters that influence the limits of this range [45].

As can be seen from Figure 2.22, scotopic luminance (L') and photopic luminance (L) build the so-called achromatic channel in the CIE brightness model [38], which is modeled by the product $(L')^{1-a} L^a$. The weighting of

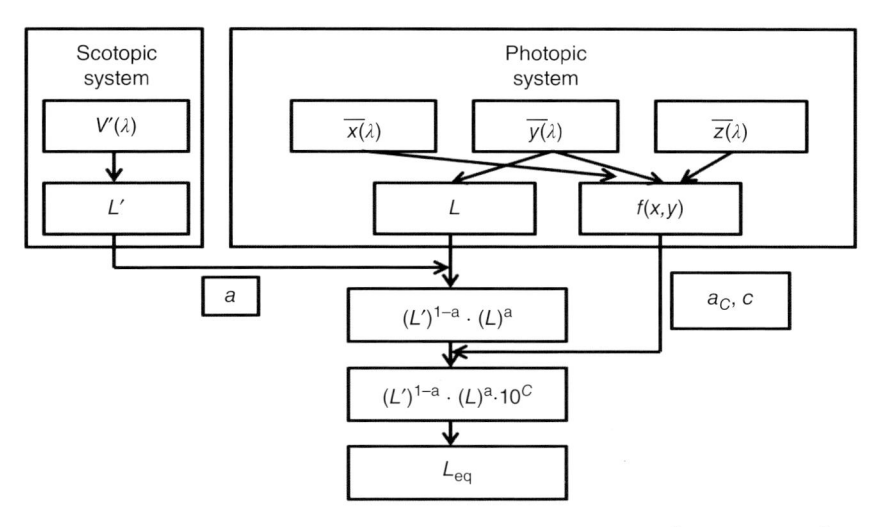

Figure 2.22 Structure of the CIE brightness model [38]. $\alpha = 0.05$ cd/m²; $\beta = 2.24$ cd/m²; $k = 1.3$, and $f(x, y)$ is a chromaticity weighting function [44]. (Khanh et al. 2014 [1]. Reproduced with permission of Wiley-VCH.)

scotopic luminance to photopic luminance is represented by the dynamic factor $a = L/(L+\alpha)$ with the parameter value $\alpha = 0.05 \text{ cd/m}^2$. Note that the latter parameter value represents only one possibility for the transition from the mesopic range to the photopic range of vision. Especially for large viewing fields when subjects assess their so-called spatial mesopic brightness impression, the influence of the rods can be more significant than predicted by this formula [1].

The contribution of the two chromatic channels (L − M channel, red-green opponency and L + M − S channel, yellow-blue opponency) is modeled by the quantity c, which appears in the exponent as 10^c. This quantity is combined with the achromatic channel in the scheme of Figure 2.22 to obtain the output of the model, a descriptor quantity of perceived brightness, *equivalent luminance* L_{eq} (as mentioned above); see Eq. (2.11).

$$L_{eq} = (L')^{1-a} \cdot L^a \cdot 10^c \quad \text{with} \quad c = a_c \cdot f(x, y) \tag{2.11}$$

In Eq. (2.11), the exponent c is the product of the chromatic adaptation parameter a_c and the function $f(x, y)$, which represents a dependence on the chromaticity coordinates x and y. The chromatic adaptation parameter a_c is an increasing function of photopic luminance; see Eq. (2.12).

$$a_c = kL^{0.5}/(L^{0.5} + \beta) \quad \text{with} \quad \beta = 2.24 \text{ cd/m}^2 \text{ and } k = 1.3 \tag{2.12}$$

The function $f(x,y)$ [44] is defined by Eq. (2.13)

$$\begin{aligned} f(x, y) = (1/2) \cdot \log_{10}\{ &-0.0054 - 0.21x + 0.77y + 1.44x^2 \\ &- 2.97xy + 1.59x^2 - 2.11\,[(1-x-y)y^2]\} - \log_{10}(y) \end{aligned} \tag{2.13}$$

2.5.3.2 The Ware and Cowan Conversion Factor Formula (WCCF)

Another, alternative model of equivalent luminance (in the above defined sense) is the so-called WCCF formula [40]. It is intended for self-luminous photopic stimuli only (not for reflecting color samples). It can be applied, for example, to compare the brightness perception of two white light sources, for example, a warm white and a cool white light source. In the first step of the computation of L_{eq} according to Ware and Cowan, the ratio (B/L) of brightness (B) to luminance (L) is computed; see Eq. (2.14). In the second step, this (B/L) ratio is multiplied by the luminance of the light source (in cd/m^2); see Eq. (2.15).

$$\log(B/L) = 0.256 - 0.184y - 2.527xy + 4.656x^3y + 4.657xy^4 \tag{2.14}$$

$$L_{eq} = (B/L)L \tag{2.15}$$

2.5.3.3 The Berman *et al.* Model

In the Berman *et al.* model, brightness is modeled by Eq. (2.16) [41, 42].

$$L_{eq} = (R_{rel}/L_{rel})^{0.5} L \tag{2.16}$$

In Eq. (2.16), L represents the conventional (photopic) luminance of the stimulus while R_{rel} is the relative signal of the rods (computed by multiplying the relative SPD of the stimulus by the relative spectral sensitivity of the rods; $V'(\lambda)$) and L_{rel} is the relative signal of the luminance channel (computed by multiplying the relative SPD of the stimulus by the $V(\lambda)$ function).

2.5.3.4 Fotios and Levermore's Brightness Model

In the Fotios and Levermore model, brightness is modeled by Eq. (2.17) [43].

$$L_{eq} = (S_{rel}/L_{rel})^{0.24} L \qquad (2.17)$$

In Eq. (2.17), L represents the conventional (photopic) luminance of the stimulus while S_{rel} is the relative signal of the S-cones (computed by multiplying the relative SPD of the stimulus by the relative spectral sensitivity of the S-cones using the so-called Smith–Pokorny 2° cone fundamentals [4]) and L_{rel} is the relative signal of the luminance channel (computed by multiplying the relative SPD of the stimulus by the $V(\lambda)$ function).

2.5.3.5 Fairchild and Pirrotta's L^{**} Model of Chromatic Lightness

If the light source illuminates colored *objects* in the interior then observers assess the perceived *lightness* of these objects because these colored objects are the so-called *related* color stimuli (i.e., the color appearance of the colored object is *related* to or compared with a reference white stimulus, for example, a white wall in the room) [46]. Similar to brightness, according to the scheme of Figure 2.21, perceived *lightness* also includes a chromatic component (Helmholtz–Kohlrausch effect) and this cannot be accounted for by the nearly achromatic CIECAM02 *J* lightness scale correctly. One possibility to model this effect is the Fairchild and Pirrotta's L^{**} formula [46], which will be described here. The formula is based on visual observations [46, 47]. Equation (2.18) defines the L^{**} formula that corrects CIELAB lightness L^* (which is completely achromatic) by adding a chromatic component.

$$L^{**} = L^* + [f(L^*) \cdot g(h°) \cdot C^*] \text{ with}$$
$$f(L^*) = 2.5 - 0.025\, L^* \text{ and} \qquad (2.18)$$
$$g(h°) = 0.116\, |\sin((h° - 90)/2)| + 0.085$$

The symbols of Eq. (2.18) have the following meaning:

$f(L^*)$: A correction function based on CIELAB lightness L^*
$g(h°)$: A correction function based on CIELAB hue angle h in degrees
C^*: CIELAB chroma; h: CIELAB hue angle in degrees.

As can be seen from Eq. (2.18), the correction term, that is, $[f(L^*) \cdot g(h°) \cdot C^*]$, increases linearly when the chroma of the stimulus (colored object) increases owing to the fact that the contribution of the two chromatic channels increases with increasing CIELAB chroma. The minimum of the hue correction function is located – not surprisingly – at the yellow hue angle $h = 90°$ [1].

2.5.4 Modeling of Color Difference Perception in Color Spaces

2.5.4.1 CIELAB Color Difference

Perceived total color differences (including lightness differences) between two color stimuli can be modeled by the Euclidean distance between them in a rectangular color space, for example, CIELAB; this quantity is denoted by ΔE^*_{ab}; see Eq. (2.19).

$$\Delta E^*_{ab} = \sqrt{(\Delta L^*)^2 + (\Delta a^*)^2 + (\Delta b^*)^2} \qquad (2.19)$$

Lightness, chroma, and hue angle differences of two color stimuli (ΔL^*, ΔC^*_{ab}, and Δh_{ab}) can be computed by subtracting the lightness, chroma, and hue angle values of the two color stimuli. Hue differences (ΔH^*_{ab}) must not be confused with hue angle differences (Δh_{ab}). Hue differences include the fact that the same hue change results in a large color difference for large chroma and in a small color difference for small chroma (i.e., in the neighborhood of the CIELAB L^* axis); see Figure 2.16. CIELAB hue difference is defined by Eq. (2.20). In Eq. (2.3.10), ΔH^*_{ab} has the same sign as Δh_{ab}.

$$\Delta H^*_{ab} = \sqrt{(\Delta E^*_{ab})^2 - (\Delta L^*)^2 - (\Delta C^*_{ab})^2} \tag{2.20}$$

2.5.4.2 CAM02-UCS Uniform Color Space and Color Difference

Unfortunately, the CIELAB color space and the color difference computed in it by using Eq. (2.19) exhibit perceptual nonuniformities depending on the region of color space (e.g., reddish or bluish colors) and on color difference magnitude (small, medium, or large color differences) [48]. To address these problems, a uniform color space (so-called CAM02-UCS) was introduced on the basis of the CIECAM02 color appearance model [35] to describe all types of color difference magnitudes. The superior performance of CAM02-UCS was corroborated in visual experiments on color rendering [1, 49, 50]. The CAM02-UCS color space [35] is defined in the following two steps on the basis of the CIECAM02 color appearance model:

Step 1: The CIECAM02 correlates lightness (J) and colorfulness (M) are transformed according to Eqs. (2.21) and (2.22). The CIECAM02 hue angle h is not transformed [1].

$$J' = \frac{1.7J}{0.007J + 1} \tag{2.21}$$

$$M' = \frac{1}{0.0228} \ln(1 + 0.0228 M) \tag{2.22}$$

Step 2: The variables a' und b' are defined by Eq. (2.23).

$$a' = M' \cos(h)$$
$$b' = M' \sin(h) \tag{2.23}$$

The variables a' and b' can be imagined as new correlates of perceived redness–greenness and yellowness–blueness (similar to CIELAB a^* and b^* in Figure 2.16). But their real importance is to constitute – together with J' – three axes of a uniform color space (a', b', J') in which perceived color differences can be quantified reliably by the Euclidean distance in this space [1]. The J' axis stands in the middle of the a'–b' plane perpendicular to this plane. In order to describe the perceived color difference between two color stimuli in the CAM02-UCS color space, the so-called CAM02-UCS color difference ($\Delta E'$ or $\Delta E_{CAM02\text{-}UCS}$ or briefly ΔE_{UCS}) shall be calculated by Eq. (2.24).

$$\Delta E' = \sqrt{(\Delta J')^2 + (\Delta a')^2 + (\Delta b')^2} \tag{2.24}$$

The advantage of the CAM02-UCS color difference metric of Eq. (2.24) is that it is based on the CIECAM02 color appearance model and hence it can be applied to all viewing conditions by adjusting the values of the CIECAM02 viewing parameters, the tristimulus values of the reference white, as well as the parameters L_A, Y_b, F, D, c, and N_c [1]. Although the CAM02-UCS color difference metric (Eq. (2.24)) is perceptually more uniform (see Section 2.2.5) than the CIELAB color difference metric (Eq. (2.19)), the perceptual uniformity of CAM02-UCS is not perfect; see Figure 5.18 (circles of the same diameter would represent perfect uniformity). A comparison of Figures 5.17 (CIELAB) and 5.18 (CAM02-UCS) shows that CIELAB's ellipses are more elongated and their size varies more than the ellipses of the CAM02-UCS color difference metric.

An alternative to Eq. (2.24) is the so-called CIEDE2000 color difference formula [51] for small to moderate color differences ($\Delta E^*_{ab} < 5.0$). This formula weights the components hue, chroma, and lightness of the CIELAB color difference ΔE^*_{ab} and uses a factor to account for the interaction between the hue and chroma components. Color difference formulae are generally important to model the color fidelity property of light sources because the fidelity property of the light source is related to the perception of color differences between the color appearance of colored test objects under the actual test light source and its reference light source. In particular, Eq. (2.24) constitutes the basis to define *advanced* color fidelity indices (i.e., numeric descriptors of the color fidelity property); see Section 2.6.

Chroma difference formulae play an important role in the definition of several types of color quality metrics (i.e., numeric descriptors that are able to model a certain aspect of color quality). A value of a chroma difference metric is an important variable to characterize the color discrimination property of the light source; see Section 7.8.4. Values of a chroma difference metric computed by the use of a chroma difference formula (e.g., CIELAB ΔC^*_{ab}) can also be used to account for the color preference, naturalness, and vividness property of the light source, either as a standalone prediction variable or better as a component of a combination of two or more color quality metrics; see Section 7.7.

As can be seen from Figure 2.14, if the chroma of the colored objects illuminated by the current test light source is less than their chroma under the reference light source (Planckian or daylight), which can be expressed by $\Delta C^*_{ab} = C^*_{ab,test} - C^*_{ab,ref} < 0$ (see Figure 2.14a), then the perceived color quality is low. Consequently, in this case, the value of a suitable color quality index shall also be low. $\Delta C^*_{ab} = 0$ corresponds to Figure 2.14b with high color fidelity. With increasing positive values of ΔC^*_{ab}, color naturalness, preference, and vividness increase; see Figure 2.14c–e. If we continue oversaturating the colored objects ($\Delta C^*_{ab} > 0$) by changing the relative SPD of the illuminating (multi-LED) spectrum, there will be a reversal of the subjective judgments at a certain maximum oversaturation level ($\Delta C^*_{ab} = \Delta C^*_{ab,max}$), at least in case of naturalness and preference; see Sections 7.7 and 7.8.

In case of a given test light source, the value of ΔC^*_{ab} strongly depends on the spectral reflectance $\rho(\lambda)$ of the colored object it is calculated for; for example, in the case of a deep red colored object with a spectral reflectance function of highly accentuated high-pass nature (e.g., beginning to reflect above $\lambda > 610$ nm like VS_1 in Figure 4.19 with a steeply increasing edge), high ΔC^*_{ab} values can be

Table 2.1 Examples for the magnitude of object oversaturation in terms of CIELAB ΔC^*_{ab} values in the case of the test color samples in Figure 4.19.

Light source	RGB33	FL34	HPS	WW-LED
VS_1	14.8	−6.5	−28.8	−3.8
VS_{14}	2.0	6.5	5.7	3.5
ΔC^*_{CQS} (mean ΔC^* for VS_1–VS_{15})	6.0	0.3	−6.8	−0.2

VS_1, VS_{14}, and also the average value of ΔC^*_{ab} for all 15 test color samples VS_1–VS_{15} in Figure 4.19 denoted by $\Delta C^*(VS_1$–$VS_{15})$ or by ΔC^*_{CQS}; for different light source spectral power distributions: RGB33–RGB LED with three accentuated local spectral maxima (peaks) at CCT = 3300 K; FL34: a triphosphor fluorescent lamp at 3380 K with high local minima and maxima and gaps in its spectrum; HPS: a yellowish high-pressure sodium lamp at CCT = 2070 K used for exterior lighting; and WW-LED: a warm-white phosphor-converted LED lamp at 2740 K.

expected. In contrast, the same test light source results in lower ΔC^*_{ab} values in the case of the less saturated (yellowish) VS_4 sample, which provides less interaction with the shape of the illuminating spectrum; see the left-hand side of Figure 2.1. Table 2.1 shows examples for the magnitude of the CIELAB ΔC^*_{ab} values in case of VS_1, VS_4, and the average value of ΔC^*_{ab} for all 15 TCSs VS_1–VS_{15} in Figure 4.19, denoted by $\Delta C^*(VS1$–$VS15)$ in the case of different light source SPDs.

As can be seen from Table 2.1, depending on the type of light source and the type of object color, different values of object color desaturation ($\Delta C^*_{ab} < 0$) or oversaturation ($\Delta C^*_{ab} > 0$) arise. The light source RGB33 that exhibits three distinct local spectral maxima (the peaks of the red, green, and blue colored LEDs) strongly oversaturates the deep red TCS VS1. In contrast, the LPS light source used for exterior lighting strongly desaturates reddish colors, resulting in a very unnatural color appearance.

2.6 Modeling of Color Quality

In this section, models of the different aspects of light source color quality (described in Section 2.4) will be presented. These models start from the SPD of the test light source and the spectral reflectance of a characteristic set of TCSs (a collection of representative colored objects; see Chapter 4) and use the quantities of color appearance models and color difference equations to define a descriptor quantity (a so-called color quality metric) for a given aspect of color quality, for example, a color preference index (CPI) to model the subjective impression of color preference. The subjective impression of a given aspect of color quality scaled by a panel of observers in a visual experiment (e.g., on a scale between 0 and 100) should correlate (in average) well with the corresponding metric for a set of different test light sources.

As mentioned in Section 2.3.4, the different visual aspects (attributes) of color quality can be associated with semantic interpretations (see Figure 7.28). The ranges of the color quality index values describing a certain aspect that corresponds to the "good" or the "very good" category shall be considered as target value ranges during the spectral optimization of the light source. The numeric limits of these ranges in terms of a suitable color quality index are very important for lighting designers and lighting practitioners to ensure the user acceptance of the light source; see Section 7.9.

If the shape of the relative SPD function of the test light source changes at a fixed white tone chromaticity (i.e., if we consider metameric white spectra of different shape, see Figure 8.35, left; CCT = 3100 K) then the color appearance of the TCSs or colored objects changes compared to a reference condition and this change can be illustrated in a color space or a color diagram, for example, the CIELAB $a^* - b^*$ diagram (see Figure 8.35, right). The computational method of color quality indices works generally in such a way that the tendency of this change (including all TCSs) is characterized by a single number, the value of the color quality metric. As can be seen from Figure 8.35, if high RGB peaks (local maxima) appear in the spectrum then the location of the TCSs in the color diagram (black polygon) becomes highly distorted compared to the reference condition (red polygon). In the latter case, the value of the color fidelity index is small but the mean value of ΔC_{ab}^* is high (for high color vividness). Note that the value of ΔC_{ab}^* can be very different for different TCSs; see Table 2.1.

The computational methods of the different color quality indices found in the literature will be described according to the following system (see Table 6.1 and Figure 2.14):

1) *to describe color fidelity (realness):* A group of color fidelity indices (representing zero oversaturation); Section 2.6.1;
2) *to describe color naturalness and color preference (attractiveness or pleasantness):* A group of color preference indices (representing a moderate level of oversaturation; less for naturalness, slightly more for color preference); Section 2.6.2;
3) *to describe color vividness:* A group of color gamut indices (with the optimization target of achieving a high level of object oversaturation); Section 2.6.3; and
4) *to describe color discrimination ability (which is related to the perception of spatial color, color shadings, continuous color transitions):* A group of color discrimination indices (which require a moderate oversaturation level); Section 2.6.4.

The huge domain of knowledge on color harmony and color harmony indices (see, e.g., [52–54]) will not be dealt with in this book nor the issue of statistical color quality indices [55].

2.6.1 Color Fidelity Indices

2.6.1.1 The CIE Color-Rendering Index

The computational method of the so-called *general* CIE color-rendering index (CRI) (R_a) [56] has the following steps [1]; see the flowchart in Figure 2.23.

Step 1: A reference illuminant is selected. This has the same CCT (T_{cp}) as the test light source. If the value of T_{cp} is less than 5000 K then a blackbody radiator (see Figures 2.8 and 2.9) of the same color temperature is the

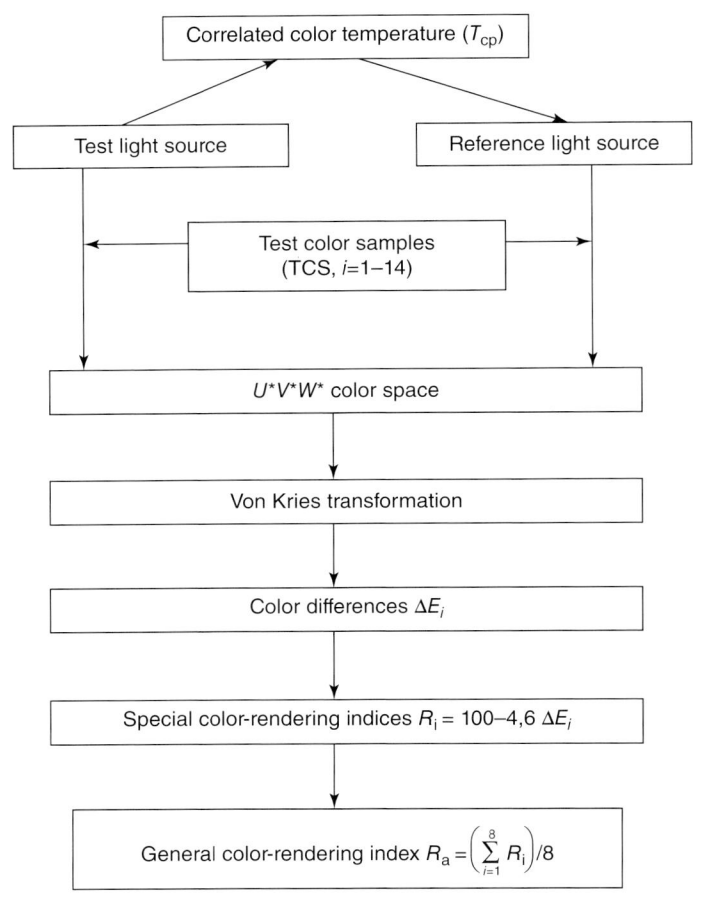

Figure 2.23 Flowchart of the computational method of the general color-rendering index CIE CRI R_a [56] (after [57]). (Khanh et al. 2014 [1]. Reproduced with permission of Wiley-VCH.)

reference illuminant. If the value of T_{cp} is 5000 K or greater than 5000 K then a phase of daylight of the same T_{cp} value is used as reference illuminant. The Euclidean distance between the test light source and the reference illuminant in the u, v color diagram shall not be greater than 5.4×10^{-3}. Otherwise, according to this method, the test light source cannot be considered a shade of white; see Chapter 3.

Step 2: Fourteen TCSs are selected from the Munsell color atlas. The first eight TCSs are used on average to compute the value of the general CRI (R_a). For every one of the 14 TCSs, 14 so-called special color-rendering indices are computed. Figure 4.15 shows the spectral reflectance functions of the spectral reflectance of these TCSs.

Step 3: The CIE 1931 tristimulus values X, Y, Z are computed for the 14 TCSs (TCS01–14) under the test light source and the reference light source. These values are transformed into CIE 1960 UCS co-ordinates (u, v) and into the currently outdated, so-called CIE 1964 U^*, V^*, W^* color space (which will not be dealt with here because it is obsolete [58]).

Step 4: In order to describe the chromatic adaptation between the white points of the test light source and the reference light source, the chromaticity of the test light source is transformed into the chromaticity of the reference illuminant by the aid of a currently outdated, so-called *von Kries transformation* [56].

Step 5: Fourteen CIE 1964 color differences (ΔE_i) are computed for each one of the 14 TCSs ($i = 1,\ldots,14$) from the differences of the U^*, V^*, W^* values under the test light source and the reference light source ($\Delta U_i^*, \Delta V_i^*, \Delta W_i^*$):

$$\Delta E_i = \sqrt{(\Delta U_i^*)^2 + (\Delta V_i^*)^2 + (\Delta W_i^*)^2} \tag{2.25}$$

Step 6: For every TCS (TCS01–14), a so-called *special* CRI is computed (R_i; $i = 1,\ldots,14$) by Eq. (2.26).

$$R_i = 100 - 4.6 \Delta E_i \tag{2.26}$$

The indices R_i are scaled according to Eq. (2.26) in the following way:
the value of $R_i = 100$ means a complete agreement between the test and reference appearances of the TCS; and
the CIE illuminant F4 [8] (a warm white fluorescent lamp) has the value of $R_a = 51$. This is established by the factor 4.6 in Eq. (2.26).

Step 7: The general CRI (R_a) is defined as the arithmetic mean of the first eight special color-rendering indices (see Figure 4.15); see Eq. (2.27).

$$R_a = (1/8) \sum_{i=1}^{8} R_i \tag{2.27}$$

The above described CRI computation method [56] (see Figure 2.23) is known to exhibit several problems. It was shown in visual experiments and in computations that the general CRI R_a is unable to describe the perceived color-rendering property of light sources correctly [59, 60]. The color rendering rank order of light sources obtained in visual experiments is often predicted incorrectly by their general CRI values (R_a). The reasons for these deficiencies [1] are described below.

1) *Choice of the TCSs*: The first eight TCSs (TCS1–8, see Figure 4.15) are unsaturated and do not represent the variety of colored natural and artificial objects (including saturated objects). An average value of the first eight special color-rendering indices (Eq. (2.27)) cannot describe the variety of the spectral reflectances of colored objects in interaction with the often discontinuous emission spectra (i.e., comprising gaps and strong local maxima) of the diverse test light sources. Chapter 4 is devoted to this issue including the choice of TCSs and the spectral reflectance and grouping of real colored objects.

2) *Nonuniform color space*: One of the most serious deficiencies of the CIE CRI color-rendering method represents the outdated color difference formula

used to predict the perceived color difference between two color appearances of the same TCS: under the test light source and under the reference light source. This color difference should be predicted by a visually relevant (i.e., perceptually uniform) color difference metric that is suitable to describe all kinds of color differences for all types of color stimuli and all magnitudes of color differences from visually just noticeable up to large differences. The outdated color difference formula of Eq. (2.25) is based on the obsolete and visually nonuniform CIE 1964 U^*, V^*, W^* color space [56]. Therefore, the color differences ΔE_i computed via Eq. (2.25) are perceptually incorrect. These problems can be solved by using the perceptually (more) uniform CAM02-UCS color space (see Section 2.5.4.2) that performs suitably both for small and for large color differences [61]. Another advantage of the CAM02-UCS color space is that it uses an advanced chromatic adaptation formula, and hence it is applicable to a wide range of the illuminant's CCTs. The updated CRI methods are based on this CAM02-UCS color space; see below.

3) *Interpretation of the values on the numeric CRI scale:* It is not easy for the user of the light source and sometimes even for lighting engineering experts to interpret the values on the CRI scale as a benchmark of color quality. Is the color-rendering property of the light source moderate, good, or very good if $R_a = 83, 87, 93$, or 97, respectively? The interpretation of the differences on the same scale is also problematic; for example, how different is the visual color-rendering judgment for $\Delta R_a = 0.5, 3.0$, or 25.0? A solution to this problem can be found in Section 7.9.1.1 (see Table 7.20).

2.6.1.2 The Color Fidelity Index of the CQS Method

The CQS (color quality scale) method of the NIST (USA) computes four different indices: Q_f (a color fidelity index, to be described in the present section), two color preference indices (Q_a and Q_p), and a color gamut index (Q_g); see Sections 2.6.2 and 2.6.3. Generally, version 9.0 (2011) of the CQS method will be dealt with, except for CQS Q_p (according to version 7.5).

The computational method of the CQS Q_f index has the following components [62]:

1) *Test color samples*: The CQS method compares the color appearance of 15 saturated test colors under a test light source and a reference light source. Figure 4.19 shows the spectral reflectance of these TCSs (the so-called VS_1-VS_{15}) chosen from the Munsell color atlas.
2) *Reference light source*: The same reference light source is used as in the CIE CRI method (Section 2.6.1.1).
3) *CIE tristimulus values*: X, Y, Z tristimulus values are computed for every test color (VS1–VS15), both under the test light source and the reference light source.
4) *Chromatic adaptation*: X, Y, Z tristimulus values under the test light source are transformed into the corresponding X, Y, Z values under the reference light source using the chromatic adaptation transformation CMCCAT2000 [63].

5) *Computation of CIELAB values*: CIELAB L^*, a^*, b^*, and C^* values are computed for every test color, both under the test light source and the reference light source.
6) *CIELAB color difference*: The value of $\Delta E^*_{ab,i}$ between the test light source and the reference light source is computed for every test color ($i = 1-15$).
7) *Root mean square*: To calculate a mean value from the CIELAB color differences ΔE^*_{ab} of the 15 test colors, the root mean square (RMS) is used instead of the arithmetic mean; see Eq. (2.28).

$$\Delta E_{rms} = \sqrt{\frac{1}{15}\sum_{i=1}^{15}(\Delta E^*_{ab,i})^2} \tag{2.28}$$

8) *Scaling*: The ΔE_{rms} value of Eq. (2.28) is scaled according to Eq. (2.29) to obtain the preliminary variable CQS $Q_{f,rms}$.

$$Q_{f,rms} = 100 - 3.0305\Delta E_{rms} \tag{2.29}$$

9) *Rescaling for the interval 0–100*: In order to avoid negative index values, the value of $Q_{f,rms}$ is transformed into the interval 0–100 to get the final value of the CQS Q_f index; see Eq. (2.30).

$$Q_f = 10\ \ln\{\exp(Q_{f,rms}/10) + 1\} \tag{2.30}$$

In the CQS Q_f method (version 9.0, 2011), special color fidelity indices are not computed.

2.6.1.3 The Color Fidelity Index CRI2012 (nCRI)

To solve the problems discussed in Section 2.6.1.1 by the introduction of an updated CRI called CRI2012 or nCRI in 2012, the following principles were considered [64]:

1) Start from a large set of representative TCSs.
2) Represent the color stimuli of these TCSs in the CAM02-UCS color space.
3) Every color stimulus of a certain TCS has two versions: one stimulus is the TCS illuminated by the reference illuminant and the other one is illuminated by the test light source. Both versions are represented in CAM02-UCS color space. The vector that originates from the reference stimulus and points toward the test stimulus is called *color shift*; see the green arrow in Figure 2.16.

Starting from the relative spectral radiance distribution of the light source, the CRI2012 method computes a general CRI, $R_{a,2012}$. The components of the CRI2012 method [64] are summarized [1] below in comparison with the CIE CRI method (Section 2.6.1.1):

1) The reference illuminant is the same illuminant as in the CIE CRI method.
2) Instead of the CIE TCS_1-TCS_{14}, there is a new set of 17 TCSs (the so-called HL17 set); see Figure 4.18, left. The starting point to define these TCSs was the so-called *Leeds 100 000* set [64] with spectral reflectance curves of more than a hundred thousand natural and artificial objects measured at the University of Leeds (England).

3) Instead of the CIE 1964 color difference calculation (Eq. (2.25)), there is a new color difference calculation in the CAM02-UCS color space.
4) Instead of Eq. (2.27), RMS (similar to Eq. (2.28)) is used to represent the mean value of the calculated color differences for the TCSs.
5) Instead of the linear scaling of the CRI, there is a new nonlinear scaling function to transform the CAM02-UCS color difference ($\Delta E'$) into the new index value $R_{i,2012}$.

The CRI2012 computation method [64] has the following steps [1]; see Figure 2.24.

1) Selection of the reference illuminant for the current test light source, the same illuminant as in the CIE CRI method (Figure 2.23).
2) Use of the HL17 set of TCSs; see Figure 4.18 (left).
3) The 10° tristimulus values X_{10}, Y_{10}, and Z_{10} are calculated for every TCS both under the test light source and under the reference light source. These

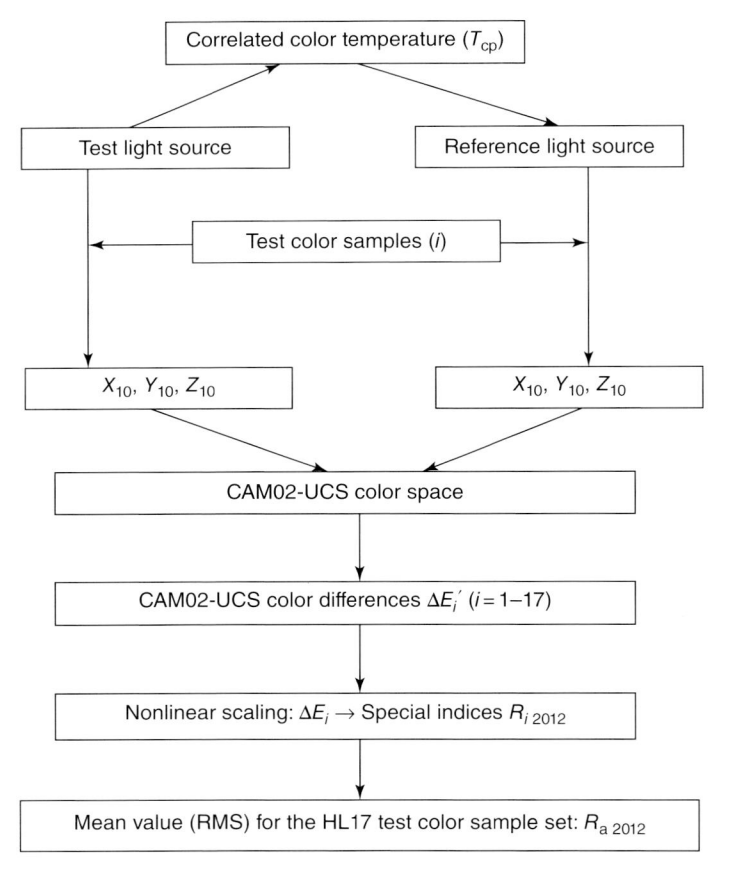

Figure 2.24 Flowchart of the CRI2012 color-rendering index computation method to obtain the special color-rendering indices $R_{i,2012}$ and the general color-rendering index $R_{a,2012}$ [64]. (Khanh et al. 2014 [1]. Reproduced with permission of Wiley-VCH.)

values are transformed into CAM02-UCS color space assuming the following CIECAM02 viewing condition parameters: $L_A = 100$ cd/m^2; $Y_b = 20$; $F = 1.0$; $D = 1.0$; $C = 0.69$; and $N_c = 1.0$.
4) CAM02-UCS color differences ($\Delta E'_i$) are computed for every TCS.
5) These color differences ($\Delta E'_i$) are scaled in a nonlinear manner to obtain the CRI values ($R_{i,2012}$, $R_{a,2012}$).
6) The value of the general CRI $R_{a,2012}$ is obtained as an RMS value calculated from the 17 $\Delta E'_i$ values of the 17 TCSs of one of the three different sets, the so-called HL17 set. It is not suggested to use these 17 TCSs to compute special color-rendering indices.

To compute *special* color-rendering indices, a set of 210 *real test colors* was defined in the CRI2012 computation method [64]. This set contains the following test colors that cover the whole hue circle at all saturation levels approximately uniformly:

1) Ninety test colors whose color stimuli change only a little if the spectral emission of the test light source changes (the so-called *high color constancy samples* or HCC);
2) Ninety test colors whose color stimuli change a lot if the spectral emission of the test light source changes (the so-called *low color constancy samples* or LCC); and
3) Ten typical artist's colors and 4×5 different skin tones (African, Caucasian, Hispanic, Oriental, and South Asian). Every skin tone was taken four times in order to increase its weight within the set.

The CRI2012 method calculates special color-rendering indices for the 210 real test colors; and, as an option, also for the so-called *Leeds 1000* test color set (containing 1000 representative colors selected from the Leeds 100 000 set). As an optional supplementary information, in addition to the value of $R_{a,2012}$, the *worst* special color-rendering indices of the subsets of the 210 test colors can be specified.

In the CRI2012 method, the CAM02-UCS color difference is computed for every one of the 17 test colors of the HL17 set ($\Delta E'_i, i = 1 - 17$) between the two color appearances of the test color under the test and reference light sources. Then, the RMS value of the set of these 17 CAM02-UCS color differences is calculated; see Eq. (2.31).

$$\Delta E'_{rms} = \sqrt{\frac{\sum_{i=1}^{17} (E'_{ii})^2}{17}} \quad (2.31)$$

The advantage of the RMS formula (Eq. (2.31)) is that if a certain color difference value ($\Delta E'_i$) is high (i.e., if there is a very poorly rendered TCS) then it has more influence than in case of the arithmetic mean. This corresponds to the visual effect that if one test color is rendered very poorly then this deteriorates the visual color-rendering property significantly. This is why the RMS value is more suitable to condense the color rendering property of a light source than an arithmetic mean value [1].

The scale of the general CRI $R_{a,2012}$ [64] is defined by the aid of the nonlinear scaling of Eq. (2.32).

$$R_{a,2012} = 100 \left[\frac{2}{1 + e^{[(1/55)(\Delta E'_{rms})^{1.5}]}} \right]^2 \quad (2.32)$$

The special indices $R_{i,2012}$ of the *Leeds 1000* set and the *210 real* set as functions of the $\Delta E'_i$ color difference values shall be computed by using Eq. (2.32), substituting the symbol $\Delta E'_{rms}$ by $\Delta E'_i$. The nonlinearity introduced by Eq. (2.32) describes signal compression, a general property of the human visual system at the extremes of perceptual scales. The value (1/55) in Eq. (2.32) corresponds to the criterion that the arithmetic mean value of the $R_{a,2012}$ values for the 12 CIE illuminants F1–F12 should have the same value ($R_{a,2012,F1-F12,mean} = 75$) as the mean of the CIE CRI R_a values for the same 12 CIE illuminants F1–F12 ($R_{a,F1-F12,mean} = 75$).

2.6.1.4 The Color Fidelity Index R_f of the IES Method (2015)

The color fidelity index R_f of the IES method (2015) is the color fidelity component of a two-measure system [65] for evaluating the light sources' color quality (in other words, color rendition) properties. The other component is R_g, a color gamut index; see Section 2.6.3.6. A set of 99 TCSs (see Figure 4.21 and the subsequent discussion in Chapter 4) represent reflectance data from real samples uniformly distributed in color space and also in wavelength space "precluding artificial optimization of the color rendition scores by spectral engineering"[65]. Thus the color fidelity index R_f represents a further improvement to $R_{a,2012}$. The method was adopted by the IES (IES TM-30-2015).

The computational method of the value of R_f is similar to the CRI2012 method (see Figure 2.24), with the following exceptions:

1) "In the CIE CRI and the CRI2012 methods, the reference illuminant jumps from a blackbody radiator to a phase of daylight at a CCT of 5000 K causing a small but unwanted discontinuity in the calculated values. The IES method addresses this shortcoming by using a linear combination of blackbody and daylight spectra for CCTs in the range 4500–5500 K; specifically, the illuminant is a blackbody at 4500 K, D55 at 5500 K and a linear combination of blackbody and daylight in-between" [65];
2) The set of 99 TCSs (Figure 4.21) is used to calculate $\Delta E'_i$ ($i = 1-99$).
3) The *arithmetic* mean of $\Delta E'_i$ ($i = 1-99$) is computed instead of RMS with the following remark: "In past research, use of the RMS mean has been proposed. In our case however, the large number of samples makes the arithmetic mean a safe and simple choice. Besides, in practice, arithmetic and RMS means yield nearly identical results for most SPDs" [65]. This arithmetic mean is denoted by $\Delta E'$.
4) The output quantity R_f is computed according to Eq. (2.33).

$$R_f = 10 \ln\{\exp[(100 - 7.54\Delta E')/10] + 1\} \quad (2.33)$$

2.6.1.5 RCRI

RCRI is an ordinal scale based CRI. Based on a series of psychophysical experiments, the RCRI method first predicts the semantic category of the CAM02-UCS color differences $\Delta E'_i$ ($i = 1–17$) between each one of 17 selected TCSs (different from the HL17 set of the CRI2012 method) illuminated by the test light source and by its reference illuminant on a five-step semantic scale R (1, excellent; 2, good; 3, acceptable; 4, not acceptable; 5, very bad) [50]. The reference illuminant is determined using the same method as that of the CIE color-rendering metric. The number of "excellent" ratings (N_1) and "good" ratings (N_2) from all 17 TCSs is calculated ($N_1 + N_2 \leq 17$). The value of RCRI is computed according to Eq. (2.34).

$$\text{RCRI} = 100\,[(N_1 + N_2)/17]^{1/3} \tag{2.34}$$

2.6.1.6 Summary of the Deficiencies of Color Fidelity Metrics

In summary, the following deficiencies are worth mentioning:

1) Currently used reference illuminants with CCT ≤ 4500 K are all artificial (blackbody) illuminants whose white points and SPDs do not yield optimal color quality, for example, they result in poor color gamut.
2) Color fidelity indices, as average numbers, cannot express the color fidelity property of particular colors, for example, saturated red colors or saturated blue colors.
3) The direction of the color shifts is not represented by the color fidelity index. For demanding applications, color distortion diagrams showing the color shift vectors of all relevant TCSs or colored objects in the given application shall be used to supplement the single value of the color fidelity index.
4) The methods do not include semantic interpretations in their definitions (except for RCRI). A solution to this problem can be found in Section 7.9.1.1 (see Table 7.20).
5) For general lighting, it is not the color fidelity aspect that plays the most important role for user acceptance. It is the color preference (attractiveness, pleasantness) and the naturalness aspects that are of crucial importance and these aspects can be modeled by the different color preference indices; see Section 2.6.2 and also Chapter 7.

2.6.2 Color Preference Indices

The common feature of these indices is that they support the (moderate) object oversaturating property of the light sources.

2.6.2.1 Judd's Flattery Index

This index (called R_f but the index "f" refers to "flattery" instead of "fidelity" in this case) was suggested by Judd as a supplement to the CIE CRI, because, at that time, worries appeared in the lighting community that the CIE "color rendering index of a light source may correlate poorly with public preference of the source for general lighting purposes [66]." Judd's flattery index is based on long-term memory colors and preferred colors. But it does not use the actual memory chromaticity and preferred chromaticity of the familiar objects [67, 68]. It follows the computational steps of the CIE CRI (its earlier version) except that the chromaticities of 10 of the CIE TCSs (TCS$_1$–TCS$_8$, TCS$_{13}$, and TCS$_{14}$) under the reference illuminant are shifted in the u, v chromaticity diagram toward the preferred chromaticities of these TCSs (based on Bartleson [27], Sanders [69], and Newhall *et al.* [70] studies). These chromaticity shifts are generally in the direction of increased saturation. The chromaticity shifts actually used ($\Delta uv_i, i = 1-8; i = 13;$ and $i = 14$) equal only one-fifth of the total chromaticity shifts between the reference chromaticity and the preferred chromaticity. The weighted sum of these chromaticity shifts (Δuv) is calculated, by the general use of the weights 5% except for TCS$_2$ and TCS$_{14}$ (15%) and TCS$_{13}$ (preferred color of complexions; 35%). R_f is then calculated according to Eq. (2.35). The maximum score of a light source equals 100 and the reference illuminant is assigned the value of 90.

$$R_f = 100 - 4.6\Delta uv \tag{2.35}$$

2.6.2.2 Gamut Area Index (GAI) in Combination with CIE R_a

The Gamut Area Index (GAI) [71] represents the color stimuli of the first eight TCSs of the CIE CRI method (Figure 4.15) in the CIE 1976 u', v' chromaticity diagram (see Eq. (2.7)) under the test light source and a fixed reference illuminant, the equienergetic stimulus (E; its spectral radiance is constant for all wavelengths). The surface area (the so-called gamut area) GA subtended by these eight points in the $u' - v'$ diagram is calculated under the test light source and the reference illuminant E. Then, GAI is defined by Eq. (2.36).

$$\text{GAI} = 100 \, \text{GA(test source)}/\text{GA(E)} \tag{2.36}$$

It was suggested [67, 71] to combine GAI with CIE R_a to account for color naturalness in the following way: $(\text{GAI} + R_a)/2$.

2.6.2.3 Thornton's Color Preference Index (CPI)

Thornton's CPI [72] is similar to Judd's flattery index, except for the following differences [67]:

1) Only the first eight TCSs are used.
2) The original magnitude of the preferred chromaticity shifts is retained (i.e., the one-fifth multiplication by Judd is not applied).
3) All samples have the same weighting and the quantity Δuv means the arithmetic mean of the chromaticity shifts.
4) The maximum score of a light source is 156 and illuminant D65 is assigned a value of 100.

Equation (2.37) shows the way the value of the CPI shall be calculated.

$$R_f = 156 - 7.18\Delta uv \tag{2.37}$$

2.6.2.4 Memory Color Rendition Index R_m or MCRI

To compute the MCRI metric, a set of nine (chromatic) TCSs [73] was selected comprising the spectral radiance coefficients depicted in Figure 4.25 (see also

Figure 4.26). The memory color rendition index, R_m (also known as MCRI) is a memory color metric that models the color preference and color naturalness aspect of the light source's color quality by comparing the rendered colors of nine familiar objects (apple, banana, orange, lavender, smurf, strawberry yogurt, sliced cucumber, cauliflower, and Caucasian skin) with their actual long-term memory colors by the use of empirically derived similarity functions [73]. Similarity functions are intended to "describe the psychophysical response to a chromaticity deviation from the memory color, implicitly taking potential differences in chroma and hue tolerance into account [68]."

The long-term memory colors of the objects and the corresponding similarity functions were obtained in psychophysical experiments [74] in which each familiar object was presented in different chromaticities to subjects who had to rate the objects' color appearance on a five-point semantic rating scale (1, very bad; 2, bad; 3, neutral; 4, good; 5, very good) with reference to their long-term memory colors. For each object, "a similarity function was derived by normalizing the bivariate Gaussian model fitted to the observer ratings" [68]. Similarity functions showed "higher tolerance for deviations in chroma than for hue [68]." Also, compared to the objects' typical color under daylight, (long-term) memory colors of most familiar objects were more saturated [68] consistent with literature [27–29, 69, 70, 75], explaining why light sources that increase object saturation get (up to a certain level of oversaturation) a better color preference and color naturalness rating [68]; see also Figure 2.14.

The memory color rendition metric, R_m of the MCRI method is defined by the following calculation steps (based on the description in [68]):

1) For each one of the nine familiar objects illuminated by the test source plus a 10th object (neutral gray), the 10° chromaticity under D65 is calculated in the so-called IPT color space [76]. P and T represent the red–green and yellow–blue axes of the IPT color space. Chromatic adaptation is taken into account by the CAT02 chromatic adaptation transform (the one used in CIECAM02). The degree of adaptation D is determined by the actual luminance of the adaptation field. If the latter is unknown then $D = 0.90$ shall be used.
2) The specific degree of similarity S_i ($i = 1$–10; the tenth object being neutral gray) with the long-term memory color of a familiar object is determined by using the similarity function $S_i(P_i, T_i)$ [73], according to Eq. (2.38).

$$S_i(P_i, T_i) = \exp[(-1/2)\{a_{i,3}(P_i - a_{i,1})^2 + 2a_{i,5}(P_i - a_{i,1})(T_i - a_{i,2}) + a_{i,4}(T_{i,1} - a_{i,2})^2\}] \tag{2.38}$$

In Eq. (2.38), $a_{i,1} - a_{i,5}$ ($i = 1$–10 for the nine familiar objects plus neutral gray) are fitting parameters describing the similarity function's centroid, shape, size, and orientation. The distance from the long-term memory colors represented by Eq. (2.38) is a non-Euclidean distance called *Mahalanobis distance* [68]. Parameter values are listed in Table 1 of [73].
3) The general degree of similarity S_a is obtained as the geometric mean of the S_i values [68].

4) The 0–1 range of the general degree of similarity S_a is rescaled to the 0–100 range using the sigmoid function in Eq. (2.39).

$$R_m = 100\,[2/\{\exp(p_1|\ln(S_a)|^{p_2} + 1)\}]^{p_3} \tag{2.39}$$

In Eq. (2.39), the rescaling parameters p_1–p_3 were chosen so that the CIE illuminants F4 and D65 have R_m values of 50 and 90, respectively, and so that $S_a < 0.5$ corresponds to $R_m \sim 0$.

2.6.2.5 The Color Preference Indices of the CQS Method (Q_a, Q_p)

2.6.2.5.1 The Color Preference Index CQS Q_a

According to Figure 2.14, increasing the saturation (or increasing the chroma) of the colored objects compared to the color fidelity ($\Delta C^*_{ab} = 0$) condition (up to a certain level), color preference increases. To account for this, the computation method of the CQS color fidelity index Q_f was modified and to other metrics, the CQS Q_a metrics [62] were introduced to account for the visual impression of color preference. The CQS Q_a metric optimizes the light source's spectrum for less oversaturated colored objects while Q_p optimizes for more oversaturated colored objects. First, the CQS Q_a metric will be dealt with. To compute CQS Q_a, two further computation steps are necessary compared to Q_f (described in Section 2.6.1.2):

1) Compute *CIELAB chroma difference:* For every CQS test color VS_i ($i = 1$–15), the chroma difference is computed between the test light source and the reference light source: $\Delta C^*_{ab,i} = C^*_{ab,test,i} - C^*_{ab,ref,i}$.
2) The so-called *chroma increment factor* shall be computed. This is called *saturation factor* in the original article [62] but the present authors think that *chroma increment factor* is a better term (while the original denotation $\Delta E^*_{ab,i,sat}$ was kept) [1]. In the CIE CRI and CRI2012 methods, those test light sources that increase the chroma of a test color are penalized. As mentioned above, in the visual experiment [62] underlying the CQS method and several other visual studies, observers preferred the increase of chroma. Accordingly, the CQS method does not penalize the chroma increase of a test color when it is illuminated by the test light source instead of the reference light source, at least for the case of the main index (CQS Q_a), which is presented here. So if $\Delta C^*_{ab,i} < 0$ for a certain test color (i.e., if the CIELAB chroma of the test color is increased by the test light source compared to the situation when it is illuminated by the reference light source) then the CIELAB color difference $\Delta E^*_{ab,i}$ between the test and reference versions is transformed via Eq. (2.40) to obtain the quantity $\Delta E^*_{ab,i,sat}$. Otherwise, the value of $\Delta E^*_{ab,i}$ remains unchanged. A chroma enhancement limit is built in for $\Delta C^*_{ab,i} \geq 10$; in this case, the value of $\Delta C^*_{ab,i}$ is set to $\Delta C^*_{ab,i} = 10$. Table 2.1 shows that the magnitude of ΔC^*_{ab} depends on the type of TCS strongly and visual experiments showed smaller mean $\Delta C^*(VS_1 - VS_{15})$ limiting values (see Figure 7.51), above which the tendency of color preference is inverted (not included in Eq. (2.40)). This problem can be solved by suitable two-metric

combinations; see Chapter 7.

$$\Delta E^*_{ab,i,\text{sat}} = \sqrt{(\Delta E^*_{ab,i})^2 - 10^2} \quad \text{if } \Delta C^*_{ab,i} \geq 10$$

$$\Delta E^*_{ab,i,\text{sat}} = \sqrt{(\Delta E^*_{ab,i})^2 - (\Delta C^*_{ab,i})^2} \quad \text{if } 0 < \Delta C^*_{ab,i} < 10$$

$$\Delta E^*_{ab,i,\text{sat}} = \Delta E^*_{ab,i} \quad \text{if } \Delta C^*_{ab,i} \leq 0 \tag{2.40}$$

Equation (2.40) means that the transformed color difference ignores the chroma difference $\Delta C^*_{ab,i}$ provided that the chroma of the test color increases compared to the reference situation.

3) In the following, steps 7–9 of the CQS R_f method (Section 2.6.1.2; Eqs. (2.28)–(2.30) shall be carried out to obtain Q_a instead of Q_f with the following modifications:
   a) In Eq. (2.28), $\Delta E^*_{ab,i,\text{sat}}$ shall be used instead of $\Delta E^*_{ab,i}$.
   b) In Eq. (2.29), the factor 3.200 shall be used instead of 3.0305 to obtain the quantity $Q_{a,\text{rms}}$ instead of $Q_{f,\text{rms}}$.
   c) In Eq. (2.30), the quantity $Q_{a,\text{rms}}$ shall be used instead of $Q_{f,\text{rms}}$ to obtain the quantity Q_a instead of Q_f.

2.6.2.5.2 The Color Preference Index CQS Q_p As mentioned above, the other CPI, CQS Q_p was intended to place additional weight (compared to Q_a) on object oversaturation by the spectrum of the light source to emphasize "the notion that increases in chroma are generally preferred and should be rewarded" [62]. Hence, Q_p is calculated using the same procedure as CQS Q_a *except* that Q_p rewards light sources for increasing object chroma [62] more strongly by computing the quantity $Q_{p,\text{rms}}$ instead of $Q_{a,\text{rms}}$ according to Eq. (2.41).

$$Q_{p,\text{rms}} = 100 - 3.780 \left[\sqrt{\frac{1}{15} \sum_{i=1}^{15} (\Delta E^*_{ab,\text{sat},i})^2} - \frac{1}{15} \sum_{i=1}^{15} \Delta C^*_{ab,i} K(i) \right] \tag{2.41}$$

In Eq. (2.41), if $C^*_{ab,i,\text{test}} > C^*_{ab,i,\text{ref}}$ then $K(i) = 1$ and if $C^*_{ab,i,\text{test}} \leq C^*_{ab,i,\text{ref}}$ then $K(i) = 0$. This means that if the chroma of the test color increases under the test light source compared to the reference light source then the value of the index CQS Q_p increases more strongly than in case of CQS Q_a. In Eq. (2.30), the quantity $Q_{p,\text{rms}}$ shall be used instead of $Q_{f,\text{rms}}$ to obtain the quantity Q_p instead of Q_f. It should be noted that CQS Q_p does not appear any more in version 9 of the CQS calculation worksheet (at least in its 1 nm version); therefore, Eq. (2.41) is based on version 7.5 (2009) of the CQS worksheet.

2.6.3 Color Gamut Indices

Color gamut indices describe the volume of color space or the area of a chromaticity diagram or a projection of a color space onto the plane of the chromatic axes (e.g., the CIELAB $a^* - b^*$ diagram) covered or spanned by a representative set of object colors or TCSs (e.g., CQS $VS_1–VS_{15}$) illuminated by the test light source; see, for example, Figures 4.29 and 4.30. This volume or area is often related to the volume or area of the same object colors illuminated by the reference light source and a color gamut index is computed in this way.

2.6.3.1 The Color Gamut Index of the CQS Method (Q_g)

The CQS method defines the so-called GA scale CQS Q_g. The GA scale Q_g is calculated as the relative GA formed by the CIELAB a^*, b^* coordinates of the 15 color samples illuminated by the test light source normalized by the GA of the CIE standard illuminant D65 (the standardized phase of daylight at 6504 K) and multiplied by 100 [62]; see Figure 4.29.

2.6.3.2 The Feeling of Contrast Index (FCI)

According to Hashimoto *et al.*'s hypothesis [77], a light source that increases the feeling of contrast in the room also increases the saturation of colored objects and the color gamut. This is why this color gamut index is called *feeling of contrast index* (FCI). FCI is defined by the use of the CIELAB $a^* - b^*$ GA of four highly chromatic TCSs (red, yellow, green, and blue) under the test light source (GA_t) and the so defined GA under the D65 reference illuminant (GA_{D65}), see Figures 4.23 and 4.24 and Eq. (2.42).

$$\text{FCI} = 100\,[(GA_t)/(GA_{D65})]^{3/2} \tag{2.42}$$

2.6.3.3 Xu's Color-Rendering Capacity (CRC)

The CRC (color-rendering capacity) index [78] computes the color volume of all possible reflecting colors (resulting from a large set of all imaginable spectral reflectance curves) in CIELUV color space (mentioned in Section 2.5.1) under the test light source (V_t) and divides this value by the similar volume under the equienergetic stimulus (V_E), see Eq. (2.43).

$$\text{CRC} = V_t/V_E \tag{2.43}$$

2.6.3.4 Gamut Area Index (GAI)

The GAI was described in Section 2.6.2.2 in combination with CIE CRI R_a as a color preference measure. GAI was defined by Eq. (2.36). Here, it is mentioned again as a standalone color gamut index.

2.6.3.5 Fotios' Cone Surface Area (CSA) Index

The cone surface area (CSA) [79] index is defined as the surface area of the color cone, which has a circular base of the same area as GA(test source) in Eq. (2.36) (see Section 2.6.2.2) and a height of $w' = 1 - u' - v'$. This is expressed by Eq. (2.44).

$$\text{CSA} = \text{GA(test source)} + \pi\,[\text{GA(test source)}/\pi]^{1/2}[\text{GA(test source)}/\pi + w'^2]^{1/2} \tag{2.44}$$

2.6.3.6 The Color Gamut Index R_g of the IES Method (2015)

The color gamut index R_g of the IES method (2015) is the second (color gamut) component of the two-measure system R_f and R_g [65] for evaluating the light sources' color quality properties. The same set of 99 TCSs (Figure 4.21) is used to compute the value of R_g as used to compute the value of the color fidelity index R_f. The reference illuminant is not fixed. Instead, it has the same CCT as the test light source, similar to the R_f computation. The CAM02-UCS a' and b' values of the 99 TCSs are computed under the test light source and the reference illuminant. These 99 points are grouped into 16 hue bins of equal width in the $a' - b'$ diagram (see Figure 4.33), based on their a', b' values under the reference illuminant. In each bin, the average values of a' and b' are then computed. This results in two 16-point polygons in the $a' - b'$ plane, one under the test light source and another under the reference light source, with the areas A_{test} and A_{ref}, respectively; see Figure 4.33 (lower right image). The value of R_g is computed by Eq. (2.45).

$$R_g = 100 \times A_{\text{test}}/A_{\text{ref}} \qquad (2.45)$$

2.6.3.7 Deficiencies of Color Gamut Metrics

The above-described color gamut metrics have serious deficiencies that can be summarized as follows:

1) The choice of the TCSs of most indices is arbitrary and their number is often too low (except for the color gamut index R_g of the IES method, which uses 99 carefully chosen test colors).
2) The use of a fixed reference illuminant "tends to favor SPDs with high CCTs... however the CCT is a given specification in most lighting applications, and one wants to optimize the gamut under this constraint [65]."
3) If the TCSs become more saturated then color gamut indices tend to increase. But as color gamut indices are only amalgamated single-number descriptors of overall saturation enhancement they cannot account for the chroma change (enhancement or deterioration) of particular (and perhaps very important, like reddish) colors because they cannot describe the distortion of the shape of the distribution of the TCSs under the test light source. Color gamut indices just describe the overall change of the total area or total volume; see Section 7.6.5.
4) According to the above, it is not clear which aspect of color quality should or can be described by a (standalone) color gamut index; see also Section 2.6.4.

2.6.4 Color Discrimination Indices

Color discrimination indices (should) describe the extent to which small color differences between only slightly different TCSs can be distinguished visually under a given test light source (see Section 2.4). A CDI definition found in literature [80] assigns test light sources a better color discriminating ability if the first eight TCSs of the CIE CRI R_a calculation method (TCS$_1$–TCS$_8$) include a larger GA in the CIE 1960 $u - v$ chromaticity diagram (Eq. (2.5)). In this respect, this is a color gamut index. As mentioned above, GA, as a single number, cannot describe the entire shape of the polygon of the TCSs in the chromaticity diagram. In case of highly oversaturating light sources, strong gamut distortions might occur, and, it is possible that, although the value of the color gamut index increases, there are pairs of not very different TCSs that become visually less distinguishable; see Figures 7.34 and 7.52.

This is why the *color difference* (e.g., the CAM02-UCS color difference $\Delta E'$) between two similar, suitable TCSs under a given test light source (e.g., two adjacent desaturated color samples of Farnsworth's D-15 color vision test; see Figure 7.2, item No. 2) is a better measure of color discrimination ability. For example, if the visual task requires the color discrimination between (desaturated) reddish-orange colored objects then the color difference $\Delta E'$ between a pair of reddish-orange TCSs with only a slight difference between their spectral reflectance functions under the given test light source is a usable measure of color discrimination performance; see Figure 7.52.

2.7 Summary

In this chapter, the most important concepts of color vision, colorimetry, color appearance, and color quality were described. These concepts are necessary to understand the subsequent chapters of the book with the final aim of modeling the color quality of light sources. The long way toward the understanding of color quality models and the underlying psychophysical experiments was recapitulated in this chapter with the most important aspects for the lighting practitioner who would like to increase the color quality perceived by the user of the light source in its lit environment. According to the scheme of Figure 2.1, the light source (the subject of spectral optimization to provide excellent perceived color quality) illuminates a colored object (in the simplest situation, a homogeneous reflecting color sample), which changes the spectral composition of the light as it is reflected from its surface generating a color stimulus to be perceived by the human visual system.

The stimulus first reaches the eye media that project it onto the retina, the photoreceptor mosaic, which, as a part of the human visual brain, constitutes the interface between the brain and the (lit) environment. The visual brain processes the spatial color distributions captured by the retina and derives the different components of visual perception (motion, shape, texture, depth) including color perception. Color perception itself has three psychological dimensions: brightness, colorfulness, and hue. Between surfaces of different color stimuli, color differences are perceived. A further important issue is to describe the appearance of (nearly) achromatic color stimuli, the different shades of white or white tones, which will be dealt with in Chapter 3.

The modeling of color perception (and the influence of viewing condition parameters such as luminance level, the change of the predominating white tone in the scene, and the spatial extent or viewing angle of the colored object considered) and color difference perception constitutes the basis for the numeric evaluation and prediction of the different aspects of color quality of a light source. Specifically, the description of the color fidelity aspect needs a color difference formula to quantify the color differences between two versions of the TCSs, under the test light source and its reference light source. It will be pointed out in the coming chapters that suitable *combinations* of, for example, the chroma difference measure (ΔC^*_{ab}), CCT, the color fidelity index (R_f), the MCRI, as well as CQS Q_f result in a more accurate description of the most important color quality aspects for general lighting (color preference and color naturalness) and their semantic interpretations in terms of categories, for example, "good" color preference. The lower limit of the "good" category in terms of a suitable color quality index represents an acceptability criterion during the optimization procedure of the spectrum of the light source.

The knowledge summarized above can be used to build up the system of color quality aspects and their modeling during the reading of the following chapters of the present book in order to optimize these aspects for lighting engineering; see Figure 10.1.

References

1 Khanh, T.Q., Bodrogi, P., Vinh, Q.T., and Winkler, H. (2015) *LED Lighting-Technology and Perception*, Wiley-VCH Verlag GmbH & Co. KGaA, Weinheim.
2 Østerberg, G.A. (1935) *Topography of the Layer of Rods and Cones in the Human Retina*, Acta Ophthalmologica, vol. **6**, p. 1.
3 Sharpe, L.T., Stockman, A., Jägle, H., and Nathans, J. (1999) in *Color Vision: From Genes to Perception* (eds K.R. Gegenfurtner and L.T. Sharpe), Cambridge University Press, pp. 3–51.
4 Web database of the Color & Vision Research Laboratory, University College London, Institute of Ophthalmology, London, UK, www.cvrl.org (accessed 7 September 2016).
5 Stockman, A., Sharpe, L.T., and Fach, C.C. (1999) The spectral sensitivity of the human short-wavelength cones. *Vision Res.*, **39**, 2901–2927.
6 Stockman, A. and Sharpe, L.T. (2000) Spectral sensitivities of the middle- and long-wavelength sensitive cones derived from measurements in observers of known genotype. *Vision Res.*, **40**, 1711–1737.
7 Commission Internationale de l'Eclairage (CIE) (2005) *CIE 10 Degree Photopic Photometric Observer*, CIE, Vienna, Publ. CIE 165:2005.
8 Commission Internationale de l'Eclairage (CIE) (2004) *Colorimetry*, 3rd edn, CIE, Vienna, Publ. CIE 015:2004.
9 Bodrogi, P. and Khanh, T.Q. (2012) *Illumination, Color and Imaging. Evaluation and Optimization of Visual Displays*, Wiley-SID Series in Display Technology, Wiley-SID.
10 Stockman, A. (2004) in *The Optics Encyclopedia: Basic Foundations and Practical Applications*, vol. **1** (eds T.G. Brown, K. Creath, H. Kogelnik, M.A. Kriss, J. Schmit, and M.J. Weber), Wiley-VCH Verlag GmbH, pp. 207–226.
11 Stiles, W. and Burch, J. (1959) N.P.L. color-matching investigation: final report (1958). *Opt. Acta*, **6**, S.1–S.26.
12 Trezona, P.W. (1984) Individual Observer Data for the Stiles-Burch 2° Pilot Investigation. NPL Report QU68.
13 Carroll, J., Neitz, J., and Neitz, M. (2002) Estimates of L:M cone ratio from ERG flicker photometry and genetics. *J. Vision*, **2**, 531–542.
14 Speranskaya, N.I. (1959) Determination of spectrum color co-ordinates for twenty-seven normal observers. *Opt. Spectrosc.*, **7**, 424–428.
15 Commission Internationale de l'Eclairage (CIE) (2007) *Measurement of LEDs*, 2nd edn, CIE, Vienna, Publ. CIE 127:2007.

16 MacAdam, D.L. (1942) Visual sensitivities to color differences in daylight. *J. Opt. Soc. Am.*, **32** (5), 247–274.
17 Baer, R., Barfuß, M., and Seifert, D. (eds) (2016) *Grundlagen Beleuchtungstechnik*, 4th edn, LiTG, Deutsche Lichttechnische Gesellschaft e.V..
18 Commission Internationale de l'Eclairage (CIE) (2011) *Standard CIE S 017/E:2011 ILV: International Lighting Vocabulary*, CIE, Vienna.
19 Derefeldt, G., Swartling, T., Berggrund, U., and Bodrogi, P. (2004) Cognitive color. *Color Res. Appl.*, **29** (1), 7–19.
20 Fairchild, M.D. and Reniff, L. (1995) Time course of Chromatic adaptation for color-appearance judgments. *J. Opt. Soc. Am.*, **12** (5), 824–833.
21 Rinner, O. and Gegenfurtner, K.R. (2000) Time course of chromatic adaptation for color appearance and discrimination. *Vision Res.*, **40**, 1813–1826.
22 Humphreys, G.W. and Bruce, V. (1989) *Visual Cognition: Computational, Experimental, and Neuropsychological Perspectives*, Laurence Erlbaum Associates, Hillsdale, NJ.
23 Barsalou, L.W. (1999) Perceptual symbol systems. *Behav. Brain Sci.*, **22**, 577–660.
24 Bodrogi, P. and Tarczali, T. (2002) in *Colour Image Science. Exploiting Digital Media, Part 1*, Chapter 2 (eds L.W. MacDonald and M.R. Luo), Chichester, John Wiley & Sons, Ltd..
25 Yendrikhovskij, S.N., Blommaert, F.J.J., and de Ridder, H. (1999) Representation of memory prototype for an object color. *Color Res. Appl.*, **24**, 393–410.
26 Bodrogi, P. (1998) Shifts of short-term colour memory. PhD thesis. University of Veszprém, Veszprém.
27 Bartleson, C.J. (1960) Memory colors of familiar objects. *J. Opt. Soc. Am.*, **50**, 73–77.
28 Bartleson, C.J. (1961) Color in memory in relation to photographic reproduction. *Photogr. Sci. Eng.*, **5**, 327–331.
29 Bartleson, C.J. and Bray, C.P. (1962) On the preferred reproduction of flesh, blue-sky, and green-grass colors. *Photogr. Sci. Eng.*, **6**, 19–25.
30 Bodrogi, P., Brückner, S., Krause, N., and Khanh, T.Q. (2014) Semantic interpretation of color differences and color rendering indices. *Color Res. Appl.*, **39**, 252–262.
31 Bodrogi, P., Brückner, S., Khanh, T.Q., and Winkler, H. (2013) Visual assessment of light source color quality. *Color Res. Appl.*, **38**, 4–13.
32 Comission Internationale de l'Eclairage (CIE) (2004) *A Color Appearance Model for Color Management Systems: CIECAM02*, CIE, Vienna, Publ. CIE 159-2004.
33 Fairchild, M.D. (2013) Color Appearance Models, http://www.cis.rit.edu/fairchild/CAM.html (accessed 7 September 2016).
34 Fairchild, M.D. (1997) *Color Appearance Models*, The Wiley-IS&T Series in Imaging Science and Technology, 2nd edn, John Wiley & Sons, Ltd..
35 Luo, M.R., Cui, G., and Li, C. (2006) Uniform color spaces based on CIECAM02 color appearance model. *Color Res. Appl.*, **31**, 320–330.
36 Khanh, T.Q. and Bodrogi, P. (2013) Farbqualität von weißen LEDs und von konventionellen Lichtquellen- systematische Zusammenhänge und praktische

Bedeutungen, in *Handbuch für Beleuchtung (Lange)*, ecomed sicherheit Heidelberg.

37 Commission International de l'Eclairage (2001) Testing of Supplementary Systems of Photometry, CIE Technical Report, Wien, CIE Publ. 141–2001.

38 Sagawa, K. (2006) Toward a CIE supplementary system of photometry: brightness at any level including mesopic vision. *Ophthalmic Physiol. Opt.*, **26**, 240–245.

39 Commission International de l'Eclairage (2011) *CIE Supplementary System of Photometry*, CIE, Vienna, CIE Publ. 200:2011.

40 Commission Internationale de l'Éclairage (1995) *CIE Collection in Colour and Vision, 118/2: Models of Heterochromatic Brightness Matching*, CIE, Vienna, Publ. CIE 118-1995.

41 Berman, S.M., Jewett, D.L., Fein, G., Saika, G., and Ashford, F. (1990) Photopic luminance does not always predict perceived room brightness. *Light. Res. Technol.*, **22** (1), 37–41.

42 Berman, S.M. (1995) Implications of rod sensitivity to interior lighting practice. Proceedings of the CIE Symposium on Advances in Photometry, Vienna, December 1–3, 1994, pp. 171–176.

43 Fotios, S.A. and Levermore, G.J. (1998) Chromatic effect on apparent brightness in interior spaces II: sws Lumens model. *Light. Res. Technol.*, **30** (3), 103–106.

44 Nakano, Y., Yamada, K., Suehara, K., and Yano, T. (1999) A simple formula to calculate brightness equivalent luminance. Proceedings of the CIE 24th Session 1, pp. 33–37.

45 Stockman, A. and Sharpe, L.T. (2006) Into the twilight zone: the complexities of mesopic vision and luminous efficiency. *Ophthalmic Physiol. Opt.*, **26**, 225–239.

46 Fairchild, M.D. and Pirrotta, E. (1991) Predicting the lightness of chromatic object colours using CIELAB. *Color Res. Appl.*, **16** (6), 385–393.

47 Wyszecki, G. (1967) Correlate for lightness in terms of CIE chromaticity coordinates and luminous reflectance. *J. Opt. Soc. Am.*, **57**, 254–257.

48 Commission Internationale de l'Eclairage (CIE) (1993) *Parametric Effects in Color-Difference Evaluation*, CIE, Vienna, CIE Publ. 101-1993.

49 Li, C., Luo, M.R., Li, C., and Cui, G. (2012) The CRI-CAM02UCS color rendering index. *Color Res. Appl.*, **37**, 160–167.

50 Bodrogi, P., Brückner, S., and Khanh, T.Q. (2011) Ordinal scale based description of colour rendering. *Color Res. Appl.*, **36**, 272–285.

51 Commission Internationale de l'Eclairage (CIE) (2001) *Improvement to Industrial Color-Difference Evaluation*, CIE, Vienna, CIE Publ. 142-2001.

52 Nemcsics, A. (2012) Experimental determination of laws of color harmony. Part 6: numerical index system of color harmony. *Color Res. Appl.*, **37** (5), 343–358.

53 Szabó, F., Bodrogi, P., and Schanda, J. (2010) Experimental modeling of color harmony. *Color Res. Appl.*, **35** (1), 34–49.

54 Li-Chen Ou and Luo, M.R. (2006) A color harmony model for two-colour combinations. *Color Res. Appl.*, **31** (3), 191–204.

55 Zukauskas, A., Vaicekauskas, R., Ivanauskas, F., Vaitkevicius, H., Vitta, P., and Shur, M.S. (2009) Statistical approach to color quality of solid-state lamps. *IEEE J. Sel. Top. Quantum Electron.*, **15** (6), 1753–1762.

56 Commission Internationale de l'Éclairage (1995) *Method of Measuring and Specifying Color Rendering Properties of Light Sources*, CIE, Vienna, Publ. CIE 13.3-1995.

57 LiTG, Deutsche Lichttechnische Gesellschaft e.V (2012) LiTG-Schrift Farbwiedergabe für moderne Lichtquellen, in German: Color rendering for modern light sources, LiTG Publ. No. 28.

58 Halstead, M.B. (1977) Colour rendering: past, present, and future, in *Proceedings of the AIC Color 77*, Adam Hilger, Bristol, pp. 97–127.

59 Commission Internationale de l'Éclairage (2007) *Color Rendering of White LED Light Sources*, CIE, Vienna, Publ. CIE 177:2007.

60 Bodrogi, P., Csuti, P., Horváth, P., and Schanda, J. (2004) Why does the CIE colour rendering index fail for white RGB LED light sources? Proceedings of the CIE Expert Symposium on LED Light Sources: Physical Measurement and Visual and Photobiological Assessment, Tokyo, 2004.

61 Wang, H., Cui, G., Luo, M.R., and Xu, H. (2012) Evaluation of colour-difference formulae for different colour-difference magnitudes. *Color Res. Appl.*, **37** (5), 316–325.

62 Davis, W. and Ohno, Y. (2010) The color quality scale. *Opt. Eng.*, **49** (3), 033602.

63 Li, C.J., Luo, M.R., Rigg, B., Hunt, R.W.G., and CMC (2002) 2000 chromatic adaptation transform: CMCCAT2000. *Color Res. Appl.*, **27**, 49–58.

64 Smet, K.A.G., Schanda, J., Whitehead, L., and Luo, M.R. (2013) CRI2012: a proposal for updating the CIE color rendering index. *Light. Res. Technol.*, **45**, 689–709.

65 David, A., Fini, P.T., Houser, K.W., Ohno, Y., Royer, M.P., Smet, K.A.G., Wei, M., and Whitehead, L. (2015) Development of the IES method for evaluating the color rendition of light sources. *Opt. Express*, **23** (12), 15888–15906.

66 Judd, D.B. (1967) A flattery index for artificial illuminants. *Illum. Eng.*, **62**, 593–598.

67 Smet, K., Ryckaert, W.R., Pointer, M.R., Deconinck, G., and Hanselaer, P. (2011) Correlation between color quality metric predictions and visual appreciation of light sources. *Opt. Express*, **19** (9), 8151–8166.

68 Smet, K.A.G. and Hanselaer, P. (2016) Memory and preferred colours and the color rendition of white light sources. *Light. Res. Technol.*, **48**, 393–411.

69 Sanders, C.L. (1959) Color preferences for natural objects. *Illum. Eng.*, **54**, 452–456.

70 Newhall, S.M., Burnham, R.W., and Clark, J.R. (1957) Comparison of successive with simultaneous color matching. *J. Opt. Soc. Am.*, **47** (1), 43–54.

71 Rea, M.S. and Freyssinier-Nova, J.P. (2008) Color rendering: a tale of two metrics. *Color Res. Appl.*, **33**, 192–202.

72 Thornton, W.A. (1972) A validation of the color preference index. *Illum. Eng.*, **62**, 191–194.

73 Smet, K., Ryckaert, W.R., Pointer, M.R., Deconinck, G., and Hanselaer, P. (2012) A memory color quality metric for white light sources. *Energy Build.*, **49**, 216–225.
74 Smet, K.A.G., Ryckaert, W.R., Pointer, M.R., Deconinck, G., and Hanselaer, P. (2011) Colour appearance rating of familiar real objects. *Color Res. Appl.*, **36**, 192–200.
75 Siple, P. and Springer, R.M. (1983) Memory and preference for the colors of objects. *Percept. Psychophys.*, **34**, 363–370.
76 Ebner, F. and Fairchild, M.D. (1998) Development and testing of a color space (IPT) with improved hue uniformity. Proceedings of the IS&T 6th Color Imaging Conference, Scottsdale, AZ, 1998, pp. 8–13.
77 Hashimoto, K., Yano, T., Shimizu, M., and Nayatani, Y. (2007) New method for specifying color-rendering properties of light sources based on feeling of contrast. *Color Res. Appl.*, **32** (5), 361–371.
78 Xu, H. (1993) Colour rendering capacity and luminous efficiency of a spectrum. *Light. Res. Technol.*, **25**, 131–132.
79 Fotios, S.A. (1997) The perception of light sources of different color properties. PhD thesis. UMIST, Manchester.
80 Thornton, W.A. (1972) Color-discrimination index. *J. Opt. Soc. Am.*, **62**, 191–194.

3

The White Point of the Light Source

As mentioned in Chapter 2, the quality of the perceived *white tone* of the light source is an important aspect of color quality. The term *white point* is also widely used because the white tone of the light source is represented by a point in the chromaticity diagram [1]. Indeed, the first primary requirement of spectral engineering (of multi-LED light engines) is to assure an appropriate shade of white, the so-called *target white tone chromaticity* [1], depending on the lighting application, the type of illuminated objects, and the intended users of the light source. The chromaticity of the white tone of the light source also determines the chromaticity of achromatic or nearly achromatic objects (spectrally aselectively or nearly aselectively reflecting surfaces) in the lit environment [1]. These white or nearly white surfaces (e.g., white walls, window sills, or gray furniture) often cover a large part of the room, and hence their perceived white tone determines the state of chromatic adaptation, influences spatial brightness perception, and contributes to the assessment of color quality significantly [1]. If the quality of the white tone in the room is not acceptable (e.g., if it contains a strange shade of green, see Figure 3.1b), this fundamentally prohibits the visual acceptance of any other color quality aspect. This is why the issue of white point deserves special attention and a full chapter, and the present chapter is devoted to it.

In lighting industry, the terms *warm white*, *neutral white*, and *cool white* (sometimes also called *cold white*) are often used. Going along the blackbody locus in Figure 2.8, the CCT ranges of the sets of white points that belong to these terms should be defined. But, according to the uncertain usage of these terms, these CCT ranges often overlap. "Warm white" sometimes refers to CCTs between 2500 and 3200 K while the term *cool white* is sometimes used for CCTs ranging between 4000 and 6500 K [3]. In widespread usage, the term *neutral white* partially overlaps with "cool white"; the term *neutral white* is often used for light sources with CCTs between about 4000 and 5000 K. In the present book, we use the term *warm white* for CCTs below 3700 K; "cool white" (or, alternatively, also "cold white") for CCTs above 4500 K; and "neutral white" for CCTs between 3700 and 4500 K. Another possible grouping is to use CCT < 3500 K for warm white, CCTs between 3500 and 5000 K for neutral white, and CCT ≥ 5000 K for cool white tones [1]; see Figure 3.10. A previous visual study [4] at lower illuminance levels (50 and 200 lx) showed that the upper limit of "warm white"

Color Quality of Semiconductor and Conventional Light Sources, First Edition.
Tran Quoc Khanh, Peter Bodrogi, and Trinh Quang Vinh.
© 2017 Wiley-VCH Verlag GmbH & Co. KGaA. Published 2017 by Wiley-VCH Verlag GmbH & Co. KGaA.

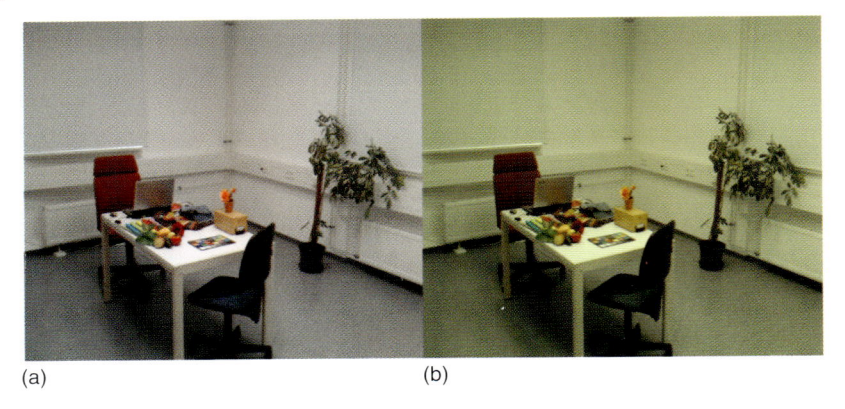

Figure 3.1 Illustration of a room illuminated by different white tones. (a) A preferred white tone in an office and (b) this illumination chromaticity is not acceptable due to the unexpected greenish shade. Image source: [2]. (Khanh *et al.* 2014 [1]. Reproduced with permission of Wiley-VCH.)

ranged between 3000 and 3700 K and the lower limit of cool white ranged between 4900 and 5400 K.

In order to obtain high color quality, the white tone illuminating the room should be *preferred* by the users in the context of the actual application; it should not only be accepted but also be found to be attractive, suitable, and pleasant. The target white tone chromaticity of the optimization of the light source's spectrum should be a *preferred* white tone. *Preferred white tones*, however, generally do not coincide with the so-called *unique white* tones (see Section 3.3), the latter having maximum perceived whiteness. Perceived whiteness is the degree to which a light source or a surface appears white without any chromatic shade, without, for example, a slight yellowish, greenish, or purplish tint [1]. Unique white tones do not have any tint or perceived hue; they are neither yellowish, nor purplish or greenish, and so on. If the perceived color of a light source or a surface does contain a chromatic shade (usually a yellowish shade, e.g., in case of a warm white LED), it is often preferred by (e.g., European and North American) light source users against a light source of greater perceived whiteness (e.g., against a cool white LED), anyway; for example, in the context of a living room (as such a yellowish shade of "white" provides a relaxing environment) [1]. In contrast, a cool white LED of high CCT (which contains a slight bluish shade) might be preferred for office lighting [1].

White tone preference is a complex cognitive issue; it depends on lighting application (e.g., office, living room, retail), gender, regional and cultural background (e.g., German vs. Chinese observers), and the characteristic color of the objects being illuminated (e.g., reddish, bluish, mixed, or white objects; see Sections 7.4 and 7.5). As mentioned above, if the amount of a certain perceived hue in the chromaticity of the illuminating spectrum exceeds a certain tolerance limit then the lit environment cannot be accepted for the purpose of general illumination at all; see Figure 3.1. In a so-called *accent lighting* application, for example, in a discotheque or a theater to achieve special emotional effects, the scene can be illuminated temporarily by any chromaticity, even saturated purples or greens [1].

One possibility to specify target white tone chromaticities follows the CIE CRI method [5]; target white chromaticities should be located at the blackbody radiators' locus (or Planckian locus) for CCT < 5000K and at the daylight locus for CCT ≥ 5000K (see Figure 2.8) with the criterion that the Euclidean distance between the light source and these loci in the u, v chromaticity diagram (Eq. (2.5)) must not be greater than $\Delta uv = 0.0054$ [1]. Note that CCT is often denoted by the symbol T_{cp}. The quantity Δuv is denoted by DC in [5]. Δuv is defined to be negative if the chromaticity of the light source is located below the Planckian locus or the daylight locus and positive above these loci. Instead of Δuv, it is more advantageous to use the quantity $\Delta u'v'$ (Eq. (2.7)), as the $u' - v'$ diagram is perceptually more uniform than u, v) in a similar way. In some studies, the denotation Duv is also used referring to the distance from the blackbody locus in the u, v diagram independent of CCT (i.e., the distance from the blackbody locus *also* for CCT ≥ 5000K; above positive, below negative).

An alternative way to specify target white tone chromaticities follows the ANSI chromaticity binning standard [6] with eight nominal CCT categories: 2700, 3000, 3500, 4000, 4500, 5000, 5700, and 6500 K and an additional category with flexible CCT between 2700 and 6500 K [1]. The centers of the target chromaticities (the so-called binning ranges) gradually shift from the Planckian locus (at low CCTs) toward the daylight locus (high CCTs) and the size of each parallelogram-shaped binning range corresponds to a so-called 7-step MacAdam ellipse (a MacAdam ellipse magnified 7 times; see Figure 2.10) in the x, y or in the u', v' chromaticity diagrams [1]. Some lighting companies use a finer grid based on 4-step or 2-step MacAdam ellipses to specify their target white points.

It will be pointed out subsequently that, in reality, the above-defined sets of possible target white chromaticities along the blackbody or daylight loci do not (necessarily) correspond to the location of preferred white tones or unique white tones in the chromaticity diagram. It should be noted that, if the luminance of the light source or of the brightest aselectively reflecting surface in the field of view of the observer is low (e.g., less than 1 cd/m^2) then there is no white perception in the lit scene; the surface (e.g., a wall) or the faint light source itself appears to be gray [1].

According to the above considerations and thinking further about the most important problems of the choice and implementation of the white point of the light source, the following questions will be discussed in this chapter:

1) At which chromaticities do unique white tones appear? Are they located at the Planckian locus, the locus of the phases of daylight, or above/below these locations in the chromaticity diagram? Where are the white tones (of different CCT) with minimal tint [3, 7] located (Section 3.1)?
2) Can the location of the unique white chromaticities be explained by retinal physiology, in terms of L − M and L + M − S signals [7] (Section 3.2)?
3) How large is the interobserver variability (an important component of human centric lighting; see Chapter 9) of white tone perception (combined with the effect of varying viewing field size) for different types of light source spectra? How can this interobserver variability be estimated based on the variability

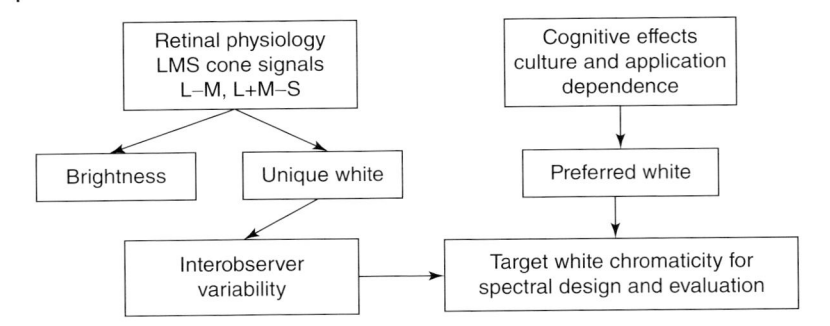

Figure 3.2 The most important issues for the choice of the white point of the light source.

of color matching functions and what are its consequences for the spectral design and evaluation of light sources (Section 3.3)?
4) How is the difference between the set of unique white chromaticities (or white chromaticities of minimal tint) and the light source users' "preferred white" tones at the different CCTs? With this question in mind, we leave the realm of retinal physiology explaining "unique white" perception to enter the level of psychological (cognitive) effects of human visual information processing to account for the light source users' white tone preference (Sections 3.4, 7.4, and 7.5).
5) Does white tone preference depend on the combination of object colors illuminated, for example, red, blue, or mixed colored objects as well as objects of different white tones (Sections 7.4 and 7.5)?
6) How is the perceived brightness of different white tones at a fixed illuminance level (Section 3.5; for strongly metameric white points, see Section 7.3)?

The block diagram of Figure 3.2 summarizes the above questions and the corresponding structure of this chapter. As can be seen from Figure 3.2, the final aim is to specify target white chromaticities and to find out the interobserver variability of their perception for the correct spectral design and evaluation of the light sources.

3.1 The Location of Unique White in the Chromaticity Diagram

The concept of *unique white* has already been defined above; in other words, it can also be defined as "a color sensation absent hue" [3]. This means that, after at least 2 min of chromatic adaptation to the unique white stimulus (a homogeneous self-luminous surface or a diffusely reflecting surface), not even a slight amount of any hue (or tint; e.g., yellowish, bluish, or greenish) can be perceived. Most practical light sources used for general illumination have a certain tint even after a long adaptation period [3] but they are still considered as "white" (e.g., "warm white" or "cool white"). "Neutral white" tones exhibit either no tint or only a slight amount of tint. It should be noted that *"source tint"* [3] should be distinguished from *"object tint"* as the white objects in the scene (e.g., cup, napkin, table cloth, chair) reflect spectrally (slightly) selectively (see Figure 7.15) providing different shades of white in the lit room even if the light source illuminates all objects homogeneously. In the above context, the following two questions arise [1]:

1) At a fixed CCT, which white points are *unique white* tones, or have maximum whiteness?
2) What is the percentage of hue (e.g., green-yellow or purple-violet) perceived and assessed by the subjects for the different white tones at the same CCT, for example, along the so-called *isotherm lines* that are perpendicular to the blackbody locus in the u, v chromaticity diagram? For lighting practice, the prediction of the type and magnitude of perceived chromatic tints in the target white tone chromaticities (on the blackbody locus, the daylight locus, above or below) is very important.

To answer the above two questions [1], the method and the results of two selected previous studies (the Rea and Freyssinier study [3] and the Smet *et al.* study [8] will be described and compared here. Several other studies have also dealt with this subject (a good summary can be found in [8]).

In the Rea and Freyssinier experiment [3], seven different spectral power distributions (SPDs) were generated at 300 lx (by using a multi-LED light source in combination with a halogen lamp; 300 lx corresponds to a luminance level of about 90 cd/m^2) for each one of the following fixed CCTs: 2700, 3000, 3500, 4100, 5000, 6500 K along the isotherm lines perpendicular to the blackbody locus in the u, v chromaticity diagram. One of the seven SPDs was always situated on the blackbody locus. Twenty observers evaluated the white tone of the interior of a viewing box (with an aselectively reflecting white paint) covering their whole field of view. Observers had to judge whether the white point was either purple/violet or green/yellow (hue choice judgments) and then also judge what percentage of that hue was present in the illuminated cube (hue magnitude judgments – 0% corresponded to unique white and 100% corresponded to maximum saturation of the given hue). Both tasks were carried out both immediately after seeing the viewing box and after 45 s of exposure in order to account for chromatic adaptation effects. Figure 3.3 [3] shows the contours of constant tint in the x, y chromaticity diagram for the "after 45 s exposure" condition (this condition is more relevant for interior lighting [1]).

As can be seen from Figure 3.3, the spectra of the same CCT but of different chromaticity (i.e., at different positions above or below the Planckian locus in the chromaticity diagram) appear to have different white tones with slight amounts of different perceived hues (e.g., having a slight yellowish or greenish tint) and they do not necessarily appear unique white (with 0%) [3, 7]. There is a line of chromaticities of minimum perceived tint in the chromaticity diagram (see the continuous curve labeled by 0%), which was found to be unique white by the subjects [3, 7]. It can also be seen from Figure 3.3 that using the white points on the blackbody locus as target white tone chromaticities above about 5000 K correspond to about 1–2% of purple-violet hue perception (see the curves labeled by −1% and −2% in Figure 3.3). Between 5000 and 6500 K, the daylight locus (not plotted in the diagram) approximately corresponds to the dash–dot line (labeled by −1%) of 1% of purple-violet. Between 5000 and 6500 K, white tones of maximum

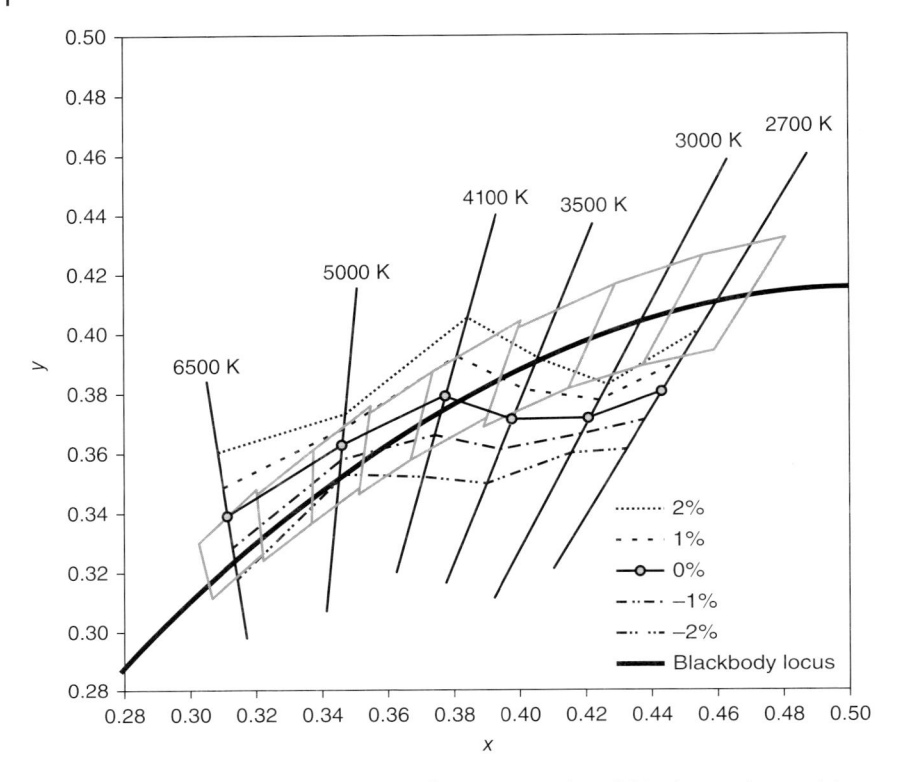

Figure 3.3 Reproduction of the contours of constant tint from [3] in the x, y chromaticity diagram. +%: percentage of green-yellow hue and −%: percentage of purple-violet hue. The u, v isotherm lines of the six different CCTs (thin lines), the blackbody locus (thick line), and the ANSI chromaticity quadrangles for solid-state lighting [6] are also shown. (Rea and Freyssinier 2011 [3]. Reproduced from Color Res. Appl. with permission of Wiley.)

whiteness (0%) are located above the blackbody locus and above the daylight locus. Between 4100 and 5000 K, the blackbody locus exhibits a purple-violet hue and the phases of daylight are perceived to be approximately achromatic. For CCT < 4100 K, white tones of zero hue (0%) are located well below the blackbody locus and the blackbody locus itself exhibits (as is well known) an increasing amount of yellow hue with decreasing CCT (this is accepted and preferred for living room lighting in Western culture [1]).

In the Smet *et al.* study [8], the chromaticity of unique white was investigated for three luminance levels (200, 1000, and 2000 cd/m^2). Thirteen observers of normal color vision had to *set* (adjust) and *rate* (in two different series of experiments) unique white at a real, small illuminated cube (side length 8.5 cm in a dark, 300 cm × 120 cm × 265 cm viewing booth) of nonselective spectral reflectance on a dark background. Both methods (rating and setting) resulted in significant interobserver variability. Considering all three luminance levels, the individual observers' unique white *settings* covered a range between 4845 and 7845 K (average: 5963 K; SE: 267 K) with a Duv (distance from the blackbody locus in the u, v diagram; positive above; negative below) range of −0.0123 and 0.0007 (average: Duv = −0.0065; SE = 0.0012). The observers' unique white *ratings* covered a range between 5331 and 6916 K (average: 5989 K; SE: 136 K) with a Duv range of −0.0111 and −0.0054 (average: Duv = −0.0085; SE = 0.0004).

Comparing the two experiments, the Smet *et al.* study [8] used a small white stimulus at typically higher luminance levels ($200 - 2000$ cd/m^2) while the Rea and Freyssinier experiment [3] used immersive viewing (which is more relevant for lighting practice) at the luminance level of about 90 cd/m^2. In the Rea and Freyssinier experiment [3], the range of unique white (0% tint) extended from 2700 K (below the blackbody locus) to 6500 K (above the blackbody locus) with an intersection point at about 4100 K. In the Smet *et al.* experiment [8], the range of unique white encompassed the range of about 5000–7000 K, always below the blackbody locus. This contradicts the results of the Rea and Freyssinier experiment [3] possibly due to the different viewing conditions.

To interpret the location of unique white points in the chromaticity diagram, it is interesting to carry out a computational cause analysis of these findings in terms of a simple hypothesis based on the $L - M$ and $L + M - S$ signals representing the visual system's chromatic channels and compare the results of this computation with the (practically more relevant) Rea and Freyssinier [3] results; see Section 3.2.

3.2 Modeling *Unique White* in Terms of $L - M$ and $L + M - S$ Signals

As described in Chapter 2, the chromatic (opponent) signals $L - M$ and $L + M - S$ are responsible for the perception of chromaticity. After the complete chromatic adaptation to the predominating white tone in the scene, the perception of any hue in the white tone should vanish and the white tone should be completely achromatic (i.e., unique white). But this is not always possible. With increasing values of the chromatic signals in the stimulus that is considered white (or achromatic) by the human visual system, there will be more and more remaining tint (hue component) even after full chromatic adaptation and the observer's perception will differ from unique white (see, e.g., Figure 3.1b). To model the perception of unique white according to the above simple hypothesis, the weighted chromatic signals $(L - aM)$ and $(L + aM - bS)$ with the weighting parameter values $a = 1.29384$; $b = 6.47002$ (the values taken from Figure 7.20) were computed for the case of blackbody radiators and different phases of daylight of varying CCT. (More complex hypotheses and models for unique white and whiteness perception can be found in literature under the keywords "color constancy" and "whiteness formula") Figure 3.4 shows the result of the present LMS based computation.

As can be seen from Figure 3.4, blackbody radiators exhibit $(L - aM) > 0$ *and* $(L + aM - bS) > 0$, which corresponds to a reddish-yellow tint for CCTs below 4000 K according to the common experience of orange-yellowish appearance of such blackbody radiators. This is also in accordance with Figure 3.3 in which positive tint percentage values indicating the perception of a yellowish tint are

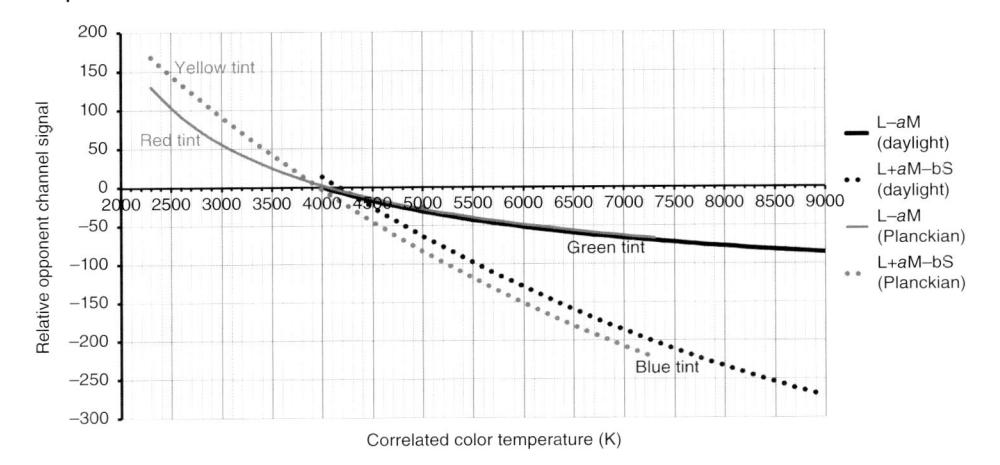

Figure 3.4 Relative weighted chromatic (opponent) signals $(L - aM)$ and $(L + aM - bS)$ with the weighting parameter values $a = 1.29384$; $b = 6.47002$ (the values are taken from Figure 7.20; an explanation of these values are given in Section 7.4.3) for Planckian (blackbody) radiators and phases of daylight of varying CCT.

depicted for this CCT range (< 4000 K) along the blackbody locus (the thick black curve in Figure 3.3). It can also be seen from Figure 3.4 that, according to the computation, both blackbody radiators and the phases of daylight (along the daylight locus) should exhibit a (slightly greenish) blue tint above 4000 K. In Figure 3.3, a violet tint can be observed (negative tint percentage values) in the same range. As violet has a reddish component (instead of a greenish component), the explanation of the Rea and Freyssinier [3] visual results in terms of these channels is not perfect; just the main tendencies could be predicted. The unique white point of the blackbody locus at about 4000 K (Figure 3.3 [3]) could be reproduced.

3.3 Interobserver Variability of White Tone Perception

As mentioned in Section 3.1, subjective unique white assessments showed considerable interobserver variability due to the interobserver variability of the perception of a white tone stimulus. To explore a possible reason, the 53 individual 10° color matching functions (interpolated between 380 and 780 nm using 1 nm steps) of the 49 observers (+4 repetitions) [9] from Figure 2.7 were used in addition to the color-matching functions of the CIE 1931 standard colorimetric observer (2°; Figure 2.5) and the color-matching functions of the CIE 1964 standard colorimetric observer (10°; Figure 2.5) to compute the $u' - v'$ chromaticity coordinates of selected white light sources (defined by their relative spectral radiance distributions between 380 and 780 nm in 1 nm steps). Thus a set of 55 different $u' - v'$ chromaticity coordinates were obtained for every light source. Figure 3.5 shows this result for the case of a warm white triphosphor fluorescent light source and its reference light source (the blackbody radiator at the same CCT).

3.3 Interobserver Variability of White Tone Perception

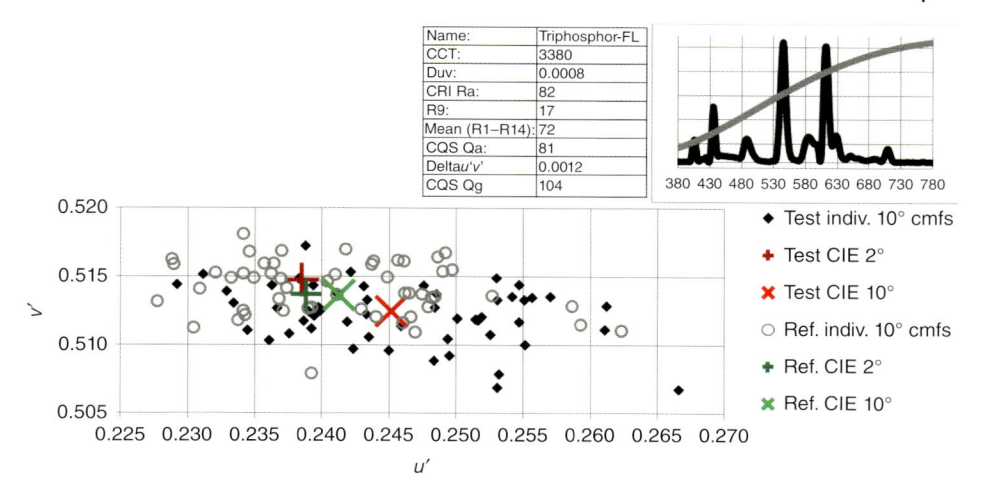

Figure 3.5 Fifty-five different $u' - v'$ chromaticity coordinates of a warm white triphosphor fluorescent light source (test light source) and its reference light source (the blackbody radiator at the same CCT). The different $u' - v'$ chromaticity coordinates were obtained by using the 53 individual 10° color-matching functions from Figure 2.7 as well as the standard 2° and 10° observers from Figure 2.5. Inset table: colorimetric data; inset diagram: relative spectral radiance (black curve: test light source; gray curve: its reference light source).

The following can be seen from Figure 3.5:

1) Using the 2° standard observer, the distance between the test (dark red plus sign) and the reference (dark green plus sign) light sources is about $\Delta u'v' = 0.001$ (see the inset table).
2) Using the 10° standard observer instead of 2° in case of the test light source (i.e., if the white surface becomes larger; red cross sign), the chromaticity of the white tone shifts about $\Delta u'v' = 0.007$ toward greater u' values and about $\Delta u'v' = 0.003$ toward smaller v' values (this chromaticity shift should be well visible).
3) The 2°–10° shift equals only about $\Delta u'v' = 0.002$ in case of the spectrally more balanced reference light source (dark green plus sign vs. green cross sign).
4) There is a huge scatter among the 53 individual 10° $u' - v'$ data for both the test and the reference light sources, with a typical diameter of $\Delta u'v' = 0.035$ (a well visible, large chromaticity difference) in both cases implying, at least at the color matching level of basic colorimetry, that a certain white stimulus can lead to individually different white tone perceptions. At higher visual processing levels, these interindividual differences are counterbalanced by the life-long adaptation (self-compensation) of the visual system to colored object scenes under familiar (spectrally balanced) light source spectra (e.g., tungsten halogen or daylight), which are often seen in everyday life. This ensures the ability of chromatic adaptation to the white tone prevailing in the actual scene. Highly structured spectra (e.g., fluorescent or especially RGB LEDs) might impair the ability of full chromatic adaptation and emphasize interobserver

variability: the target white point of the standard 2° observer might evoke the persistent perception of a strange shade of green or purple for a certain individual observer (with normal color vision) looking at a large surface.

Another set of 55 $u' - v'$ chromaticity coordinates is shown in Figure 3.6 for a special cool white multi-LED light source (test light source) optimized to desaturate purplish blue objects and yellowish-green objects and saturate red and cyan objects and its reference light source (the phase of daylight at the same CCT).

The following can be seen from Figure 3.6:

1) Using the 2° standard observer, the distance between the test (dark red plus sign) and the reference (dark green plus sign) light sources is about $\Delta u'v' = 0.001$ (according to the inset table).
2) Using the 10° standard observer instead of 2° in case of the test light source (i.e., if the white surface becomes larger; red cross sign), the chromaticity of the white tone shifts about $\Delta u'v' = 0.007$ toward smaller u' values and about $\Delta u'v' = 0.01$ toward greater v' values (this shift should be well visible).
3) The 2°–10° shift equals only about $\Delta u'v' = 0.001$ in case of the spectrally more balanced reference light source (dark green plus sign vs. green cross sign).
4) There is a huge scatter, again, among the 53 individual 10° $u' - v'$ data for both the test and the reference light sources, with a typical diameter of $\Delta u'v' = 0.030$ (a well visible chromaticity difference) in both cases.

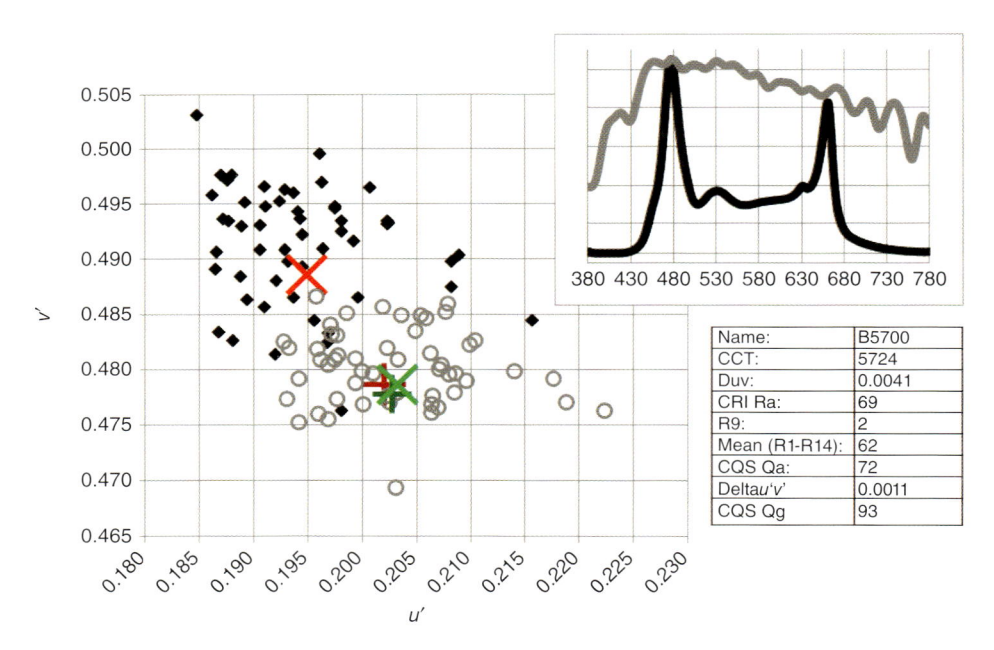

Figure 3.6 Fifty-five different $u' - v'$ chromaticity coordinates of a special cool white multi-LED light source (test light source) optimized to desaturate purplish blue objects and yellowish green objects and saturate red and cyan objects; and its reference light source (the phase of daylight at the same CCT). For legend and further explanations, see the caption of Figure 3.5.

5) Using the 10° standard observer, the distance between the test (red cross) and the reference (green cross) light sources becomes about $\Delta u'v' = 0.014$ (a well visible difference) and the two distributions (test, filled black diamonds; and reference, gray circles) of the individual $u' - v'$ white points follows the trend of the 10° standard observer. For the two light sources (test and reference) whose 2° standard observer white points (almost) match ($\Delta u'v' = 0.001$), two dissimilar and well-separated individual 10° white point distributions are apparent from Figure 3.6. This means that the instrumental matching of the chromaticities of the two white points using the 2° standard observer does not guarantee that the two white tones will match on a large surface for a given individual. The highly structured spectrum (the multi-LED test light source B5700 in Figure 3.6) will possibly not match its spectrally more balanced daylight counterpart (i.e., its reference light source) for a given individual observer. This is the so-called *white tone chromaticity matching discrepancy*.

As a third example, the set of 55 $u' - v'$ chromaticity coordinates is shown in Figure 3.7 for a warm-white hybrid multi-LED with a balanced spectrum and a high color rendering index as well as for its reference light source (the blackbody radiator at the same CCT).

The following can be seen from Figure 3.7:

1) Using the 2° standard observer, the distance between the test (dark red plus sign) and the reference (dark green plus sign) light sources is only $\Delta u'v' = 0.0003$ (see the inset table) because the spectrum of the multi-LED light source

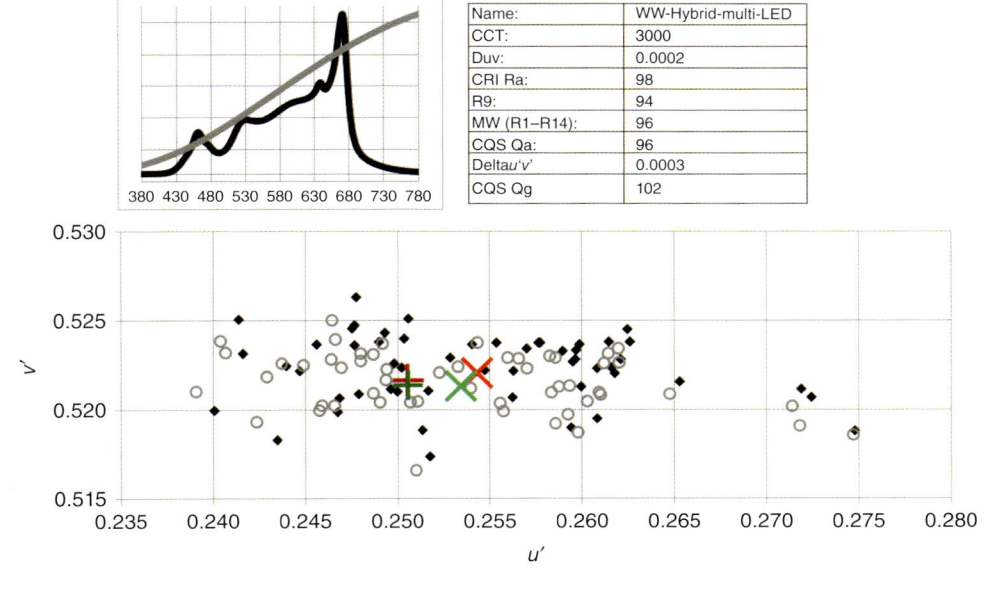

Name:	WW-Hybrid-multi-LED
CCT:	3000
Duv:	0.0002
CRI Ra:	98
R9:	94
MW (R1–R14):	96
CQS Qa:	96
Deltau'v'	0.0003
CQS Qg	102

Figure 3.7 Fifty-five different $u' - v'$ chromaticity coordinates of a warm-white hybrid multi-LED with a balanced spectrum and a high color-rendering index as well as for its reference light source (the blackbody radiator at the same CCT). For legend and further explanations, see the caption of Figure 3.5.

was optimized to be located at the blackbody locus by using the 2° standard observer.

2) Using the 10° standard observer instead of 2° in case of the test light source (i.e., if the white surface becomes larger; red cross sign), the chromaticity of the white tone shifts about $\Delta u'v' = 0.004$ toward greater u' values (this shift is well visible).
3) The $2° - 10°$ shift of the reference light source (dark green plus sign vs. green cross sign) has a similar magnitude ($\Delta u'v' = 0.003$).
4) There is a large scatter among the 53 individual 10° $u' - v'$ data for both the test and the reference light sources, with a typical diameter of $\Delta u'v' = 0.035$ (a very well visible chromaticity difference) in both cases.
5) Using the 10° standard observer, the distance between the test (red cross) and the reference (green cross) light sources remains small ($\Delta u'v' = 0.001$; approximately at the visibility threshold) and the two distributions (test, filled black diamonds; and reference, gray circles) overlap. This means that, in tendency, there is less white tone chromaticity matching discrepancy in case of the balanced high-color rendering test light source (WW-Hybrid-multi-LED) and its blackbody radiator reference light source.

A similar order of magnitude of white tone matching discrepancies between highly structured, narrowband spectra (RGB LEDs) and broadband spectra was found in visual studies [10, 11] ($\Delta u'v' = 0.002 - 0.04$). The latter range of matching discrepancies generally corresponds to a "bad" visual agreement between the two white tones (in terms of semantic categories); see Figures 7.58 and 7.59. In literature, it was found that "differences between visual and instrumental matches increase as one moves in the chromaticity diagram from yellowish white lights toward greenish and bluish lights [10]." This finding is in agreement with the results of the computational approach in Figures 3.5 and 3.7 (warm white, yellowish, less matching discrepancy between narrowband test and broadband reference) and in Figure 3.6 (cool white, bluish; more matching discrepancy between narrowband test and broadband reference).

The *mean* white tone matching discrepancy (among all observers) between the instrumental and visual matching of broadband and narrowband white tones can be *reduced* by the use of alternative color-matching functions instead of the standard CIE 2° or 10° color matching functions [10]. These alternative color-matching functions include:

1) the CIE's 2° and 10° "physiologically relevant" color matching functions [12], which are linear transformations of the Stockman and Sharpe (2000) 2° and 10° cone fundamentals [13] (see Chapter 2);
2) the new color-matching functions of the Technische Universität Ilmenau (2006) for 2° and for 10° obtained by shifting the spectral sensitivity of the S cones' fundamental spectral responsivity function by 3 nm toward shorter wavelengths [14]; and
3) the new color-matching functions of the University of Pannonia (2006) for 2° and for 10° obtained by shifting the spectral sensitivity of the S-cone's fundamental spectral responsivity function by 6 nm toward shorter wavelengths [15].

Finally, two *alternative* approaches, intended to reduce the above-described white tone matching discrepancies, should be mentioned:

1) Oicherman *et al.* [16] attributed the above discrepancies to post-receptoral adaptation of the blue-yellow chromatic channel and proposed a framework to describe this effect; and
2) Sarkar and Blondé [17] proposed eight different observer categories (instead of a single standard observer) with characteristic color-matching functions in every group.

3.4 White Tone Preference

As suggested by the block diagram of Figure 3.2, the target chromaticities for the optimization of the white point of LED light sources should be the preferred white points that do not necessarily coincide [18] with unique white tones (exhibiting maximum perceived whiteness) [1]. In contrast to perceived unique white tones, white tone *preference* includes several subjective and cognitive factors [19]: lighting application (e.g., home, office, or retail lighting), the choice of the characteristic colors of the illuminated object scene (e.g., white objects, reddish or bluish objects, or objects of various different colors; see Sections 7.4 and 7.5) as well as the subjects' characteristics (e.g., region of origin, culture, or gender) [19]. The issue of white tone preference is usually divided into two sub-questions, CCT preference (warm white, neutral white, or cool white; see Sections 7.4 and 7.5) and the preference of the distance from the blackbody locus or the daylight locus expressed by the quantities described above ($\Delta u'v'$ or Duv).

Kruithof's experimental data [20] showed that the lower and upper limits of the comfortable illuminance for general indoor lighting depend on CCT: for low CCTs, lower illuminances, and for high CCTs, higher illuminances are appropriate [19]. In another study [21] on illuminance and CCT preference in a living room, illuminance values varied between 100 and 800 lx and CCTs ranged between 3000 and 6700 K with an average color-rendering index of $R_a = 88$. Warm white CCT was preferred at all illuminance levels and 400 lx was the most preferred [21]. Thus Kruithof's results [20] could not be confirmed.

Further studies [22, 23] also found results that contradicted the Kruithof rule: Boyce and Cuttle [22] found that the assessment of the lighting of a small room across different CCTs and illuminances did not depend on CCT but there was a significant dependence on illuminance. Bringing fruit and flowers into the room improved the subjective impressions evoked by the light source spectra [22]. Davis and Ginthner [23] also found that illuminance influenced the preference for the atmosphere in an experimental room but there was no influence of CCT in their study.

Visual preference for various illuminating spectra (multi-LED spectra and fluorescent spectra) was investigated in a 1 : 6 scale model of an office built into a light booth [24]. Brightness, colorfulness (of the mockup objects), and pleasantness of the office model illuminated at 500 lx between 2855 and 6507 K by preset multi-LED spectra of five LED channels and one fluorescent spectrum (at 3750 K) were assessed. In another series of experiments [24], observers chose their preferred spectra by adjusting the five LED channels – either as a free choice or a choice within an illuminance range of 450 and 550 lx.

Concerning the preset judgments [24], the lowest and the highest CCTs had lower ratings than the middle CCT conditions, similar to the findings described in Sections 7.4 and 7.5. The free and illuminance-constrained lighting choices did not yield different results. The adjusted, preferred CCTs were located in the broad range between 2850 and 14 000 K and the preferred white tone chromaticities were distributed significantly below the blackbody locus (typically Duv = −0.02 in average) [24].

In another study, the naturalness of white light chromaticity was investigated in a simulated interior room environment with fruits and vegetables on the table using the NIST Spectrally Tunable Lighting Facility [25]. Twenty-one subjects compared pairs of spectra of different Duv values (i.e., different chromaticity shifts above and below the Planckian locus). The illuminating spectra with about Duv = −0.015 appeared most natural for all investigated CCTs, 2700, 3500, 4500, and 6500 K. As Duv was negative, these preferred white points were located below the Planckian locus, in accordance with the previous study [24]).

Whitehead [18] compared the subjective assessments of unique white and preferred white (for workplace illumination) resulting from two different studies [26, 27]. This comparison is reproduced in Figure 3.8.

As can be seen from Figure 3.8, although both distributions (unique white and preferred white) are widespread indicating large interpersonal variability, their shape is different: below the Planckian locus, the distribution of preferred white points (Figure 3.8b) is more condensed at the Planckian locus while the distribution of unique white points is more widespread below the Planckian locus (Figure 3.8a).

The most important issues described in Section 3.4 can be summarized as follows:

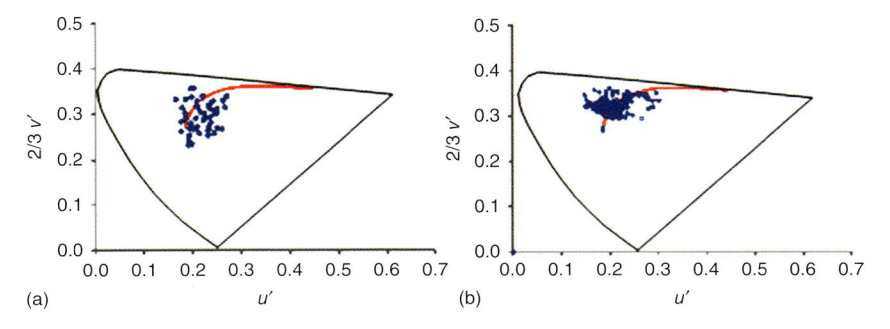

Figure 3.8 Comparison [18] of the subjective assessments of unique white [26] (a) and preferred white for workplace illumination [27] (b). (Whitehead 2012 [18]. Reproduced from Color Res. Appl. with permission of Wiley.)

1) White tone preference has two aspects: CCT preference (warm white, neutral white or cool white, see Sections 7.4 and 7.5) and the distance of the white point from the Planckian or daylight loci in the chromaticity diagram.
2) Preferred white points exhibit large interindividual scatter and are generally different from unique white points.
3) Preferred white points depend on lighting context and user characteristics.
4) Kruithof's rule (for low CCTs, lower illuminances and for high CCTs, higher illuminances are preferred) could not be corroborated in recent experiments.
5) There are studies contradicting the general tendency of preferred white points being located below the blackbody locus; further studies are necessary to find out the final answer.

3.5 The White Tone's Perceived Brightness

As mentioned in Section 2.5.3, the brightness impression of different white stimuli of the same luminance vary with their chromaticity (the so-called brightness–luminance discrepancy, see Figure 2.10) and this can be described by different brightness models. In this section, the Ware and Cowan Conversion Factor formula (WCCF; Section 2.5.3.2) was used to compute the equivalent luminance of different blackbody radiators and different phases of daylight of varying CCT at a fixed luminance, with (100 cd/m^2) as an example. The result of this computation is shown in Figure 3.9.

As can be seen from Figure 3.9, the equivalent luminance L_{eq} computed by the WCCF formula of Eq. (2.15) (the quantity intended to characterize the brightness impression of the stimulus) increases monotonically with increasing CCT for both blackbody (Planckian) radiators and the phases of daylight. This means that, at least according to the WCCF prediction, neutral white is perceived to

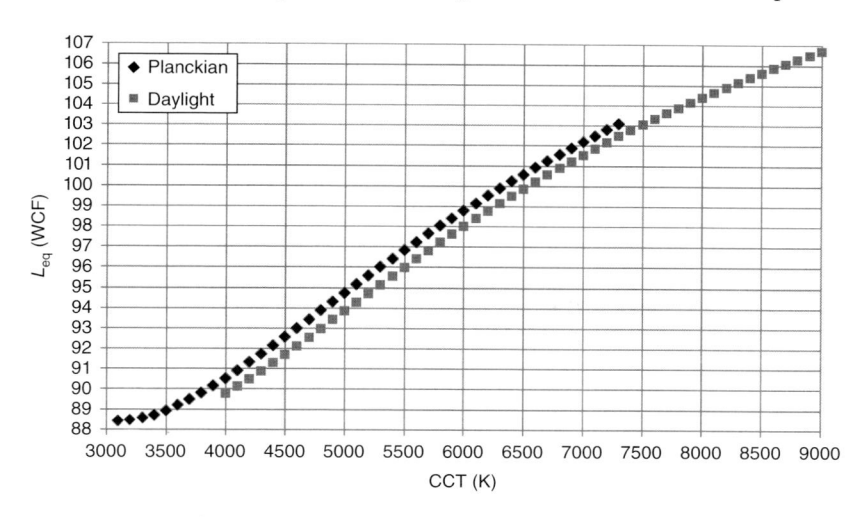

Figure 3.9 Equivalent luminance (L_{eq}) according to the Ware and Cowan conversion factor formula (WCCF; Section 2.5.3.2) of different blackbody radiators and different phases of daylight of varying CCT at the fixed luminance of 100 cd/m^2.

86 | *3 The White Point of the Light Source*

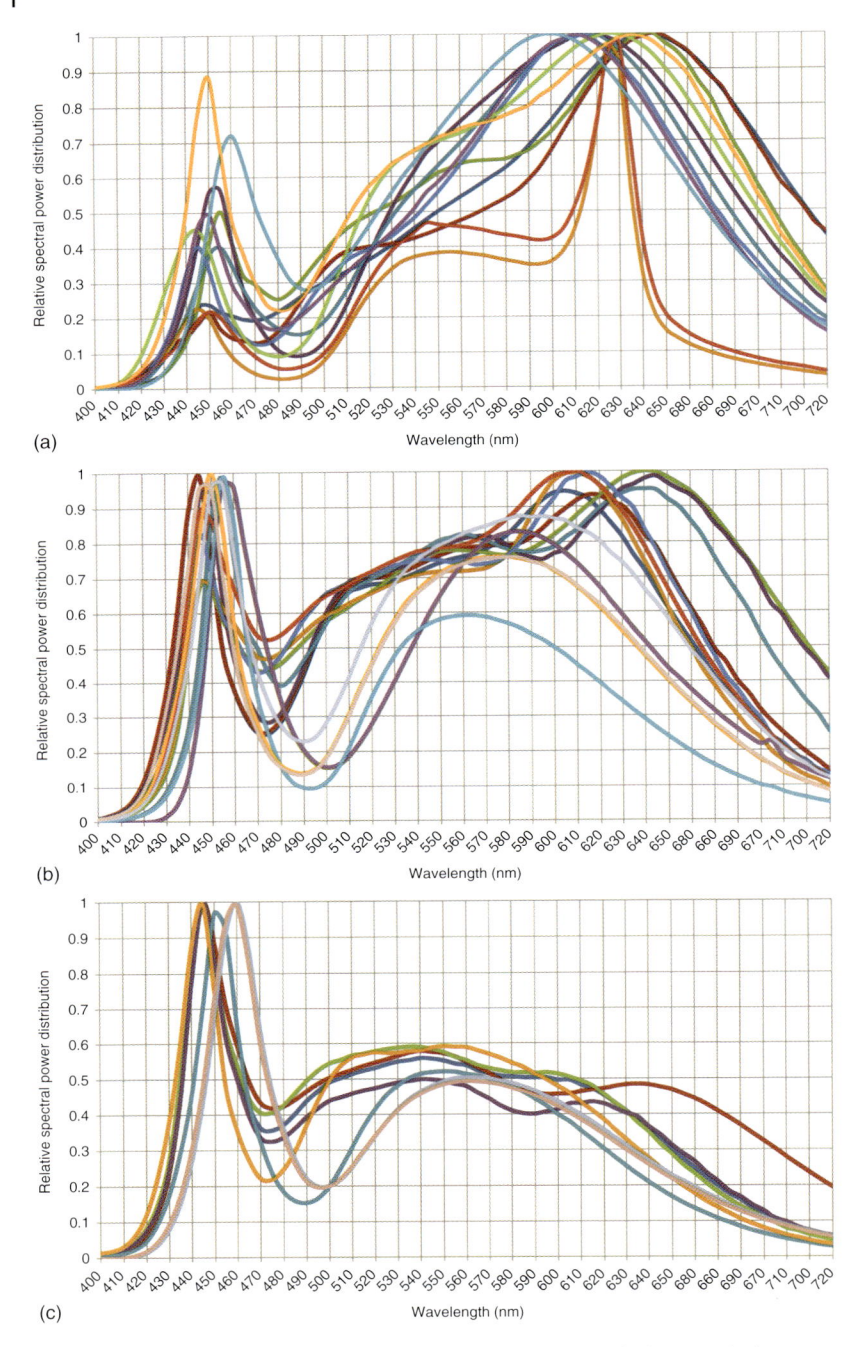

Figure 3.10 Relative spectral power distribution of 34 selected white LED light sources including some LEDs with high-quality phosphor mixtures, multiple phosphor-converted LEDs, and two warm white LEDs with an additional red semiconductor LED. (a) WW (warm white) with CCT < 3500 K; (b) NW (neutral white) with 3500 K ≤ CCT < 5000 K; and (c) CW (cool white) with CCT ≥ 5000 K. (Khanh *et al.* 2014 [1]. Reproduced with permission of Wiley-VCH.)

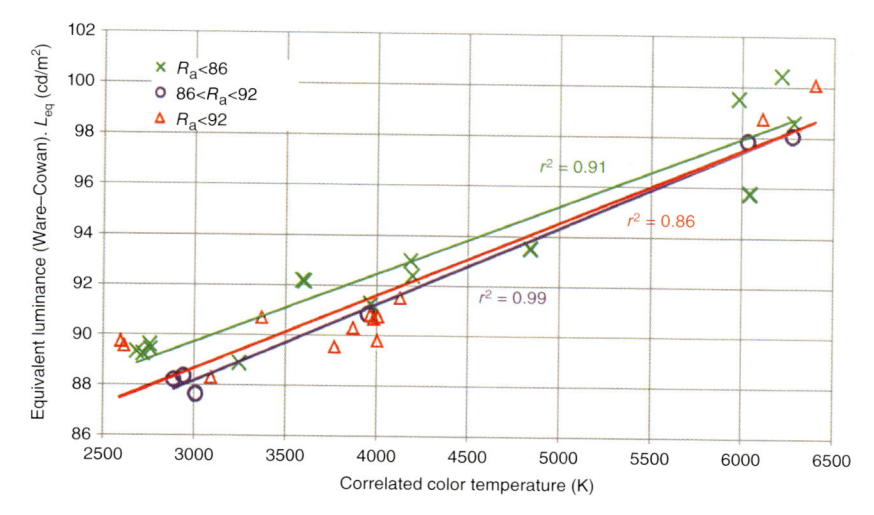

Figure 3.11 WCCF L_{eq} values (see Section 2.5.3.2) for the 34 white LEDs of Figure 3.10 at a fixed luminance of $L_v = 100$ cd/m², as a function of CCT (abscissa). Light sources are grouped according to their general color-rendering index (CRI R_a) level. (Khanh et al. 2014 [1]. Reproduced with permission of Wiley-VCH.)

be brighter than warm white, and cool white (or cold white) is perceived to be brighter than neutral white at a fixed luminance level. This agrees well with the everyday visual experience of brightness in a room illuminated at a *fixed* luminance level of $L_v = 100$ cd/m² first by a warm white LED (2700 K; providing a dark environment) and then by a cold white LED (6500 K; providing a bright environment). It should be noted that the scale of the quantity L_{eq} is not intended to describe the absolute amount of perceived brightness (as mentioned in Section 2.5.3), just the correct ascending order of perceived brightness. The perceived brightness difference between, for example, $L_{eq} = 88$ and $L_{eq} = 89$, is not necessarily the same as the perceived brightness difference between, for example, $L_{eq} = 100$ and $L_{eq} = 101$.

In a second example, WCCF L_{eq} values were also computed for 34 selected white LED spectra [1] (see Figure 3.10) at the fixed luminance of $L_v = 100$ cd/m² again, as a function of CCT. The result is depicted in Figure 3.11.

As can be seen from Figure 3.11, L_{eq} values increase with increasing CCT, as in Figure 3.9; for example, a warm white LED at $L_v = 100$ cd/m² will provide a darker lit environment than a cool white LED at the same luminance level.

3.6 Summary and Outlook

An important constraint of spectral engineering is to guarantee the appropriate white point of the light source. This should be a white point that is not only accepted but also *preferred* by the intended users of the light source. Preferred white is, in general, different from unique white (the one that does not contain any tint). White tone preference (with its two components, CCT preference and the distance of the white tone's chromaticity from the Planckian or daylight loci) depends on the lighting application, the type of objects being illuminated (e.g., white, red, blue. or mixed objects; this issue will be discussed in Sections 7.4 and 7.5), and also on the user's characteristics (age, gender, and cultural background [19]). Because the users of lighting installations generally have only a limited knowledge on lighting quality, Rea and Freysinnier proposed [7] a "class A color" designation for light sources for general illumination with the following three requirements: the illumination should be "white" (in the sense of having no tint); and two color quality requirements are CRI $R_a \geq 80$ and $80 \leq GAI \leq 100$ (see Chapter 7).

As mentioned above, preferred white tones exhibit large interobserver variability due to the cognitive effects that determine *preference judgments.* This variability is complicated by the similarly large effect of the interobserver variability of color-matching functions on white tone *perception.* The latter effect results in white tone perception artifacts and white tone matching discrepancies among different observers and different field sizes. Strongly metameric white tones will not match visually despite the matching of their x, y chromaticity coordinates computed by the aid of the color-matching functions of the standard 2° observer. This effect was quantified in a computational approach re-analyzing the Stiles–Burch individual 10° color-matching dataset. This discrepancy is especially serious for the case of highly structured spectra with conspicuous local spectral maxima (e.g., RGB LED spectra) to be matched with broadband spectra and less obvious in case of matching a broadband spectrum (e.g., multiphosphor LEDs of high color-rendering property) with another broadband spectrum.

The use of new (large field) color-matching functions or the use of different observer groups with associated different color-matching functions might reduce this discrepancy. The white tone matching discrepancy might threaten, for example, the acceptability of the white points of multi-LED spectra intended to oversaturate certain object colors by the aid of their high local maxima with the intention of enhancing their color quality (see Chapter 7). Such highly structured spectra with narrowband details might diminish the *interindividual agreement* (an important issue for human centric lighting, see Chapter 9) on perceived white tone quality for the practically relevant case when multiple users assess the lighting quality of a given lighting system at the same time.

Today's definition of the target white point, the $\Delta uv = \pm 0.0054$-wide stripe along the locus of the blackbody radiators and the phases of daylight (with a breakpoint at 5000 K), does not (necessarily) coincide with the unique white points or the preferred white points resulting from visual studies. Rea and Freyssinier [3] found that the target white points of the above definition contain shades of purple/violet or green/yellow (a cause analysis in terms of LMS cone signals in this chapter was partially able to explain these hues). To determine the level of user acceptance and user preference of different tints perceived in the white tone, further visual experiments are necessary. The perceived brightness of different white tones was also investigated in this chapter at a fixed luminance level: white tones of higher CCTs turned out to be brighter. The brightness issue will be further analyzed in Section 7.3 in case of strongly metameric white points.

References

1. Khanh, T.Q., Bodrogi, P., Vinh, Q.T., and Winkler, H. (2015) *LED Lighting-Technology and Perception*, Wiley-VCH Verlag GmbH & Co. KGaA, Weinheim.
2. Pepler, W., Böll, M., Bodrogi, P., and Khanh, T.Q. (2012) *Einfluss unterschiedlicher Beleuchtungskonzepte und Lampenspektren in der Innenraumbeleuchtung* (The Effect of Different Illuminating Concepts and Lamp Spectra for Interior Lighting), Conference Licht 2012, Deutsche Lichttechnische Gesellschaft e.V. (LiTG), Germany, (In German).
3. Rea, M.S. and Freyssinier, J.P. (2011) White lighting. *Color Res. Appl.*, **38** (2), 82–92.
4. Davis, W., Weintraub, S., and Anson, G. (2013) Perception of correlated color temperature: the color of white. Proceedings of the 27th Session of the CIE, Sun City, 2013, pp. 197–202.
5. Commission Internationale de l'Éclairage (1995) Method of Measuring and Specifying Color Rendering Properties of Light Sources, Publ. CIE 13.3-1995.
6. ANSI C78.377-2008 (2008) *American National Standard for Electric Lamps: Specifications for the Chromaticity of Solid State Lighting Products*, National Electrical Manufacturers Association.
7. Rea, M.S. and Freyssinier, J.P. (2012) The class a color designation for light sources, in *Proceedings of Experiencing Light 2012: International Conference on the Effects of Light on Wellbeing* (eds Y.A.W. de Kort, W.A. IJsselsteijn, M. Aarts, A. Haans, D. Lakens, K.C.H.J. Smolders, F. Beute, and L. van Rijswijk), Eindhoven University of Technology, The Netherlands.
8. Kevin, A.G., Smet , Deconinck, G., and Hanselaer, P. (2014) Chromaticity of unique white in object mode. *Opt. Express*, **22** (21), 25830–25841.
9. Stiles, W. and Burch, J. (1959) N.P.L. color-matching investigation: final report (1958). *Opt. Acta*, **6**, 1–26.
10. Csuti, P. and Schanda, J. (2008) Colour matching experiments with RGB-LEDs. *Color Res. Appl.*, **33** (2), 108–112.
11. Bieske, K., Csuti, P., and Schanda, J. (2007) Colour appearance of metameric lights and possible colorimetric description. CIE Expert Symposium on Visual Appearance, 2007.
12. CIE (2006) Fundamental Chromaticity Diagram with Physiological Axes. Parts 1 and 2. Technical Report 170-1, Central Bureau of the Commission Internationale de l' Éclairage, Vienna.
13. Web database of the Color & Vision Research Laboratory, University College London, Institute of Ophthalmology, London, UK www.cvrl.org.
14. Polster, S. (2014) Neue Spektralwertfunktionen für die korrekte Bewertung von LED-Spektren (New color matching functions for the correct evaluation of LED spectra). 3. Darmstädter farbwissenschaftliches Expertenseminar (The 3rd Darmstadt Expert Symposium for Color Science) (in German).
15. Csuti, P. and Schanda, J. (2010) A better description of metameric experience of LED clusters. *Light Eng.*, **18**, 44–50.

16 Oicherman, B., Ronnier Luo, M., Rigg, B., and Robertson, A.R. (2009) Adaptation and colour matching of display and surface colours. *Color Res. Appl.*, **34** (3), 182–193.

17 Sarkar, A. and Blondé, L. (2013) Colorimetric observer categories and their applications in color and vision sciences. CIE Centenary Conference, April 2013.

18 Whitehead, L. (2013) Interpretation concerns regarding white light. *Color Res. Appl.*, **38** (2), 93–95.

19 Bodrogi, P., Lin, Y., Xiao, X., Stojanovic, D., and Khanh, T.Q. (2015) Intercultural observer preference for perceived illumination chromaticity for different coloured object scenes. *Light. Res. Technol.* doi: 10.1177/1477153515616435

20 Kruithof, A.A. (1941) Tubular luminescence lamps for general illumination. *Philips Tech. Rev.*, **6**, 65–96.

21 Nakamura, H. and Karasawa, Y. (1999) Relationship between illuminance/colour temperature and preference of atmosphere. *J. Light Visual Environ.*, **23** (1), 29–38.

22 Boyce, P.R. and Cuttle, C. (1990) Effect of correlated colour temperature on the perception of interiors and colour discrimination performance. *Light. Res. Technol.*, **22** (1), 19–36.

23 Davis, R.G. and Ginthner, D.N. (1990) Correlated color temperature, illuminance level and the Kruithof curve. *J. Illum. Eng. Soc.*, **19** (1), 27–38.

24 Dikel, E.E., Burns, G.J., Veitch, J.A., Mancini, S., and Newsham, G.R. (2014) Preferred chromaticity of color-tunable LED lighting. *LEUKOS: J. Illum. Eng. Soc. North Am.*, **10**, 101–115.

25 Ohno, Y. and Oh, S. (2016) Vision experiment II on white light chromaticity for lighting. Proceedings of CIE 2016 "Lighting Quality and Energy Efficiency", Melbourne, Australia, March 3–5, 2016.

26 Werner, J.S. and Schefrin, B.E. (1993) Loci of achromatic points throughout the life span. *J. Opt. Soc. Am. A*, **10**, 1509–1516.

27 Veitch, J.A., Dikel, E.E., Burns, G., and Mancini, S. (2011) Individual control over light source spectrum: effects on perception and cognition. 27th Session of the Commission Internationale de l'É clairage, Sun City, South Africa, 2011, pp. 213–218.

4

Object Colors – Spectral Reflectance, Grouping of Colored Objects, and Color Gamut Aspects

4.1 Introduction: Aims and Research Questions

If an observer considers a scene with colored and neutral objects (see Figure 4.1), the chain of the cognitive imaging and color perception process begins with the spectral radiant power distribution of the light source $L(\lambda, t, x, y, z)$ with a certain correlated color temperature (CCT), chromaticity, and B/R ratio (blue to red ratio); see Figure 4.2. After a short time of chromatic adaptation of about 2 min, if illuminating in the photopic luminance range of vision, the colored objects appear in front of the observer with a certain brightness (achromatic and chromatic brightness), colorfulness, and hue owing to the spectral reflectance $R(\lambda, t, x, y, z)$ of each object in the visible wavelength range. Besides the color perception of the individual colored objects, the spatial structure of each object (e.g., fine and rough structure of textiles, oil paintings, or flowers) and the color contrast within the color scene evoke a perception of color differences between the objects.

The perceived color difference is constituted by the differences in brightness, saturation, and hue caused by the color difference of adjacent colored objects by illumination of a light source simultaneously or by two different light sources successively illuminating the same color objects. While color fidelity or the perception of color differences is a physiological property of human vision related to a comparison procedure between a test and a reference light source, the rating (i.e., visual assessment) of color preference, naturalness, or vividness is another, higher level process based on the observer's expectations, viewing context, lighting application, or visual experience of the observer. Therefore, color preference, naturalness, and vividness should have an absolute psychological scale with an anchor value depending on the individual observers. In this cognitive process, each individual can find a semantic scale for the specific color scene with the ratings *"excellent, very good, good, tolerable, bad,* or *unusable."* The semantic evaluations of a panel of observers of different age, cultural background, profession, and ethnic background establish a representative assessment of an illuminated scene with a multitude of colored objects.

The task of color quality science is to determine the scale and the relationship between the perceptual issues *"color preference, color fidelity, naturalness,* and *vividness"* and the colorimetric properties of the illuminated colored objects.

Color Quality of Semiconductor and Conventional Light Sources, First Edition.
Tran Quoc Khanh, Peter Bodrogi, and Trinh Quang Vinh.
© 2017 Wiley-VCH Verlag GmbH & Co. KGaA. Published 2017 by Wiley-VCH Verlag GmbH & Co. KGaA.

Figure 4.1 A tabletop with colored and structured objects (® Technische Universität Darmstadt).

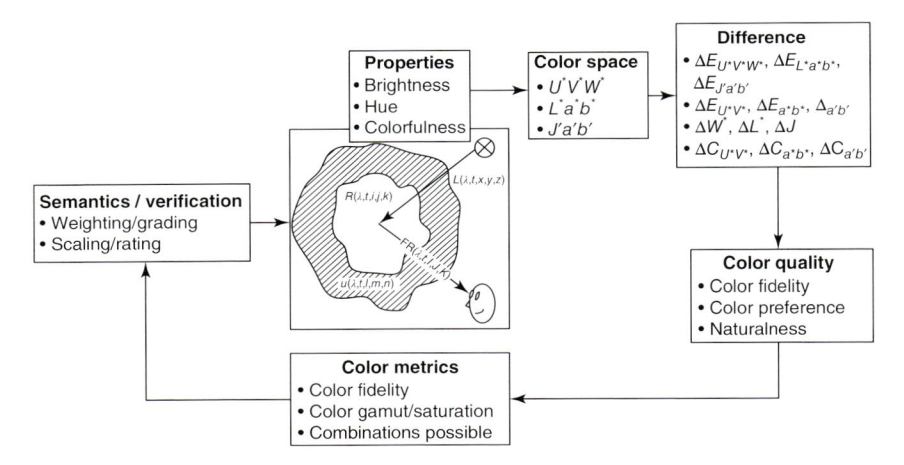

Figure 4.2 Process of color perception, evaluation, and rating (i.e., visual assessment).

This issue will be dealt with in Chapter 7 in more detail. If these relationships, their metrics, and the corresponding semantic scales are known, then the spectral power distributions, chromaticities, and illuminance levels of the light sources can be optimized for different lighting applications and lighting contexts.

Based on the above description, it can be recognized that the spectral reflectance of the colored objects, their meaning for the design of an illuminated space, and the cultural identification of the users with the colors are of vital importance. For a reasonable optimization of light source spectra, a representative collection of test colors or test color objects has to fulfill the following requirements.

- The set of test color objects should be representative for a specific lighting application or many typical lighting applications (e.g., TV and film studios, shop lighting, lighting of oil paintings in a gallery, general museum lighting, food lighting, the lighting of living rooms, etc.).
- The colored objects should be homogeneously distributed over typical ranges of lightness (expressed, e.g., as CIELAB L^*), hue angle H (red, orange, skin tones, yellow, green, blue, or violet), and saturation degree C^* (strongly saturated colors or desaturated colors).
- The spectral reflectance of the collected test colors should be uniformly distributed over the visible wavelength range between 400 and 700 nm with their typical band-pass, short-pass, and long-pass characteristics.

In the history of color science, several color collections have been proposed and used to calculate color quality parameters. They include (see also Chapter 2)

- the 14 CIE test color samples (TCSs) for the color-rendering index 1965 and 1995
- the 15 CQS TCSs for the CQS proposal of NIST (USA)
- the 4 TCSs of the FCI definition
- the 9 TCSs of the memory color rendering index (MCRI) definition
- the 17 artificial test colors for the color fidelity formula of CIE CRI 2012
- the 99 TCSs for the definition according to IES TM-30 (2015).

There are both advantages and disadvantages to the above test color collections in the evaluation of the color quality of a light source. If the color collection is not suitable for a certain color quality metric or for a specific lighting application, then the value of the color quality metric (e.g., CIE CRI $R_a = 83$) will not be able to be used to evaluate the illuminated color scene under consideration, or it will cause a wrong selection of the light source for a dedicated lighting application. In both cases, the acceptance of the users or decision makers for the specific lighting system should be in the focus of the selection of a TCS set and the establishment of the corresponding color quality metric.

However, the aim of the present chapter will not be to propose any new color collection. Instead, in the first part, the spectral and colorimetric characteristics of numerous natural and artificial colors under different light source spectra (e.g., D65 for daylight or tungsten halogen light with a CCT of 3200 K) will be analyzed. In the second part, the spectral and colorimetric properties of the well-known color collections mentioned above will be analyzed and their advantages and disadvantages will be pointed out. In the third part, general principles and various methods of color classification will be presented for different applications accompanied by two important examples, one with museum oil painting colors and another with the IES test colors. The aim of these two examples is to point out how to apply the principles and methods of color classification.

The chapter presents results of several spectral reflectance measurements with different color object groups such as skin tone, flowers, fruits, foods, textiles, or art painting colors implemented in the lighting laboratory of the authors;

Figure 4.3 Color object groups spectrally measured in the Lighting Laboratory of the Technische Universität Darmstadt.

see Figure 4.3. In addition, external databases of color collections, for example, the one of the University of Leeds (United Kingdom) and others, will also be analyzed.

4.2 Spectral Reflectance of Flowers

Flowers constitute an important group of natural objects. Their various shapes and colors have influenced the culture and social life of mankind since many centuries because they represent a main element in the decoration of living spaces for different applications, for example, festivals, parties, or in representative rooms; they are strongly identified with the character of the acting persons. Numerous types of flower from different seasons and geographical areas were measured in the laboratory (see Figure 4.4; measured areas are marked on the surface of the flower by small open circles) and their spectral reflectance was analyzed together with the other spectral data of international databases (see Figure 4.5).

Based on the two-dimensional hue circle (CAM02-UCS $a' - b'$) in Figure 4.6, it can be recognized that the most analyzed flowers are concentrated with a very high density in the hue angle ranges between 0° and 120° (red, orange, yellow) and from 210° to 360° (blue, violet), and most of the colorful flowers can be found at the red ($H = 25°$) and yellow hue tones ($H = 90°$). A deeper analysis of their spectral reflectance (or spectral radiance coefficient; see Figure 4.5) shows that most blue-violet flowers have a maximum of reflectance in the wavelength range between 425 and 455 nm. Yellow and red flowers exhibit long-pass spectral reflectance behavior with the edge between 510 and 650 nm. In order to truly render and reproduce the hue and colorfulness of these flowers, optical radiation power between 500 and 700 nm (as in case of daylight and tungsten halogen light) must be therefore available in reasonable amounts. Lightness values

Figure 4.4 Measurement of the spectral reflectance of flowers.

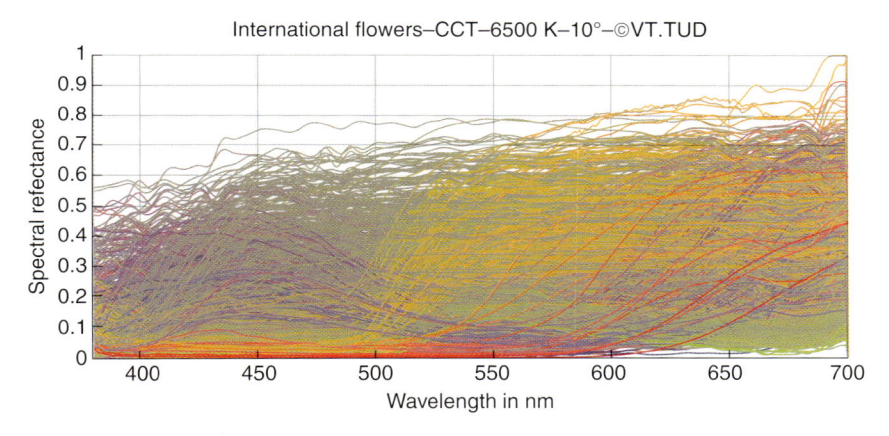

Figure 4.5 Spectral reflectance (spectral radiance coefficient) of 3968 measured flowers (Data Source: [1] FReD of Sarah E. J. Arnold *et al.* and the added database of TU Darmstadt [2]).

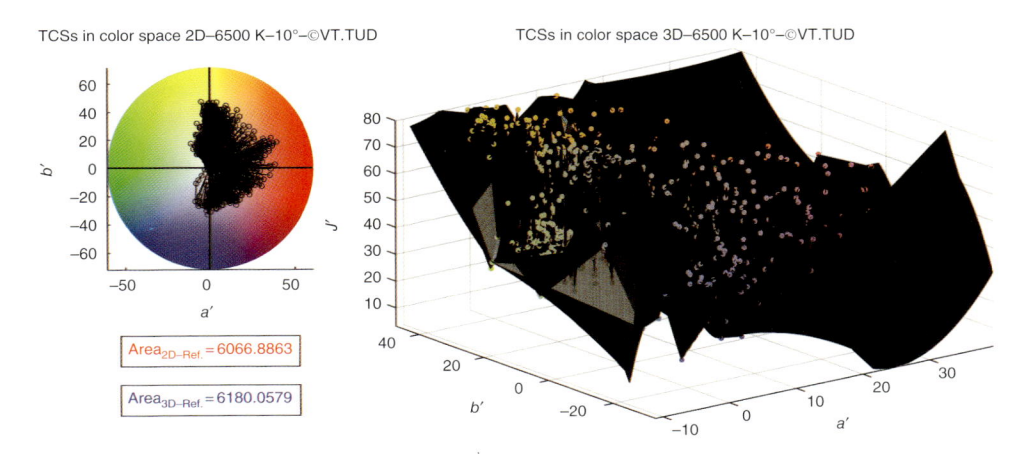

Area$_{2D-Ref.}$ = 6066.8863
Area$_{3D-Ref.}$ = 6180.0579

Figure 4.6 Colorimetric data of the 3968 analyzed flowers illuminated by the D65 illuminant.

(CAM02-UCS J') range between 20 for blue-violet flowers and about 80 for the yellow flowers.

4.3 Spectral Reflectance of Skin Tones

Skin tones and excellent skin tone illumination play an essential role in photography, TV and film production, fashion shows, painting art, in the beauty industry, and generally in daily-life situations such as social talks, meetings, and communication. Skin tones do not only have a physiological meaning but they also represent important memory colors because they are related to the aim to be beautiful and attractive. For this purpose and in order to optimize the spectra of light sources, the face skins of 64 young persons (students) coming from different countries of Africa, Asia, and Europe were spectrally measured in the laboratory of the present authors. These persons did not use skin cosmetics before and during the measurements. In Figure 4.7, the spectral reflectance of these skin tones are shown.

Some of the skin tones of the African subjects showed strong absorbance and their spectral reflectance curves did not show a significant structure of spectral peaks or spectral bands. The spectral reflectance curves of the European and Eastern Asian persons, however, exhibit strong variability with a maximum at 520 nm. From 575 nm on, the spectral curves strongly increase with a strong slope up to 630 nm and stay at high spectral reflectance values up to 780 nm. The spectral and colorimetric data of the analyzed skin tones illuminated with the D65 illuminant are illustrated in Figure 4.8. The hue angles are at about 50° independent of the origin of the subjects and the chroma values are at about 15 in the CAM02-UCS $a' - b'$ diagram.

Under the optical radiation of thermal radiators (e.g., halogen tungsten light) with CCT = 3200 K, the saturation increases until a chroma value of about 20 (see Figure 4.9) while the lightness J'-values range between 50 and 70.

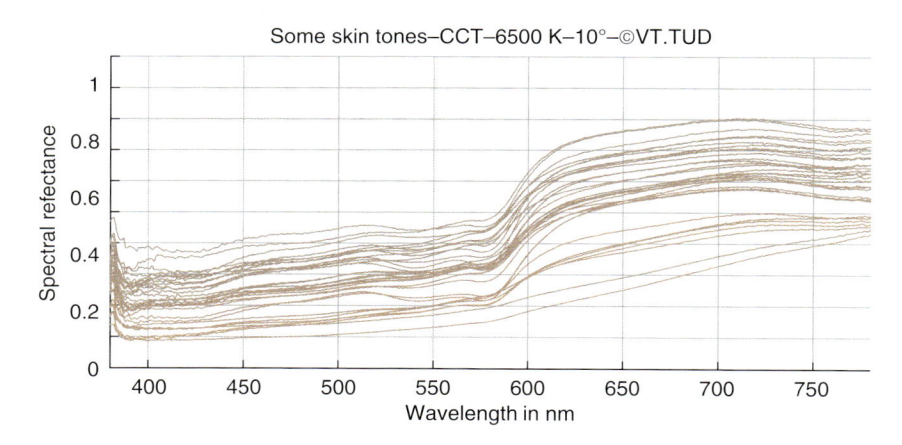

Figure 4.7 Spectral reflectance (spectral radiance coefficient) of the measured skin tones of 64 persons (® TU Darmstadt, [2]).

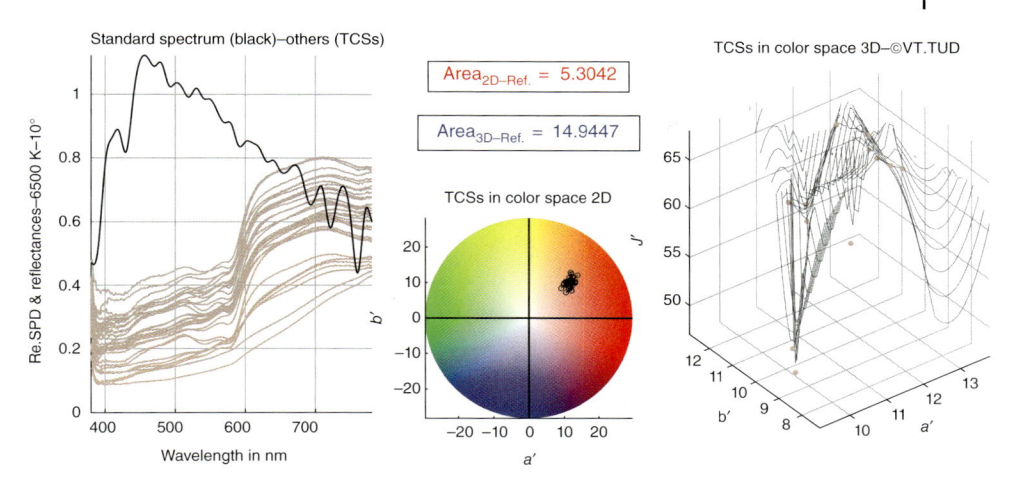

Figure 4.8 Spectral and colorimetric data of the analyzed skin tones illuminated by D65.

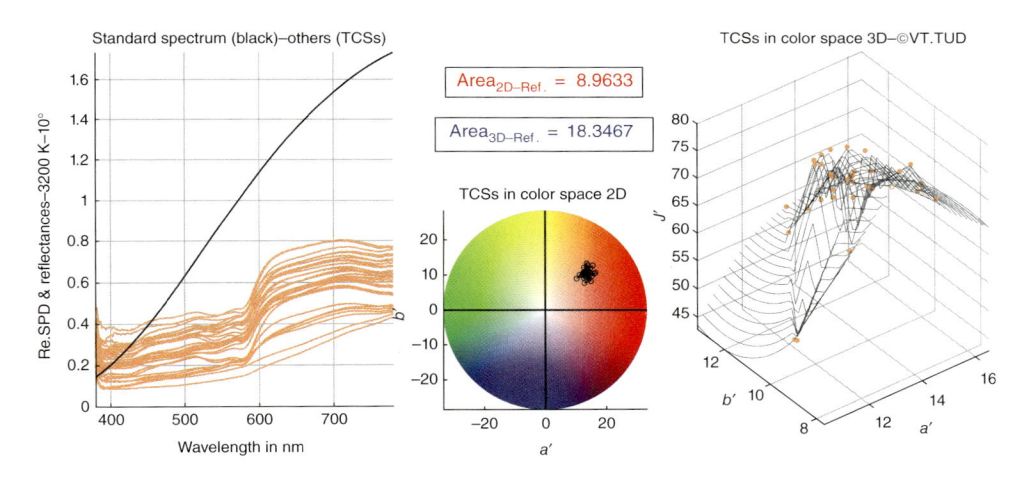

Figure 4.9 Spectral and colorimetric data of the analyzed skin tones illuminated by the Planckian radiator at 3200 K.

4.4 Spectral Reflectance of Art Paintings

Art paintings consist of different color pigments, and special mixtures of many hue tones and saturation degrees can be created allowing the artists to have large freedom in designing and realizing their ideas. The paintings in their final form in the museums and galleries are cultural goods of historical, educational, and social importance so that illuminating spectra with very good lighting quality and color quality should be ensured. As the recent introduction of LED (and OLED) technology allows for the use of light sources with high color quality and variable chromaticity (variable CCTs), the spectral reflectance of painting materials was systematically collected and measured in the authors' laboratory at the TU

Figure 4.10 Examples of measured art painting materials (® TU Darmstadt).

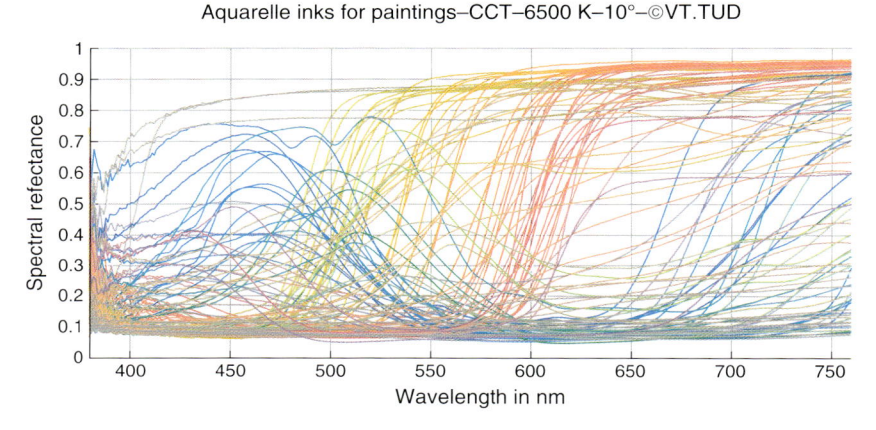

Figure 4.11 Spectral reflectance (spectral radiance coefficients) of some measured painting materials (® TU Darmstadt, [3]).

Darmstadt (see Figure 4.10). An example of the spectral radiance coefficient of measured painting materials is shown in Figure 4.11.

Compared to the spectral reflectance of flowers or skin tones, there are several spectral reflectance curves of painting materials that contain a high slope in the range between 480 and 700 nm making the use of saturated colors possible. Parallel to this fact, there is also a variety of spectral reflectance curves with a strong band-pass characteristic with maximal wavelengths between 420 and 540 nm. This is the condition for the generation of saturated violet, blue, cyan, and green painting colors that can be seen in the two-dimensional hue circle $a' - b'$ in Figure 4.12. However, in the range of the hue angles between 120° and 180° and between 245° and 360°, many gaps can be seen indicating missing higher saturations in these ranges.

4.5 The Leeds Database of Object Colors

The color scientists of the University of Leeds (UK) collected and measured the spectral reflectance of many subcollections of surface colors (see Figures 4.13 and

4.5 *The Leeds Database of Object Colors* | 99

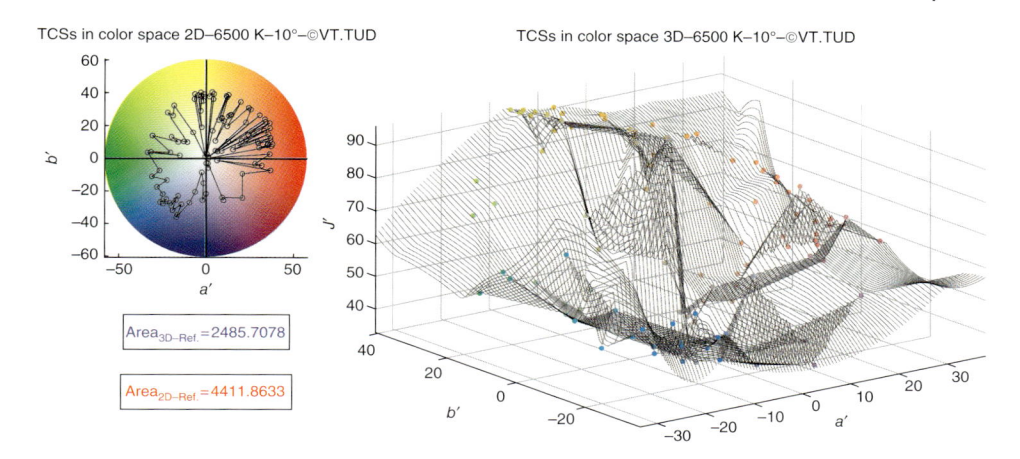

Figure 4.12 Colorimetric data of the analyzed painting materials illuminated by D65.

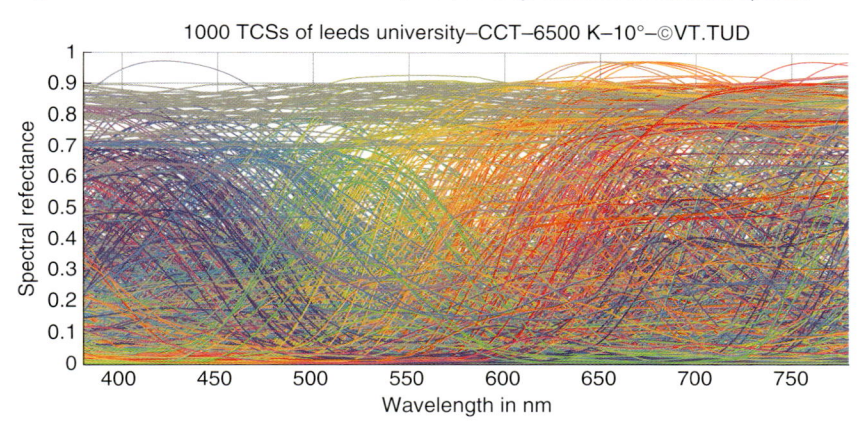

Figure 4.13 Spectral reflectance (spectral radiance coefficient) of object colors in the database of the University of Leeds (UK).

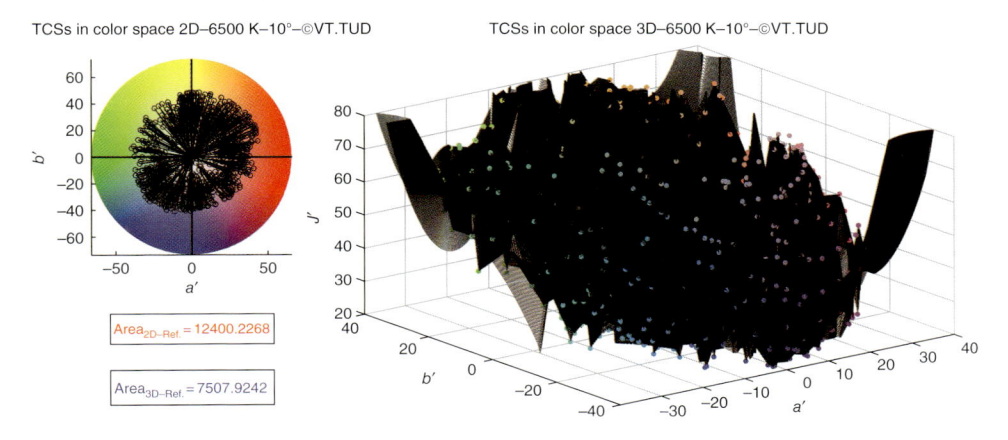

Figure 4.14 Colorimetric data of the analyzed Leeds color materials illuminated by D65.

4.14; ® Database (R. Luo, Personal communication on the Leeds's color object data base, 2010)).

These spectral reflectance curves include the following:

- Red curves with strong slope from 580 nm and yellow curves from 510 nm on.
- Blue and green curves are spectral radiance coefficients of blue and green colored objects with band-pass characteristics of low spectral reflectance outside the band-pass range guaranteeing saturated colors with chroma values ranging up to 50 (see Figure 4.14). The violet curves contain the spectral reflectance both in the range of 380–470 nm and from 570 nm on.
- Many gray curves indicate the spectral reflectance of (nearly) neutral colored objects with spectral reflectance values going up to 0.9 (90%).

These four color collections of flowers, skin tones, art paintings, plus the Leeds database represent important examples of many different material types studied and measured in several color research laboratories, among them, the Technische Universität Darmstadt. These spectral and colorimetric data give an impression on the plurality of colorimetric features of natural and artificial object colors that have to be taken into account if color researchers would like to correctly define a color quality metric for a wide range of lighting applications. The above description also shows that many collected object colors are uniformly distributed at different CIELAB L^* (or CAM02-UCS J') levels of lightness, over the whole hue circle and they contain different degrees of saturation. The task of modern color science is to select a limited number of samples from these object colors in order to reveal or represent the color fidelity, color gamut, and color (over-) saturation potential and color preference enhancement potential of the light source.

4.6 State-of-the-Art Sets of Test Color Samples and Their Ability to Evaluate the Color Quality of Light Sources

In the last century, several object colors and color groups have been selected for use in art (e.g., Munsell colors for the painting art), RAL colors in technology (in Germany) or in color science (e.g., the Oswald colors or the NCS colors). To characterize the color quality of light sources, the following groups of color samples can be defined according to three perceptual attributes of color quality:

1) For the definition of "*color fidelity metrics*": for example, the 8 TCSs of the CIE 1965 for the definition of R_a, 14 TCSs of the CIE 1965 for an extended R_a definition for higher color requirements; 17 TCSs for nCRI (CIE CRI 2012); 15 CQS test color samples for the definition of the CQS Q_f metric; and the 99 IES TCSs for the definition of the R_f color fidelity metric.
2) For the definition of "*color gamut metrics*": for example, the 15 CQS test colors for the Q_g definition, 4 colors to calculate FCI, 8 colors for GAI; and the 99 IES test colors to compute R_g.
3) For the definition of "*color preference*": the 15 CQS TCSs to define CQS Q_p or the "*memory color*" samples, that is, the 9 TCSs used in the MCRI definition.

The criteria for the selection of the colored objects or TCSs can be described from today's point of view as follows.

- If the colors are typical for a specific application (e.g., for flower shops, for a book exhibition, for a gallery of a famous artist with his special oil painting colors, or for a well-defined decoration of a specially featured cinema film), the color quality parameters have to be optimized for these specific colors that can also be selected by the shop owner, the architect or the producer, or a creative artistic manager of the cinema film.
- Generally, most light sources are optimized for a plurality of applications (e.g., conference halls, schools, offices, meeting rooms, hospitals) at the same time and if the color fidelity of the illumination system (i.e., the similarity to a reference light source; see Chapter 2) is the criterion to be fulfilled then the selected colors have to be the representatives for numerous different contexts, different lightness levels, hue tones, and saturation levels. Because the number of the selected colors is limited (in the order of maximal 100–120 colors), these colors have to be found by a reasonable grouping method that will be roughly described in Section 4.7.
- In contrast to the criterion "color fidelity," color gamut is, by definition, a feature of a specific light source enabling the rendering of the hue and the saturation of highly saturated object colors. If the rendering capacity for saturated colors is fulfilled, then it is also acceptable for desaturated colors. Therefore, to characterize color gamut, object colors have to exhibit higher saturation, also depending on the lightness level expressed by CIELAB L^* or CAM02-UCS J'.
- In most cases, color gamut criteria can be optimized for certain light sources only and for a certain group of colors (such as red-orange colors, green colors, or violet colors) only and not for all colors at the same time, which are distributed around the hue angle at a certain lightness level. It is therefore not reasonable to define a big volume of colors at many levels of lightness, covering the full hue angle at maximal saturation. Instead, the optimization aims of *"large color gamut"* or *"color saturation enhancement"* should be limited to a certain, specified smaller area or volume of the whole color gamut.
- If this philosophy is consequently realized then the color collections for color fidelity and color gamut should not be the same.
- If the color preference, naturalness, or vividness feature of a specific light source has to be evaluated and/or optimized then the color collection for this evaluation has to contain saturated and moderately saturated colors and therefore, the color collection of color fidelity should be expanded by some more saturated colors with suitable contextual, social, cultural, or memory characteristics. Memory color metrics or color preference metrics cannot contain saturated or desaturated colors only.

The aim of this section is to characterize the color collections mentioned above. They are approved by the CIE or published in the drafts of CIE technical committees or in scientific literature.

The color collection of the CIE, established in 1965 (see [4]), has already been mentioned in Chapter 2. The spectral reflectance (spectral radiance coefficient)

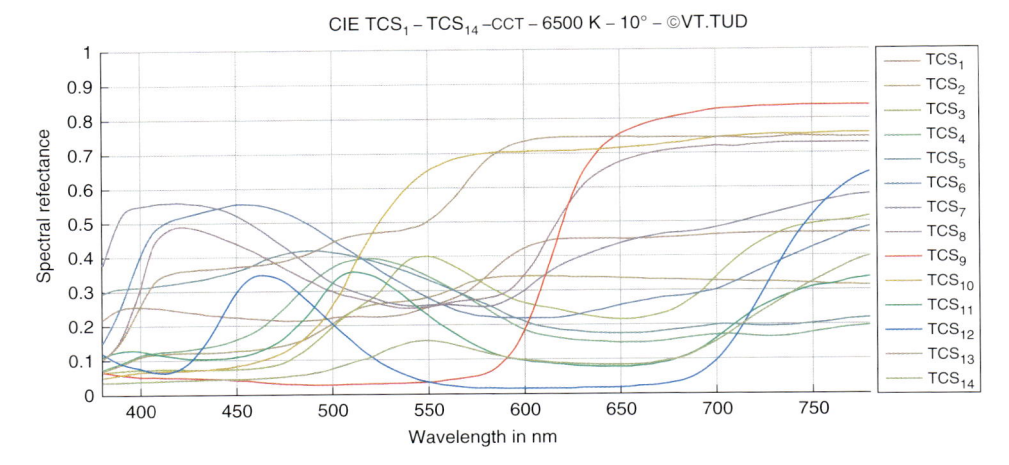

Figure 4.15 Spectral reflectance (spectral radiance coefficient) of the 14 test color samples defined for the color-rendering index by the CIE in 1965.

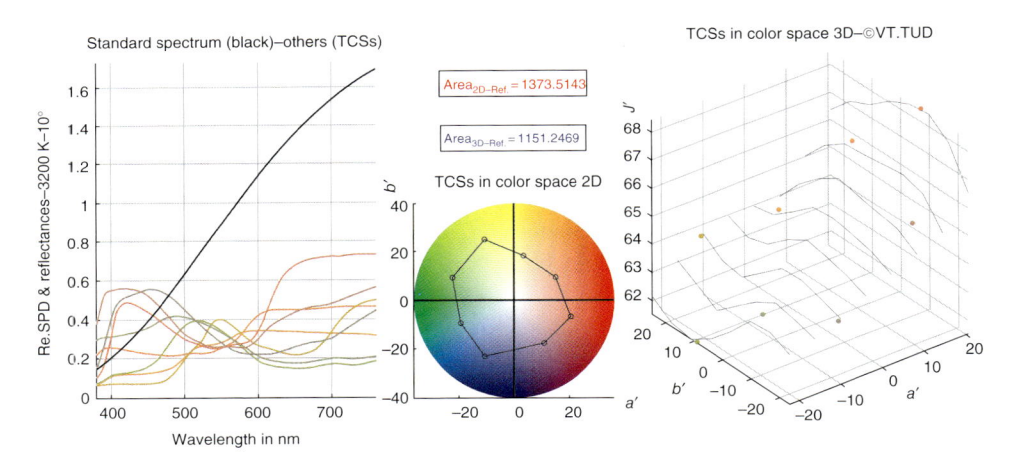

Figure 4.16 Spectral and colorimetric data of the eight CIE test color samples for the definition of CIE CRI R_a. Samples were illuminated by a Planckian radiator at 3200 K.

of these 14 TCSs is shown in Figure 4.15. Only two saturated colors (TCS_{10} for yellow and TCS_9 for red) have a long-pass characteristic at 510 and 610 nm. There are no other TCSs with similar spectral behavior between 510 and 610 nm and at longer wavelengths between 610 and 670 nm.

The colorimetric characteristics of the first eight TCSs used for the definition of the general color rendering index R_a (to model the color fidelity property of a light source) are illustrated in Figure 4.16. The lightness value J' is only between 62 and 68. Generally, the chroma values are below 22 and this means low saturation. Obviously, the saturation in the red–orange–yellow-range is very low, although most important natural and artificial colors can be found in this hue range (e.g., flowers, foods, cosmetics, textiles). The definition of the color gamut

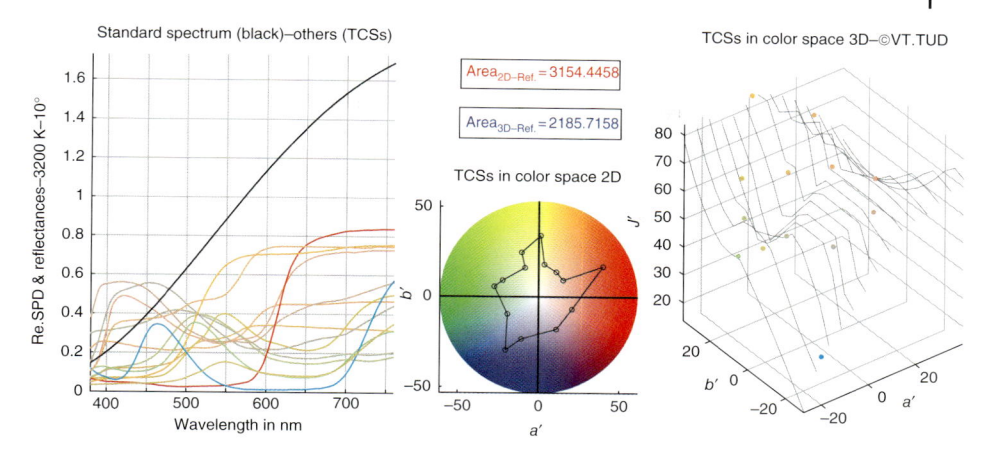

Figure 4.17 Spectral and colorimetric data of the 14 CIE test color samples of the CIE CRI method illuminated by a Planckian radiator at 3200 K.

metric GAI ([5]; see also Chapter 2) with these eight colors is therefore not an optimal solution to characterize the color gamut property of light sources. The colorimetric data of the 14 CIE TCSs including the 6 saturated colors are shown in Figure 4.17. The saturation enhancement for red (by TCS_9), yellow (by TCS_{10}), and blue (by TCS_{12}) are visible. The lightness value of the blue color is about 20 and those of the green and yellow colors in the order of about 60 and about 80 for the saturated yellow color sample TCS_{10}.

In 2012, the technical committee of the CIE responsible for color fidelity proposed 17 colors defined by their spectral reflectance functions in order to compute a new color rendering index to describe the color fidelity property. These 17 colors are mathematically defined functions of band-pass character (see Chapter 2). The core idea is to uniformly divide the peak wavelengths of these 17 band-passes over the visible wavelength range between 380 and 750 nm. In contrast to most of the blue and violet real colors whose spectral reflectance curves have a portion of reflectance in the range between about 600 and 700 nm, these mathematical curves are pure band-pass functions. Violet and blue colors do not reflect in the range longer than 600 nm at all. In Figure 4.18, the colorimetric data of these 17 colors are presented. Although these 17 functions are uniformly distributed over wavelength, their reflectance in the blue–red–violet hue range is very low. This is associated with small saturation in the violet range due to the reasons mentioned above. Different lightness levels are also underrepresented by this set.

Concerning the 15 CQS TCSs defined by NIST (USA) [6], the uniformity of hue distribution combined with a reasonable saturation level was achieved; see Figures 4.19 and 4.20. These TCSs are selected for a multipurpose definition: color fidelity (Q_f), color gamut (Q_g), and color preference (Q_a, Q_p).

The selection of these TCSs (VS1–VS15; Figure 4.19) is reasonable. The first color VS_1 is saturated red (CQS version 9.0; 2011); it exhibits long-pass filter characteristics with an increasing branch around 620 nm. The entire set of 15 test

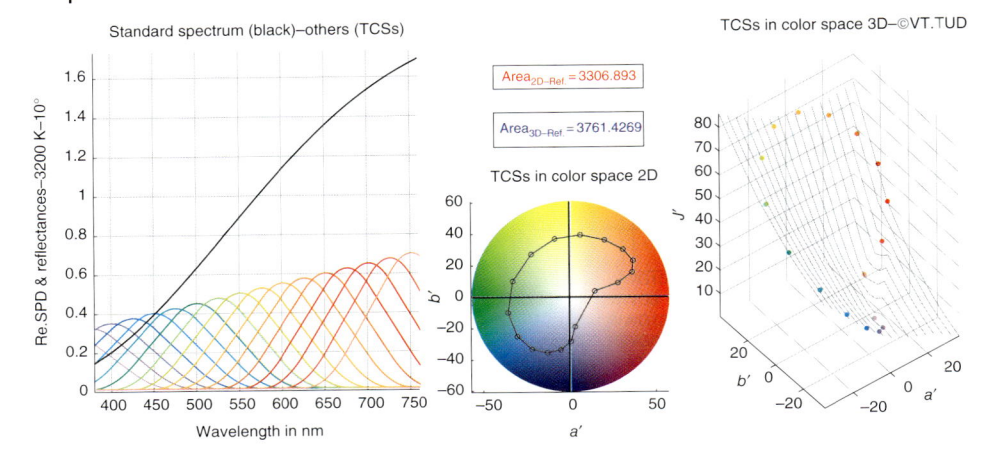

Figure 4.18 Spectral and colorimetric data of the 17 test color samples in the CRI2012 (or nCRI) definition. Illuminant: Planckian radiator at 3200 K.

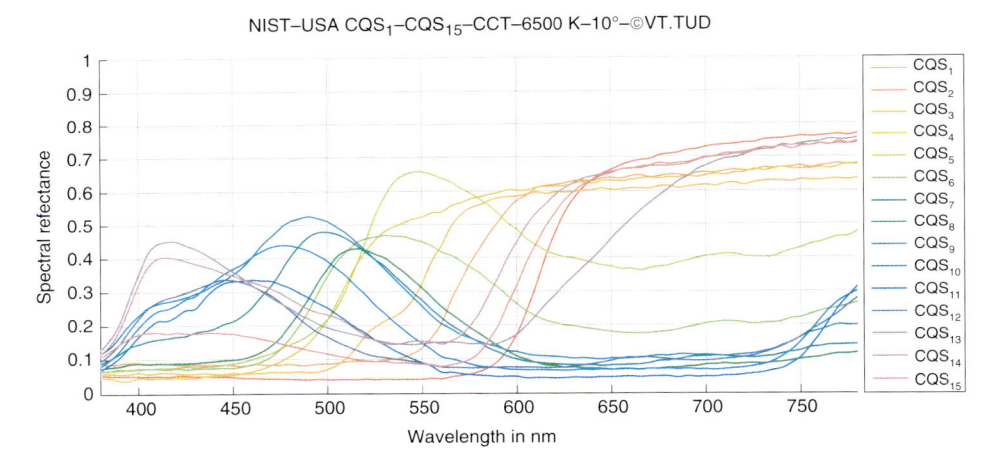

Figure 4.19 Spectral reflectance (spectral radiance coefficient) of the 15 test color samples in the CQS system of NIST (USA) according to CQS version 9.0 (2011).

colors covers the whole rainbow with the corresponding long-pass, short-pass, or band-pass filter characteristics, which can be observed from Figure 4.19. In contrast to the 14 TCSs of the CIE (1965) or the 17 colors of CRI2012 method, the advantages of these 15 CQS colors (VS1–VS15) can be seen from Figure 4.20: the distribution of the colors around the hue circle is nearly uniform and, at the same time, their chroma is high. These 15 TCSs are situated at different lightness levels between 40 and 80.

It can be concluded that the numbers of 15 colors (CQS), 17 colors (CIE CRI 2012), or 14 colors (CIE 1965) are not high enough to be distributed well on a representative number of lightness levels. If someone assumes that 10–15 TCSs should be enough for a hue plane at a certain fixed lightness level and totally 6 or

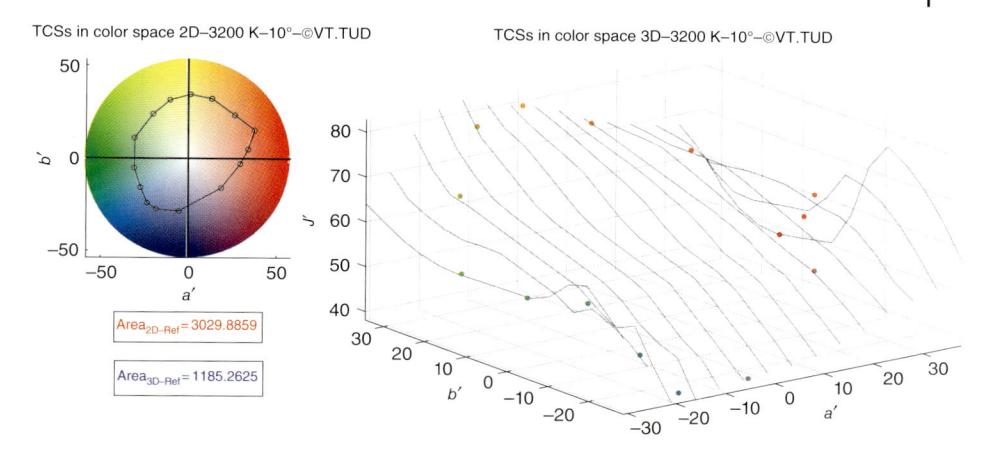

4.6 State-of-the-Art Sets of Test Color Samples and Their Ability to Evaluate | 105

Figure 4.20 Colorimetric properties of the 15 CQS test color samples illuminated by a Planckian radiator at 3200 K.

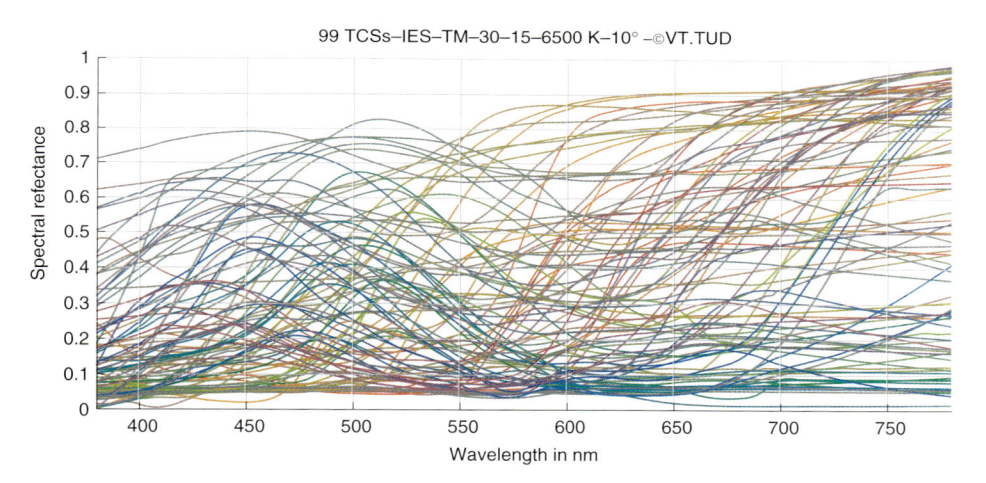

Figure 4.21 Spectral reflectance (spectral radiance coefficients) of the 99 test colors in the IES-TM-30 system.

7 lightness levels are needed, then it can be calculated that about 60–105 colors should be selected from all important colored objects of nature and in technical products. A color collection proposed in 2015 by the IES (USA, TM-30, [7]) for the definition of color fidelity (R_f) and color gamut (R_g) includes 99 colors whose spectral radiance coefficients are illustrated in Figure 4.21.

Some advantages already mentioned in the above discussion of the 15 CQS colors can be repeated in the case of Figure 4.21 for these 99 colors. With these 99 colors, a reasonable distribution of colors along the wavelength axis can be realized, and, in the most important hue range between 0° and 90° for red and orange colors, enough colors have been selected. The disadvantage is, on one hand, a small number of long-pass spectral reflectance curves between 500 and 600 nm for yellow colors. On the other hand, most spectral reflectance curves between

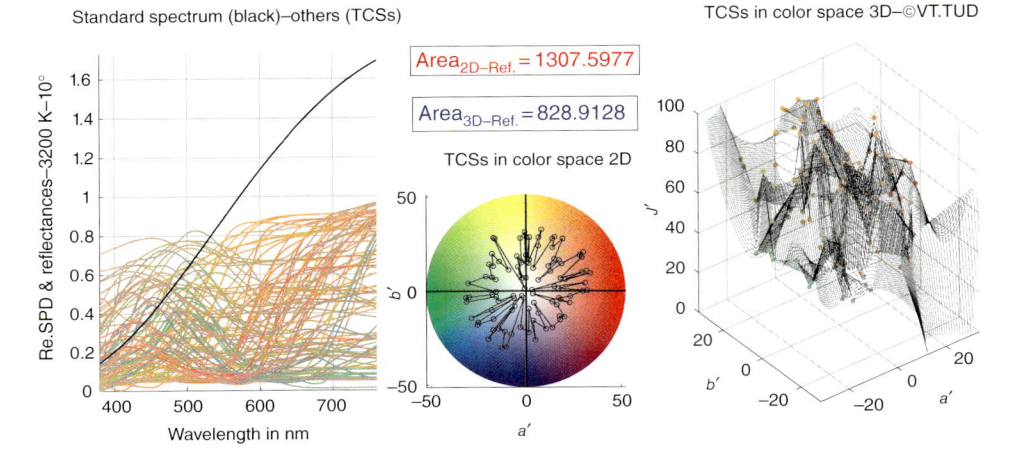

Figure 4.22 Spectral and colorimetric data of the 99 IES test colors illuminated by a Planckian radiator at 3200 K.

500 and 680 nm with long-pass characteristics do not have a high slope. The consequences of these two disadvantages can be seen from Figure 4.22 in the hue circle in the $a' - b'$ diagram.

Taking a closer look at Figure 4.22, it can be seen that

1) in the yellow hue angle range between 90° and 120° only colors with rather low saturation are available;
2) totally, the chroma of most colors is not higher than 40 – rather with a maximum at the chroma of 35. This should be enough regarding the definition of color fidelity but it is relatively low in order to analyze the color gamut property of semiconductor light sources, for example, hybrid LEDs (i.e., RGB plus phosphor-converted white LEDs).

Concerning the description of color gamut, the definition of FCI ([8], see also Chapter 2) is often used in the lighting industry. To model color gamut, four saturated object colors are taken into consideration whose spectral radiance coefficients are shown in Figure 4.23.

A blue and a green color with band-pass forming spectral reflectance curves with a low reflectance at wavelengths longer than 600 nm establish two saturated colors. The other two TCSs are yellow and red defined by long-pass spectral reflectance functions with half-value edges at about 515 and 620 nm, respectively. The small number of only four colors intended to characterize the color gamut rendering potential of the test light source is obvious; see also Figure 4.24.

As can be seen from Figure 4.24, in the two-dimensional hue circle diagram (CAM02-UCS $a' - b'$), showing only four colors, big gaps arise in the hue angle range between about 255° and 380° and between 90° and 170° (yellow-green). From the distribution of the colors in the three-dimensional hue volume (right diagram of Figure 4.24; $J' - a' - b'$), it can be clearly recognized that these four colors are situated at four very different lightness levels with the blue color at around $J' = 20$ and the yellow color at $J' = 80$.

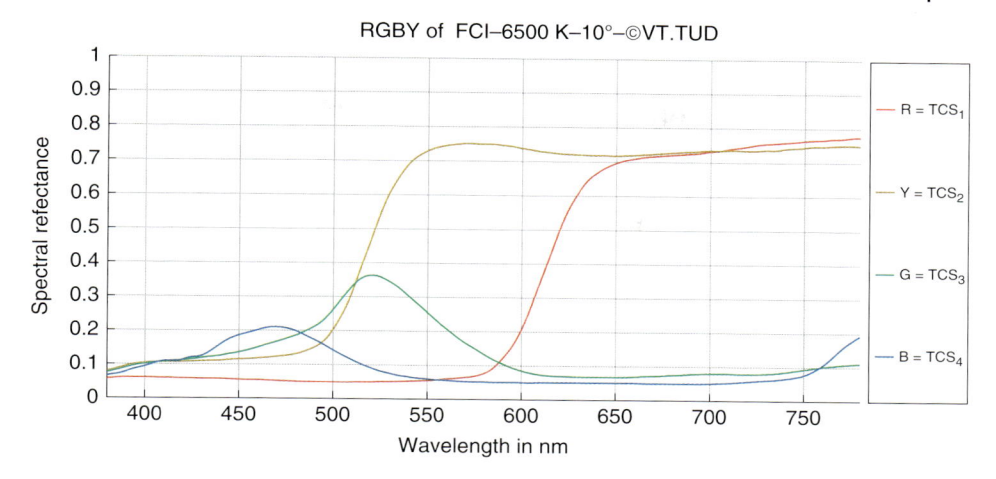

Figure 4.23 Spectral reflectance (spectral radiance coefficients) of the four test color samples of the FCI definition.

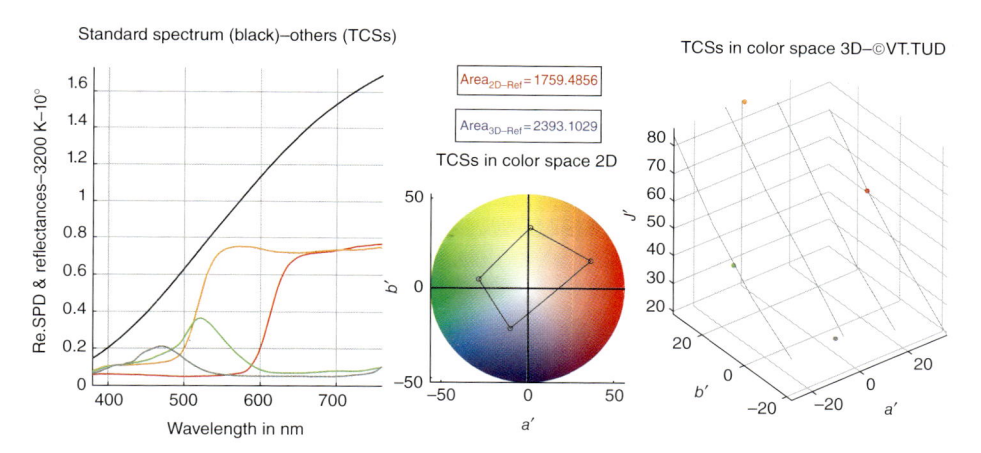

Figure 4.24 Spectral and colorimetric data of the four FCI test color samples illuminated by a Planckian radiator at 3200 K.

Besides the color collections for color fidelity, color gamut, and color preference (in the context of CQS Q_p), a further color collection is being used in the context of memory colors [9] based on the definition of the *MCRI* metric (see Chapter 2). For this purpose, a set of nine TCSs was selected with the spectral radiance coefficients shown in Figure 4.25.

A can be seen from Figure 4.25, the TCSs selected to compute MCRI contain typical spectral reflectance distributions for desaturated surface colors with a wide half-width for band-pass curves and very flat slopes in case of the spectral reflectance curves with long-pass characteristics. The colorimetric data of these nine colors illuminated under the optical radiation of a Planckian radiator at a CCT of 3200 K are shown in Figure 4.26.

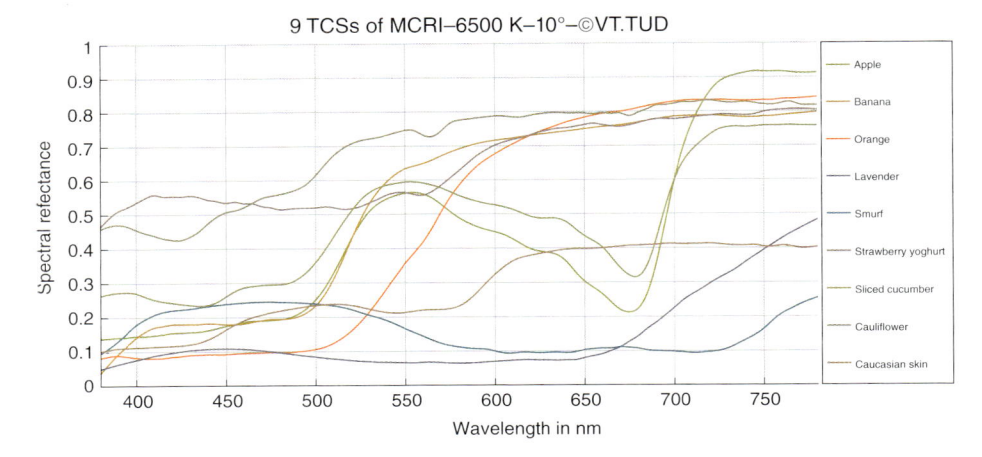

Figure 4.25 Spectral reflectance (spectral radiance coefficient) of nine object colors in the MCRI definition.

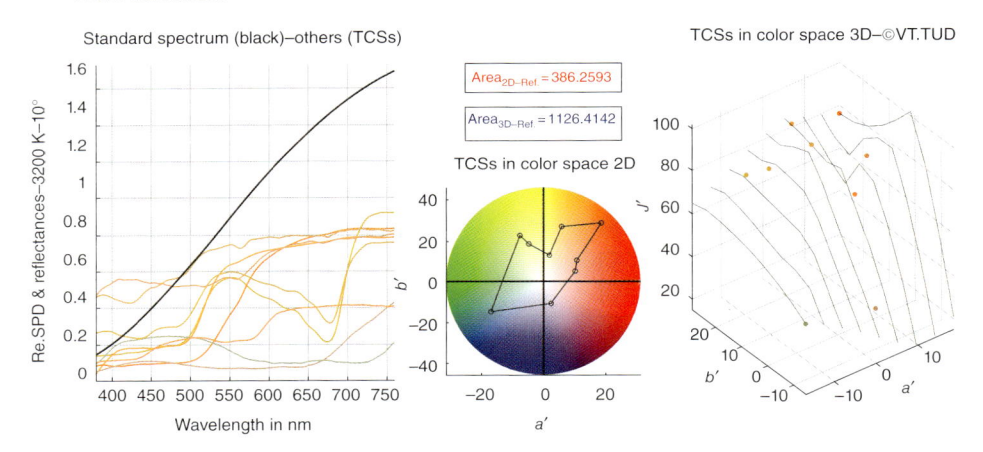

Figure 4.26 Spectral and colorimetric data of the nine MCRI test color samples illuminated by a Planckian radiator at 3200 K.

As can be seen from Figure 4.26, most MCRI colors contain a chroma value of about 15–20, with an exception of the one orange-red color. In the hue angle range between 120° and 200° (yellow/green–green–cyan) and from 200° to 280° (cyan–blue), no MCRI TCSs have been selected.

Methodologically, there exist different ways to characterize color collections used to compute color fidelity indices or color gamut indices. Scientists might consider the characteristics of the colors alone or in the general context of the metric's definition, or also by looking at their relationship with the light source's properties. In any case, the 2° or 10° color-matching functions of the standard observer shall be used as a starting point. The methods to characterize these color collections include the following:

1) Analysis of the spectral reflectance behavior and colorimetric data of the color collections under consideration (distributions of the lightness, hue, or saturation of the colors illuminated by daylight or by Planckian radiators of different CCTs). This method was used in the above discussions of the present chapter.
2) Sample calculations of the values of the metrics that belong to the three groups (color fidelity, color gamut, and color preference) with their own color spaces, color collections, their own scaling factors, and for a large number of typical light source spectra. Based on these calculations, the values of the metrics in each group can be compared and the correlation between them can be analyzed. This is a universal method to point out the practical relevance of the metric and its value range and also to analyze its own color collection. This method will be the main focus of Chapter 6.
3) To compute the values of every color gamut metric or color fidelity metric, scientists can also use several different color collections in order to check the sensitivity of the metric values if different object colors are used that cover different volumes in color space. For example, the color fidelity definition of the CIE (1965) does not necessarily have to be applied to the eight desaturated colors (see Figure 4.16). Other collections such as the 99 colors of IES or the 15 colors of the CQS system or a collection of art painting colors can also be used. Such calculations will not be shown in this book.
4) Color scientists can also use a mixture between methods (1) and (2) and vary the test light sources in a wide range (starting from the reference light sources of the color fidelity or color gamut metrics and considering all possible different test light sources with a high variability in their spectral distributions) and observe the behavior of each test color of the color collection. This mixed method will be the subject of the description below.

In Figure 4.27, the spectra of Planckian radiators (2700–4500 K) and the ones of phases of daylight (5000–6500 K) can be seen. Planckian radiators generally contain less blue radiation in the range between 380 and 480 nm while daylight sources have a good balance between the optical radiation in the range up to 500 nm and between 500 and 780 nm. The typical color quality metrics and their values for these reference light sources are listed in Table 4.1.

Based on the data in Table 4.1, the following aspects can be concluded:

- Because most color quality metrics use these light sources as reference light sources, color quality index values usually equal 100.
- FCI values continuously decrease if the CCTs of the reference light sources increase from 2700 to 6500 K. In the context of color gamut in the sense of FCI, a tungsten light source with CCT = 2700 K has a greater color gamut than the one of a daylight radiator at 6500 K.
- The color preference CQS Q_p values increase steadily if CCT increases from 2700 to 5000 K with the maximal value at 5000 K and decrease for the reference light sources with higher color temperatures (higher than 5000 K). A similar tendency can also be seen for MCRI for the color temperatures up to 4000 K and higher than 4000 K. It means that the FCI values and the Q_p/MCRI values exhibit a different course. This is caused by the metrics' definitions or/and by the usage of different color collections and, as a consequence, researchers,

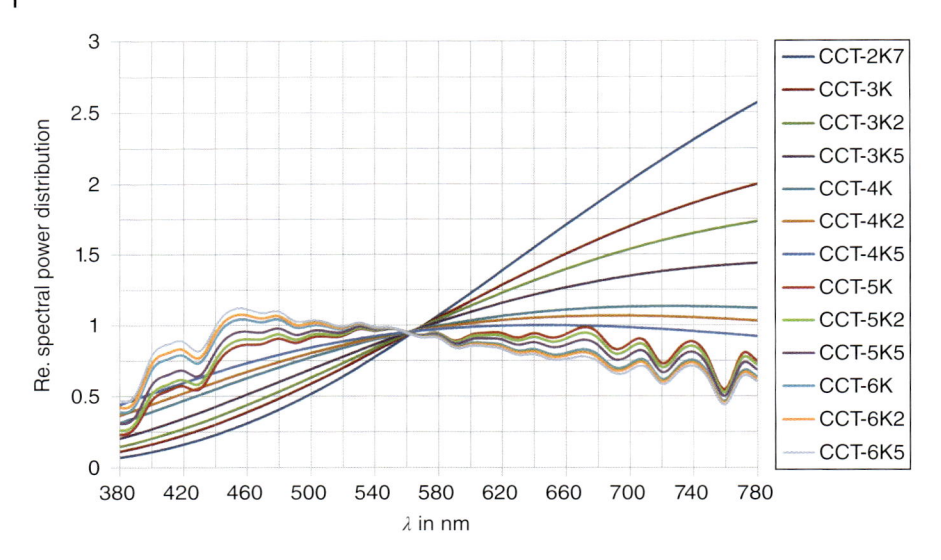

Figure 4.27 Relative spectral power distributions of Planckian radiators and daylight phases of different correlated color temperature.

Table 4.1 Color quality metrics and their values for the reference light sources in Figure 4.27.

Name	CCT-2K7	CCT-3K	CCT-4K	CCT-5K	CCT-5K5	CCT-6K	CCT-6K5
CCT	2700	3000	4000	5000	5500	6000	6500
R_a	100	100	100	100	100	100	100
$R_{1,14}$	100	100	100	100	100	100	100
GAI_{rel}	100	100	100	100	100	100	100
Q_a	100	100	100	100	100	100	100
Q_f	100	100	100	100	100	100	100
Q_g	100	100	100	100	100	100	100
Q_p	87.83	90.33	93.78	**94.39**	93.84	93.20	92.54
CRI_{2012}	100	100	100	100	100	100	100
MCRI	89.26	90.05	**90.78**	90.51	90.33	90.16	90.01
FCI	124	121	114	108	105	102	100
R_f	100	100	100	100	100	100	100
R_g	100	100	100	100	100	100	100

lighting engineers, and light source users will experience a conflict between the different metrics.

Because the reference light sources automatically result in the value of 100 for several different metrics, the strategy can be to use light sources with such spectra that rapidly change their spectral power within a small range of wavelengths.

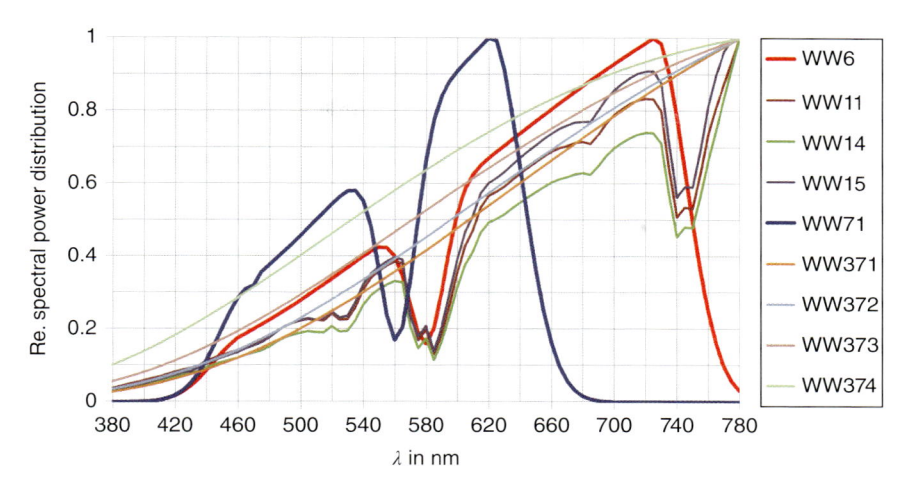

Figure 4.28 Spectra of Planckian radiators (WW371 − WW374) compared with filtered tungsten halogen lamps (WW6 − WW71).

Examples for these light sources are RGB LEDs or the filtered halogen tungsten lamps often used in the past in shop lighting; see Figure 4.28.

The lamps labeled by WW371 − WW374 are Planckian radiators similar to tungsten halogen lamps while the lamps labeled by WW6 − WW71 are tungsten halogen lamps whose lamp glass is coated with a special coating that changes the spectral emission in the range between 540 and 620 nm. The spectral emission of WW71 has a maximum at 620 nm and the other filtered lamps WW6 − WW15 exhibit similar spectral distributions. The colorimetric properties of these lamp types are shown in Table 4.2.

The following conclusions can be drawn from Table 4.2:

- Because the thermal radiators WW371 − WW374 are reference lamps, many metrics deliver the values of 100 and the FCI and Q_p/MCRI metric show different tendencies.
- The values of the color fidelity metrics R_a or $R_{1,14}$ continuously increase from the lamp type WW6 to WW71. In contrast to this phenomenon, the color fidelity values R_f, Q_f, and CRI_{2012} increase from WW6 to WW11 and WW14 and decrease in case of WW71. This tendency can also be seen in case of the color gamut metrics R_g, FCI, and Q_g and of the color preference metrics Q_p and MCRI.
- In comparison to the thermal radiators such as WW372 with $Q_p = 88.78$, $R_g = 100$, and $Q_g = 100$, the filtered tungsten halogen lamp WW14 with similar color temperature contains $Q_p = 102.53$, $R_g = 108.88$, and $Q_g = 114.32$. This increase of the color gamut values of the different metrics and of color preference is a reason for the production of these lamp types in order to illuminate products with more vividness. In Figure 4.29, the color gamut of the lamp WW14 and its corresponding reference lamp are shown. The difference between the two lamp types in the range around 560–600 nm and in the red range between 620 and 750 nm is the reason for the gamut increase

Table 4.2 Colorimetric data of the thermal radiators WW371 – WW374 and the filtered halogen tungsten lamps WW6 – WW71.

Name	WW6	WW11	WW14	WW15	WW71	WW371	WW372	WW373	WW374
CCT	2887	2869	2826	2756	3003	2700	2800	3000	3300
$\Delta u'v'$	2.60E−04	5.28E−03	5.45E−03	4.82E−03	3.58E−05	2.91E−06	2.81E−06	2.64E−06	2.42E−06
R_9	21.29	9.65	11.54	15.47	46.95	100	100	100	100
R_{13}	70.45	73.88	73.79	73.46	84.89	100	100	100	100
R_a	75.59	76.76	76.88	76.93	81.11	100	100	100	100
$R_{1,14}$	71.62	72.61	72.74	72.93	76.49	100	100	100	100
Q_f	80.11	83.20	83.32	83.46	82.15	100	100	100	100
Q_g	106.92	114.10	114.32	113.89	93.93	100	100	100	100
Q_p	89.79	102.25	102.53	101.89	73.05	87.83	88.78	90.33	91.98
CRI_{2012}	87.26	91.26	91.40	91.55	79.09	100.00	100.00	100.00	100.00
MCRI	91.52	93.21	93.10	92.73	86.31	89.26	89.59	90.05	90.45
FCI	142.52	144.82	145.02	145.17	113.20	123.72	122.97	121.37	118.92
R_f	80.98	85.45	85.51	85.56	77.66	100	100	100	100
R_g	103.15	108.78	108.88	108.57	92.66	100	100	100	100

C.S. $U^*V^*W^*$–CIE TCS$_{1-14}$ of WW14

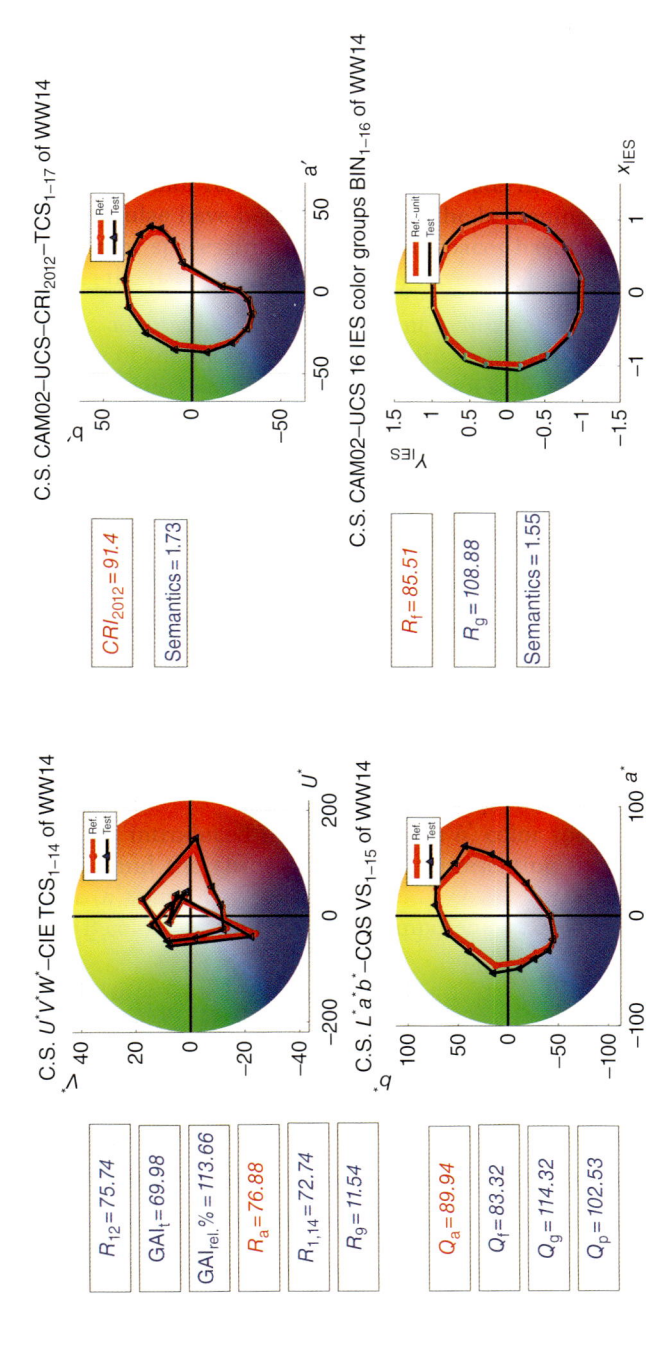

C.S. CAM02–UCS–CRI$_{2012}$–TCS$_{1-17}$ of WW14

C.S. $L^*a^*b^*$–CQS VS$_{1-15}$ of WW14

C.S. CAM02–UCS 16 IES color groups BIN$_{1-16}$ of WW14

$R_{12} = 75.74$
$GAI_t = 69.98$
$GAI_{rel.}\% = 113.66$
$R_a = 76.88$
$R_{1,14} = 72.74$
$R_9 = 11.54$

$CRI_{2012} = 91.4$
Semantics = 1.73

$Q_a = 89.94$
$Q_f = 83.32$
$Q_g = 114.32$
$Q_p = 102.53$

$R_f = 85.51$
$R_g = 108.88$
Semantics = 1.55

Figure 4.29 Color gamut and other color parameters of the lamp type WW14 and their corresponding reference illuminant in their own color spaces and using their own color sample definitions.

of WW14 in the hue angle range of 0–210° (red–orange–yellow–green–cyan) if the 15 CQS colors are considered as the optimal color collection for this gamut comparison.

Considering Figure 4.29 together with Figure 4.30 and analyzing the diagrams for IES BIN$_{1-16}$ (taken from the 99 IES colors) in Figure 4.30 and comparing them with CIE CRI 2012 and CQS VS1–VS15, it is clear that the tendency of having more color gamut in the red–orange and in the green–cyan range is reproduced by all three color collections. However, this tendency is not obvious when regarding the gamut graphic of the 14 CIE TCSs (CIE $R_{1,14}$; see the upper left diagram of Figure 4.30).

In the context of color differences, if the lamp type WW14 and its corresponding reference lamp (thermal radiator with CCT = 2826 K) are compared and if the three color collections CIE TCS$_{1-14}$, CQS TCS$_{1-15}$, and IES-16 binning groups are taken into account then the greatest color differences can be observed for R_4, R_5, R_6 (green, green–yellow), and R_9 (red) in the case of CIE TCS$_{1-14}$. In the case of CQS VS$_{1-15}$, large differences can be found in the case of VS$_1$, VS$_2$, VS$_{14}$, VS$_{15}$ (orange–red), and VS$_7$ (green). In the case of the 16-bin group of the IES method, the groups TCS-BIN$_1$, TCS-BIN$_2$ (red), TCS-BIN$_{16}$ (violet), TCS-BIN$_7$ (green), and TCS-BIN$_9$ (cyan) show the most obvious color differences between the test and reference appearance of the individual TCSs; see also Figures 4.31–4.33. As can be seen from these figures, spider web diagrams represent a very good tool to visualize color appearance and color gamut comparisons.

4.7 Principles of Color Grouping with Two Examples for Applications

As mentioned above, the choice of the set of TCSs is very important to set up color metrics for the evaluation of color quality for modern lighting systems. If the set of TCSs covers the hue circle and the color volume well enough, for example, the real colors to be found in nature, then the evaluation of the color quality of light sources based on the color set shall yield satisfactory results. If there are missing colors and gaps in the gamut volume then the quality of the lighting system will not be satisfactory, accurate, and/or objective. Therefore, the size of the color set should be large enough but, unfortunately, this issue causes problems by itself when the TCSs should be presented visually for light source designers or users. Consequently, it is necessary to minimize the size of the color set so that the set should be convenient in use with about 14–20 representative color samples. However, these samples must be able to represent the complete set of all TCSs within a given group of TCSs. It means that these representative color samples must have the predominating properties of all TCSs belonging to the given color group.

As a result, color grouping or color classification becomes an important subject for the establishment of color metrics. In color science, there are several methods and approaches to carry out this task. In Figure 4.34, these methods are classified into two types including modern and classic approaches.

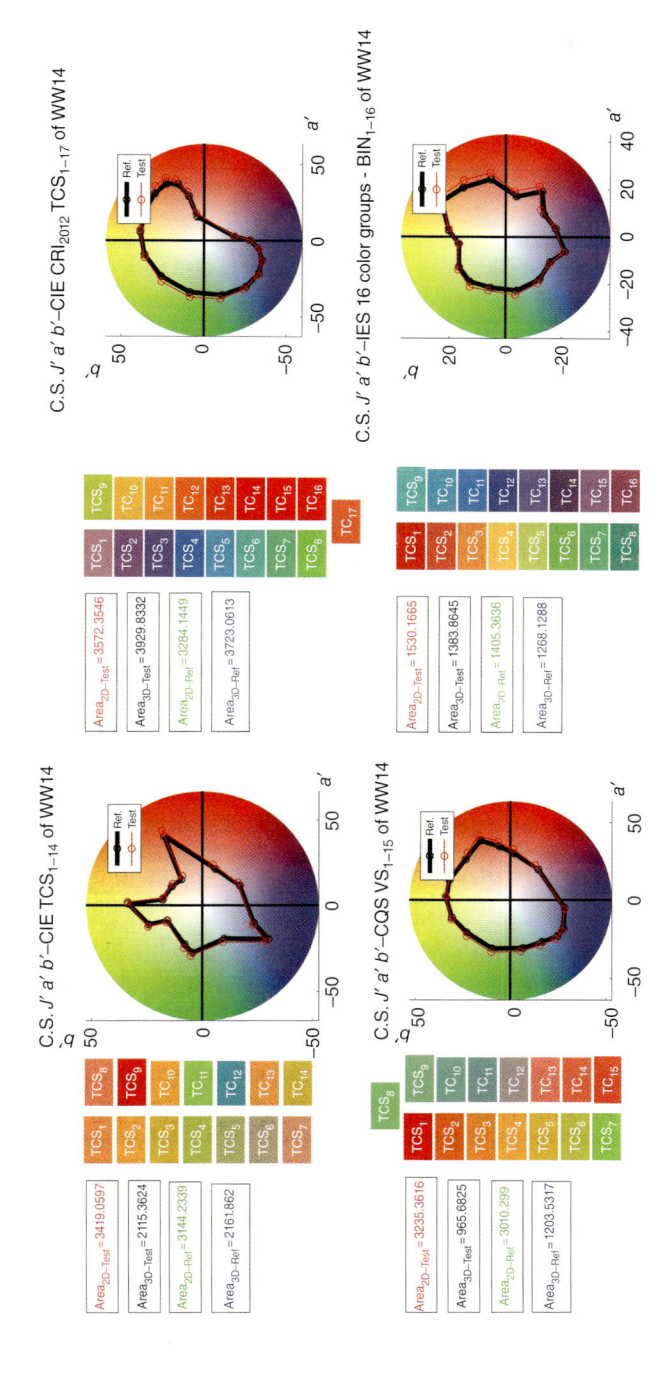

Figure 4.30 The 2D and 3D color gamut area of the lamp type WW14 and of their corresponding reference under the same color space CIE CAM02 UCS and color definitions.

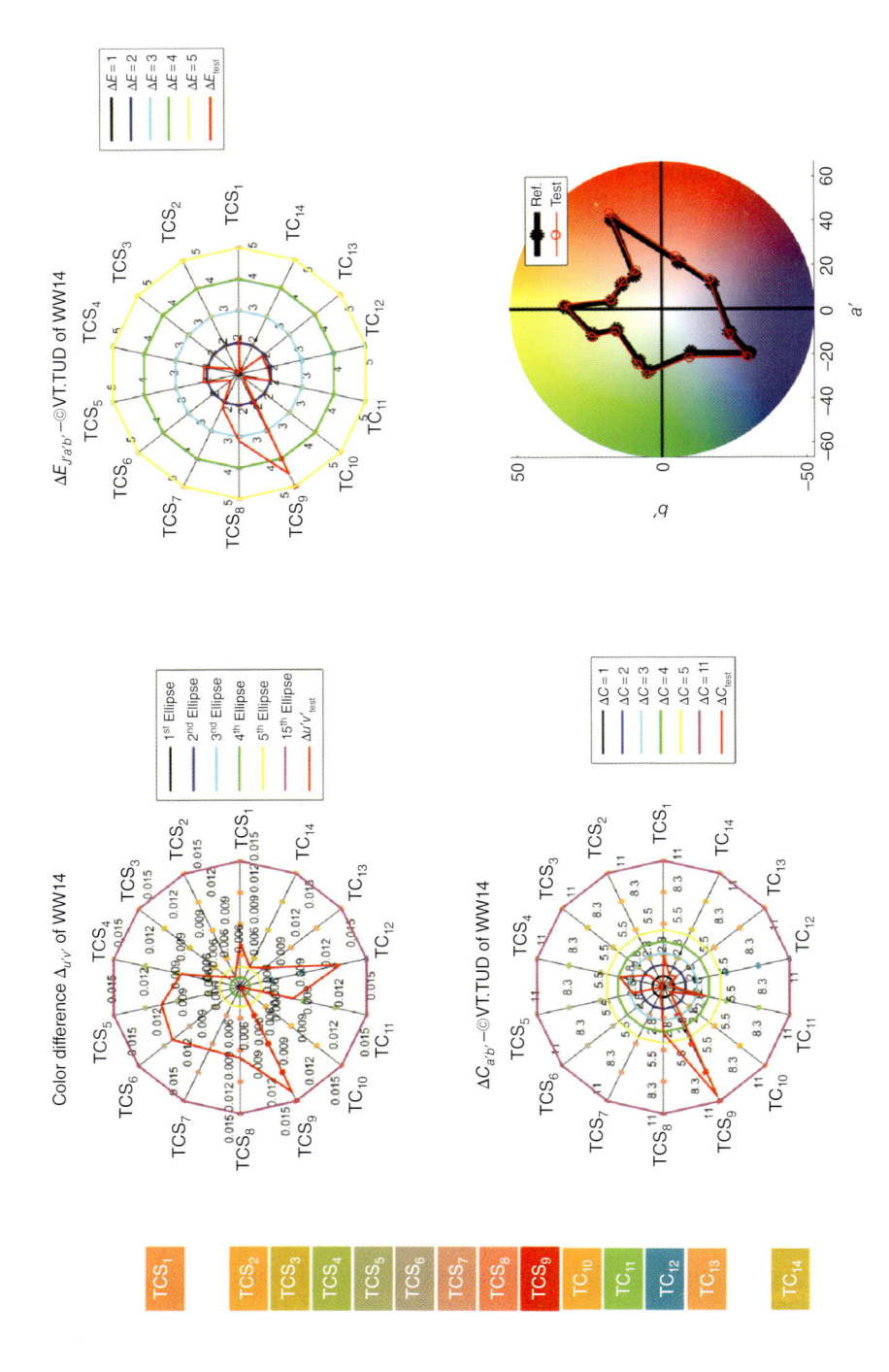

Figure 4.31 Spider web diagrams of color difference and chroma difference; hue circle diagram ($a' - b'$): comparison between the lamp type WW 14 and its reference light source for the 14 CIE test color samples.

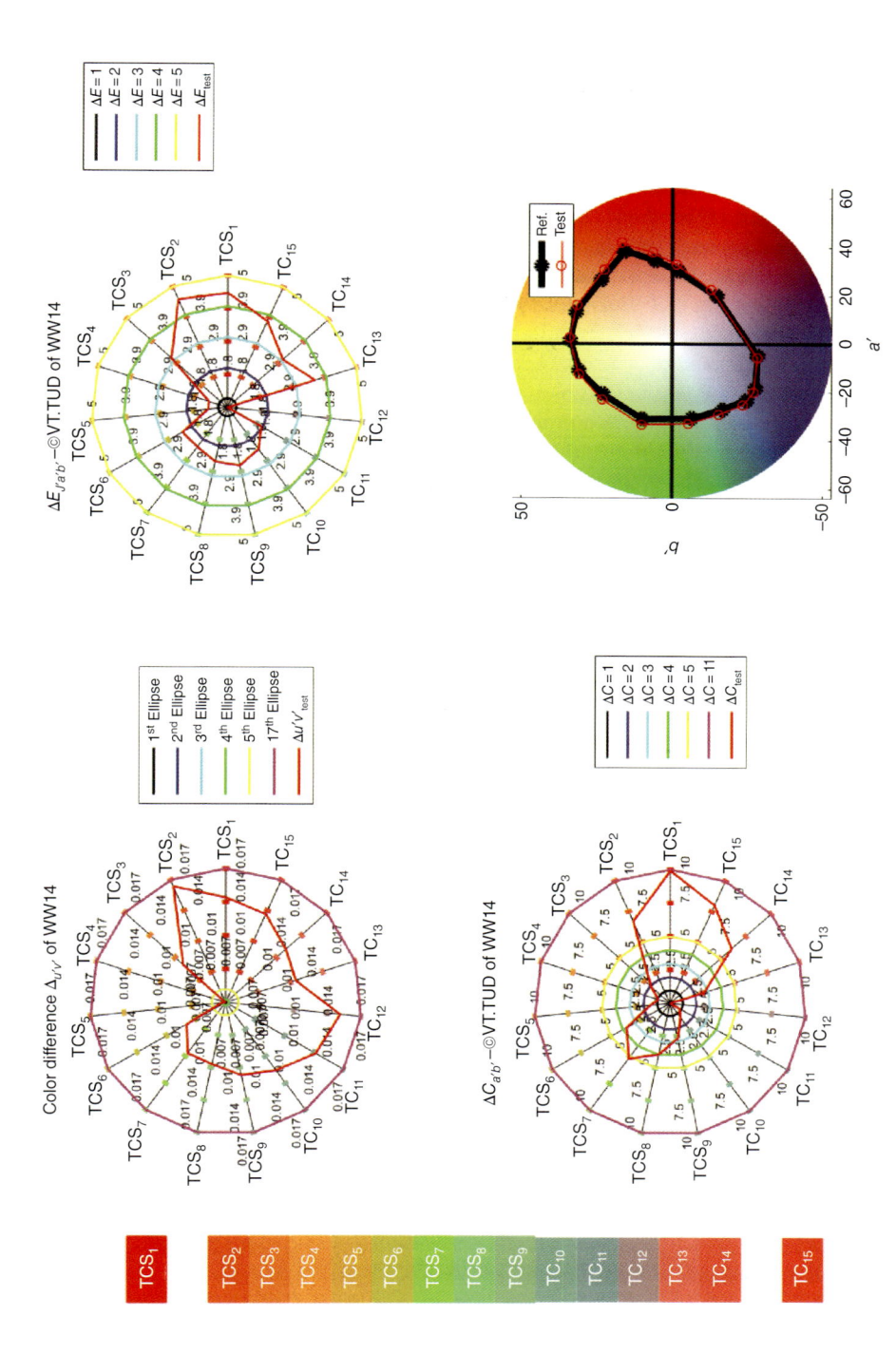

Figure 4.32 Spider web diagrams and hue circle diagram for the 15 CQS TCSs (NIST/USA, 2010).

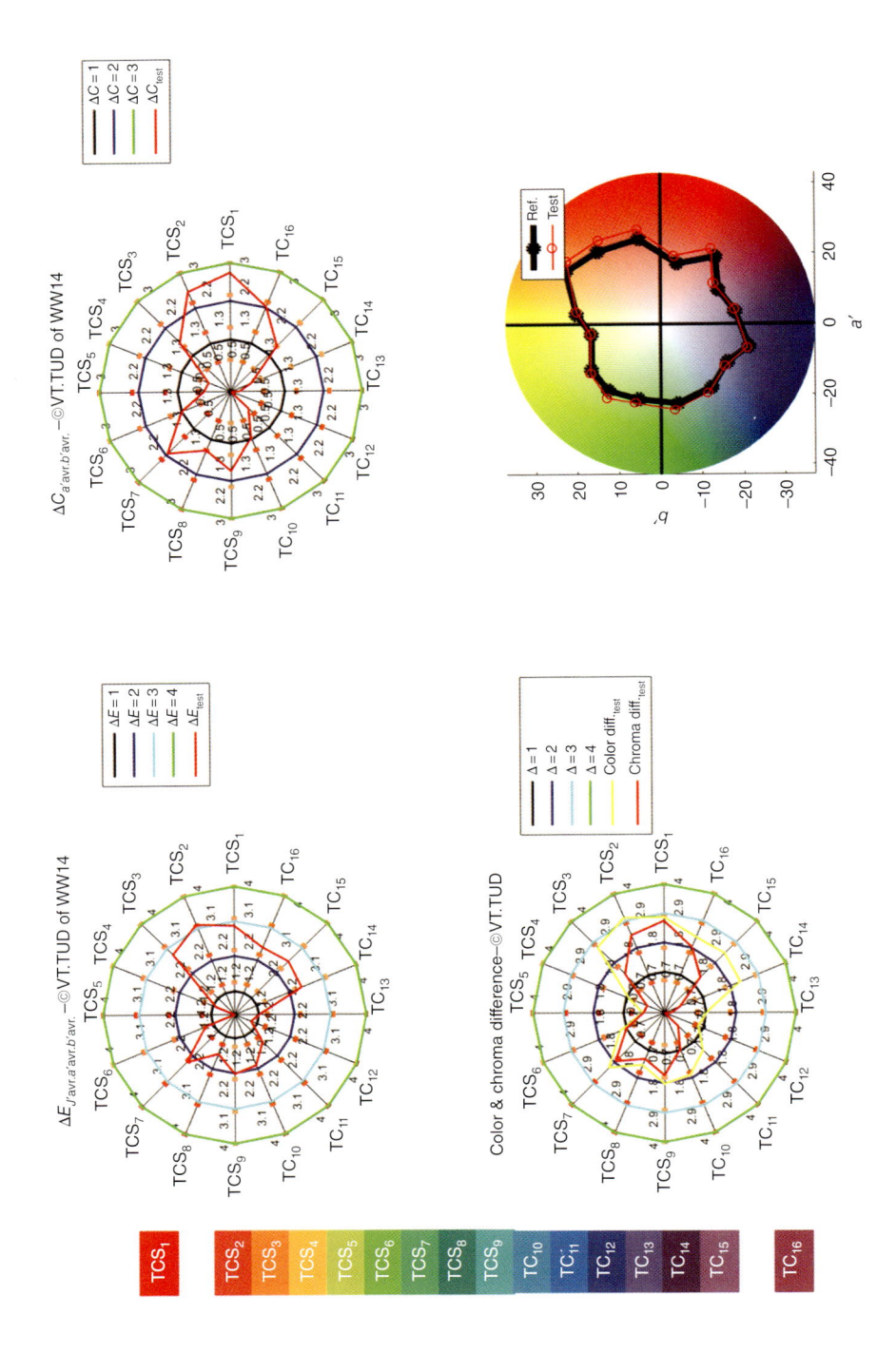

Figure 4.33 Spider web diagrams and hue circle diagram for the 16 color binning groups of the IES colors (2015).

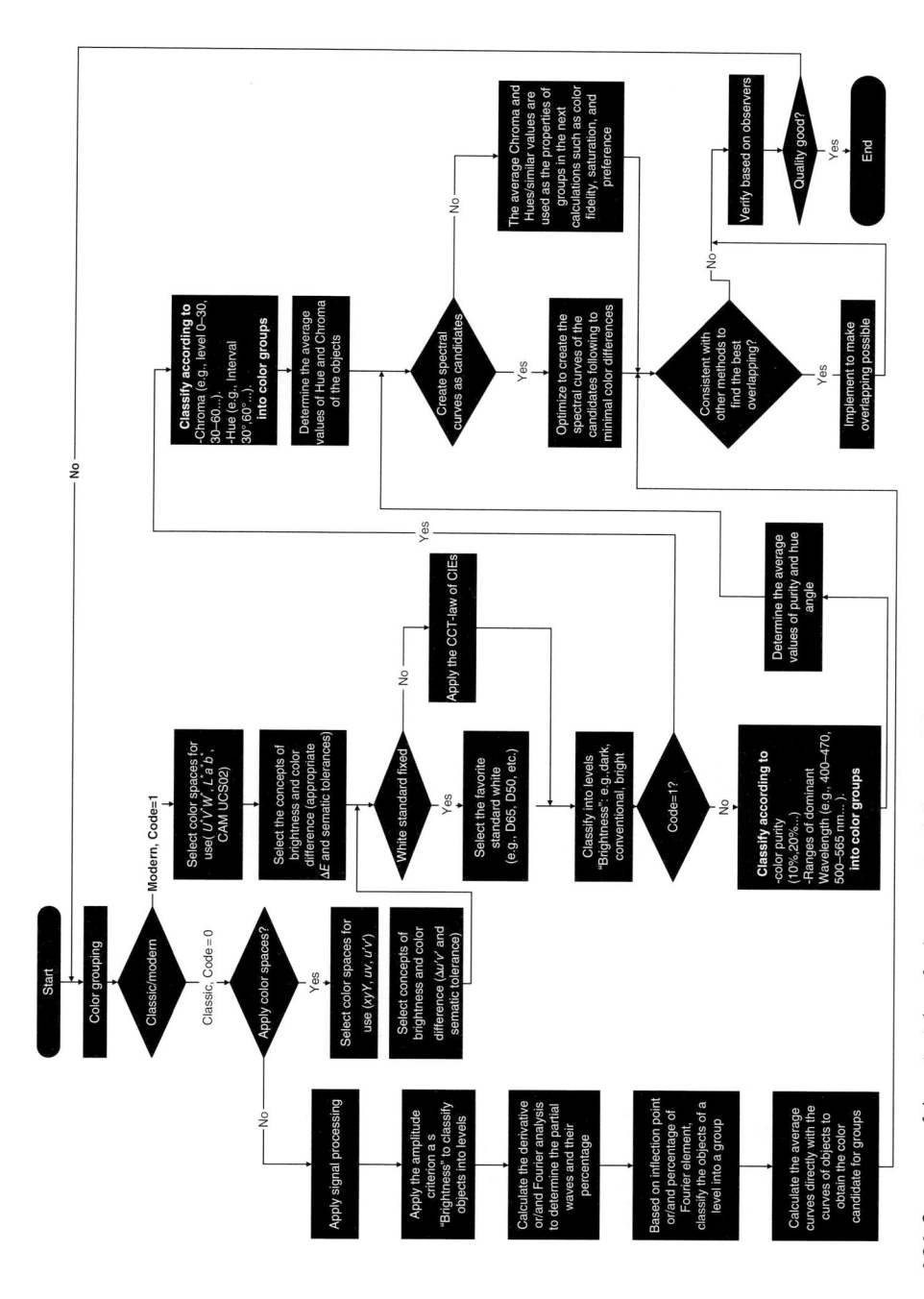

Figure 4.34 Summary of the principles of color grouping following different methods and approaches.

As can be seen from Figure 4.34, in the modern approach, the classic color concepts from the years between 1930 and 1960 such as *xyY*, *uv*, or *u′v′* should not be applied. Then, the concept of color difference such as $\Delta u'v'$ is not necessary. In contrast, in the classical approaches, these classical concepts and parameters are the tools to be used for color classification. In addition, in the classical approach, if a color space should not be applied then the theory of signal processing becomes essential for color grouping. Although the latter method is quite strange for color experts, it is very useful and objective for color classification because it does not need a subjective visual color model to establish the candidate set of TCSs. In order to analyze the approaches and methods, the following issues should be explained and solved in the process of color grouping:

- Should a color space or a more complicated visual color model be applied?
- If a color space is appropriate, which color space shall be used with which color difference definition? How are semantic interpretation categories and their tolerances defined?
- Shall a white point be fixed or should the CCT be varied in the progress (to apply the so-called "CCT-law" in the flowchart of Figure 4.34)?
- Should the spectral curves of the TCS candidates be created one by one or shall just the average values of chroma and hue within a group be taken?
- Should a combination of several methods be implemented?
- Should the finished color classification be verified by test persons?

The answers for the above questions lead to the establishment of three methods (Methods 1–3).

4.7.1 Method 1 – Application of the Theory of Signal Processing in the Classical Approach

In this method, no visual color model is applied. Instead, the spectral reflectance curves of the TCSs are considered as signals and/or signal spectra. Then, these signal spectra will be analyzed by their derivatives or Fourier series in the time domain or frequency domain. Because there is no visual color model, the amplitude of the spectra must be used as the criterion for the classification of "*brightness*" in order to group the TCSs into different brightness levels such as "*dark*," "*middle*," or "*bright*." Then, the calculation of derivatives or Fourier analysis helps determine the partial wave elements and their percentages. Based on the inflection points or/and the percentage of Fourier elements, color objects are classified into different groups of brightness level. If the spectral reflectance curves of the representative color samples are needed then the calculation of the average curves of the investigated colored objects in the color groups will help find out the desired spectral reflectance curves. Based on this method, the absence of the application of a visual color model helps avoid possible subjective error factors. Also, the natural properties of test color objects are investigated and generalized objectively, without any human factors or the application of actual light source spectra. Therefore, no problem arises concerning the variation of colorimetric parameters under different light sources and different subjects belonging to different cultures, age, or educational groups, and the concept of color differences and their tolerances and semantics can be avoided. Hence this method of color grouping is simple, effective, and powerful.

4.7.2 Method 2 – the Application of a Visual Color Model in the Classical Approach

In Method 2, the classical but conventional and effective concepts of colorimetric tools and their parameters are applied for color classification. Specifically, chromaticity xy, uv, and $u'v'$ in the color diagrams xy, uv, or especially $u'v'$ and their semantics and tolerances have been applied in many studies. Typically, MacAdam's ellipses and subsequent color research have been important to define the limits of color difference in various color areas (see [4, 10–12]). In this method, the parameter Y is assumed to be the "*Brightness*" of color objects under a specific light source so that colored objects are grouped into different brightness levels. However, previously, a desired standard white light source such as the black body radiator at 3000 K (the favorite white standard in museum lighting) or the daylight D65 or D50 (the favorite white standard in conventional color systems) shall be fixed or the CCT of the light source shall be changed. Because the classical color concept is very simple, color classification shall apply color purity and the range of dominant wavelengths for the grouping. Colorimetric purity represents the degree of color saturation, and the range of dominant wavelengths or/and color difference $\Delta u'v'$ represents the hue factor of the TCSs. Finally, depending on demand, the spectral reflectance curves can be generated. In the case of applying a fixed white light source, the spectral reflectance curves must be generated one by one simply. But in the case of applying the rule of varying CCTs, each set of spectral reflectance curves is only correct for a specific CCT value. For a different CCT, a new set must be established. The advantage of this method is that it is simple, in line with the classical concepts and their parameters. Dominant wavelength and colorimetric purity are objective descriptors to represent hue and saturation. However, the quantities intended to describe "*Brightness*" (Y) or "*Color difference*" ($\Delta u'v'$) are not accurate enough to correlate with human perception in view of today's lighting technology and the recent results of color science.

4.7.3 Method 3 – the Application of Visual Color Models in the Modern Approach

In recent years, Method 3 has been the favorite one among color experts. Many color researchers have applied this method in their studies (see [11–13]) by applying a color space such as $U^*V^*W^*$, $L^*a^*b^*$, or CAM02-UCS ($J'a'b'$).

The color space $U^*V^*W^*$ is the conventional one in the establishment of the CIE color-rendering index with the factor "*Color difference*" $\Delta E_{U^*V^*W^*}$ and "*Brightness*" W^* while the color space CIELAB $L^*a^*b^*$ is frequently used in the color evaluation of digital images and displays/monitors. Although the color space CAM02-UCS has just appeared for about 10 years, it has played an important role with its parameters "*Color difference*" $\Delta E' = \Delta E_{J'a'b'}$ and "*Brightness*" J' in certain color quality metrics such as CIE CRI2012 in 2012 or IES TM15-15 in 2015 as the earliest international application.

Compared with Method 1, Method 3 applies a modern color space. And compared with Method 2, Method 3 is similar at many points. Specifically, it must also face the problem of a fixed standard white light source or the application of the method of varying CCTs. As well, it must apply a parameter assumed to describe "*Brightness*" to classify the test color objects into different brightness levels. However, several differences appear in this modern method. In detail, Method 3 does not apply the classical concept of color purity or the range of dominant wavelengths. Therefore, as a color saturation parameter, it uses chroma in order to group the test color objects by the use of different distance intervals (expressed as $\Delta U^* V^*$, $\Delta a^* b^*$, or $\Delta a' b'$), for example, 0–10, 10–20, 20–30, or other distance intervals. Concerning hue, it must apply a hue angle to classify the samples such as $\alpha = \text{atan}(U^*, V^*)$, $\alpha = \text{atan}(a^*, b^*)$, or $\alpha = \text{atan}(a', b')$ in the ranges 0–30°, 60–90°, …, 330–360° or by the use of other hue angle intervals. It can be recognized that this calculation is more systematic. However, its accurateness depends on the visual color models or color spaces used. If the accurateness of the used color spaces is not good enough then the resulting color evaluation or color grouping will not be appropriate. In addition, if the semantics of "*Color difference*" ($\Delta U^* V^* W^*$, $\Delta a^* b^* L^*$, or $\Delta a' b' J'$) or/and the one of "*Brightness*" (W^*, L^*, or J') is not associated with the visual tolerance of the subjects correctly then it will become difficult to specify the correct scaling factor to establish a correct color grouping. As a result, the combination of different methods is necessary in order to unify all advantages of the above-mentioned methods and to avoid or reduce their disadvantages.

Two typical examples will be described below (see the Sections 4.7.4 and 4.7.5) for the application of the above methods of color grouping with a specific lighting system and the use of a suitable color metrics.

4.7.4 First Example of Color Grouping with a Specific Lighting System Applying Two Methods

The author in [13] used the combination of Method 1 and Method 2 for color grouping of 79 oil colors in the context of museum lighting. Here, the 79 oil colors were analyzed in the color space CAM02-UCS. First, the chroma distance $\Delta a' b'$ was applied to classify the oil colors into two groups including saturated oil colors and desaturated oil colors. Then, the hue angle $\alpha = \text{atan}(a', b')$ was used to group the oil colors into smaller color groups and the spectral reflectance curves were checked and classified again with Method 1. If the overlapping results of the two methods were acceptable then the color grouping was finished. The 79 original spectral reflectance curves and the 18 grouped oil color representatives are shown in Figures 4.35 and 4.36, respectively. The achieved results of the 18 representative color samples of oil colors correspond to the application of a white light source at 3000 K often used for museum lighting. If the method of varying CCTs is applied then these resulting spectral reflectance curves must be recomputed for every CCT.

The quality of color grouping can be observed in Figure 4.37 for the 79 original oil colors and the 18 representative oil color samples. It can be seen that the chromaticity distribution of the 79 original oil colors and the 18 representative

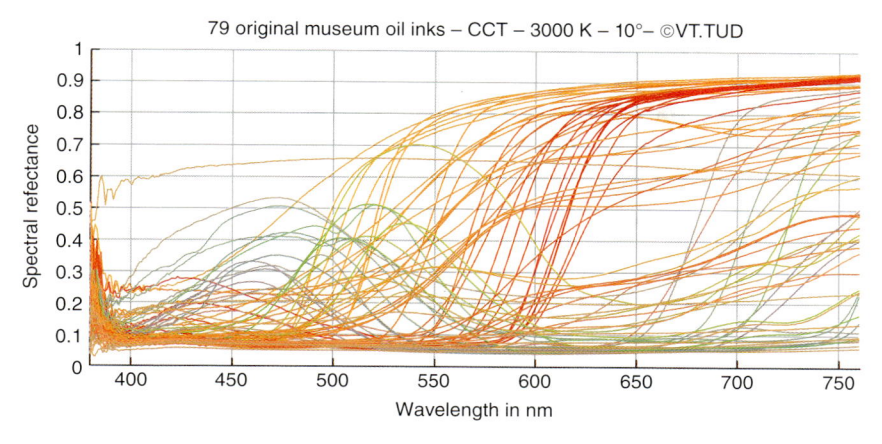

Figure 4.35 79 original oil colors before the color grouping in the case of CCT = 3000 K for museum lighting [3].

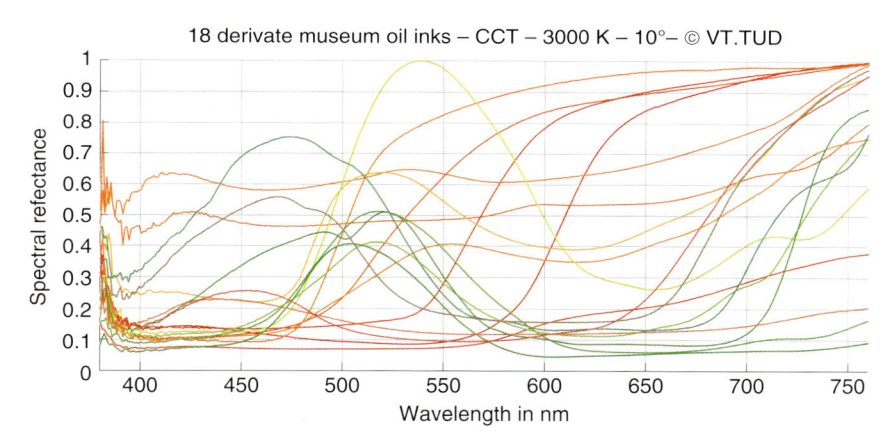

Figure 4.36 18 representative oil color samples after color grouping in the case of CCT = 3000 K for museum lighting [13].

oil colors in the color diagrams (xy, uv, $u'v'$, $a'b'$, and $J'a'b'$) are comparable. The saturation and hue distributions of the original samples and the representative samples are similar. The representative spectral reflectance curves are able to characterize the full set of the original ones in a simpler manner and with a smaller number of TCSs. The weak point of this example [13] is that the author did not carry out any visual experiment with test subjects to verify color grouping.

4.7.5 Second Example of Applying Method 3 by Using Modern Color Metrics

IES TM30-15 was published in 2015 as the new color quality metrics with the 99 TCSs and new parameters including R_f for the evaluation of color fidelity and R_g for color gamut. In Section 4.6, the advantages and disadvantages of this standard have already been discussed based on the aspects of TCSs and the definition

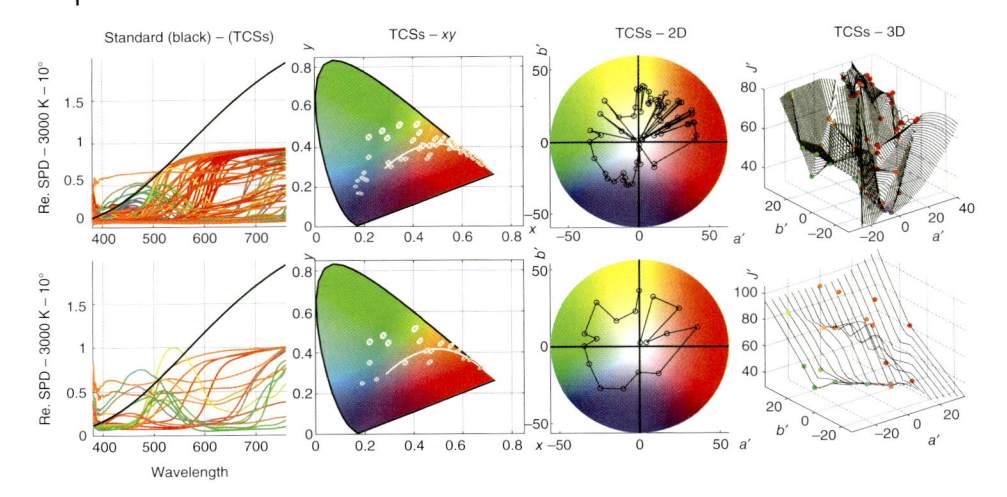

Figure 4.37 79 original and 18 representative oil colors before (upper diagrams) and after (lower diagrams) color grouping.

of its colorimetric concepts. Here, the color grouping according to this standard will be described and discussed. This second example applies Method 3 by varying CCT.

Similar to the first example, the color space CAM02-UCS is applied for all calculations with the IES TM-30-15 samples. The "*Brightness*" J' is not used for color classification but "hue angle" is calculated according to Eqs. (4.1)–(4.4).

$$v = \text{rem}((a\tan 2(b', a')), 2 \cdot \pi) \quad \text{if } v < 0 \text{ then } v = 2 \cdot \pi + v \tag{4.1}$$

$$\text{Nr}_{\text{BIN}} = \text{floor}\left(\left(\frac{\frac{v}{2}}{\pi}\right) \cdot 16\right) + 1 \tag{4.2}$$

$$a'_j, b'_j, J'_j \text{ classified into group } i(\text{Nr}_{\text{BIN}} = i = 1 \to 16) \tag{4.3}$$

$$\text{Angle}_{\text{BIN}} = -\frac{\pi}{16} + \frac{\pi}{8} \cdot \text{Nr}_{\text{BIN}} \tag{4.4}$$

According to Eqs. (4.1)–(4.4), while the binning number (Nr_{BIN}) runs from 1 to 16, the a'_j, b'_j, and J'_j values are grouped into the 16 color binning groups. Because the values a'_j, b'_j, and J'_j vary depending on the actual value of CCT according to the method of varying CCTs, for every CCT, there will be a new set of the a'_j, b'_j, and J'_j values, and, consequently, 16 new representative color samples will arise; see Figure 4.38 (for CCT = 6500 K). These spectral reflectance curves can be established by the use of the set of the a'_j, b'_j, and J'_j values of the TCSs. These curves are generated based on an optimization algorithm so that they have minimal color differences ($\Delta a'b'J'$) compared with the average value of the a'_j, b'_j, and J'_j of the test color objects belonging to a color group; see Eq. (4.5).

$$\text{minimize } \Delta J'a'b'_{\text{candidate}-J'a'b'_{\text{average}}} \to \text{reflectance curve}_{\text{color group Nr}_{\text{Bin}}} \tag{4.5}$$

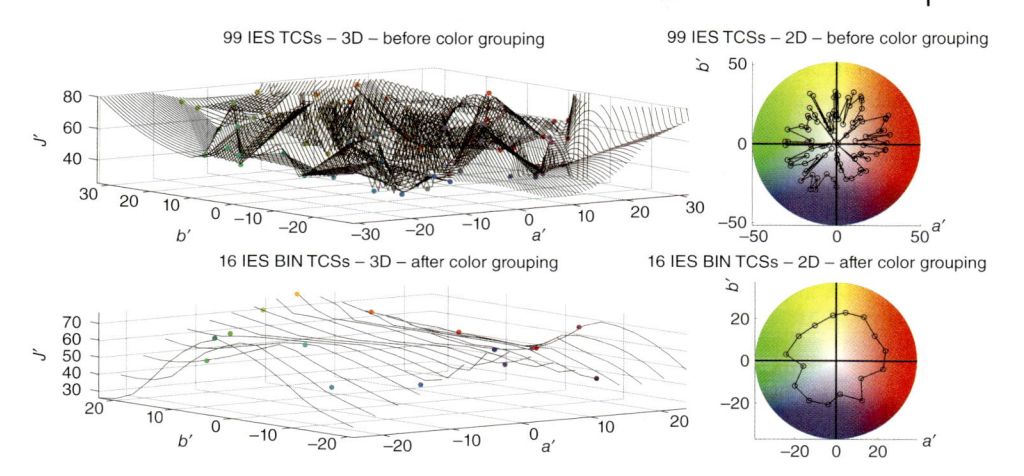

Figure 4.38 99 original and 16 derivate TCSs before (upper diagrams) and after (lower row) color grouping in case of CCT = 6500 K.

Note that the weak points of this application are that the classification after brightness is not considered, the combination of several methods is not implemented, and the spectral reflectance curves are not generated but only the average values of the a'_j, b'_j, and J'_j of the TCSs are used in the calculation. The method has not been verified in subjective visual experiments.

4.8 Summary and Lessons Learnt for Lighting Practice

After a comprehensive analysis of the color collections used by the CIE or in publications of national organizations (e.g., IES) or from different color scientists, the requirements to be fulfilled by the object colors were formulated:

1) The colored objects should be representative for a specific or many typical lighting applications (e.g., TV and film studios, shop lighting, oil paintings for galleries and museum lighting, food lighting, living rooms, etc.).
2) The colored objects should be homogeneously distributed over many typical ranges of lightness L^*, hue angles H (red, orange, skin tones, yellow, green, blue, or violet), and saturation degree C^* (strongly saturated or desaturated colors).
3) The spectral reflectance of the collected colors should be uniformly distributed over the visible wavelength range between 400 and 700 nm with their typical band-pass, short-pass, and long-pass characteristics.

The analysis of the object colors for flowers, skin tone, art painting, and the Leeds database has shown that many colors with moderate and high saturation, positioned at different lightness levels and distributed well around the hue tone circle, are available and have to be grouped in order to find out representative color collections used to define the metrics for color fidelity and color gamut, and also for the analysis of the object oversaturating property of light sources.

Regarding the TCS collections of current color quality metrics, the 15 CQS TCSs come close to the idea of an ideal color collection for color gamut and saturation enhancement. Using this philosophy of the CQS system and expanding it to several lightness levels, the designers and the users of the light sources should have a reasonable collection of test colors for color gamut analysis. For color fidelity purposes, the IES color collection is nearly optimal if it is expanded by a few saturated colors that have to be distributed around the hue circle uniformly. For the evaluations of color preference, naturalness, and vividness as a *multiple color quality metric* (see Chapter 7), a simple combination of the 99 IES and the 15 CQS TCSs should be the first step in the right direction before a general final step would be done in order to have about 120 colors that fulfill the requirements mentioned above.

References

1 Arnold, E.J., Faruq, S., Savolainen, V., McOwan, P.W., and Chittka, L. (2010) FReD: the floral reflectance database-a web portal for analyses of flower colour. *PLoS One*, **5** (12), e14287–e14289.
2 Krause, N., Bodrogi, P., and Khanh, T.Q. (2012) Bewertung der Farbwiedergabe-Reflexionsspektren von Objekten unter verschieden weißen Lichtquellen. *Zeitschrift Licht*, Heft 5, 62–69.
3 Pepler, W. and Khanh, T.Q. (2013) Museumsbeleuchtung-Lichtquellen, Reflexionsspektren, optische Objektschädigung, Teil 1. *Zeitschrift Licht*, Heft 1–2, 70–77.
4 Commission Internationale de L'Éclairage (1995) *Method of Measuring and Specifying Colour Rendering Properties of Light Sources*, CIE Publication 13.3-1995, CIE, Vienna.
5 Rea, M.S. and Freyssinier-Nova, J.P. (2008) Color rendering: a tale of two metrics. *Color Res. Appl.*, **33**, 192–202.
6 Davis, W. and Ohno, Y. (2010) Color quality scale. *Opt. Eng.*, **49**, 033602.
7 David, A., Fini, P.T., Houser, K.W., and Whitehead, L. (2015) Development of the IES method for evaluating the color rendition of light sources. *Opt. Express*, **23**, 15888–15906.
8 Hashimoto, K., Yano, T., Shimizu, M., and Nayatani, Y. (2007) New method for specifying color-rendering properties of light sources based on feeling of contrast. *Color Res. Appl.*, **32**, 361–371.
9 Smet, K.A.G., Ryckaert, W.R., Pointer, M.R., Deconinck, G., and Hanselaer, P. (2012) A memory colour quality metric for white light sources. *Energy Build.*, **49**, 216–225.
10 Bieske, K., Wolf, S., and Nolte, R. (2006) Wahrnehmung von Farbunterschieden von Licht-und Körperfarben. Vortrag auf der Lichttagung der Deutschsprachigen Länder, Bern, Schweiz, September 10–13, 2006, pp.63–64.
11 Frohnapfel, A. (2008) Psychophysiche und physiologische Aspekte der Farbwiedergabe für LED Anwendungen. Master thesis. Technische Universität Darmstadt, pp. 57–60.

12 Brückner, S. (15 December 2014) Farbdifferenz-Skalierung zur Farbqualitäts-beurteilung von Halbleiter-Lichtquellen. PhD dissertation. TU Darmstadt.

13 Quang Vinh, T. (10 October 2013) Characterization, optimization and stabilization of the lighting quality aspects of high qualitative hybrid LED-lamps by development of transient LED models. PhD dissertation. TU Darmstadt, Chapter 7, pp. 140–160.

5

State of the Art of Color Quality Research and Light Source Technology: A Literature Review

This chapter plays the role of a review chapter on the color quality of conventional light sources and solid-state light sources. To fulfill this aim, general thoughts on the meaning of color quality metrics and the approaches to find out available metrics to describe the different aspects of color rendition will be described, analyzed, and discussed in a comprehensive manner. The reason is that new knowledge can only be presented meaningfully in the subsequent chapters of the book if the reader is first given a state-of-the-art update of the research from literature and from light source technology.

5.1 General Aspects

The general task of luminaire developers, lighting designers, and building managers is to develop luminaires and lighting systems for the users with high energy efficiency, long-term stability, and high reliability with smart characteristics to achieve high user acceptance. The final product of a lighting system is artificial light with high lighting quality and color quality such as a little amount of glare, a reasonable luminance distribution in the viewing field, high object contrast, and the possibility to illuminate the objects in lit space with a high degree of color naturalness, color vividness, visual comfort, and attractiveness, allowing light source users in the room to ensure good concentration on their working task and their clear identification of the room's arrangement and the colored objects' structure. In this manner, work quality and life quality of the users of the lighting system can be improved.

For the development of lighting systems, a well-defined selection of the optical, thermal, and electrical components of the luminaire is important. In order to maximize human acceptance over a long time, lighting research has to determine the parameters influencing the perception and well-being of the users under different viewing conditions, their interaction and interdependence and the range of these parameters for an appropriate semantic interpretation of lighting quality and color quality. All these qualitative and quantitative relationships between the perceptual aspects and components should be described, modeled, and experimentally proven in realistic visual tests so that lighting engineers and designers

130 | 5 State of the Art of Color Quality Research and Light Source Technology

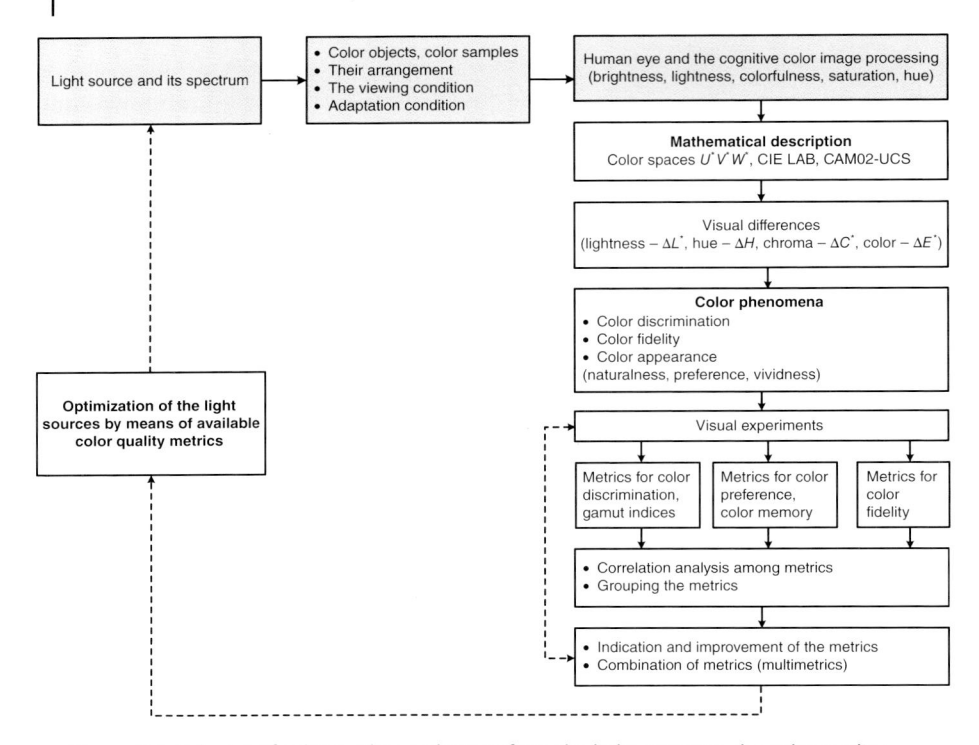

Figure 5.1 Scheme of color quality evaluation from the light source to the color quality metrics.

obtain objective criteria for the configuration and design of the lighting system in order to fulfill the user's demands in practical applications. The scheme to systematize the available color quality metrics corresponding to the different perceptual attributes is shown in Figure 5.1.

The aim of the systematic description of Figure 5.1 is to answer the question how color quality and its metrics can be systematized and processed appropriately corresponding to their roles and importance. From the scheme of Figure 5.1, it can be recognized that the color quality of light sources and illuminating systems can only be evaluated and designed properly if the aspects of the light source's spectrum, object colors, and humanly perceptual characteristics are analyzed in a synoptic view under different viewing conditions. Then, perceived color attributes (brightness, hue, and chroma or colorfulness) are expressed in various color spaces (such as $U^*V^*W^*$, $L^*a^*b^*$, CAM02-UCS) in which perceived color attributes and perceived color differences are modeled correctly. These attributes are processed by the methods of color science to arrive at a correct description of color quality attributes.

The human perception of visual differences of object colors under artificial illumination, the establishment of their modeling, and expression of their categorical (semantic) acceptability limits constitutes the basis for the definition of all color quality metrics because its color differences represent the core of the description of color fidelity, color discrimination, color saturation, color memory, and color preference metrics. Color difference, in turn, includes various attributes (to be described by different formulae) such as lightness difference (ΔL^*), hue difference (ΔH), chroma difference (ΔC^*), the two-dimensional chromaticity difference ($\Delta u'v'$ or ΔE_2^*), and the three-dimensional (or total) color difference (ΔE_3^*)). The so-called color gamut ratio ($G_{\text{test}}/G_{\text{ref.}}$) should also be mentioned at this point. These differences (plus the color gamut ratio) are conventionally constructed from the visual color attribute differences between a test and a corresponding reference light source of the same correlated color temperature (CCT in kelvin). These differences can have various degrees of correlation under each other.

Indeed, while color fidelity and its related metrics are created by the use of the three-dimensional color difference ($\Delta E^*_{|U^*V^*W^*\text{test-ref.}|}$, $\Delta E^*_{|L^*a^*b^*\text{test-ref.}|}$, or $\Delta E^*_{|J'a'b'\text{test-ref.}|}$) of the test color samples (TCSs) under the test and the reference light sources in a three-dimensional color volume, chroma difference is built from the radius difference projected to a constant hue plane ($\Delta C^*_{|U^*V^*|\text{test-ref.}}$, $\Delta C^*_{|a^*b^*|\text{test-ref.}}$, or $\Delta C^*_{|a'b'|\text{test-ref.}}$), and color gamut ratio is usually formulated by the ratio of color polygon areas on a two-dimensional hue circle. Hue differences ($\Delta H^*_{\angle U^*V^*\text{test-ref.}}$, $\Delta H^*_{\angle a^*b^*\text{test-ref.}}$, or $\Delta H^*_{\angle a'b'\text{test-ref.}}$) of adjacent colored objects (an example is shown in Figure 5.2) in a two-dimensional hue circle can be applied to describe color discrimination ability.

The description of color differences shall be combined with semantic (categorical) interpretation in visual experiments to be able to find out categorical acceptance limits for lighting design. In the same manner, when subjects with their human visual system's properties are considered in the experiments, their

Figure 5.2 Color arrangement of the 100 Farnsworth–Munsell Hue Test[R] to test color discrimination ability.

comfortable color difference perception and color memory or color preference characteristics shall be investigated and described by grouping and weighting the test colors to help establish usable color quality models for color memory, color preference, color naturalness, color vividness, and other relevant attributes.

In the last decades, many color quality metrics have been defined based on different color spaces, TCSs, and psychophysical scales. All aspects of these color perceptual issues and color quality metrics should be described, grouped, and verified in appropriately designed visual experiments in which correct scientific methods and suitable viewing conditions shall be used. In this chapter, the corresponding literature will be reviewed and analyzed to summarize the state-of-the-art of color research. Note that all color quality metrics applied in this chapter have already been defined in Chapter 2.

5.2 Review of the State of the Art of Light Source Technology Regarding Color Quality

The first electrical light sources were tungsten incandescent lamps invented by Thomas Edison at the end of the nineteenth century. Early at the beginning of the twentieth century, tubular low-pressure mercury lamps with different phosphor systems were developed allowing for the illumination of industrial halls, offices, and public buildings. Since 1959, the first tungsten halogen lamps as thermal radiators have found the way to the market and in the years around the 1980s and 1990s, several generations of compact fluorescent lamps and high-pressure metal halide lamps were introduced.

From the point of view of color and lighting architecture, the white points and the CCTs in kelvin are the first criteria and important discussion points (see Figure 7.14). Herewith, the emotional atmosphere and human feelings are essential aspects in the design and application of lighting systems. These factors need a connection with the research of the human eye's adaptation and the cognitive and psychological orientation of the test subjects in order to evaluate their visual assessments and categorical acceptability judgments correctly.

In Table 5.1, color temperatures are classified into three categories, warm white, neutral white, and cold white (sometimes also called *cool white* but the present authors believe that "cold white" is a more appropriate term) light sources typically used in current lighting technology. In Figure 5.3, the spectra of selected tungsten incandescent lamps and fluorescent lamps in the warm-white range are

Table 5.1 Classification of color temperatures into three ranges: warm white, neutral white, and cold white light sources.

Category	Correlated color temperature (K)
Warm white	2000–3500
Neutral white	3500–5000
Cold white	>5000

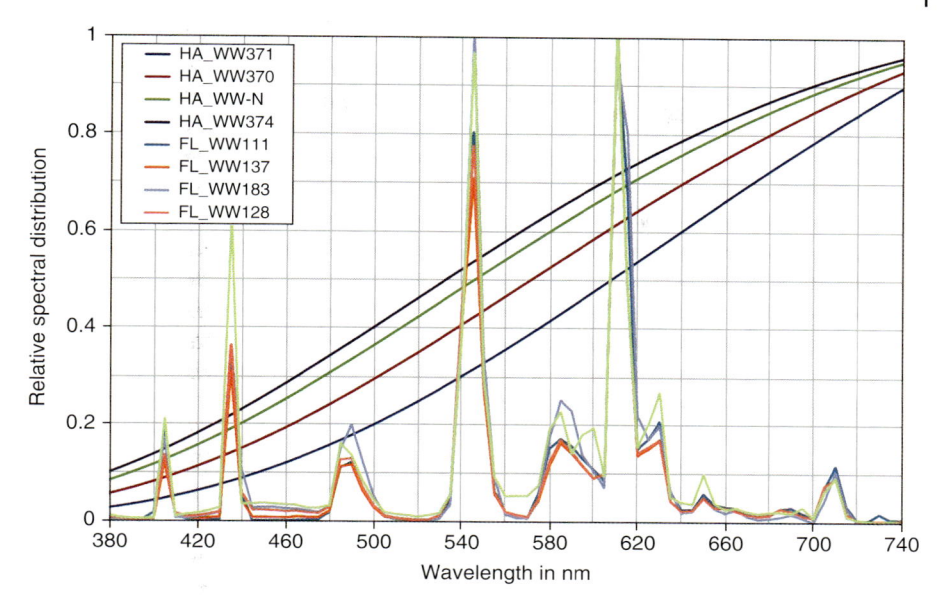

Figure 5.3 Spectra of selected tungsten incandescent lamps and fluorescent lamps in the warm white range.

shown exposing many "*emission gaps*" of the fluorescent lamps between 440 and 480, 500 and 535, 560 and 570, and from 640 nm on. Their corresponding color quality values are listed in Table 5.2.

The tungsten halogen lamps listed in Table 5.2 have a CCT range between 2700 K (optimal for home illuminations) and 3300 K (very good for a photo gallery or TV studios) with the highest color quality described by CIE color-rendering index (CRI), color quality scale (CQS) Q_a-Q_f-Q_g, CRI_{2012}, R_f-R_g from TM-30-15 or by gamut area index (GAI) (for all TCSs). Values of 100 occur because these lamp types are the reference radiators or very close to them in the definitions of most metrics. The values of color preference or color memory according to CQS Q_p and memory color-rendering index (MCRI) are in the region of 90. But the selected tubular fluorescent lamps achieve the values for the color-rendering (color fidelity) index CIE CRI $R_a = 82$–84 and color preference values of CQS $Q_p = 75$–80 and MCRI = 77–83 only.

The color gamut values R_g for the two lamp types are in the range of 100–103, being similar to the color gamut index and color area of the corresponding reference light sources. The gamut areas of these two light source types are illustrated in Figure 5.4.

As can be seen from Figure 5.4, tungsten halogen lamps have a more extended color gamut in the red and cyan–blue range while fluorescent lamps exhibit a greater gamut area in the orange–green and blue–violet range. This analysis is an early illustration to show that the color gamut area and/or gamut area-based metrics such as feeling of contrast index (FCI), GAI, Q_g, or R_g cannot reflect, indicate, represent, and evaluate the color gamut properties of light sources in specific hue ranges in some cases because they just characterize the size

Table 5.2 Color quality values of selected warm white light sources.

Name	Ha.WW$_{371}$	Ha.WW$_{370}$	Ha.WW$_N$	Ha.WW$_{374}$	FL.WW$_{111}$	FL.WW$_{137}$	FL.WW$_{183}$	FL.WW$_{128}$	FL.WW$_{48}$
CCT (K)	2700	3000	3200	3300	2700	2802	2929	3003	3101
$\Delta u'v'_{CCT}$	2.9E−06	4.6E−06	0.0E+00	2.4E−06	3.7E−03	3.4E−03	2.8E−04	1.4E−03	3.0E−03
R_9	99.99	99.99	100.00	99.99	−7.22	−8.90	−3.59	−1.34	−4.19
CIE R_a	**100.00**	**100.00**	**100.00**	**100.00**	**82.26**	**82.74**	**84.27**	**82.57**	**80.39**
$R_{1,14}$	100.00	100.00	100.00	100.00	70.56	71.24	73.53	71.53	69.92
GAI	100.00	100.00	100.00	100.00	94.31	95.19	93.98	95.00	90.39
Q_a	100.00	100.00	100.00	100.00	80.19	81.34	82.79	81.36	77.77
Q_f	100.00	100.00	100.00	100.00	79.21	80.30	81.04	79.17	74.99
Q_g	100.00	100.00	100.00	100.00	100.98	101.43	101.84	103.69	103.96
Q_p	**87.83**	**90.33**	**91.50**	**91.98**	**72.94**	**75.17**	**78.27**	**80.05**	**78.63**
CRI$_{2012}$	100.00	100.00	100.00	100.00	70.58	72.10	73.69	73.14	71.27
MCRI	**89.26**	**90.05**	**90.35**	**90.45**	**77.70**	**79.82**	**82.74**	**81.27**	**79.68**
FCI	123.72	121.38	119.74	118.92	115.79	115.58	113.39	116.16	111.91
R_f	100.00	100.00	100.00	100.00	71.74	73.94	76.60	74.15	71.08
R_g	100.00	100.00	100.00	100.00	101.98	102.08	101.48	103.13	103.12
ΔC^*_{CQS}	1.29E−04	5.59E−04	2.19E−06	4.49E−05	−1.03	−0.81	−0.38	−0.05	−0.30

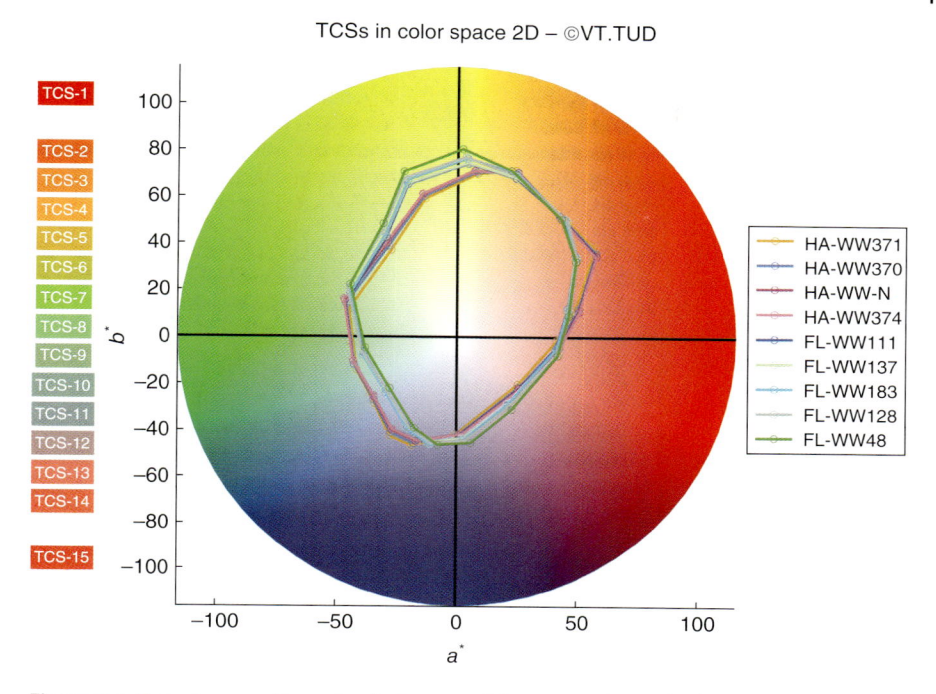

Figure 5.4 Gamut areas of tungsten halogen and fluorescent lamps in the warm white range in the CIELAB a^*–b^* diagram.

of the overall gamut area. A more detailed analysis of this issue will follow in Chapter 7.

From the end of the twentieth century, with the introduction of the first high-power white LEDs up to now, many cyan, green, yellow, orange, and red luminescent material systems have been developed and applied in modern, efficient white LEDs. In Figure 5.5, the spectra of a selected number of warm white LEDs are shown as typical demonstration examples for two different luminescent material configurations. In the case of one conventional luminescent material system, an orange luminescent material layer such as YAG or Bose with the maximum emission wavelength around 590–605 nm and in case of two luminescent material systems, a combination of green phosphors (540–560 nm such as LuAG, LuGaAG, or mint luminescent materials) and red phosphors (such as red Bose or other luminescent materials of the phosphor family Eu^{2+}) with a maximum emission wavelength in the range of 620–640 nm (e.g., pcLED_WWXCEX2) are realized in a package with blue LED chips with peak wavelengths in the range of 440–460 nm.

The general CRI CIE R_a of the spectra runs between 80 and 97. The special index R_9 for the saturated red TCS R_9 has a value range between 2 and 97.6. The color preference values CQS Q_p range between 68 and 100. The special white phosphor-converted LED (pcLED) $WW_{EX,4}$ has very high color-rendering indices R_9 and R_a; and $Q_p = 100.14$ and a gamut index R_g according to TM-30-15 of 108.94 at the CCT of 2969 K. This LED type also yields the best MCRI value

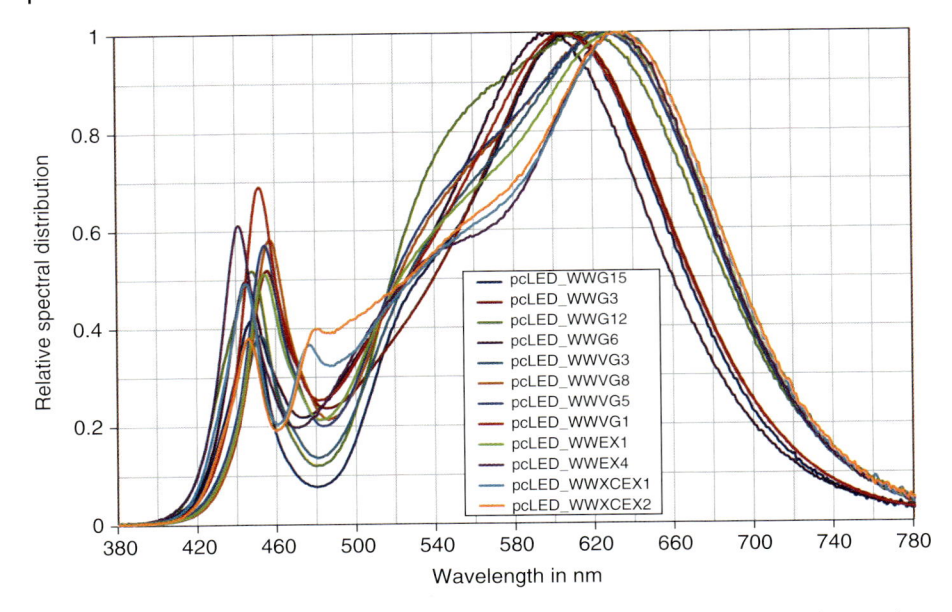

Figure 5.5 Spectra of a selected number of warm white LEDs with different phosphors and blue chip types.

of 91.88. The price for this excellent feature is a remarkable deviation from the Planckian locus of $\Delta u'v' = 0.00981$ that should be checked and compared with the user's visual white point acceptance (see Chapter 3) (Table 5.3).

In the neutral white range from 3500 K on, thermal radiators do not play a practical role in lighting engineering. In most modern offices, industrial halls, and public buildings such as hospitals or schools, fluorescent lamps and neutral white LEDs are often used. Consequently, the relative spectral distributions of selected and representative fluorescent lamps and white LEDs are shown in Figure 5.6. The typical *"emission gaps"* of the fluorescent phosphor layers beside typical emission lines of mercury atoms in the low-pressure state and the relatively continuous spectral curves of white pcLEDs can be observed clearly in Figure 5.6. Values of the corresponding color quality metrics are listed in Table 5.4.

The selected fluorescent lamps in Figure 5.6 illustrate the case of typically moderate general color-rendering indices R_a of 80–82 and very low special color rendering indices of the red color R_9 of 14–25. The color preference values are in the range of 78–81 in the case of Q_p and near 82–84.5 in the case of MCRI. The color gamut index R_g is around 99–100 similar to most reference light sources of the same CCT. The presented white LEDs illustrate the achievement of higher values for the color-rendering indices R_a or $R_{1,14}$, the color memory index MCRI, and also for the color gamut index R_g. Similar to the case of warm white light sources, most neutral white light sources that are commercially available on the international market have negative or slightly positive chroma difference (i.e., oversaturation or desaturation of object colors by the test light source compared to the reference light source) values. Consequently, most of the light sources in Table 5.4

Table 5.3 Color quality values of selected warm white LED light sources.

Name	pcLED WW$_{G,15}$	pcLED WW$_{G,3}$	pcLED WW$_{G,12}$	pcLED WW$_{G,6}$	pcLED WW$_{VG,3}$	pcLED WW$_{VG,8}$	pcLED WW$_{VG,5}$	pcLED WW$_{VG,1}$	pcLED WW$_{EX,1}$	pcLED WW$_{EX,4}$	pcLED WW$_{XC,EX1}$	pcLED WW$_{XCEX,2}$
CCT (K)	2708	2801	3081	3134	2782	2998	3026	3105	2946	2969	2973	2994
$\Delta u'v'_{CCT}$	2.59E−04	5.26E−03	2.22E−03	6.02E−04	1.44E−03	1.29E−03	5.78E−04	5.99E−03	1.72E−03	9.81E−03	6.51E−03	1.15E−03
R_9	9.91	23.65	35.94	2.22	53.70	61.74	61.08	29.55	73.20	97.48	97.66	94.06
CRI R_a	79.80	84.05	81.32	80.27	88.09	90.56	90.09	85.64	93.75	94.41	96.42	97.39
$R_{1,14}$	73.38	80.21	75.09	74.11	83.73	86.80	86.18	81.55	91.10	94.32	96.12	96.75
GAI	86.29	87.46	88.83	85.84	92.77	93.30	93.37	88.00	95.58	98.43	98.98	100.46
Q_a	79.51	83.58	81.63	81.27	88.21	89.06	88.99	83.84	92.11	91.34	95.47	98.63
Q_f	78.37	82.50	80.84	81.59	87.89	88.44	88.32	82.63	91.05	85.69	91.92	98.16
Q_g	98.34	97.43	98.52	93.20	98.79	99.45	99.80	98.81	101.48	112.04	106.95	100.58
Q_p	70.97	74.63	74.70	68.47	77.89	81.32	81.69	77.83	85.92	100.14	97.13	90.46
CRI$_{2012}$	81.88	87.01	85.70	85.07	92.04	92.61	92.82	87.95	94.77	93.28	97.21	97.84
MCRI	83.31	87.99	84.28	84.94	87.53	89.55	89.33	88.94	90.21	91.88	91.66	90.55
FCI	111.56	110.98	114.18	103.23	119.30	119.71	119.58	111.11	122.31	128.20	125.22	120.99
R_f	76.47	83.07	78.83	81.28	86.70	87.57	87.70	83.65	90.64	87.68	93.84	96.41
R_g	100.02	96.79	100.84	96.04	100.23	99.31	100.08	98.91	100.94	108.94	104.80	100.27
ΔC^*_{CQS}	−0.69	−0.27	−0.67	−1.75	−0.58	−0.09	−0.12	−0.15	0.31	2.54	1.55	0.22

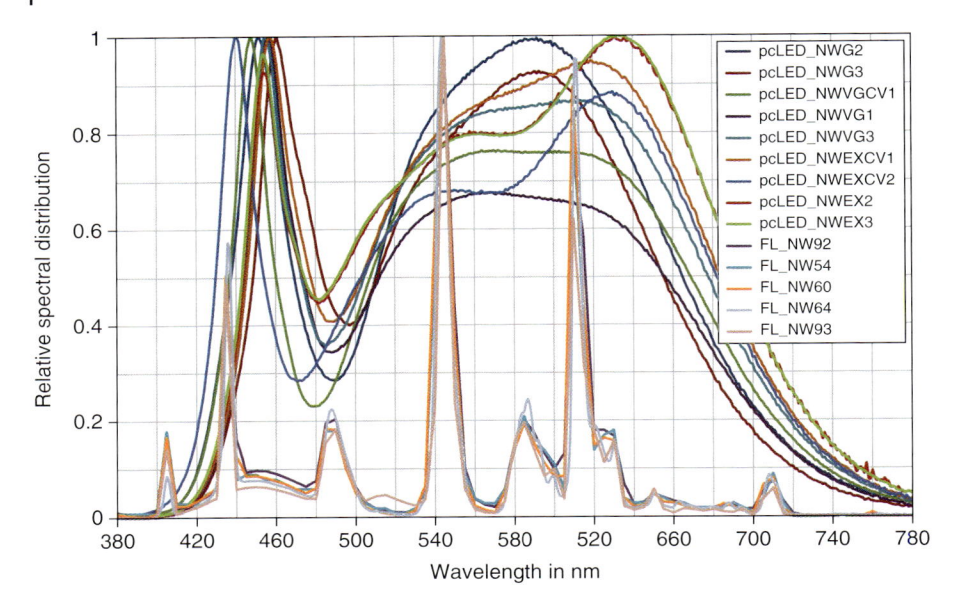

Figure 5.6 Spectra of a variety of neutral white light sources (LEDs and fluorescent lamps).

and Figure 5.6 are slightly desaturating or cause a small chroma enhancement only; see Figure 5.7.

The philosophy discussed for several decades in color research journals and in the international commissions for lighting research and technology as well as in the subsequent chapters of this book is that a positive effect on human color preference, vividness, and memory colors should be established and achieved based on the acceptable amount of color saturation enhancement beside the allowable lower limit of color fidelity. This desire has been impossible to realize in case of conventional light source technology. But with the appearance of multichannel LED systems that allow previously unknown spectral flexibility, it is necessary and feasible to intensify the development process of future lamps and luminaire designs with solid-state lighting devices (i.e., LEDs). In this respect, the design potentials of conventional light sources (limited) and solid-state light sources (with previously unknown flexibility) are compared in Figures 5.7 and 5.8.

All the above findings for neutral white light sources are also true for commercially available cold white LEDs. In Figure 5.8, the values of the CRI R_f and color gamut index R_g according to the TM-30-15-proposal (see Chapter 2) for all light source technologies in the warm white, neutral white, and cold white ranges are shown. The R_f values extend between 70 and 100 but the R_g values are concentrated at the value range around 100 with a maximum of about 105. Light sources manufactured by conventional and solid-state lighting technologies that are commercially available with a color gamut index higher than 105 are rare. Solid-state light sources with a much higher chroma difference and color gamut area combined with moderate and reasonable color fidelity index values constitute therefore the main objective of light source optimization strategies; see Chapter 8.

Table 5.4 Color quality values of selected neutral white LED light sources and fluorescent lamps.

Name	pcLED $NW_{G,2}$	pcLED $NW_{G,3}$	pcLED $NW_{VG,CV1}$	pcLED $NW_{VG,1}$	pcLED $NW_{VG,3}$	pcLED NW_{EXCV1}	pcLED NW_{EXCV2}	pcLED $NW_{EX,2}$	pcLED $NW_{EX,3}$	FL. NW_{92}	FL. NW_{54}	FL. NW_{60}	FL. NW_{64}	FL. NW_{93}
CCT (K)	3981	4116	4194	4614	4057	3906	3942	3890	3924	4001	4009	4044	4116	4195
$\Delta u'v'_{CCT}$	2.10E−03	2.10E−03	9.68E−04	3.31E−05	7.76E−04	1.06E−03	9.36E−03	1.80E−03	2.76E−03	1.47E−03	2.41E−03	1.19E−03	1.64E−04	3.78E−03
R_9	25.52	23.96	57.82	64.01	62.19	71.91	94.43	96.75	93.38	24.95	14.10	19.11	18.04	17.24
CIE R_a	82.74	84.65	87.47	90.91	90.58	92.96	93.09	98.05	97.72	82.86	80.64	80.79	81.49	82.89
$R_{1,14}$	76.27	79.08	83.07	86.88	86.38	89.77	92.96	96.69	96.18	73.21	70.37	70.61	71.69	73.85
GAI	86.97	88.64	92.14	93.14	93.43	94.69	98.66	99.72	99.99	94.05	92.81	93.91	92.50	94.31
Q_a	80.76	82.90	86.42	88.23	88.93	90.48	91.81	95.98	96.10	81.07	80.21	79.60	80.39	83.65
Q_f	80.34	83.61	85.35	88.38	88.99	90.50	87.70	95.47	95.32	79.13	78.59	77.54	78.60	82.50
Q_g	96.02	90.10	100.30	96.05	96.90	97.62	109.02	101.30	102.05	101.58	100.06	102.12	100.27	99.58
Q_p	73.82	68.17	83.77	79.55	80.89	82.94	98.97	92.41	93.85	80.56	78.02	80.27	78.74	80.19
CRI_{2012}	86.47	86.28	91.16	91.70	92.73	93.13	93.78	96.05	96.21	78.99	77.81	76.71	78.29	79.94
MCRI	87.49	88.09	89.20	90.61	90.18	90.98	92.44	92.27	92.57	84.57	82.18	82.47	83.87	83.48
FCI	105.45	100.83	110.72	107.23	110.86	113.25	120.68	118.83	119.36	110.28	107.62	110.10	106.42	105.13
R_f	80.71	80.56	84.85	85.78	87.04	87.91	88.13	92.65	92.84	77.75	76.53	75.45	77.12	79.78
R_g	97.08	90.70	101.33	95.83	97.11	97.32	107.86	100.34	100.80	100.96	99.97	101.44	100.10	100.10
ΔC^*_{CQS}	−0.74	−2.21	−0.17	−0.77	−0.65	0.44	2.08	0.45	0.64	0.21	−0.20	0.33	−0.07	−0.41

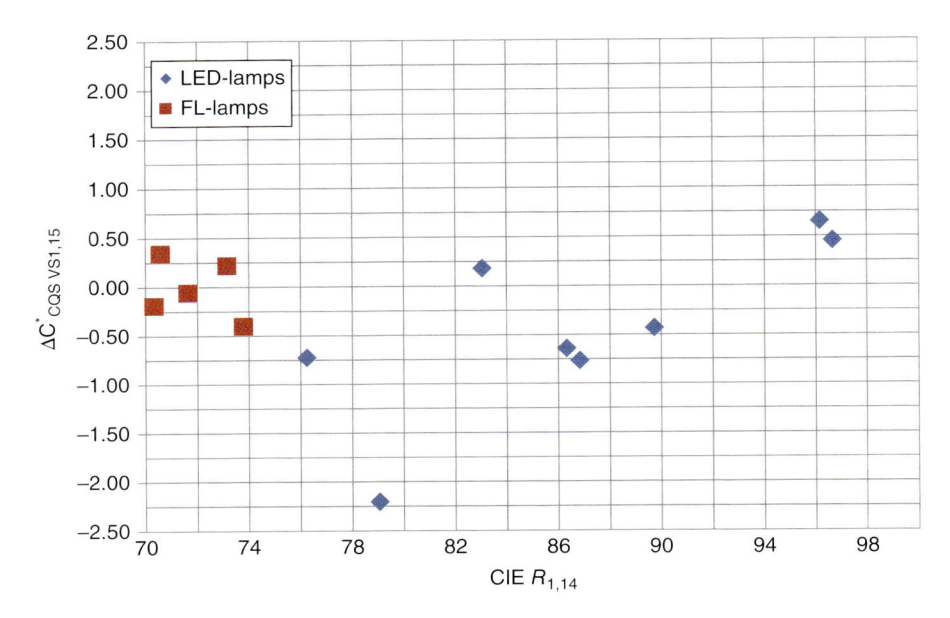

Figure 5.7 Possible chroma differences ΔC^* (average for the 15 CQS TCSs; see Chapter 4) and color-rendering indices $R_{1,14}$ (CIE 1995) of the neutral white LEDs and fluorescent lamps in Figure 5.6.

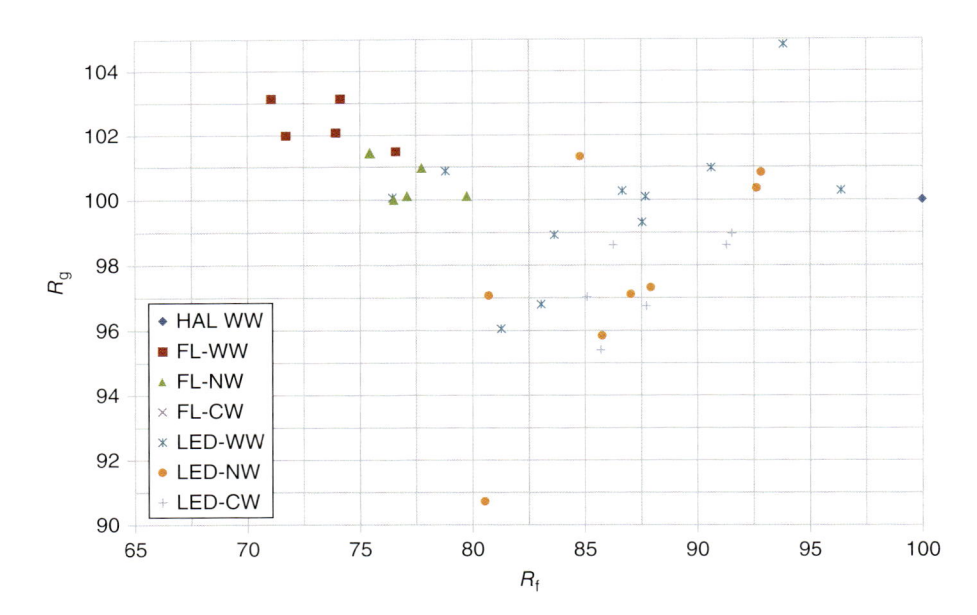

Figure 5.8 Achievable limits of the color-rendering index R_f and color gamut index R_g for all warm white, neutral white, and cold white light sources in the examples of this chapter. FL: fluorescent lamp.

5.3 Review of the State of the Art of Colored Object Aspects

Because of the importance of colored objects and object selection in the process of the definition of color quality metrics and the visual and numeric evaluation of illuminating practice, a comprehensive analysis and interpretations of different prominent color object collections used in the well-known color quality metrics have been implemented in Chapter 4. Some important aspects should be repeated and discussed in depth here in form of a short summary (see Figures 4.1–4.4 and 4.10):

- Colored objects of a collection for a certain color metric definition should cover the whole hue circle and different saturation degrees uniformly. Their spectral reflectance should have a uniform coverage over the visible wavelength range between 380 and 780 nm (or, alternatively, 720 nm). In this manner, the value of the considered color quality metric can be representative for many practical applications and illuminating situations (e.g., museums, shops for textiles or cosmetics, film and TV production, printing, and car industry) and an inappropriate optimization for a specific color quality metric with a small number of colored objects can be avoided.
- According to the analysis in [1], a given light source can provide a very good color quality for non-saturated colored objects but poor color quality for saturated samples. Anyway, the reverse formulation is impossible. Therefore, a well-thought out selection of colored objects with a reasonable mixture of non-saturated and saturated color samples should be implemented. Based on the analysis in Chapter 4, a mixture of 99 TCSs of IES (TM-30-15) and 15 saturated CQS–TCSs should be a right step from the viewpoint of current color research.
- Planar, homogeneous surface color patches (e.g., color cards or the Munsell colors) are useful for visual experiments with color difference metrics in their focus. For those visual experiments with absolute internal, personality-dependent color assessments such as the experiments on color memory, color preference, naturalness, and attractiveness, a complex cognitive process related to the cultural identification, visual and professional experience, expectations, and education of the observers is activated by the absolute level and chromaticity (CCT) of the illumination as well as the color, texture, geometrical shape, and arrangements of the familiar objects in the observed scene. Therefore, a complex system of real colored objects should be chosen for the appropriate psychophysical experiments in order to establish and verify any color quality metric (see Figures 7.21, 7.22, 7.25, 7.36, and 7.45).
- Blue objects contain a relatively higher spectral reflectance in the wavelength range between 400 and 470 nm and red objects have a higher spectral reflectance from 580 nm on. It is meaningful from the point of view of lighting design and color research to consider what white tone of light sources with suitable CCTs are visually appropriate for blue or bluish objects (e.g., blue jeans in a clothes shop), red objects (e.g., cosmetics and makeup products),

and for a mixture of colored objects of all possible colors. A visual study in this context is discussed in Section 7.5.
- In most color quality metrics, a well-defined collection of color samples is used and the metric is formulated based on the average value (geometric mean) of color differences of this color collection assuming that all used color objects have the same meaning and importance. This approach is carried on for the establishment of all color fidelity indices and also in the case of MCRI. The open question is what color object group (red, yellow, green, or blue tones) should have a higher or smaller weighting factor in a certain color arrangement (e.g., foods, fruits, and flowers on the table of a party; see Figure 7.25) in case of a certain illuminating chromaticity (such as warm white at 3000 K or cold white at 5500 K). According to this approach, Judd in 1967 (in [2]) tried to weight the TCSs (e.g., the preferred color of the complexion was 35%, the one of butter equaled 15%, and the one of green foliage was 15%) without any visual experiment or any other explanation. Consequently, intensive visual experiments in different contexts and color arrangements are essential to bring light into this research issue.

5.4 Viewing Conditions in Color Research

Viewing conditions have a high influence on the quality, correctness, and reliability of the results of visual experiments and the resulting color quality metrics and they should correspond to the real viewing conditions in practical illumination situations (such as shopping in a supermarket, selection of cosmetics in a makeup shop, observing an art product in a museum). In detail, the time course of chromatic adaptation, the level of luminance in the viewing field, the size of the adaption field, and interpersonal color perception differences are important factors that determine the experimental results.

In visual experiments (e.g., ranking experiments) to define color quality metrics, different light sources with different spectra and illumination chromaticities (CCTs) at the same luminance level are successively presented to illuminate a group of TCSs and then the experimental results are matched with the desired visual color quality attribute (such as vividness, naturalness, or color preference; see Figure 5.9).

In general, test subjects shall be adapted for the illuminating chromaticity of the light source used. When a test light source is turned on, the adaptation phase should take enough time for the test subject to complete chromatic adaptation. Fairchild and Reniff [3] studied the time course of chromatic adaptation for color appearance judgments and found two stages of adaptation including one rapid stage within a few seconds and the other one slower with approximately 1 min. Chromatic adaptation at a constant luminance was 90% complete after approximately 60 s so that the experiment designers have to account for approximately 1.5 min for a reliable visual evaluation before the subject starts the visual assessment of the colored objects.

In visual experiments, in order to have a good chromatic adaptation course and reliable test results, the luminance level of the scene of the test objects must

Figure 5.9 Arrangement of colored objects under a certain light source in a visual experiment. (Photo source: The Lighting Laboratory of the Technische Universität Darmstadt, Germany.)

be high enough allowing for photopic vision and good visual comfort during the experiment. Thus, in [4] Boyce and Simons implemented two different series of color experiments with 300 and 1000 (series 1) and 400, 800, and 1200 lx (series 2). Although the results are only for one age group, the increase of illuminance level has a significant effect on visual performance. However, Boyce and Simons also mentioned in [4] about the finding that the difference of the visual test's results were unnoticeable between the luminance level of 300 and 12 000 lx for the case of fluorescent lamps and young test persons. Boyce and Simons found a noticeable conclusion about color differences, color discrimination ability, and illuminance [4] and formulated an important hint that should be taken into consideration in visual experiments: *"it is possible that for color discrimination tasks in which the color differences are much smaller and/or in which prolonged work is involved, particularly by older age groups, illuminance could become an important factor in determining the level of performance."*

Concerning another aspect, in [6] Schierz found that the age group around 65 years needs 200% more light entering the eye in order to maintain retinal illuminance on the same level as the young age group of 20–25 years experiences. Consequently, a level of about 600 lx and more is needed for visual tests with the elderly age group. A comprehensive analysis of the age-dependent demand of light and color is discussed in Section 9.4.

Based on different mesopic color experiments in the Lighting Laboratory of the present authors (not published at the time of writing the present book), photopic color appearance perception generally begins at a luminance value of about 40 cd/m^2 hence, computing with a diffuse reflectance factor of 18–20% (0.18–0.2) of the colored object, the minimum necessary illuminance level on the objects should be 697 lx (i.e., about 700 lx).

In contrast to color discrimination experiments in which only the comparative (i.e., relative) mechanisms of the human visual system are activated, visual experiments on color rendition (in other words, color quality) such as color preference,

Table 5.5 Illuminance levels in visual experiments in the literature.

Literature	Test criterion	Illuminance (lx)
[7]	Preference, fidelity, vividness, naturalness, attractiveness	248 lx ± 2%
[8]	Attractiveness, naturalness	230 lx ± 3%
[9]	Preference, vividness, naturalness, liveness	530 lx
[10]	Preference	200 lx
[11]	Color discrimination	371–374 lx

color vividness, and memory colors stimulate a more complicated response of the observers that comes from the cognitive level of signal processing including mental behavior. This involves higher level mechanisms depending on the individual's personality related to the terminology used on the questionnaire, *"brilliant, excellent, very good, I like it, bad, and so on."* For these visual tests, the level of illuminance on the objects or the object luminance is very important in order to receive the most reliable ratings.

In high-quality shops in big cities, the illuminance level on the goods (e.g., textiles, jewelry) is often beyond the value of 1200 lx. In Table 5.5, the illuminance levels in some visual experiments in the literature are listed. In Chapter 7, visual experiments with the test criterion *"color preference, color vividness, and naturalness"* were implemented at illuminance levels between 470 and 2300 lx (see Sections 7.4–7.7) in our Lighting Laboratory of the Technische Universität Darmstadt. An intensive visual experiment analyzing the influence of illuminance level on achromatic visual acuity and chromatic perception is described in Section 7.2. Also, in Section 7.4, a significantly different assessment of the test subjects in an office-like room at two different illuminance levels 500 and 1200 lx was obtained.

For the consideration of the viewing conditions, not only the chromatic adaptation and the level of luminance of the colored objects but also their surroundings are important. In almost all color viewing contexts in reality such as reading books in a library, buying flowers in a shop, observing a painting in an art gallery, or monitoring product quality in printing industry, the adaptation field of more than 10° has to be taken into consideration. In most visual experiments in the literature, test subjects were positioned in front of a color arrangement with a white or neutral gray background and had to adapt to a viewing field of more than 10°; see, for example, Figure 5.9.

Therefore, the color-matching functions $x_{10}(\lambda)$, $y_{10}(\lambda)$, and $z_{10}(\lambda)$ for 10° viewing angle [12] should be applied. On one hand, these functions were carefully determined by Stiles and Burch in their time at the NPL (National Physical Laboratory of the United Kingdom/UK). On the other hand, based on the important physiological phenomenon of human eyes, because of the known receptor distributions on the retina, the use of the 10° color-matching functions can deliver more similar color assessment results among a panel of test subjects and lighting users than the use of the 2° functions of the CIE (1931). With these two main facts in mind, recent color quality metrics such as the CRI (color fidelity index) CIE CRI_{2012}, or MCRI apply the 10° color-matching functions. The calculated tristimulus values are then converted into different color spaces with the corresponding chromatic adaptation formulae.

5.5 Review of the State-of-the-Art Color Spaces and Color Difference Formulae

In the history of color research from 1931 to date, many color spaces and color difference formulae have been used. Since the beginning of color science until 1976, most color spaces are mathematical spaces having the task of arranging and grouping the tristimulus values. They have the following disadvantages:

- Chromatic adaptation to the illuminant's chromaticity and/or the predominating tone of the more or less chromatic colored objects (e.g., furniture or walls) in the main viewing field was not or not sufficiently considered (e.g., the chromatic adaptation transformations of Judd or Kries).
- The tristimulus values X, Y, and Z or x, y, and Y (luminance) are not directly related to the color attributes describing the daily color impression of the users such as brightness (bright–dark), hue (red–yellow–green–blue) and saturation (desaturated, saturated, colorful). This deficit brought some problems into the communication between technicians, artists, and lighting designers.
- The color spaces up to 1976 are not uniform enough. Equidistant color differences between two arbitrarily selected color coordinates do not provide equal color difference perceptions. The consequences of this deficit can be seen from the experiments of MacAdam [13] (see Figure 5.10) with the ellipses for the threshold value of color difference with the criterion *"just noticeable."* The ellipses exhibit different sizes for the same criterion and show different orientations in different areas in the color diagram.

In 1976, CIE introduced the CIELAB $L^*a^*b^*$ color space in order to predict lightness (L^*) and the chromatic values representing the red–green (a^*) and yellow–blue (b^*) opponent channels.

In CIELAB color space, lightness is computed by Eq. (5.1).

$$L^* = 116 \cdot \sqrt[3]{\frac{Y}{Y_n}} - 16 \tag{5.1}$$

Red–green content is described by Eq. (5.2).

$$a^* = 500 \cdot \left(\sqrt[3]{\frac{X}{X_n}} - \sqrt[3]{\frac{Y}{Y_n}} \right) \tag{5.2}$$

Finally, yellow–blue content is modeled by Eq. (5.3).

$$b^* = 200 \cdot \left(\sqrt[3]{\frac{Y}{Y_n}} - \sqrt[3]{\frac{Z}{Z_n}} \right) \tag{5.3}$$

In all three formulae for L^*, a^*, and b^*, no human visual system relevant and eye-physiologically correct chromatic adaptation transformation has been included. An indirect, imperfect chromatic adaptation attempt has been adopted

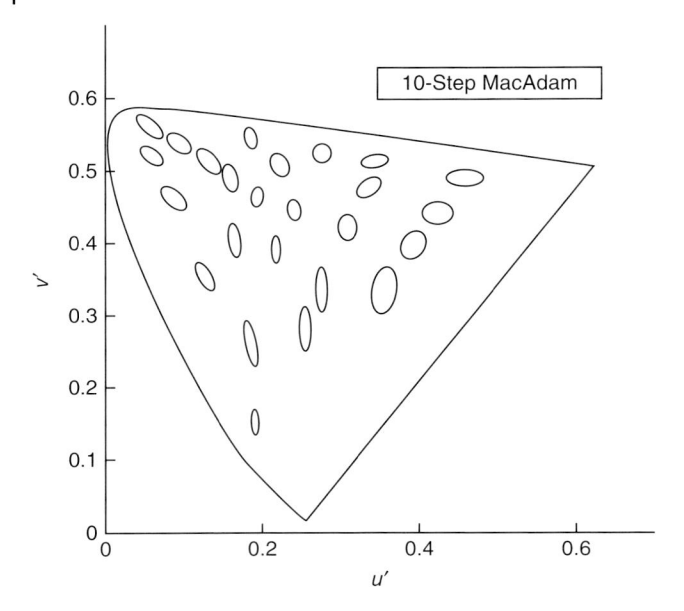

Figure 5.10 Threshold values of chromaticity differences for the criterion "just noticeable" according to MacAdam [13]. (Image source: reproduced with permission from Dr. G. Kramer; Büchenbach-Germany.)

by referring the tristimulus values X, Y, and Z of the actual color to the tristimulus values X_n, Y_n, and Z_n of the white point. A main deficit of this CIELAB $L^*a^*b^*$ color space is the distortion of hue characteristic. In the polar coordinate system such as the a^*b^* plane, the colors with constant perceived hue tones have to be in the same hue angle or stay on the same radial line from the origin of the coordinate system. It is not the case in the CIELAB $L^*a^*b^*$ system according to the research result of Hung and Berns [14]. The distortion of the hue line in CIELAB $L^*a^*b^*$ is illustrated in Figure 5.11. Most distortions can be found in the blue range between approximately 260° and 320° and in the chroma range $C^* > 50$.

The color space CIE $L^*a^*b^*$ is the basis of the well-known CQS-system of color quality [1]. The main research activity of color science since 1976 up to now has been the establishment and definition of new color spaces based on the most recent knowledge on the color processing chain from the retina to the visual brain with correct *LMS* cone receptor spectral sensitivity functions and a proper chromatic adaptation transformation (CMCCAT2000, CAT02, see [16, 17]) numerically providing the perceptive attributes brightness, lightness, colorfulness, saturation, and hue. The newest color appearance model of the CIE (CIECAM02) is the product of this research activity.

Once a color space has been defined, a color difference formula shall also be derived. Most color quality metrics for color fidelity (R_a of CIE1965 and 1995, CRI_{2012}, CQS-Q_f, and CQS-Q_a; R_f according to TM-30-15), for color preference (CQS-Q_p, Judd's flattery index, and Thornton's color preference index) are based on the establishment of a suitable arithmetic color difference between the test

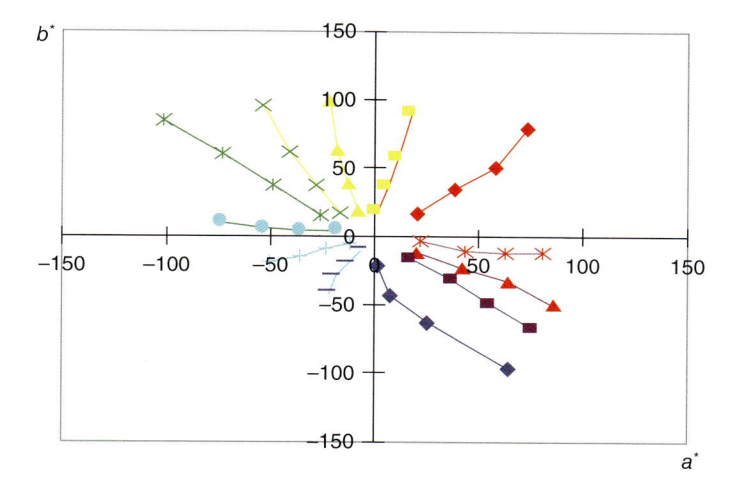

Figure 5.11 Lines of constant perceived hue plotted in CIELAB $L^*a^*b^*$ color space according to the result of Hung and Berns (see [15]).

light source and the corresponding reference light source at the same CCT. In its long history, color science has followed two pathways to find an appropriate color difference formula corresponding to human color difference perception:

- *First pathway:* Selection of an appropriate color space (e.g., CAM02-UCS) or DIN99 and use of a Euclidean formula for determining the distance between two color points in the space.
- *Second pathway:* The use of an imperfect color space (e.g., CIELAB $L^*a^*b^*$) and then, the introduction of correction factors for the components of lightness, hue, and chroma. The prominent example for the color difference formula in this approach is the CIEDE2000 formula [18].

After the color space and the color difference formula have been found, a validation procedure has to be carried out in order to answer the following questions:

- How well do the different color difference formulae correspond to visually perceived color difference evaluations of the test subjects? The answer to this question should help find the best color difference formula.
- How can the color differences (e.g., $\Delta E = 1, 2,\ldots,7$) be semantically categorized and interpreted on an ordinal rating scale (e.g., excellent, good, acceptable, not acceptable, bad, or very bad color similarity) in order to implement the descriptions or evaluations of perceived color differences as accurate, as convenient, or as comfortable as possible in the information exchange between the users, designers, or color experts in technical, marketing, and other activities of human life.

Based on comprehensive visual experiments, the above two questions can be answered [19]. A double-chamber viewing booth was constructed; see Figure 5.12. Inside the chamber, the walls were painted white with a spectrally nonselective $BaSO_4$ coating. This white surface had a luminance of $240\,cd/m^2$

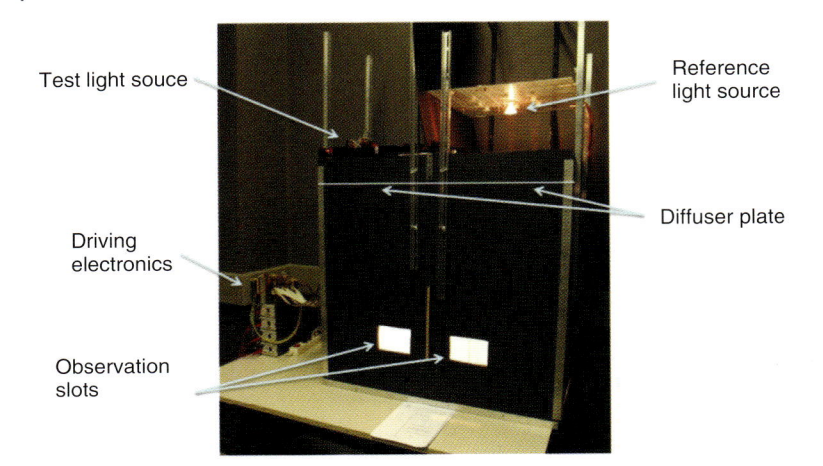

Figure 5.12 Double-chamber viewing booth. (Bodrogi 2010 *et al*. [19]. Reproduced with permission of Wiley.)

(±2%) for all light sources at the bottom of the booth. The test light source illuminated the left chamber and the reference light source illuminated the right chamber. The light went through a diffuser plate in order to provide a uniform lighting distribution on the ground of the chambers. Spectral measurements were carried out by a well-calibrated, high-end spectroradiometer. The spectral power distributions (SPDs) of the light sources were measured on a white standard placed on the bottom of the chambers. Colorimetric properties of the light sources (such as CCT; chromaticity coordinates x, y; and CIE color rendering indices R_a, see Table 5.6) could be determined simply from these SPDs.

Two groups of test and reference light sources were investigated, one at approximately 2700 K and the other one at approximately 4500 K. The 2700 K group consisted of two types of warm white pcLED (HC3L and C3L), an RGB-LED cluster (RGB27), and two warm white fluorescent lamps (FL627 and FL927) as test light sources, and a tungsten halogen plays the role of a reference light source (TUN). The 4500 K group consisted of two types of white phosphor LED (HC3N and C3N), an RGB-LED cluster (RGB45), and two cool white fluorescent lamps (FL645 and FL945) as test light sources, and a HMI plays the role of a reference light source. The relative SPDs of the 2700 K and the 4500 K light sources are shown in Figures 5.13 and 5.14, respectively.

Two identical copies of matte color papers (pairs of identical uniform standalone TCSs subtending about 4° × 3° viewed from the observation slot, one under the test light source and the other one under the reference light source) were observed on a gray background ($L = 59$ cd/m^2). Seventeen pairs were observed altogether, after each other. Each pair had a different color. A gray scale color difference anchor helped scale the perceived color differences between the test and the reference chamber of the viewing booth; see Figure 5.15.

Twelve of the 17 TCSs were taken from the Macbeth Color Checker Chart (Nos. 1–12) and 5 from the NIST color set (Nos. 13–17). A majority of these colors therefore has a relatively high saturation. The SPD of the light reflected

Table 5.6 Correlated color temperatures (CCTs), chromaticity coordinates (x, y), and CIE color-rendering indices (R_a) of the light sources used in the experiment [19].

Light source	Description	CCT (K)	x	y	R_a
HC3L	T; white phosphor LED	2798	0.448	0.401	97
C3L	T; white phosphor LED	2640	0.476	0.432	67
RGB27	T; RGB LED	2690	0.462	0.414	17
FL627	T; a 4000 K fluorescent lamp filtered by a Rosco color filter plus the diffuser plate to get close to 2700 K	2786	0.456	0.415	64
FL927	T; a 3000 K fluorescent lamp filtered by the diffuser plate to get close to 2700 K	2641	0.466	0.414	90
TUN	R; tungsten halogen	2762	0.460	0.419	97
HC3N	T; white phosphor LED	4869	0.349	0.355	95
C3N	T; white phosphor LED	4579	0.363	0.393	69
RGB45	T; RGB LED	4438	0.361	0.355	22
FL645	T; the same 4000 K fluorescent lamp as by FL627 filtered by another Rosco color filter and the diffuser plate to get close to 4500 K	4423	0.365	0.371	69
FL945	T; a 5400 K fluorescent lamp filtered by the diffuser plate to get close to 4500 K	4391	0.366	0.372	92
HMI	R; a gas discharge light source filtered by the diffuser plate to get close to 4500 K	4390	0.362	0.353	92

T: test light source and R: reference light source.

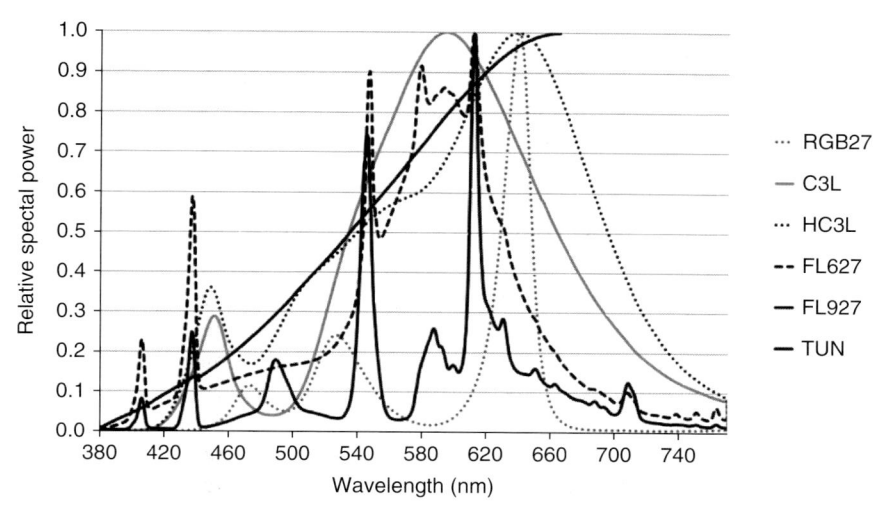

Figure 5.13 Relative spectral power distributions of the light sources of the 2700 K group. (Bodrogi et al. 2010 [19]. Reproduced with permission of Wiley.)

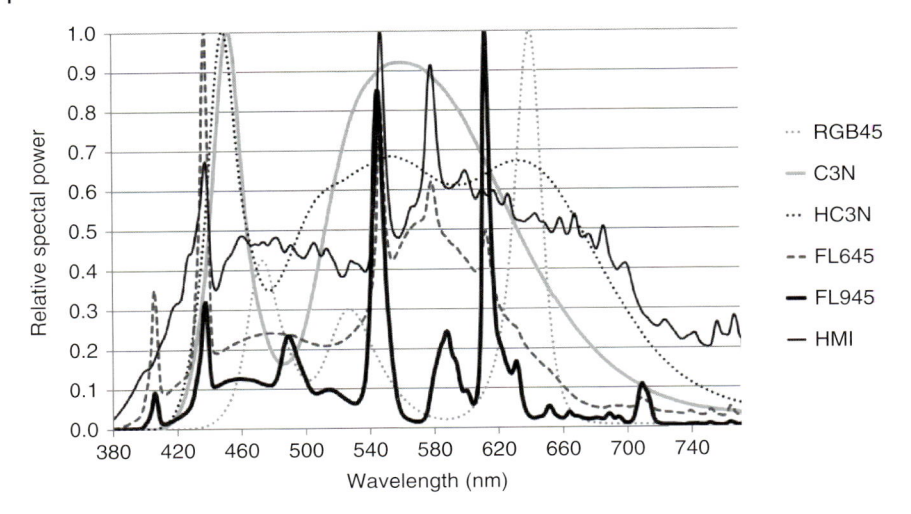

Figure 5.14 Relative spectral power distributions of the light sources of the 4500 K group. (Bodrogi et al. 2010 [19]. Reproduced with permission of Wiley.)

Figure 5.15 Reference side of the viewing booth with a 4° × 3° square test color sample (matte paper) on a gray background and the color difference grayscale anchor that helped scale color differences (one gray scale step was equal $\Delta L^* = 2$). The observer saw this scene when looking into the reference (right) chamber through the observation slot (see Figure 5.12). (Bodrogi et al. 2010 [19]. Reproduced with permission of Wiley.)

from each TCS under each light source was measured in situ by the above mentioned spectroradiometer. Eight observers of normal color vision (tested by Farnsworth's D-15 test) completed the experiment by evaluating each color difference. Totally, 1318 observations were evaluated. Before the observation, subjects were trained to get familiar with assessing color attributes by using NCS color training material. After this training phase, each of the 1318 color differences was evaluated visually in the following way.

5.5 Review of the State-of-the-Art Color Spaces and Color Difference Formulae | 151

Übereinstimmung links und rechts bewerten

| 1 | 2 | 3 |

Tadellos

☐ 1 Tadellos
☐ 2 Gut
☐ 3 Tolerierbar
☒ 4 Nicht annehmbar
☐ 5 Sehr schlecht

Variable R

Grauskala 8

Variable ΔE_{vis}

Kein Unterschied

Dunkler   −6 −5 −4 −3 −2 −1 0 +1 +2 +3 +4 +5 +6   Heller

Grüner   −6 −5 −4 −3 −2 −1 0 +1 +2 +3 +4 +5 +6   Roter

Blauer   −6 −5 −4 −3 −2 −1 0 +1 +2 +3 +4 +5 +6   Gelber

Variable P

Blasser −6 −5 −4 −3 −2 −1 0 +1 +2 +3 +4 +5 +6   Kräftiger

Sehr schlecht

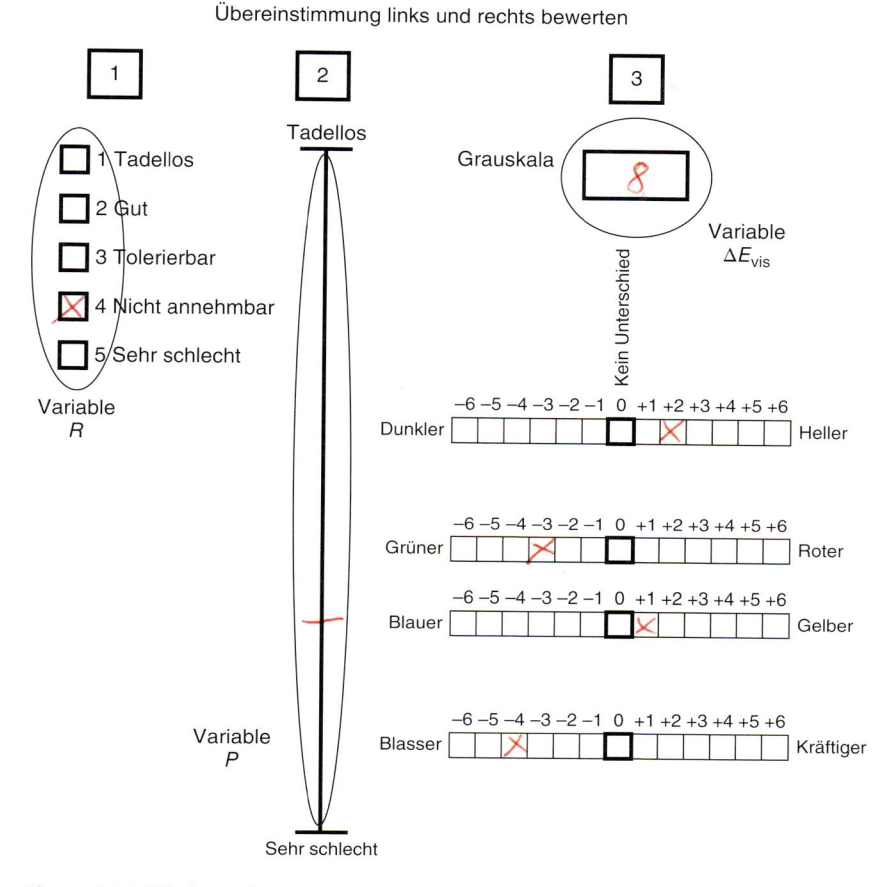

Figure 5.16 Filled sample questionnaire to evaluate the perceived color difference between the test and the reference light sources. Variable R: rating scale for the perceived color difference with 1, excellent (in German: tadellos); 2, good (gut); 3, acceptable (tolerierbar); 4, not acceptable (nicht annehmbar); 5, very bad (sehr schlecht); variable P: graphical scale to assess the degree of similarity between the test and the reference side; in German: "Uebereinstimmung links und rechts bewerten" top, excellent; bottom, very bad; observers had to put a cross on the interval; and variable ΔE_{vis}, scaled color difference using the gray scale anchor "Grauskala" of Figure 5.15. There are four supplementary scales below ΔE_{vis}. (Bodrogi et al. 2010 [19]. Reproduced with permission of Wiley.)

First, observers had to rate the color difference on a five-step ordinal rating scale: 1, excellent; 2, good; 3, acceptable; 4, not acceptable; and 5, very bad. This constituted the dependent variable *R*. Second, observers had to assess the degree of similarity between the test and the reference side by putting a cross on a graphical rating scale (independent variable *P*). This scale was used to check the consistency of the observers' judgments and to help observers scale the total visual difference (ΔE_{vis}) in their third task by the aid of the gray scale color difference anchor (see Figure 5.16).

In Figure 5.16 a filled sample questionnaire is illustrated. The supplementary scales below the ΔE_{vis} data helped the observer obtain the value of ΔE_{vis}: observers had to tell whether the TCS was brighter or darker (in German: heller/dunkler), more or less saturated (in German: kräftiger/blasser) under the test source than under the reference source, and also whether it contained more or less red (roter), green (grüner), blue (blauer), or yellow (gelber). Observers usually filled these supplementary scales first. They proved to be very helpful to find the gray scale equivalent of the perceived color difference. But the values themselves were not used in the analysis of the results.

Observers were taught to look into one of the chambers through one of the two observation slots binocularly by placing their head always directly to the slots for total immersion. The observation through the slots was used to achieve complete immersion and adaptation. At the beginning, observers had to look into the reference chamber for 10 min to adapt. In the observation phase, they were taught to look into the reference chamber and, after at least 2 s, change their line of sight into the test chamber, and then, in turn, after at least 2 s, look into the reference chamber again. Observers had to repeat this procedure until they could assess the color difference and fill the questionnaire (Figure 5.16).

For comparison with theoretical predictions, color differences were computed for each of the 10 test light sources and each of the 17 test colors by using 6 color difference formulae: CIE LAB (ΔE_{ab}^a), CIE DE2000 (ΔE_{2000}, $k_L = k_C = k_H = 1$), the Euclidean difference in CIE CAM02 J-a_C-b_C space (ΔE_{02}) as well as CAM02-LCD, CAM02-SCD, and CAM02-UCS [20–22].

The existence of well-delimited categories (in terms of ΔE_{vis}) showed the way to the idea of *predicting* the subjective rating categories based on the prediction of the visual color differences related to each category. The best predictor for ΔE_{vis} was selected from the set of six color difference formulae described above. The calculated predictors of color differences are denoted by ΔE_{calc}. The correlation coefficients between ΔE_{vis} and ΔE_{calc} were computed from the whole evaluated dataset of 1318 observations. Table 5.7 shows this result.

Based on the results in Table 5.7, it can be recognized that CIELAB and CIEDE2000 are significantly poorer than CAM02-UCS and also, CAM02-LCD and CAM02-SCD are insignificantly poorer than CAM02-UCS. This means that all of the latter three CIE CAM02-based formulae performed almost equally well in predicting perceived color differences but the CAM02-UCS metric was chosen owing to its slightly better performance compared to LCD and SCD.

Table 5.7 Pearson's correlation coefficients (r) between visual color differences ΔE_{vis} (and their z-scores $z\Delta E_{vis}$) and ΔE_{calc} for the whole dataset of 1318 observations.

r	CIELAB	CIEDE2000	CIECAM02	CAM02-LCD	CAM02-SCD	CAM02-UCS
ΔE_{vis}	0.596	0.609	0.645	0.651	0.650	0.654
$z\Delta E_{vis}$	0.647	0.660	0.698	0.704	0.702	0.706

Source: Reproduced with permission from Color Research and Application.
All correlations were significant at $p = 0.01$ (two-sided) [19].

Based on this result and on the results of further experiments, recent color quality metrics use CAM02-UCS as a basis for the new definitions of color fidelity indices (CIE CRI_{2012}, R_f of TM-30-15).

The question is how the relatively moderate Pearson's correlation coefficients (r) between visual color differences ΔE_{vis} and ΔE_{calc} (in the order of $r = 0.65$) can be explained. In the research work of Luo, several studies have been implemented in this direction and a comparison between the color space CIELAB $L^*a^*b^*$ and CAM02-UCS has been carried out. This is illustrated in Figures 5.17 and 5.18.

From Figures 5.17 and 5.18, it can be recognized that the CIELAB color space has a potential problem at the a^* and b^* values greater than 40 in the hue ranges between 0° and 180° (red–yellow–green) and between 225° and 285° (cyan–blue–violet). All these problems have strongly been reduced in the CAM02-UCS color space (note the different value scales in the two figures). However, CAM02-UCS is not the perfect color space either, because the ellipses for the same threshold of color difference perception are also different in size and orientation at different positions on the $a'b'$-diagram and they are not circular.

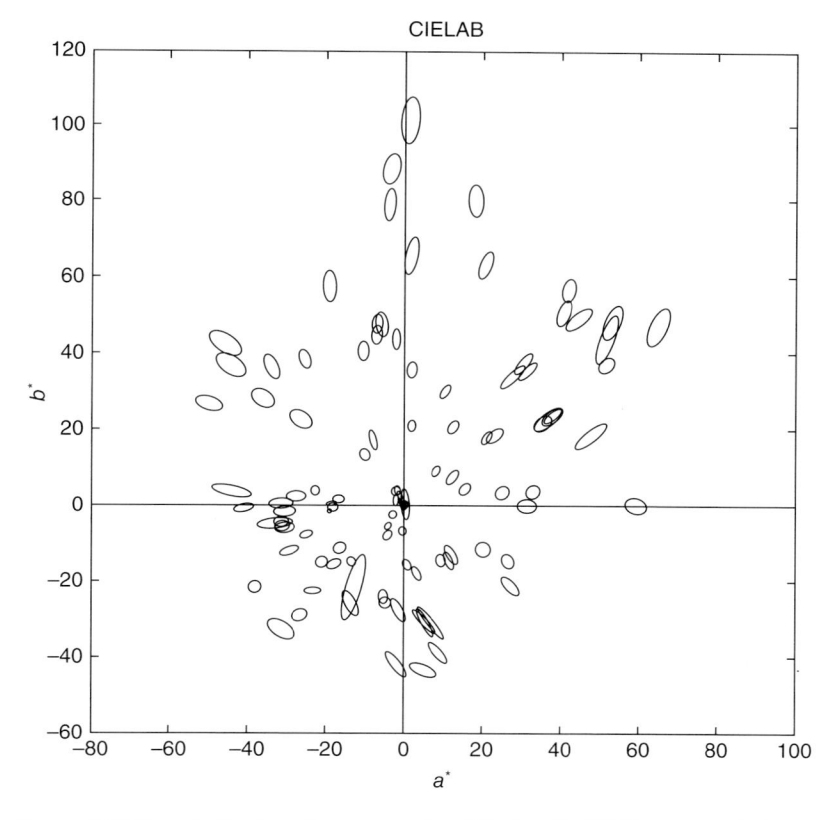

Figure 5.17 The distribution of areas over the a^*b^*-plane in CIELAB space with the same color difference thresholds. (Reproduced with kind permission of Professor Luo, personal communication.)

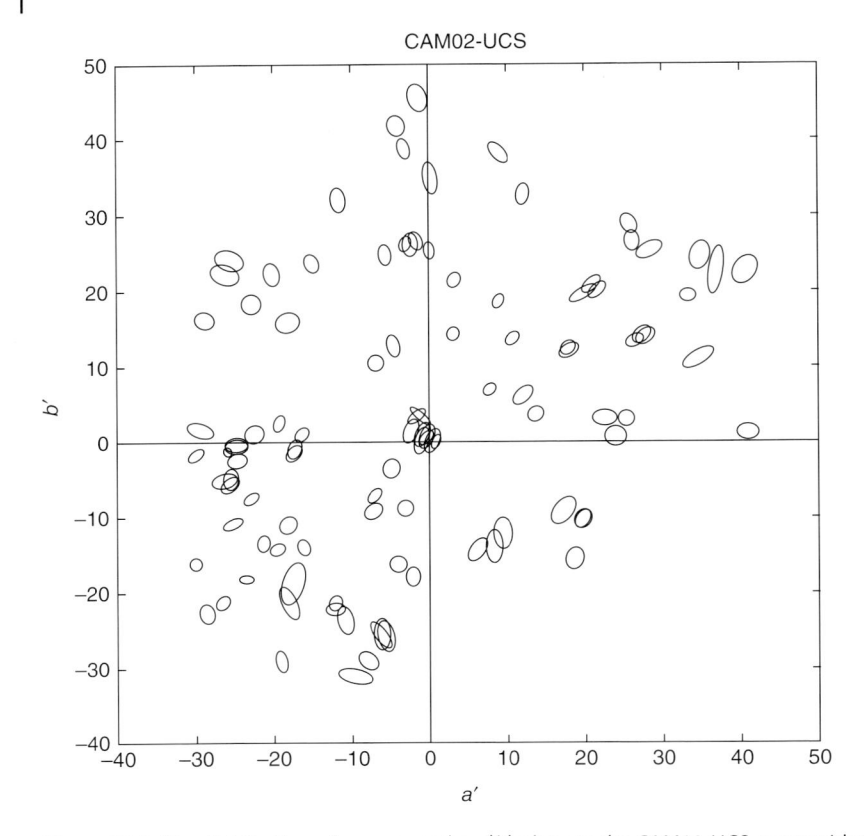

Figure 5.18 The distribution of areas over the $a'b'$-plane in the CAM02-UCS space with the same color difference thresholds. (Image source: reproduced with permission from R. Luo, personal communication.)

5.6 General Review of the State of the Art of Color Quality Metrics

The definitions and the colored objects related with the color quality metrics have been described in Chapters 2 and 4. From Figure 5.1, it can be seen that there are three main groups of color quality metrics including color fidelity, color discrimination/gamut, and color preference/color memory.

Although the experiments on color preference and color memory were carried out early in the time around 1957–1960 [23–26], the first color quality metric adopted by the CIE is the color fidelity index R_a in 1965 with some minor corrections in 1974 and 1995 [27]. It is also the only color quality metric officially endorsed by the CIE. With the introduction of white LEDs and the explosive development of LED applications in current lighting technology, some improved color fidelity indices have been proposed in the time between 2010 and 2015 (such as CQS-Q_f, CIE CRI$_{2012}$, or TM-30-15 R_f). These modern color fidelity indices are listed in more detail in Table 6.1.

Following the philosophy of *color fidelity*, good light sources are the ones that can render object colors with the most similar color appearance (color impression) compared with those under the illumination of a corresponding reference light source of the same CCT and possibly, at the same brightness level. Any color difference (color shift) in chroma, hue, or lightness from the reference light source causes a reduction of the special color-rendering indices R_i (Eq. (5.4)) and the general CRI R_a (Eq. (5.5)).

$$R_i = 100 - 4.6 \cdot \Delta E_i (U^* V^* W^*) \qquad (5.4)$$

$$R_a = \frac{1}{8} \cdot \sum_{i=1}^{8} R_i \qquad (5.5)$$

Gradually over time and in the course of applications, it has been recognized that the CRI R_a has the deficits of a nonuniform color space $U^* V^* W^*$ with an obsolete chromatic adaptation transformation of von Kries type and the use of only eight desaturated TCSs.

Fortunately, all these deficits have been improved by the definitions of the new proposals such as CQS Q_f, CRI_{2012}, and the TM-30-15 R_f definition with new color spaces, a higher number of color samples, and a visually relevant chromatic adaptation model (CAMCAT02). However, there has been still an unsolved issue for the generation rule of reference light sources in the range of CCT $\leq$ 5000 K because the thermal radiators as reference light sources according to the Planckian law always have only a small spectral amount in the blue wavelength range (see Figure 5.3) and a small gamut area in the color diagram (see Figure 5.4). Consequently, optimized light sources in the last decades with the target of color fidelity in the whole CCT range have still faced the problem of low chroma and small color gamut in the specific color area although their color-rendering indices have sometimes achieved the near perfect level.

On the other hand, the concept of *color gamut indices* is usually based on the vision that a light source with its SPD should render as many colors in the direction of the longer radius (i.e., more chroma) at all hue angles in a hue circle as possible. With such a light source, the volume of all presentable colors in a three-dimensional color space or the area in a two-dimensional color plane (e.g., $a^* b^*$-plane, see Figure 5.4) should be large in comparison to that of a fixed absolute reference light source (e.g., the Illuminant D65 or C) or a correspondingly generated reference light source with the same CCT (e.g., Planckian radiators for CCT $\leq$ 5000 K and daylight illuminants for CCT > 5000 K). For a long time period, most scientists have assumed that, although a light source should achieve a high color gamut (see Figures 4.29 and 4.30 of the light source WW14 with a high gamut area in Chapter 4), it still has to ensure an acceptable level of color discrimination ability at all hue angles. This means that the enhancement of chroma and/or color gamut by a light source is only permitted up to a limited value.

The size of the color gamut (area or volume), its shape, and its colorimetric relevance are strongly dependent on the selection of the colored objects (or TCSs). This is demonstrated in Figure 4.29 for the special light source WW14 with four different well-known color collections (14 CIE TCSs for the CRI 1995, 17 TCSs for CRI$_{2012}$, the 15 TCSs in the CQS system, and 16 binning groups of the 99 IES TCSs). The TCSs in the color collection for a color gamut metric should cover the hue circle in fine hue steps and should have enough saturation. In this context, the 15 CQS-colors and the 99 IES-colors can be recommended. The definition of the color gamut indices FCI [5] (four saturated colors) and GAI [28] (eight desaturated colors) is therefore not optimal.

In the last years, some new color gamut indices have been defined (such as Q_g in the CQS-system, R_g in the TM-30-15 system, FCI and GAI). Researchers generally believe that the color gamut index is able to describe an important aspect of color quality. The ability of a light source to enhance color saturation up to the allowable limit of color discrimination is not a new subject but it has a long history. Indeed, the first definition of a color gamut index has been announced by Thornton [29]. In Table 5.8, this definition and others are also listed in their development history.

Each definition of the conventional or modern color gamut metrics includes a fundamental problem. The color gamut index is a ratio of the gamut (an area or a volume) of a number of TCSs under a test light source to that under a reference light source. Therefore, the exact shape of the gamut at every hue angle and the maximal chroma values at every hue angle are not reflected by this definition. It is only the size of the entire gamut that is reflected mathematically. In real applications, however, the color rendition of some specific colors, for example, the reddish colors in a butcher's shop, and their color rendition are very important and only the light sources fulfilling this special color rendition should come into play. The evaluation for the applicable light sources or/and the technical support for designing and planning an appropriate light source are difficult in the case of such specific applications with colored objects of specific hue ranges when the above-mentioned definition of color gamut is used.

Table 5.8 Color gamut indices established before 2009.

Parameter	Reference light source	Color space	Metric calculation
CDI (color discrimination index)	Standard illuminant C	CIE 1960 (U, V, W)	Area enclosed by the polygon connected by the coordinate points of the eight TCSs used for determining R_a in the CIE 1960 uv-diagram [29]
GAI (gamut area index)	Equal-energy-spectrum GAI = 100	CIE 1976 UCS diagram (u', v')	Area enclosed by the polygon connected by the coordinate points of the eight TCSs used for determining R_a in the CIE 1976 $u'v'$-diagram [28]
CRC93 (color rendering capacity)	Equal-energy-spectrum	CIE 1976 UCS diagram (u', v')	[30]
CSA (cone surface area)	No reference illuminant	CIE 1976 (L^*, u^*, v^*)	[31]

5.6 General Review of the State of the Art of Color Quality Metrics

Beside color fidelity and color gamut (or color saturation of certain specific TCSs or specific colored objects), the essence of the memory- and preference-based color quality metrics has also been considered. It was pointed out that the color rendition of the object colors by a light source should be improved when the color appearance of familiar objects come close to the appearance that the observer expects (according to long-term memory colors stored in the brain) or prefers (according to esthetic considerations). In this respect, the individual internal *"preference mechanism"* of each observer has to be considered according to professional and personal experience, cultural roots, living location, and social environment. It is difficult to separate the two components *"memory color"* and *"preferred color."* However, the context of the two phenomena will be explained below as follows:

1) *Memory colors*: during the lifetime of every person, many familiar objects have often been observed. They are outdoor objects such as blue sky, foliage, trees, flowers, grass, parks, hills, clouds, house façades, fruits, sea water, or sand on the beach in different facets, forms, and colors that are dependent on season, weather, type of sunlight, and geographical location (Asia, Africa, Australia, America, or Europe). Indoor objects are objects in a room, at school, or in the family such as foods (butter, bread, and meat), furniture, textiles, decorations, books, and cosmetics. The frequent observation and mental contact with the objects and their colors initiate a process of identification (positive direction) or rejection (negative direction). The positively recognized objects with their colors and appearance form a mental system of familiar objects. It is a process being developed for a long time and its growth and establishment depends on personal, social, historic, and cultural conditions. If the light source under consideration is able to render the object colors – according to the actual mental state of the subject – similar to a familiar appearance thus generating an acceptable (perhaps intimate) atmosphere for the user, then the probability of acceptance of the illumination is high.

2) *Preferred colors*: In the additional role of color memory as its fundamental effect, the preference for certain objects with their shape, material texture, and colors is a product of social and technological development. In the era of technological progress, many new technical products with new features and possibilities have been used and then preferred (e.g., smartphones, internet, social media, cars, sport shoes, or some types of clothes according to fashion). Similar to the new possibilities, new streams, and new products, new color trends and new color acceptance tendencies have also appeared from time to time (e.g., new color for new sport shoes from well-known shoe manufacturers). Some colors became suddenly preferred by the young generation but they do not have a long tradition associated with a deep color memory effect.

Anyway, the mechanism of memory colors and preferred colors can be stimulated by the use of suitable light sources with a high color rendition capacity. The color quality metrics for color preference and color memory (i.e., the rendering of colors according to long-term memory colors stored in the observer's brain) have the task to analyze and describe these mechanisms in order to give an available

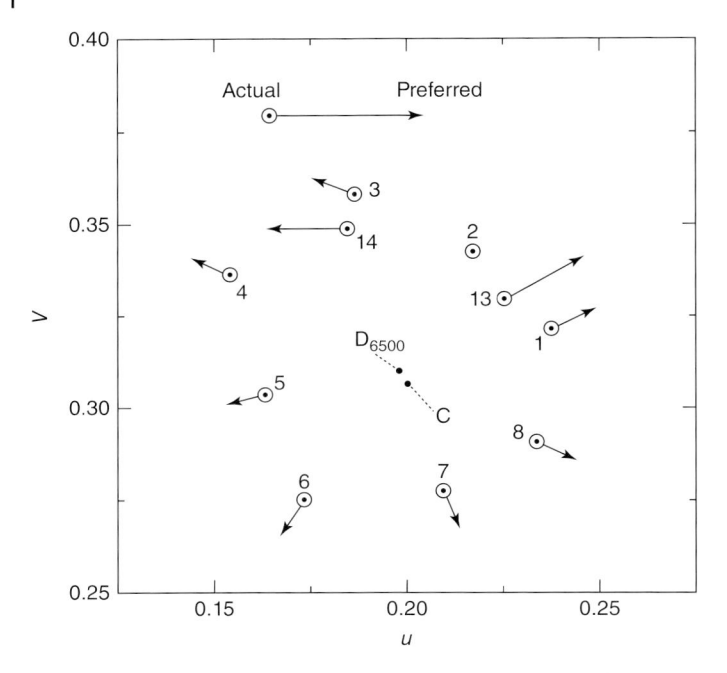

Figure 5.19 Judd's preferred chromaticity shifts (full color shift magnitudes). (Judd 1967 [2]. Reproduced with permission of Illuminating Engineering Society.)

measure and tool for developing and evaluating lighting systems for well-defined specific applications.

Modern color preference and color memory metrics (such as CQS Q_a and CQS Q_p as well as MCRI) are listed in Table 6.1. In the history of color quality research, the definition and description of color preference metrics, especially the interesting color preference metrics of Sander (Sander's preferred color index R_p in [32]), Judd's Flattery index (R_f, [2]), and Thornton' preference index CPI [33], have been discussed since 1957 (see also in [23–26]). The essence of preferred colors can be illustrated by the data of Judd [2] in Figure 5.19 for the case of the eight CIE R_a-colors and the CIE colors No. 13 (skin tone-TCS13) and No. 14 (green leaf-TCS14). As can be seen, the preferred chroma shift is related to increased saturation. As an example, the skin tone (No. 13) is shifted toward the direction of more chroma for red and the green leaf would contain more of the green hue after the color shift. It means that a chroma shift in the direction of more color saturation (plus a hue shift) plays an important role in this context for both colors TCS13 and TCS14.

With a variety of color metrics in different categories in mind (some of them are in a mixed form), the following essential scientific and practical questions arise:

1) How can the color metrics be categorized in different groups with a similar meaning?
2) Do the groups of metrics represent different aspects of color quality (color rendition)?

3) Which metric corresponds to the results of visual experiments designed to analyze the perceptual attributes color fidelity, color preference, color naturalness, vividness, or attractiveness?

According to the above discussions, Figure 5.20 summarizes the methodology of color quality science.

As can be seen from Figure 5.20, the methodology of color quality science can be dived into five steps:

- *First step:* Classification and grouping the color quality metrics by means of a correlation analysis
- *Second step:* Visual experiments with several colored objects, a number of light sources, and relevant perceptual attributes
- *Third step:* Finding the correlation among the perceptual attributes (e.g., color preference to naturalness)
- *Fourth step:* Finding the correlation between the color quality metrics and the perceptual attributes
- *Fifth step:* Modeling relevant perceptual attributes with relevant metrics.

A successful color quality research should do all these five steps appropriately in order to achieve both numeric and psychophysical results. In this way, applicable definitions of new color quality metrics can be established and applied for technical exchange, research, and the optimization, design and evaluation of real lighting applications (such as lamps, luminaires, or systems of luminaires). In the history of color quality research, not all steps have been implemented appropriately. Therefore, a literature review and much more analytical calculations and visual psychophysical experiments are necessary to implement them properly.

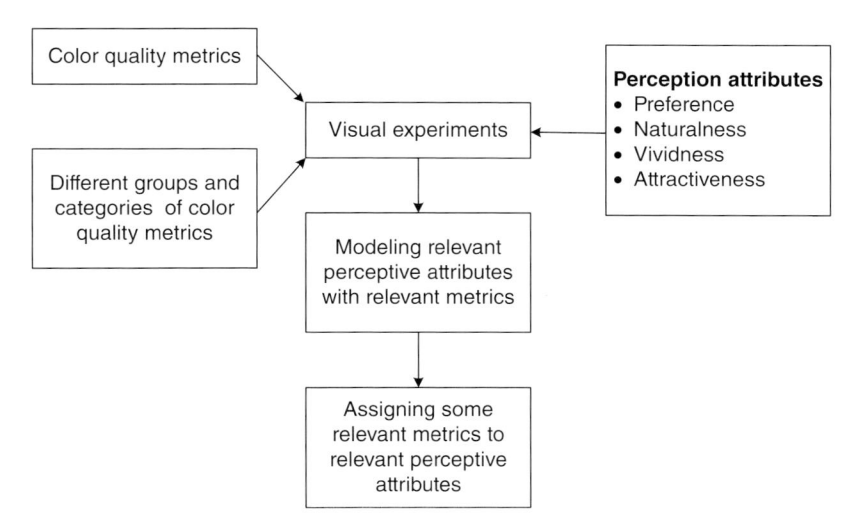

Figure 5.20 Methodology of color quality science.

5.7 Review of the Visual Experiments

In Chapter 7, a series of visual experiments in the Lighting Laboratory of the present authors will be described as the illustrations of the steps 2–5 mentioned above. Here, the following aspects will be considered in detail:

1) *Test subjects:* They should have good concentration at the time of the visual experiment and should not have any color deficiency to be tested by a widely used color deficiency test (such as the Ishihara test or the Farnsworth–Munsell test). Some tests and questionnaires, for example, on sleeping quality and on concentration, should also be carried out (optionally). Test subjects should have a training of 10–12 min before the beginning of the experiment. They should be informed about the meaning and procedure of the experiment. In many cases, it is desirable to invite test subjects from different age groups with enough number of male and female subjects in order to increase the relevance and usability of the test results. In contrast to several published color studies using naive subjects or subjects without any knowledge on color science or lighting engineering, the test subjects in the experiments of Chapter 7 are lighting engineering students, Ph.D. students for lighting engineering and color science, and developing engineers in lighting companies in order to increase the sharpness, decisiveness, and cognitive relevance of the test response.

2) *Test color objects:* The test color selection shall fulfill the context and the intention of the experiment. In many cases, the colored objects used are real familiar objects representing a wide range of hue tones and chroma. The spectral radiance coefficient of each colored object should be measured in order to calculate the values of the color quality metrics coming into consideration.

3) *Test light sources:* The light sources should have different spectra of different white tones (warm white, neutral white, and cold white) with the spectral design and distribution being relevant for the research questions to be answered by the experiment. The number of light sources shall be defined according to the duration of the experiment. In order to avoid the fatigue of the test subjects, the test duration should be generally no longer than 1 h. The light distribution on the color objects shall be in most cases diffuse and homogeneous.

4) *Test procedure:* The chromatic adaptation time of 1.5 min on a white wall of the test chamber or test room after changing the light source should be enforced before starting the observation and the assessment. Some pretrials (without evaluation) should be conducted in a first training phase in order to teach the test subjects the method and prepare his or her visual cognition for performing the main experiment. By the use of the arrangement of the actual colored objects under two or three different light sources to be presented in the pretest, the concepts to be assessed visually, for example, color preference, color naturalness, or color vividness shall be explained to the test subjects.

All light sources should be presented randomly to the test subject. Optionally, one or two light sources can be repeated within the light source sequence in order to check the stability and understanding of the test subject and the physical

Figure 5.21 A test arrangement with three copies of the same test objects and three different light sources with different spectra at three similar correlated color temperatures.

properties of the test procedure during the experiment. The following methods can be used concerning the sequence of the different light sources:

- **Either** each light source should be tested by an available questionnaire designed correspondingly with an appropriate hypothesis such as the one in Chapter 7 with separate scales indicated by two individual vertical axes: one for numeric scaling (0–100) and the other one for the semantics (e.g., very bad… very good) (see the Figures 7.28 and 7.40). In many visual experiments published in the literature, the scale can have five categories, seven categories, nine categories, or a continuous scale from 0 to 100. For each light source to be tested, the average rating of a large number of test subjects (more than 20 test persons), its standard deviation as well as statistical significance tests shall be carried out and evaluated.
- **Or** the set of light sources to be assessed should be divided into pairs and each pair of the light sources shall be compared and then evaluated in order to rank the light sources based on the visual assessment results.

Two or three sets of the same colored objects with similar color and reflectance characteristics can also be arranged in two or three adjacent chambers illuminated by different light source spectra at the same luminance level and similar CCTs (see Figure 5.21). Because color preference and color naturalness are those visual color quality attributes that do not need any reference light source, the use of only one chamber is most frequent.

5.8 Review of the State-of-the-Art Analyses about the Correlation of Color Quality Metrics of Light Sources

In this section, a literature review dealing with the first step *"classification and grouping of the color quality metrics by means of correlation calculations"* (as mentioned above) is discussed. Further consequences and color research potentials will be discussed in the subsequent chapters of the book.

The analytical classification of color quality metrics, the research on the similarity and limitation of each color quality metric, and the consequences of using a multimetric system were described in two fundamental articles of Guo *et al.* [34] and of Houser *et al.* [35] in 2013. In nearly one decade from 2004 to 2013, the technological progress of white LEDs was accelerated and the number and quality of the proposed color rendition metrics was improved. Methodologically, each metric is defined with a number of TCSs and assumed reference light sources. In order to point out the behavior and prediction potential of each metric, a multitude of light sources with various spectral distributions from different conventional and modern (solid-state-lighting) technologies and CCTs were used, and the value of each metric was calculated. In this manner, each light source provided many color rendition (color quality) metric values and then, in turn, a correlation, ranking, and grouping analysis was performed.

In the publication [34], 10 color quality metrics for color gamut (CDI, CRCO, CRC, and CSA), color fidelity (R_a and R_aO), color preference (R_f and CPI, Pointer index), and CCT were analyzed with 34 light sources (thermal radiators, fluorescent lamps, metal halide lamps). In 2004, it was recognized that different color quality metrics ranked the light sources differently. Correlation computations showed that the color gamut indices cone surface area (CSA), color discrimination index (CDI), CRCO, and color-rendering capacity (CRC) did have high correlation among each other. This was also true for the color fidelity indices R_a, R_aO, and R_f and indicated the fact that the flattery index R_f of Judd exhibits a behavior similar to color fidelity index. Thornton's CPI correlated moderately with the color fidelity indices (Pearson's correlation coefficient was in the order of 0.598–0.797) and it did not correlate with the color gamut metrics.

The authors' approach (2004) with a composite z–score, which equally weights and sums all indices for color gamut, color preference, and color fidelity issues, is interesting. From today's point of view, it is a methodologically equivalent approach to considering a multimeasure system in order to describe the color quality of light sources. But the authors also remarked that *"the z-score sum, however, does not have a rigorous theoretical basis."* They also recognized the importance of color temperature: *"CDI, CSA, CRC, and Pointer's index all have a high correlation with CCT. This may be an indication that higher color temperature will produce larger color gamut."*

In 2013, Houser *et al.* [35] carried out an analysis using similar methods for 22 color quality measures plus CCT and with 401 light sources and illuminants including incandescent lamps, LEDs, pcLEDs, mixed (hybrid) LEDs, fluorescent lamps, high-intensity discharge lamps, and some theoretical illuminants. Besides the 10 metrics analyzed in 2004, more recent color gamut metrics (such as GAI, FCI), color fidelity metrics (CRI_{2012}), the four CQS-metrics (Q_f, Q_a, Q_g, and Q_p) and the well-known MCRI have been added to their analysis. The matrix of Spearman's rank correlation coefficients is shown in Table 5.9.

As can be seen from Table 5.9, there are three separate groups of metrics (a group of color fidelity indices, a group of color gamut/color discrimination measures, and a group a *"preference-based"* metrics). Based on the data in Table 5.9,

Table 5.9 Matrix of Spearman's rank correlation coefficients that also illustrate blocks of similarity from the MDS scaling solution (see the next section).

	R_aO	R_f	R_a	R_9	Q_a	Q_f	$RCRI_a$	R_a12	CPI	FCI94	FCI02	Q_g	Q_p	MCRI	CDI	FMG	CRC84	CRC93	CSA	GAI	PI	FSCI	CCT
	1965	1967	1974	1974	2010	2010	2010	2012	1974	2007	2007	2010	2009	2010	1972	1977	1984	1993	1997	2004	1986	2004	
R_aO	1.00																						
R_f	0.901**	1.00																					
R_a	0.946**	0.87**	1.00																				
R_9	0.792**	0.766**	0.843**	1.00																			
Q_a	0.921**	0.930**	0.937**	0.838**	1.00																		
Q_f	0.950**	0.894**	0.952**	0.826**	0.979**	1.00																	
$RCRI_a$	0.875**	0.842**	0.891**	0.813**	0.906**	0.925**	1.00																
R_a12	0.815**	0.750**	0.530**	0.787**	0.885**	0.883**	0.837**	1.00															
CPI	0.588**	0.799**	0.607**	0.701**	0.714**	0.610**	0.614**	0.497**	1.00														
FCI94	0.266**	0.434**	0.326**	0.480**	0.405**	0.315**	0.347**	0.272**	0.655**	1.00													
FCI02	0.300**	0.453**	0.372**	0.541**	0.461**	0.374**	0.412**	0.376**	0.643**	0.977**	1.00												
Q_g	0.716**	0.824**	0.757**	0.803**	0.872**	0.794**	0.789**	0.785**	0.841**	0.623**	0.685**	1.00											
Q_p	0.332**	0.551**	0.378**	0.483**	0.468**	0.345**	0.356**	0.284**	0.821**	0.802**	0.790**	0.739**	1.00										
MCRI	0.581**	0.723**	0.646**	0.746**	0.751**	0.685**	0.748**	0.696**	0.778**	0.637**	0.710**	0.891**	0.638**	1.00									
CDI	0.016	0.105*	0.024	0.100*	0.061	0.013	0.038	0.079	0.300**	0.153**	0.131**	0.255**	0.272**	0.240**	1.00								
FMG	0.003	0.0816	0.000	0.075	0.034	0.011	0.015	0.050	0.272**	0.189**	0.169**	0.216**	0.237**	0.209**	0.996**	1.00							
CRC84	0.060	0.0537	0.059	0.005	0.039	0.081	0.044	0.099*	0.266**	0.175**	0.189**	0.134**	0.242**	0.145**	0.943**	0.953**	1.00						
CRC93	0.044	0.084	0.053	0.123*	0.094	0.071	0.097	0.195**	0.179**	0.236**	0.186**	0.243**	0.116**	0.259**	0.945**	0.946**	0.874**	1.00					
CSA	0.032	0.0089	0.041	0.0057	0.023	0.046	0.028	0.0259	0.135*	0.351**	0.326**	0.102**	0.0636	0.097**	0.966**	0.977**	0.927**	0.951**	1.00				
GAI	0.016	0.105*	0.023	0.099*	0.061	0.013	0.037	0.078	0.300**	0.154**	0.133**	0.254**	0.272**	0.238**	1.00**	0.996**	0.944**	0.944**	0.966**	1.00			
PI	0.579**	0.539**	0.570**	0.494**	0.553**	0.573**	0.572**	0.537**	0.444**	0.012*	0.032	0.539**	0.244**	0.495**	0.0591**	0.0575**	0.514**	0.602*	0.553**	0.591**	1.00		
FSCI	0.319**	0.204	0.322**	0.253**	0.307**	0.332	0.26**3	0.489**	0.009	0.382**	0.308**	0.221**	−0.125*	0.0765	0.485**	0.487**	0.341**	0.594**	0.548**	0.485**	0.460**	1.00	
CCT	−0.081	−0.12*6	−0.110*	0.102**	−0.134**	−0.111**	−0.102*	−0.057	−0.104**	−0.564**	−0.548**	−0.126*	−0.240**	−0.102**	0.790**	0.816**	0.777**	0.818**	0.908**	0.791**	0.471**	0.543**	1.00

Source: Bodrogi et al. 2010 [19]. Reproduced with permission of Wiley.

The upper left shading in orange identifies a cluster that can be called *preference-based measures*, the middle shading in green identifies a cluster that can be called *fidelity-based measures*, the lower right shading in blue identifies a cluster that can be called *gamut based* (i.e., discrimination) measures. The year of the literature reference is also provided.

** (*) indicates that the correlation is significant at the 0.01 (0.05) level (two-tailed).

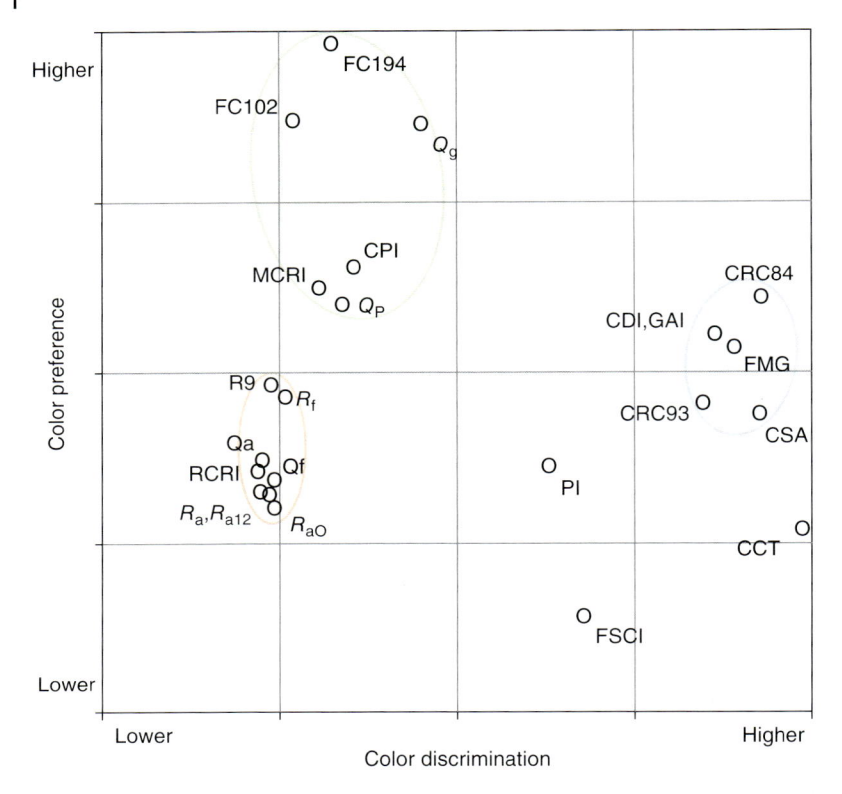

Figure 5.22 Two-dimensional Euclidian distance MDS solution for 22 color measures based on 401 light source spectra. (Houser *et al.* 2013 [35] Reproduced from Optics Express with permission from OSA.)

it can also be seen that the group of gamut indices does not or only very weakly correlates with the group of color fidelity indices. This means that these two types of color quality metric can be optimized independently.

Houser *et al.* [35] also carried out a multidimensional scaling (MDS) in order to identify groups (clusters) of similarity and the underlying mechanisms concerning all 23 metrics and the 401 light sources. In Figure 5.22, the two-dimensional Euclidian distance MDS solution for 22 color measures based on 401 light source spectra is shown. As can be seen, there are three groups of metrics. The group of color fidelity metrics contains low color discrimination and moderate color preference; the group of color gamut indices exhibits higher color discrimination and moderate color preference. The three indices FCI_{02}, FCI_{94}, and Q_g result in the highest color preference and the "*color preference based*" metrics MCRI, Q_p, and CPI appear to have higher color preference and moderate color discrimination.

Owing to the fact that several color gamut indices do have a fixed reference light source (e.g., GAI with the equal-energy spectrum as a reference), the authors [35] calculated the color gamut indices with the reference light source having the same CCT as the light source under consideration. This is depicted in Figure 5.23, which is an upgradation of Figure 5.22.

5.8 Review of the State-of-the-Art Analyses about the Correlation | 165

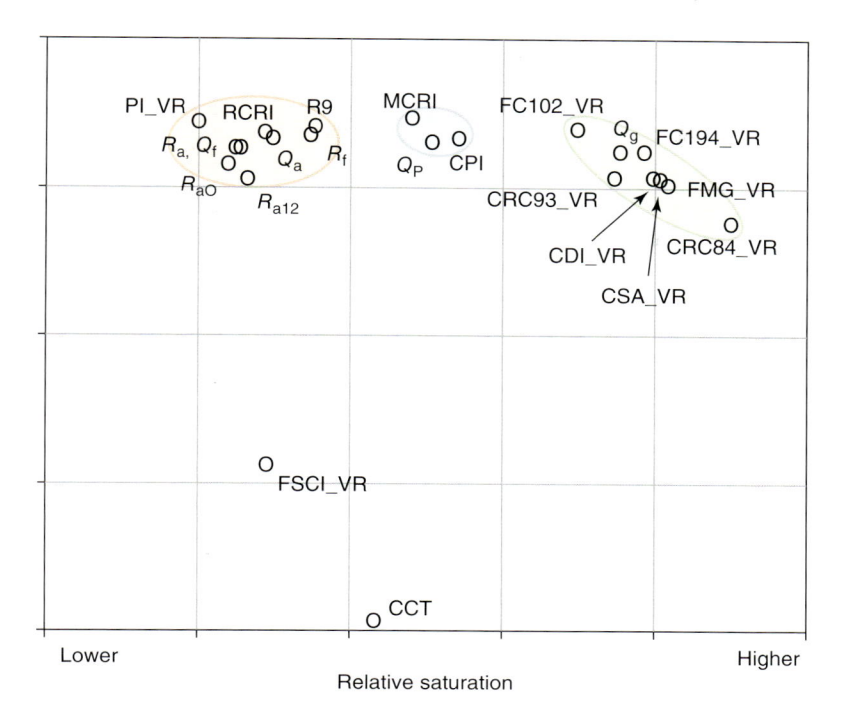

Figure 5.23 Two-dimensional Euclidian distance MDS solution for 22 color quality measures based on 401 SPDs. All measures were computed by using a reference illuminant at the same CCT as the test illuminant. (Houser et al. 2013 [35] Reproduced from Optics Express with permission from OSA.)

With the above new calculation method, color quality research reached a new milestone in the analysis and in conception. It can also be seen from Figure 5.22 that three groups of metrics (fidelity, gamut/discrimination, and color preference) can be identified. In each group, the correlation between the metrics is relatively high. The group of the color fidelity indices represents the property of lower object saturation and the group of all color gamut indices embodies the highest object saturation. The group of the three color preference indices MCRI, Q_p, and CPI is situated in the middle of Figure 5.23.

The combined view of Figures 5.22 and 5.23 suggests that a light source with an acceptable color preference/color memory property should have high enough color fidelity (e.g., $R_a > 80–83$ or $85–86$) and, additionally, it should also have – as a bit of compromise – an appropriate enhancement of color saturation indicated by the increase of chroma difference or color gamut area compared with the corresponding reference light source. In other words, color preference/color memory represents a compromise between color fidelity and color (over)saturation (i.e., more color gamut or a positive chroma difference) but color discrimination ability shall always remain within its acceptable limits as a constraining factor or boundary condition. Color discrimination as an essential color quality aspect plays an important role in the optimization strategy of light source spectra. This will be further discussed in Chapter 8.

A further mathematical analysis showed that the color preference index (Q_p) and the MCRI include both attributes (color fidelity and color saturation/color gamut). This idea should also motivate color scientists and analysts to model color preference metrics in terms of the synthesis of two metrics (e.g., a fidelity metric and a chroma difference or color gamut metric). Such combination methods will be shown in Chapters 6 and 7 including modern color quality metrics such as the recently proposed TM-30-15 color quality metrics.

5.9 Review of the State-of-the-Art Analysis of the Prediction Potential and Correctness of Color Quality Metrics Verified by Visual Experiments

Generally, a color quality metric describes either the color differences between the color appearance of an arrangement of color objects illuminated with a test and a reference light source of the same CCT (color fidelity) or between the color appearance of the objects illuminated by a test light source and an expected and/or preferred color appearance (color preference, color memory). The color gamut index is a mathematical tool to be used to quantify how big the volume or area of a defined color collection illuminated by a test light source is, in comparison to the volume or area of the same color collection illuminated by a reference light source. It is the task of color research to find out, by means of visual experiments, how these mathematical constructions (color quality metrics) correlate with each other and with the (mean) visual ratings of a panel of test subjects.

The correlation calculation described in Section 5.8 can only help group and classify the relatively big number of the currently proposed color quality metrics. The identification of the underlying dimensionality of the metrics by the MDS technique shows the order of the three groups of color fidelity, color preference, and color gamut indices qualitatively along the object (over)saturation axis. But this analysis in Section 5.8 does not answer the question of how well the prediction of each metric corresponds to the visual appreciations and ratings of the test subjects for a given set of TCSs illuminated under a test light source.

In the last years, especially from the beginning of the twenty-first century until now, many visual experiments have been carried out in order to clarify the meaning of the perceptual color quality attributes (such as color fidelity, color preference, naturalness, or vividness) and determine the prediction potential and correctness of these metrics referring to these attributes. These studies were published in [7–10, 28, 36–46].

Visual experiments always include the chain "light sources–colored objects–test subjects." The light sources used in the visual experiments can be conventional light sources (such as thermal radiators, fluorescent lamps, high-pressure metal halide lamps) or solid-state light sources (such as mixed LED systems or pcLEDs) with well-defined spectra. These light sources shall be grouped into blocks of the same CCT. Most of the published experiments investigated the warm white range between 2600 and 3200 K. The set of colored test objects included TCSs (e.g., Munsell colors) or colored objects that comprise some

Table 5.10 Average correlation coefficients r(av) for the visual color quality attributes preference/attractiveness [47].

Metrics	Average correlation coefficients r(av)	Remarks
CIE R_a	0.17	Color fidelity indices
$R_{a,CAM02\text{-}UCS35}$	0.22	
RCRI	0.23	
CQS Q_f	0.20	
CQS Q_a	0.29	CQS-metrics
CQS Q_p	0.58	
FCI	0.42	Color gamut indices
GAI	0.40	
CSA	0.26	
S_a	0.88	Color memory rendering index
R_f	0.29	Color preference metrics
CPI	0.39	
GAI-R_a	0.41	Combined metric

contextual meaning (such as cosmetics, fruits, flowers, skin, complexion, or other familiar objects). Test subjects are regarded as "*response-givers*" with their physiological, cognitive, and social characteristics. For an arrangement of colored objects and light sources, subjects are asked to give their judgment on a set of dedicated perceptual color quality attributes defined on the questionnaire.

From all publications in the period 2007–2015 before the time of writing of this book, there have been two comprehensive studies of Smet *et al.* [47, 48] dealing with the correlations between the metrics and the visual scaling results systematically. In [47], the color metrics of color fidelity (R_a, $R_{a,CAM02\text{-}UCS35}$, RCRI), the CQS metrics (CQS Q_a, CQS Q_f, CQS Q_p), the color gamut indices (FCI, GAI, CSA), the preference and memory based metrics (S_a/MCRI, R_f, CPI), and the combined metric GAI-R_a were studied based on the analysis of nine visual experiments [7–10, 36, 37, 49]. Based on the spectral distribution of the light sources used in the experiments, the metrics mentioned above could be calculated and based on the visual ratings of the test subjects, Spearman's correlation coefficients between the metrics' predictions and the visual ratings on perceived color preference and naturalness were calculated. For each metric, the average correlation coefficients r(av) were calculated by weighting the individual correlation coefficients by the number of the test subjects in each visual experiment. In Table 5.10, the average correlation coefficients r(av) are listed for the visual color quality attributes color preference/attractiveness.

The following can be seen from Table 5.10:

- All color fidelity metrics poorly correlate with the visual ratings according to the criterion "*color preference and attractiveness.*" To some extent, CQS-Q_a can also be regarded as a type of special color fidelity index that does not penalize saturation enhancement. This metric also does not correlate with visual ratings. Beside the weaknesses of the color space, the chromatic adaptation transformation, the reference illuminant (CCT < 5000 K), and the selection of the colored objects, it should be emphasized that these color fidelity indices are not intended, according to their definition, to represent the perceived attributes color preference and attractiveness. These metrics are defined to indicate the similarity of color appearance between a test light source and its reference light source. Therefore, the discussion on the deficits of color fidelity metrics is not necessary.

- The same formulation is also true for the color gamut indices CSA, FCI, and GAI. This means that a higher or a smaller gamut of the light sources in the context of their definition does not necessarily provide a higher or a lower preference or attractiveness of the illuminated scene. The metrics CPI and R_f have also shown the same weakness.
- The color memory-rendering index S_a of Smet et al. correlates well with the perceived color quality attributes. But the combined metric GAI-R_a does not correlate with visual ratings.

In Table 5.11, the average correlation coefficients r(av) for the perceived color quality attribute "*naturalness*" are shown.

It can be seen from Table 5.11 that the color gamut indices also do not correlate with the visually scaled values of "*naturalness.*" Color fidelity indices show a somewhat better correlation in the order of about 0.63–0.65 implying that "*naturalness*" contains a remarkable content of color fidelity (color realness). The combined metric GAI-R_a exhibited the best correlation with r(av) = 0.85. This finding was tested by other visual experiments; see Chapter 7.

Table 5.11 Average correlation coefficients r(av) for the visual color quality attribute naturalness [47].

Metrics	Average correlation coefficients r(av)	Remarks
CIE R_a	0.65	Color fidelity indices
$R_{a,\ CAM02\text{-}UCS35}$	0.65	
RCRI	0.64	
CQS Q_f	0.67	
CQS Q_a	0.63	CQS-metrics
CQS Q_p	0.56	
FCI	−0.17	Color gamut indices
GAI	0.06	
CSA	0.30	
S_a	0.45	Color memory index
R_f	0.68	Color preference metrics
CPI	0.53	
GAI-R_a	0.85	Combined metric

The authors of [47] wrote *"Finally, predictive performance in terms of naturalness was found to be roughly negatively correlated with the predictive performance for preference (r = −0.44, p = 0.13). Therefore, a metric that performs well for one aspect of color quality will not perform well for the other. This confirmed the finding of Rea and Freyssinier [49] that a complete description of the color quality of a light source will probably require more than one metric."*

Five years after writing this publication [47], Smet *et al.* continued this analysis by using a similar method. Meanwhile, the technological development of white LEDs and their applications in indoor lighting has achieved many new evolutionary steps and the need for a system of correct color quality metrics has strongly increased. New color quality metric definitions, improvements of the previously defined metrics, and several new studies on the optimal test color set have emerged parallel to an increased number of visual color quality experiments [9, 10, 36–44]. These experiments constitute the basis for the research presented in [48] at the end of 2014 and at the beginning of 2015. In a similar manner to [47], correlation calculation and MDS analysis were implemented [48]. Table 5.12 shows the weighted average (artifact-corrected) Spearman correlation coefficients $r(av)$ between the metrics and the visually scaled values.

As can be seen from Table 5.12, color fidelity indices and CQS Q_a are not able to account for the perceived attribute *"visual appreciation"* (which can be considered as a synonym of "color preference"). The color gamut indices, the combined metric GAI-R_a, and the color preference metrics (with an exception of the R_p index of Sanders) do have a moderate correlation in the order of 0.67–0.77. The best correlation was achieved by the MCRI ($r = 1.0$). The four CQS metrics Q_f, Q_a, Q_g, and Q_p (listed in the order of aiming at an increasing level of object

Table 5.12 Average correlation coefficients $r(av)$ for the visual color quality attributes color preference/attractiveness [48].

Metrics	Average correlation coefficients $r(av)$	Remarks
CIE R_a	0.09	Color fidelity indices
CRI_{2012}	0.41	
CQS Q_f	0.16	
CQS Q_a	0.38	CQS-metrics
CQS Q_p	**0.81**	
FCI	0.67	Color gamut indices
GAI	0.74	
CQS Q_g	0.76	
R_m (MCRI)	1.00	Color memory rendering index
R_f	0.57	Color preference metrics
R_p	−0.26	
CPI	0.77	
GAI-R_a	0.77	Combined metric

oversaturation) resulted in increasing correlation coefficients in the same order ($r_{qf} < r_{qa} < r_{qg} < r_{qp}$).

Comparing Table 5.12 with Table 5.10, it can be seen that several color gamut indices and color preference indices result in a higher correlation coefficient although the methods of calculation did not change essentially. Seemingly, the quality of the results from the visual experiments used for this correlation analysis (Table 5.12) influenced the correlation results.

In the analysis in [48], the authors formulated some important thoughts worth citing at this point: *"Obviously, there will be an upper limit to the visually allowed chroma enhancement, as oversaturation is known to have a negative impact on perceived color quality. Regarding the gamut-expansion (FCI, GAI, and Q_g) and chroma enhancement (Q_a, Q_p, and GAI-R_a) based metrics, the former fail to account for such a limit. On the other hand, the latter and metrics such as R_p, R_f, CPI, R_m, and GAI-R_a do include such a limit: either by setting up a reference chromaticity with increased saturation (e.g., R_p, R_f, CPI, and R_m), or by explicitly setting a limit to the allowed chroma enhancement (e.g., Q_a and Q_p), or by the implicit counterbalance introduced by the changes in hue that are generally associated with increases in saturation (all former metrics, but especially GAI-R_a). In fact, the GAI-R_a counter-balances increase in gamut area (cfr. GAI) with full color differences with respect to unsaturated reference chromaticity (cfr. CIE R_a). Therefore, increases in chroma or gamut are always associated with increases in color difference."*

From the above cited thoughts, some ideas will be implemented in this book to show the consequences of the above-described research and for the visual experiments to be presented in Chapter 7, and also for the optimization of light source color quality in Chapters 8 and 9. These ideas can be summarized as follows:

- All color gamut indices have the same or similar constructions referring to a certain color collection. They represent the volume or area of the object colors illuminated by the test and reference light sources. This volume/area is an average value that cannot describe color difference distortions and the reduction or increase of chroma in *specific* hue ranges.
- The color preference metrics R_f, R_p, and CPI have problematic definitions (concerning their color space, their color sample set, their chromatic adaptation transform, and the description of color shift magnitudes). Therefore, they should not be taken into consideration in future works. Those metrics of color preference that take saturation limits into account are the CQS Q_p and the MCRI metrics that exhibit a useable correlation level with visual assessment results. These two metrics should be further analyzed in visual experiments under realistic viewing conditions. This will be included in Chapter 7.
- Generally, GAI and R_a are confronted with structural problems (color space, color samples, and also the chromatic adaptation transformation they use). As a single metric, they do not perform well to describe the color rendering (R_a) or the color gamut feature (GAI). The idea to combine these metrics into a new metric to account for both color fidelity and color gamut in the same color

space and with the same colored objects in order to describe the perceived color quality attributes naturalness, color appreciation, or vividness should be extended to the index pairs (Q_f, Q_g) of the CQS system and to the index pairs (R_f, R_g) of the TM-30-15's proposal [50]. These multimetrics will be investigated in Chapter 7.
- Because of the multifaceted nature of color rendition (color quality), a multi-index measure is needed. If this multi-index measure would be found then further analytical computations and also some fundamental psychophysical experiments shall be carried out in order to develop a semantic interpretation scale (e.g., very good, good, acceptable, moderate, poor, bad, very bad) for the considered aspect of color quality and to find the perceptually acceptable limits of object oversaturation. This issue will also become an interesting subject in Chapter 7.

References

1 Davis, W. and Ohno, Y. (2010) Color quality scale. *Opt. Eng.*, **49** (3), 033602–033616.
2 Judd, D.B. (1967) *A flattery index for artificial illuminants. Illum. Eng.*, **62**, 593–598.
3 Fairchild, M.D. and Reniff, L. (1995) *Time course of Chromatic adaptation for color-appearance judgments. J. Opt. Soc. Am. A*, **12** (5), 824–833.
4 Boyce, P.R. and Simons, R.H. (1977) *Hue discrimination and light sources. Light. Res. Technol.*, **9**, 125–136.
5 Hashimoto, K., Yano, T., Shimizu, M., and Nayatani, Y. (2007) *New method for specifying color-rendering properties of light sources based on feeling of contrast. Color Res. Appl.*, **32** (5), 361–371.
6 Schierz, Ch. (2008) Licht für die ältere Bevölkerung–Physiologische Grundlagen und ihre Konsequenzen. Lichttagung der deutschsprachigen Länder, Ilmenau-Germany, September 11, 2008.
7 Smet, K.A.G., Ryckaert, W.R., Pointer, M.R., Deconinck, G., and Hanselaer, P. (2010) *Memory colors and color quality evaluation of conventional and solid-state lamps. Opt. Express*, **18** (25), 26229–26244.
8 Jost-Boissard, S., Fontoynont, M., and Blanc-Gonnet, J. (2009) Perceived lighting quality of LED sources for the presentation of fruit and vegetables. *J. Mod. Opt.*, **56** (13), 1420.
9 Szabó, F., Csuti, P., and Schanda, J. (2009) Color preference under different illuminants–new approach of light source color quality, in *Light and Lighting Conference with Special Emphasis on LEDs and Solid State Lighting, Budapest, Hungary, May 27–29*, CIE, Vienna, pp. 27–29.
10 Narendran, N. and Deng, L. (2002) Color rendering properties of LED light sources. Solid State Lighting II: Proceedings of SPIE, 2002.
11 Royer, M.P., Houser, K.W., and Wilkerson, A.M. (2012) *Color discrimination capability under highly structured spectra. Color Res. Appl.*, **37** (6), 441–449.

12 Stiles, W. and Burch, J. (1959) N.P.L. color-matching investigation: inal report (1958). *Opt. Acta*, **6**, S.1–S.26.
13 MacAdam, D.L. (1942) *Visual sensitivities to color differences in daylight*. *J. Opt. Soc. Am.*, **32** (5), S. 247–S. 273.
14 Hung, P. and Berns, R.S. (1995) Determination of constant hue loci for a CRT gamut and their predictions using color appearance spaces. *Color Res. Appl.*, **20** (5), 285–295.
15 Braun, G.J., Ebner, F., and Fairchild, M.D. (1998) Color gamut mapping in a hue-linearized CIELAB color space. IS&T/SID 6th Color Imaging Conference, Scottsdale, 1998, pp. 163–168.
16 Li, C., Ronnier, L.M., Rigg, B., and Hunt, R.W.G. (2002) *CMC 2000 chromatic adaptation transform: CMCCAT2000*. *Color Res. Appl.*, **27** (1), 49–58.
17 CIE (2004) *A Review of Chromatic Adaptation Transforms*, CIE 160:2004, CIE. ISBN: 978 3 901906 30 5
18 Luo, M.R., Cui, G., and Rigg, B. (2001) The development of the CIE 2000 color–difference formula: CIEDE2000. *Color Res. Appl.*, **26**, 340–350.
19 Bodrogi, P., Brueckner, S., and Khanh, T.Q. (2011) *Ordinal scale based description of color rendering*. *Color Res. Appl.*, **36** (4), 272–285.
20 Guan, S.S. and Luo, M.R. (1999) A color-difference formula for assessing large color differences. *Color Res. Appl.*, **24**, 344–355.
21 Guan, S.S. and Luo, M.R. (1999) Investigation of parametric effects using large color differences. *Color Res. Appl.*, **24**, 356–368.
22 Guan, S.S. and Luo, M.R. (1999) Investigation of parametric effects using small color differences. *Color Res. Appl.*, **24**, 331–343.
23 Newhall, S.M., Burnham, R.W., and Clark, J.R. (1957) *Comparison of successive with simultaneous color matching*. *J. Opt. Soc. Am.*, **47**, 43–54.
24 Bartleson, C.J. (1960) *Memory colors of familiar objects*. *J. Opt. Soc. Am.*, **50**, 73–77.
25 Bartleson, C.J. (1961) Color in memory in relation to photographic reproduction. *Photogr. Sci. Eng.*, **5**, 327–331.
26 Sanders, C.L. (1959) *Color preferences for natural objects*. *Illum. Eng.*, **54**, 452–456.
27 CIE 13.3 (1995) *Method of Measuring and Specifying Color Rendering Properties of Light Sources*, CIE.
28 Rea, M.S. and Freyssinnier-Nova, J.P. (2008) *Color rendering: a tale of two metrics*. *Color Res. Appl.*, **33** (3), 192–202.
29 Thornton, W.A. (1972) *Color-discrimination index*. *J. Opt. Soc. Am. A:*, **62** (2), 191–194.
30 Xu, H. (1993) *Color-rendering capacity and luminous efficiency of a spectrum*. *Light. Res. Technol.*, **25** (3), 131–132.
31 Fotios, S.A. (1997) The perception of light sources of different color properties. PhD thesis. UMIST, United Kingdom.
32 Sanders, C.L. (1959) Assessment of color rendition under an iIlluminant using color tolerances for natural objects. *J. Illum. Eng. Soc.*, **54**, 640–646.

33 Thornton, W.A. (1974) *A validation of the color-preference index. J. Illum. Eng. Soc.*, **4** (1), 48–52.
34 Guo, X., Houser, K.W., and Akashi, Y. (2004) *A review of color rendering indices and their application to commercial light sources. Light. Res. Technol.*, **36**, 183–199.
35 Houser, K.W., Wei, M., David, A., Krames, M.R., and Shen, X.S. (2013) Review of measures for light-source color rendition and considerations for a two measure system for characterizing color rendition. *Opt. Express*, **21**, 10393–10411.
36 Schanda, J. and Madár, G. (2007) Light source quality assessment, in *Proceedings of the CIE 26th Session, Beijing*, CIE, Vienna, pp. D1-72–D1-75.
37 Vanrie, J. (2009) Appendix 4: technical report to the user committee of the IWT-TETRA project (80163): the effect of the spectral composition of a light source on the visual appreciation of a composite object set. Diepenbeek, Belgium: PHL, 2009 June 11, 2009. Report No. 42.
38 Tsukitani, A. and Lin, Y. (2014) Research on FCI. Presented at the Meeting of TC1-91, Kuala Lumpur, Malaysia, April 23–26, 2014.
39 Imai, Y., Kotani, T., and Fuchida, T. (2012) A study of color rendering properties based on color preference in adaptation to LED lighting, in *CIE2012 Lighting Quality & Energy Efficiency, Hangzhou, China*, CIE, Vienna, pp. 369–374.
40 Houser, K.W., Tiller, D.K., and Hu, X. (2005) Tuning the fluorescent spectrum for the triChromatic visual response: a pilot study. *Leukos*, **1**, 7–23.
41 Smet, K.A.G., Ryckaert, W.R., Pointer, M.R., Deconinck, G., and Hanselaer, P. (2012) *Optimization of color quality of LED lighting with reference to memory colors. Light. Res. Technol.*, **44**, 7–15.
42 Ohno, Y. and Davis, W. (2010) *Visual Evaluation Experiment on Chroma Enhancement Effects in Color Rendering of Light Sources, Report submitted to TC1-69*, National Institute of Standards and Technology.
43 Wei, M., Houser, K.W., Allen, G.R., and Beers, W.W. (2014) *Color preference under LEDs with diminished yellow emission. Leukos*, **10**, 119–131.
44 Imai, Y., Kotani, T., and Fuchida, T. (2013) A study of color rendering properties based on color preference of objects in adaptation to LED lighting, in *CIE Centenary Conference "Towards a New Century of Light", Paris, France*, CIE, Vienna, pp. 62–67.
45 Dangol, R., Islam, M.S., Hyvärinen, M., Bhusal, P., Puolakka, M., and Halonen, L. (2013) Subjective preferences and color quality metrics of LED light sources. *Light. Res. Technol.*, **45** (6), 666–688.
46 Islam, M.S., Dangol, R., Hyvärinen, M., Bhusal, P., Puolakka, M., and Halonen, L. (2013) *User preferences for LED lighting in terms of light spectrum. Light. Res. Technol.*, **45** (6), 641–665.
47 Smet, K.A.G., Ryckaert, W.R., Pointer, M.R., Deconinck, G., and Hanselaer, P. (2011) *Correlation between color quality metric predictions and visual appreciation of light sources. Opt. Express*, **19** (9), 8151.

48 Smet, K.A.G. and Hanselaer, P. (2016) *Memory and preferred colors and the color rendition of white light sources. Light. Res. Technol.*, **48** (4), 393–411.
49 Rea, M.S. and Freyssinier, J.P. (2010) *Color rendering: beyond pride and prejudice. Color Res. Appl.*, **35** (6), 401–409.
50 David, A., Fini, P.T., Houser, K.W., Ohno, Y., Royer, M.P., Smet, K.A.G., Wei, M., and Whitehead, L. (2015) *Development of the IES method for evaluating the color rendition of light sources. Opt. Express*, **23** (12), 15888.

6

Correlations of Color Quality Metrics and a Two-Metrics Analysis

6.1 Introduction: Research Questions

As mentioned in Chapter 1, the task of color quality research is to develop color rendition metrics that correlate well with the visual assessments of the users of lighting systems and the designers of light sources and luminaires regarding the criteria *"color preference,"* *"color naturalness,"* and *"vividness."* The chain of color quality evaluation and objective metric establishment can be seen in Figure 6.1.

In order to optimize the spectrum of the light sources, developers should know the lighting context, that is, the character of the application, for example, museum, shop, theater, or office with specific working conditions, the required lighting quality as well as human and social requirements in the illuminated areas that contain colored objects, for example, textiles, fruits, or art objects. Based on the input information, an appropriate color quality metric and its formulae can be applied so that the simulated spectra of the light source fulfills the color quality criteria on a semantic scale being "good" or "very good." Therefore, objective color quality metrics must be developed based on the modern knowledge of color science and verified by comprehensive visual color quality experiments.

In the history of color science, several color quality metrics have been developed that can be divided into three groups:

- Color fidelity/color realness
- Color gamut/color saturation/vividness
- Color preference/naturalness.

In the last decades, some color fidelity metrics have been defined in different color spaces with various test color objects (TCSs). In the last few years, new definitions of color gamut metrics have also been developed in order to compare the area or volume the colored objects cover in a chromaticity diagram or in color space when they are illuminated by the reference and test light sources at the same correlated color temperature. The terminology *"color gamut"* refers to a collection of defined test color samples and must have a relation with the requirements for the saturation enhancement of colored objects under a certain illumination condition. The color preference and naturalness attributes describe the positive acceptance of the light source users immersed into a specific illuminated scene

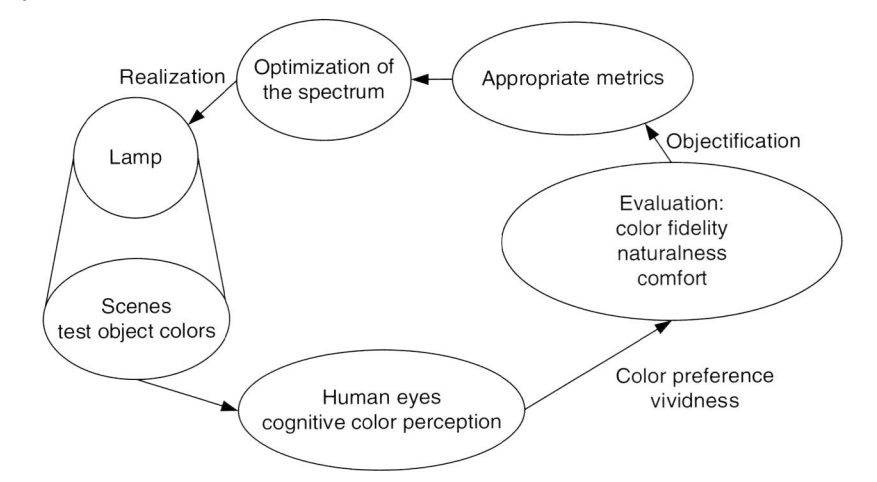

Figure 6.1 Chain of color quality evaluation and objective metric establishment.

during a short-term or a long-term observation period of the colored objects. These attributes can be analyzed in the context of the cultural background, life experience, and the age of the users or in the context of the type of the observed scene. Color quality metrics of practical importance in current international discussions are listed in Table 6.1.

Table 6.1 Frequently used and internationally discussed color quality metrics.

Color quality metrics	Symbol and year of definition	Test color samples (TCS)	Color space
Color fidelity	CIE $R_a(1-8)$, 1965, 1995	8 desaturated surface colors	$U^*V^*W^*$
	CIE $R_1 - R_{14}$, 1965, 1995	8 desaturated and 6 saturated surface colors	$U^*V^*W^*$
	CQS Q_f (CQS), 2010	15 saturated surface colors	$L^*a^*b^*$, 1976
	CIE CRI2012 (nCRI)	17 mathematically defined colors	CAM02-UCS
	R_f (IES, 2015)	99 surface colors	CAM02-UCS
Color gamut	Q_g (CQS), 2010	15 saturated surface colors	$L^*a^*b^*$, 1976
	FCI	4 saturated surface colors	CAM02-UCS
	GAI	8 desaturated surface colors	$U^*V^*W^*$
	R_g (IES, 2015)	99 surface colors	CAM02-UCS
Color preference	MCRI, 2010	9 memory surface colors	IPT-color space
	Q_a (CQS), 2010	15 saturated surface colors	$L^*a^*b^*$, 1976
	Q_p (CQS), 2010	15 saturated surface colors	$L^*a^*b^*$, 1976

From the scientific viewpoint, there are numerous color metrics that have some potential for researchers to compare and correlate them numerically or with results of visual experiments under different light sources at various brightness levels. However, regarding the need of lighting engineers and designers, the number of the color metrics should be united into one single parameter describing the color quality attributes color fidelity, color gamut/saturation, and color preference so that the comparison, definition, and evaluation of the lighting quality in lighting practice can be implemented conveniently. In order to reduce the number of metrics, the following questions shall be answered by means of a mathematical analysis.

- *How do the color metrics in the same group (Table 6.1, first column) correlate with each other?* Color metrics with a good correlation should be considered as the best representations of the group and the others should be analyzed to find out the reason for the bad correlation.
- *How is the correlation of the color metrics of this group with the ones from other groups?* The degree of similarity or dissimilarity of the color metrics of the three groups mentioned above, which will be described mathematically by the coefficient of determination (R^2) in this chapter, will expose the dimension of the underlying components. This important information will foster the establishment of more complex evaluations such as *"color preference"* or *"naturalness."*

In Section 6.2, more than 900 semiconductor and conventional light sources will undergo such a mathematical analysis to find out potential correlations. Then, in Section 6.3, the state-of-the-art color quality metrics of color preference (MCRI and CQS Q_p) will be constructed mathematically as the linear or quadratic combinations of color fidelity and color gamut metrics. Finally, an outlook and a list of research aspects will be formulated in Section 6.4 for further visual studies.

6.2 Correlation of Color Quality Metrics

This section is divided into two subsections in order to investigate the potential differences of the correlations in case of warm white (CCT < 3500 K) and cold white light sources (CCT > 5000 K).

The dataset of 618 relative spectral power distributions with 1 nm spectral resolution between 380 and 780 nm was prepared for this computational analysis. This set includes both real light sources (RGB LEDs, tungsten halogen lamps, neodymium lamps, hybrid multi-LED lamps, and phosphor-converted LEDs with one or more phosphor components) and standard illuminants such as CIE illuminants A, D, E, and F. Then, it is divided into two groups including 346 warm white light sources (CCT = 2000–3672 K) and 272 cool white light sources (CCT = 4500–10 000 K). The CCT in the range between 3673 and 4499 K was not investigated here. In Table 6.2, the parameters and group names of color quality metrics are shown in the first and second columns, respectively, followed by their minimum, mean, and maximum values in the evaluation of the warm white and cold white light sources and for the entire set.

Table 6.2 Parameters and group names of the color quality metrics with their min., max., and mean values in the evaluation of 618 different light source spectral power distributions.

Parameter of color metrics	Group of color metrics	Warm white (346 light sources)			Cool white (272 light sources)			All (618)
		Min.	Mean	Max.	Min.	Mean	Max.	Mean
CCT (K)	Colorimetric	2000	2966	3672	4500	5949	10000	4168
Δuv	Colorimetric	0	1.9E−3	1.5E−2	0	3.3E−3	3.5E−2	2.5E−3
CRI R_a [1]	Fidelity	73.2	91.2	100.0	70.2	87.1	100.0	87.4
CRI $R_{1,14}$ [1]	Fidelity	63.2	87.7	100.0	58.9	82.5	100.0	82.8
GAI [2]	Gamut	80.2	97.0	111.6	76.0	93.5	122.0	94.9
CQS Q_a [3]	Preference	14.1	86.3	100.0	22.2	85.3	100.0	84.4
CQS Q_f [3]	Fidelity	15.7	85.1	100.0	23.3	84.3	100.0	83.1
CQS Q_g [3]	Gamut	85.0	100.9	111.4	84.9	98.1	116.7	99.7
CQS Q_p [3]	Preference	52.8	90.9	101.2	57.0	89.9	100.8	89.6
$R_{a,2012}$ [4]	Fidelity	61.3	90.1	100.0	65.6	88.6	100.0	87.7
MCRI [5]	Preference	74.8	88.4	93.4	70.8	86.4	94.9	86.9
FCI [6]	Gamut	96.0	121.5	148.3	70.1	98.9	140.9	111.8
IES R_f [7]	Fidelity	69.0	87.8	100.0	66.5	85.6	100.0	85.2
IES R_g [7]	Gamut	79.4	100.6	118.2	78.8	97.3	111.7	99.4

In Table 6.2, four groups of color quality descriptors, colorimetric, color fidelity, color preference, and color gamut, are listed [1]. In the case of MCRI, the value of the general memory color-rendering index (R_m) is computed by using the "*degree of adaptation*" of 0.9. To compute the values of the IES color fidelity measure R_f [7] and color gamut measure R_g [7], the following viewing condition parameters were used: $L_A = 60$ cd/m², $Y_b = 20$, $F = 1.0$, $c = 0.69$, and $N_c = 1.0$. The value of D was computed from F and L_A according to the CIECAM02 equation.

6.2.1 Correlation of Color Metrics for the Warm White Light Sources

In this section, 346 light sources (231 LED spectra, 95 spectra of fluorescent lamps, and 20 thermal radiators) were investigated. In Figures 6.2–6.4, the relative spectral power distributions of these three light source groups are illustrated.

It can be seen from Figure 6.2 that the warm white LEDs have a blue LED chip with the peak wavelength in the range of 435–470 nm and luminescent materials (converting phosphors) with the peak wavelength of their spectral emissions in the range between 600 and 670 nm. The spectral range of the green–yellow luminescent materials is 510–560 nm and that of red luminescent materials is between 620 and 660 nm. The general color-rendering index R_a of these white LEDs is always higher than 70.

In Table 6.3, the coefficients of determination R^2 for the various color quality metrics are listed for the investigated LED types. The color quality metrics are calculated by using the LED spectra (Figure 6.2) and the own test color samples of the metrics.

6.2 Correlation of Color Quality Metrics | 179

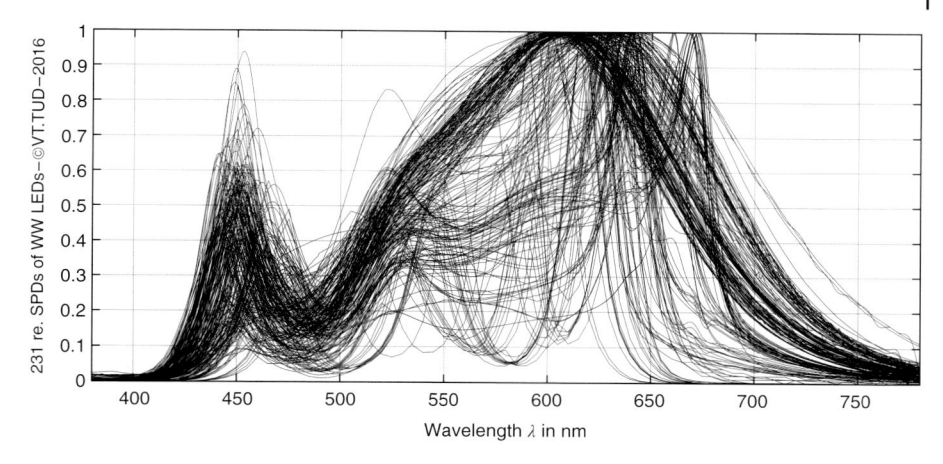

Figure 6.2 Spectral characteristics of the 231 investigated LED light sources in the warm white range.

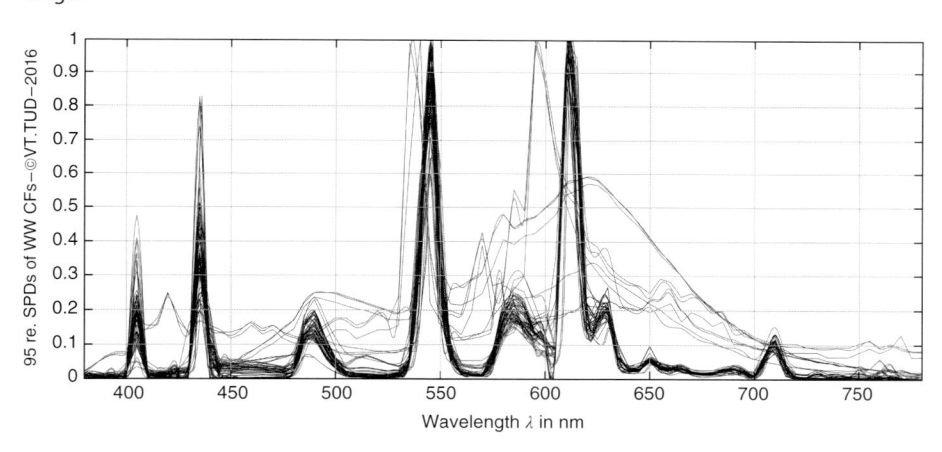

Figure 6.3 Spectral characteristics of the 95 fluorescent lamps in the warm white range.

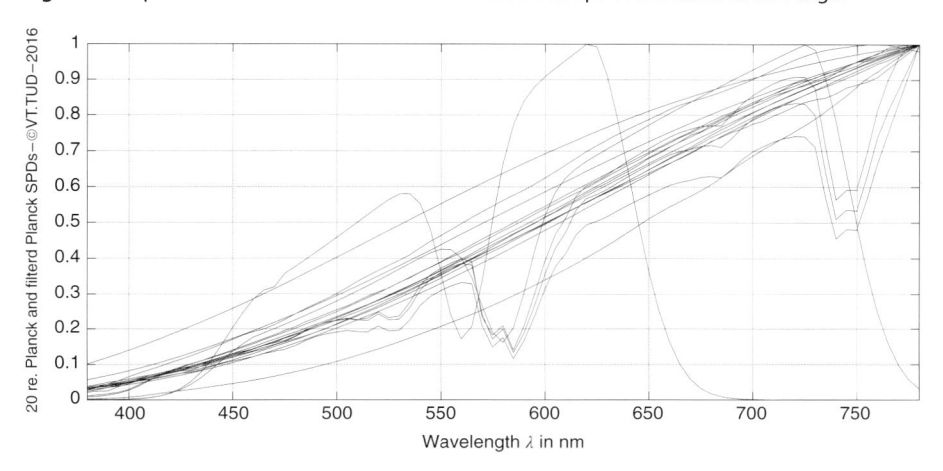

Figure 6.4 Spectral characteristics of the 20 thermal radiators (pure and filtered) in the warm white range.

Table 6.3 Correlation of the color quality metrics for the 231 warm white LEDs (Figure 6.2).

R^2	R_a	$R_{1,14}$	$GAI_{abs.}$	$GAI_{rel.}$	Q_a	Q_f	Q_g	Q_p	CRI_{2012}	MCRI	FCI	R_f	R_g
R_a	1.00	0.97	0.01	0.14	0.29	0.32	0.03	0.26	0.54	0.39	0.09	0.78	0.01
$R_{1,14}$	0.97	1.00	0.02	0.21	0.30	0.33	0.06	0.30	0.58	0.48	0.15	0.81	0.02
$GAI_{abs.}$	0.01	0.02	1.00	0.11	0.01	0.02	0.35	0.01	0.01	0.32	0.21	0.01	0.30
$GAI_{rel.}$	0.14	0.21	0.11	1.00	0.11	0.07	0.54	0.34	0.10	0.61	0.79	0.22	0.26
Q_a	0.29	0.30	0.01	0.11	1.00	0.98	0.03	0.88	0.30	0.15	0.06	0.42	0.01
Q_f	0.32	0.33	0.02	0.07	0.98	1.00	0.00	0.79	0.30	0.12	0.02	0.42	0.00
Q_g	0.03	0.06	0.35	0.54	0.03	0.00	1.00	0.23	0.10	0.46	0.81	0.11	0.88
Q_p	0.26	0.30	0.01	0.34	0.88	0.79	0.23	1.00	0.33	0.33	0.30	0.45	0.15
CRI_{2012}	0.54	0.58	0.01	0.10	0.30	0.30	0.10	0.33	1.00	0.30	0.11	0.82	0.13
MCRI	0.39	0.48	0.32	0.61	0.15	0.12	0.46	0.33	0.30	1.00	0.65	0.49	0.24
FCI	0.09	0.15	0.21	0.79	0.06	0.02	0.81	0.30	0.11	0.65	1.00	0.18	0.51
R_f	0.78	0.81	0.01	0.22	0.42	0.42	0.11	0.45	0.82	0.49	0.18	1.00	0.08
R_g	0.01	0.02	0.30	0.26	0.01	0.00	0.88	0.15	0.13	0.24	0.51	0.08	1.00

The following can be seen from Table 6.3:

- R_a and $R_{1,14}$ correlate with each other at the coefficient of determination level "*good*" and with the color-rendering index R_f from IES TM-30 at the "*relatively good*" level; see also Figure 6.5. R_a and $R_{1,14}$ do not correlate well with the color-rendering index CRI_{2012}, with the color quality scale (CQS) based color fidelity indices Q_f and with the color preference metrics MCRI and Q_p. The color fidelity index R_f does not correlate well with the CQS-based color quality metrics.
- The two color preference metrics MCRI and CQS Q_p do not have an acceptable correlation. It can be explained by the difference in the test color samples (9 desaturated colors for MCRI and 15 saturated CQS colors for Q_p) and by the difference between the two methods to calculate the values of MCRI and Q_p, respectively.
- The color gamut metric feeling of contrast index (FCI) correlates well with other gamut parameters Q_g and gamut area index (GAI). But it does not have a good correlation with R_g.

The spectra of the 95 fluorescent lamp types (see Figure 6.3) contain numerous peak wavelengths at 435.8, 546.1 nm, and at about 620 nm. They exhibit low spectral intensities in the ranges 450–480, 500–530, 560–570, and longer than 630 nm. The coefficients of determination of the color metrics for these lamp types are shown in Table 6.4.

Generally, a similar general tendency can be seen from Tables 6.3 and 6.4. For the fluorescent lamps, the correlation between MCRI and all color fidelity indices (R_a, $R_{1,14}$, CIE CRI_{2012}, R_f) is significantly higher than the one in case of the warm white LEDs. On the other hand, the correlation between MCRI and the color gamut indices such as FCI, GAI, and Q_g in the case of the fluorescent lamps is worse than that in the case of the LED spectra.

Table 6.4 Correlation of color quality metrics for the set of 95 warm white fluorescent lamps (Figure 6.3).

R^2	R_a	$R_{1,14}$	$GAI_{abs.}$	$GAI_{rel.}$	Q_a	Q_f	Q_g	Q_p	CRI_{2012}	MCRI	FCI	R_f	R_g
R_a	1.00	0.89	0.08	0.22	0.29	0.33	0.01	0.26	0.62	0.60	0.05	0.76	0.00
$R_{1,14}$	0.89	1.00	0.13	0.19	0.30	0.33	0.01	0.27	0.81	0.69	0.07	0.84	0.01
$GAI_{abs.}$	0.08	0.13	1.00	0.00	0.51	0.50	0.22	0.53	0.19	0.44	0.00	0.16	0.16
$GAI_{rel.}$	0.22	0.19	0.00	1.00	0.01	0.02	0.04	0.00	0.03	0.21	0.65	0.09	0.00
Q_a	0.29	0.30	0.51	0.01	1.00	0.99	0.08	0.98	0.41	0.38	0.03	0.39	0.13
Q_f	0.33	0.33	0.50	0.02	0.99	1.00	0.05	0.96	0.45	0.40	0.04	0.44	0.10
Q_g	0.01	0.01	0.22	0.04	0.08	0.05	1.00	0.13	0.04	0.01	0.19	0.05	0.86
Q_p	0.26	0.27	0.53	0.00	0.98	0.96	0.13	1.00	0.36	0.35	0.01	0.34	0.18
CRI_{2012}	0.62	0.81	0.19	0.03	0.41	0.45	0.04	0.36	1.00	0.56	0.00	0.89	0.01
MCRI	0.60	0.69	0.44	0.21	0.38	0.40	0.01	0.35	0.56	1.00	0.06	0.70	0.00
FCI	0.05	0.07	0.00	0.65	0.03	0.04	0.19	0.01	0.00	0.06	1.00	0.00	0.05
R_f	0.76	0.84	0.16	0.09	0.39	0.44	0.05	0.34	0.89	0.70	0.00	1.00	0.03
R_g	0.00	0.01	0.16	0.00	0.13	0.10	0.86	0.18	0.01	0.00	0.05	0.03	1.00

Table 6.5 Correlation of the color quality metrics for the 20 warm white thermal radiators (Figure 6.4).

R^2	R_a	$R_{1,14}$	$GAI_{abs.}$	$GAI_{rel.}$	Q_a	Q_f	Q_g	Q_p	CRI_{2012}	MCRI	FCI	R_f	R_g
R_a	1.00	1.00	0.41	0.92	0.97	0.97	0.54	0.71	0.75	0.30	0.61	0.91	0.34
$R_{1,14}$	1.00	1.00	0.41	0.91	0.97	0.97	0.52	0.73	0.77	0.28	0.59	0.93	0.32
$GAI_{abs.}$	0.41	0.41	1.00	0.43	0.41	0.44	0.49	0.16	0.17	0.69	0.34	0.26	0.42
$GAI_{rel.}$	0.92	0.91	0.43	1.00	0.85	0.88	0.73	0.45	0.51	0.48	0.82	0.73	0.53
Q_a	0.97	0.97	0.41	0.85	1.00	0.99	0.54	0.77	0.73	0.28	0.58	0.88	0.35
Q_f	0.97	0.97	0.44	0.88	0.99	1.00	0.61	0.71	0.67	0.34	0.65	0.85	0.42
Q_g	0.54	0.52	0.49	0.73	0.54	0.61	1.00	0.10	0.09	0.81	0.95	0.26	0.96
Q_p	0.71	0.73	0.16	0.45	0.77	0.71	0.10	1.00	0.92	0.01	0.14	0.89	0.02
CRI_{2012}	0.75	0.77	0.17	0.51	0.73	0.67	0.09	0.92	1.00	0.01	0.14	0.95	0.01
MCRI	0.30	0.28	0.69	0.48	0.28	0.34	0.81	0.01	0.01	1.00	0.68	0.09	0.84
FCI	0.61	0.59	0.34	0.82	0.58	0.65	0.95	0.14	0.14	0.68	1.00	0.33	0.87
R_f	0.91	0.93	0.26	0.73	0.88	0.85	0.26	0.89	0.95	0.09	0.33	1.00	0.11
R_g	0.34	0.32	0.42	0.53	0.35	0.42	0.96	0.02	0.01	0.84	0.87	0.11	1.00

In Figure 6.4, the spectra of the 20 thermal radiators are mostly continuous from 380 to 780 nm with the exception of the two filtered tungsten lamp types with a maximum in the green range, at 530 or 555 nm, and in the red range between 620 and 725 nm. Regarding the list of coefficients of determination of the color quality metrics for the 20 thermal radiators in Table 6.5, several differences to Tables 6.3 and 6.4 can be observed.

Table 6.6 Coefficients of determination (R^2) of some selected pairs of metrics for warm white light sources.

Number	Combination	LEDs	Fluorescent lamps	Thermal radiator
1	$R_g - Q_g$	0.88	0.86	0.96
2	$R_f - \text{CRI}_{2012}$	0.82	0.89	0.95
3	$R_{1,14} - R_f$	0.81	0.84	0.93
4	$\text{MCRI} - R_f$	0.49	0.7	0.09
5	$\text{MCRI} - R_g$	0.24	0.00	0.84
6	$\text{FCI} - R_g$	0.51	0.05	0.87
7	$\text{FCI} - Q_g$	0.81	0.19	0.95
8	$\text{FCI} - R_a$	0.09	0.05	0.61
9	$\text{GAI} - R_a$	0.14	0.22	0.92
10	$Q_p - R_f$	0.45	0.34	0.89

As can be seen from Table 6.5, the color gamut metric R_g has a very good correlation with the metrics MCRI and FCI. But this is not true in the case of the LEDs and the fluorescent lamps. In Table 6.6, the coefficients of determination of some selected pairs of metrics are listed.

The following can be seen from Table 6.6:

1) The color fidelity metric R_f (IES 2015 [7]) correlates well with the conventional color fidelity indices $R_{1,14}$ (for 14 CIE TCSs) and CIE CRI_{2012} (for 17 own TCSs).
2) The coefficients of determination between the color preference metric MCRI and the two metrics R_g and R_f, between FCI and Q_g, R_a, or R_g, between GAI and R_a, between Q_p and R_f are highly dependent on the type of light source (white LED, fluorescent lamp, or thermal radiator).

In [2], Rea et al. proposed a new metric for color preference as a combination of R_a (color fidelity) and GAI (color gamut); see Eq. (6.1).

$$\text{CP} = 0.5 \ (R_a + \text{GAI}) \tag{6.1}$$

It can be seen from Table 6.6 that R_a and GAI have a poor correlation with each other in case of warm white LEDs and fluorescent lamps. It means that R_a and GAI can be independently varied by using suitable light source spectra. For thermal radiators, a very good correlation, or in other words, a linear relationship between GAI and R_a should allow a consequence that CP (color preference according to Eq. (6.1)) is only dependent on either GAI or R_a.

Considering all warm white LEDs, fluorescent lamps, and thermal radiators as a single group of 346 light source spectra, the relationships between R_a and R_f, $R_{1,14}$ and R_f, and CIE CRI_{2012} and R_f are illustrated in Figures 6.5–6.7. The correlation between CRI_{2012} and R_f is very good because of the same color space (CAM02-UCS) and the selection of test color samples that are uniformly distributed around the hue circle in the $a'-b'$-plane. Note that the definitions of R_a or $R_{1,14}$ are based on the collection of only 8 or 14 test color samples whose color appearance is evaluated in the visually nonuniform color space $U^*V^*W^*$. The

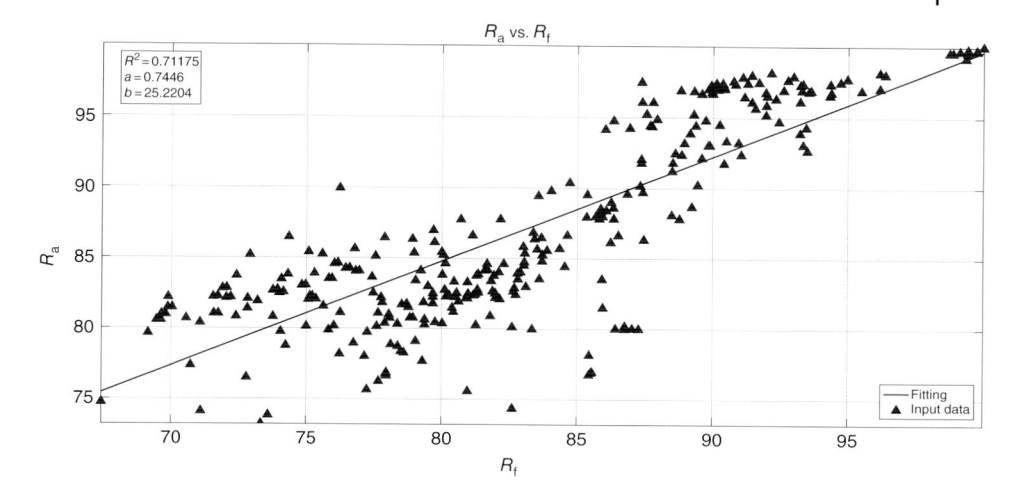

Figure 6.5 Correlation between the two color fidelity indices R_a and R_f for 346 warm white light sources.

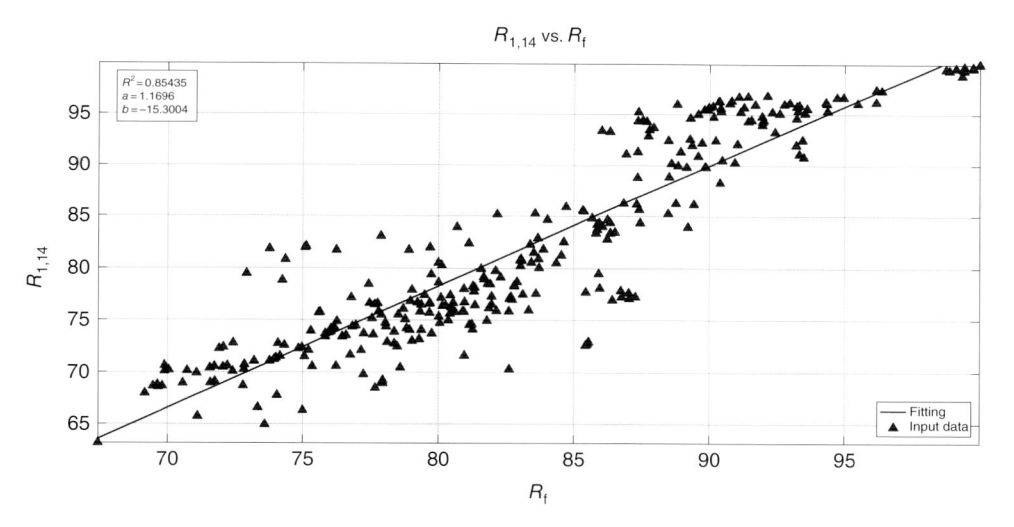

Figure 6.6 Correlation between the two color fidelity indices $R_{1,14}$ and R_f for 346 warm white light sources.

very good correlation between the Q_g metric for 15 saturated colors [3] and R_g for 99 IES colors [7] can be seen from Figure 6.8.

In Figure 6.9, the poor correlation between R_f (color fidelity index) and R_g (color gamut) with $R^2 = 0.0014$ indicates the independence of these two metrics for this set of warm white light sources. In other words, one can formulate that the color fidelity value R_f can be kept constant while the color gamut of the light sources can be varied over a broad range by the conscious design of suitable lamp spectra.

In Table 6.7, the coefficients of determination of all 346 warm white light sources are listed. It can be seen that MCRI does not have a sufficient correlation with the group of color fidelity indices (R_a, $R_{1,14}$, Q_f, Q_a) and with the color gamut

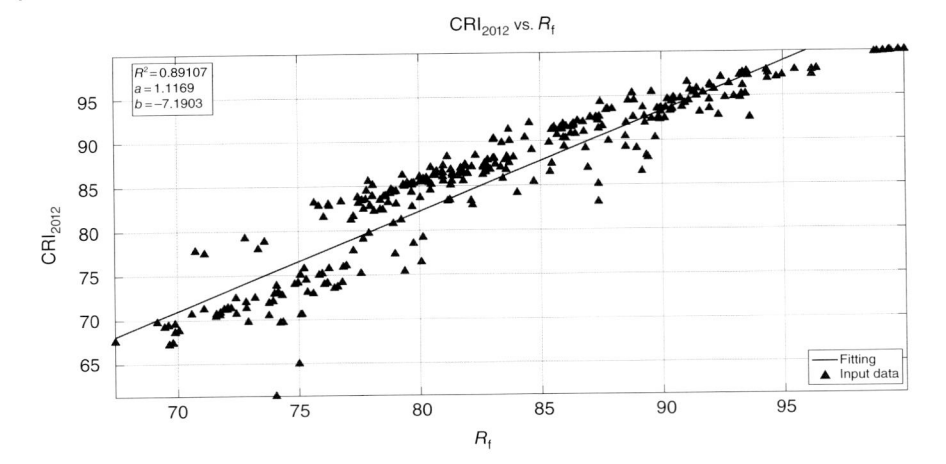

Figure 6.7 Correlation between the two color fidelity indices CRI_{2012} and R_f for 346 warm white light sources.

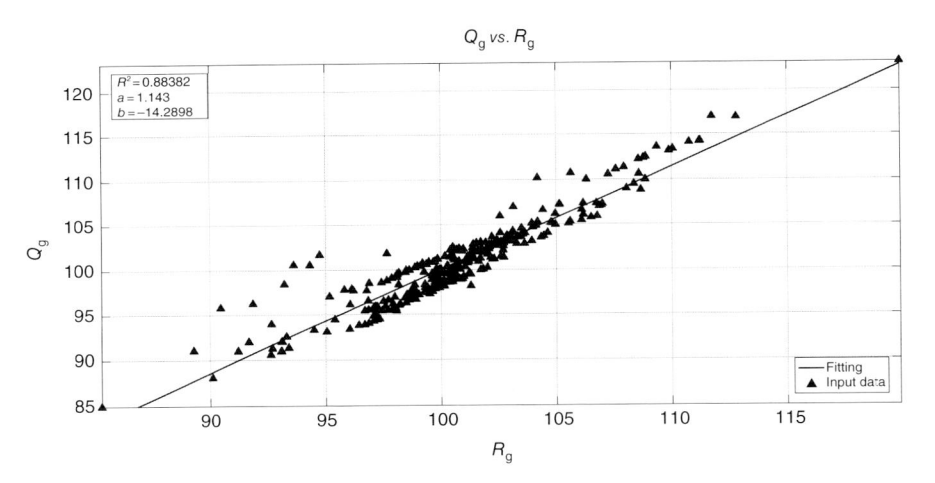

Figure 6.8 Correlation between the two color gamut indices R_g and Q_g for 346 warm white light sources.

indices (Q_g, GAI, FCI, and R_g). The poor correlation in this context confirms that a variation of the metrics for color fidelity and color gamut mentioned above does not change the MCRI values in a substantial manner. However, unfortunately, the color preference metric CQS Q_p also does not correlate well with the color fidelity indices (R_a, $R_{1,14}$, CIE CRI_{2012}, R_f) and with the color gamut metrics (GAI, FCI, R_g, Q_g) in the case of warm white light sources.

6.2.2 Correlation of Color Quality Metrics for Cold White Light Sources

In this section, 272 cold white spectra (130 LEDs, 113 fluorescent lamps, and 29 daylight spectra) are considered in the computational analysis. The spectra of

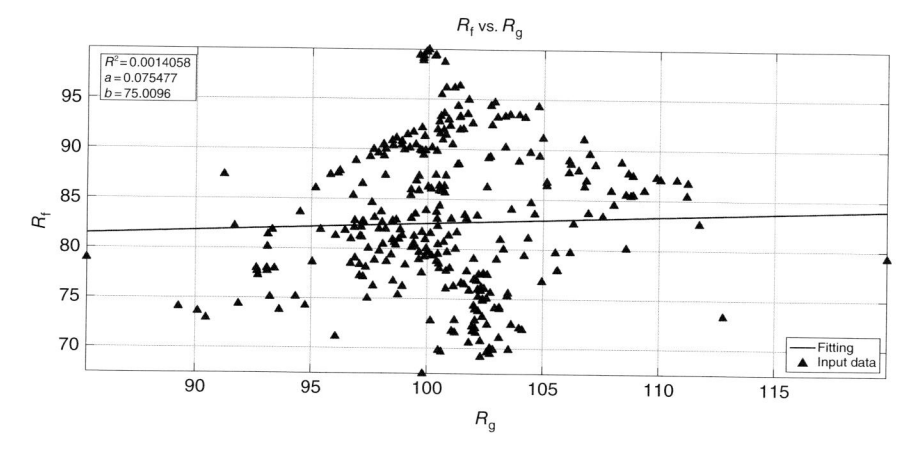

Figure 6.9 Relationship between the color gamut index R_g and the color fidelity index R_f for 346 warm white light sources.

Table 6.7 Correlation of color metrics for all 346 warm white light sources.

R^2	R_a	$R_{1,14}$	GAI$_{abs.}$	GAI$_{rel.}$	Q_a	Q_f	Q_g	Q_p	CRI$_{2012}$	MCRI	FCI	R_f	R_g
R_a	1.00	0.93	0.00	0.11	0.38	0.42	0.00	0.33	0.50	0.33	0.09	0.71	0.00
$R_{1,14}$	0.93	1.00	0.02	0.16	0.43	0.47	0.00	0.39	0.70	0.54	0.16	0.85	0.00
GAI$_{abs.}$	0.00	0.02	1.00	0.07	0.01	0.01	0.25	0.05	0.03	0.23	0.13	0.02	0.21
GAI$_{rel.}$	0.11	0.16	0.07	1.00	0.09	0.05	0.45	0.27	0.09	0.31	0.76	0.16	0.21
Q_a	0.38	0.43	0.01	0.09	1.00	0.98	0.00	0.90	0.43	0.29	0.07	0.52	0.00
Q_f	0.42	0.47	0.01	0.05	0.98	1.00	0.00	0.83	0.45	0.27	0.04	0.55	0.00
Q_g	0.00	0.00	0.25	0.45	0.00	0.00	1.00	0.12	0.00	0.07	0.60	0.00	0.88
Q_p	0.33	0.39	0.05	0.27	0.90	0.83	0.12	1.00	0.40	0.36	0.26	0.50	0.08
CRI$_{2012}$	0.50	0.70	0.03	0.09	0.43	0.45	0.00	0.40	1.00	0.64	0.16	0.89	0.00
MCRI	0.33	0.54	0.23	0.31	0.29	0.27	0.07	0.36	0.64	1.00	0.43	0.64	0.02
FCI	0.09	0.16	0.13	0.76	0.07	0.04	0.60	0.26	0.16	0.43	1.00	0.18	0.36
R_f	0.71	0.85	0.02	0.16	0.52	0.55	0.00	0.50	0.89	0.64	0.18	1.00	0.00
R_g	0.00	0.00	0.21	0.21	0.00	0.00	0.88	0.08	0.00	0.02	0.36	0.00	1.00

the 130 LEDs are illustrated in Figure 6.10 where the peak wavelengths of the blue chips are in the range 435–470 nm, the ones of the phosphor emissions in the range around 520 and 560 nm and the ones of the red semiconductor LEDs around 600, 620, and 670 nm. The spectra of the 113 fluorescent lamp types with several line emissions and continuous contributions are shown in Figure 6.11. In Figure 6.12, the spectra of some daylight illuminants containing a strong continuous spectral distribution between 380 and 750 nm are also represented.

In addition, the coefficients of determination between the color quality metrics for all cold white light source spectra are listed in Table 6.8.

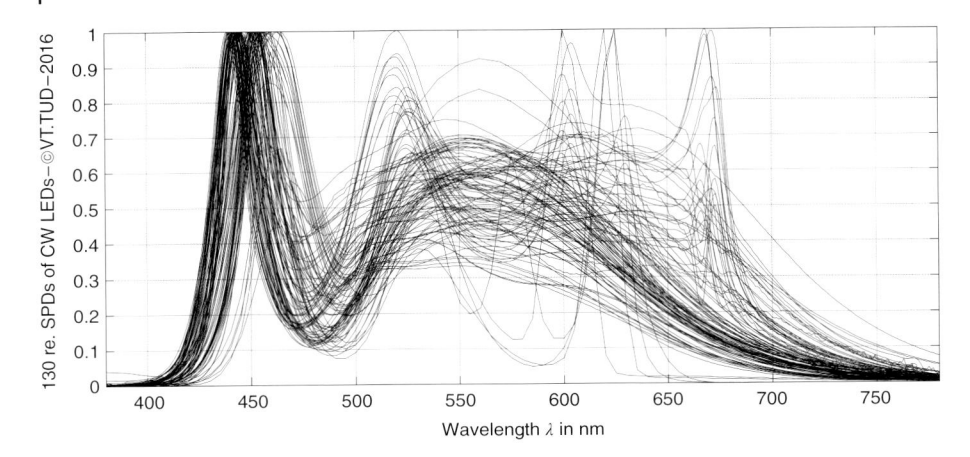

Figure 6.10 Spectral characteristics of the 130 LED light sources in the cold white range.

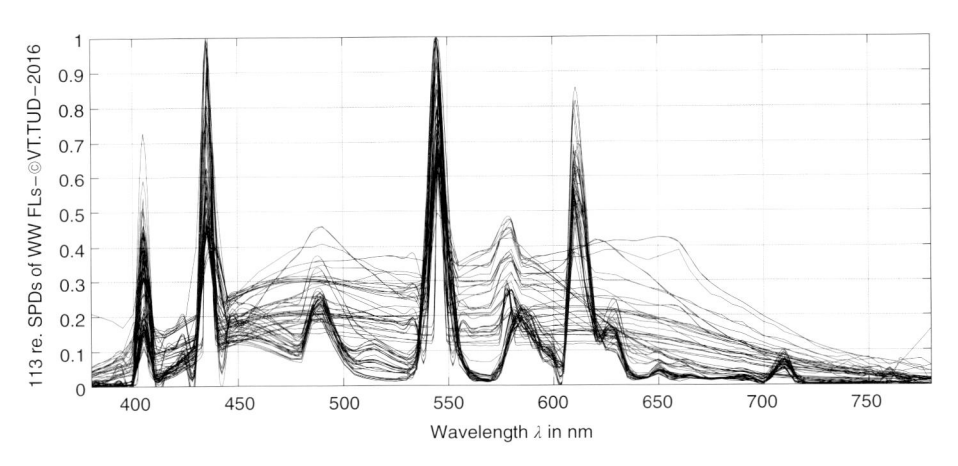

Figure 6.11 Spectral characteristics of the 113 fluorescent lamps in the cold white range.

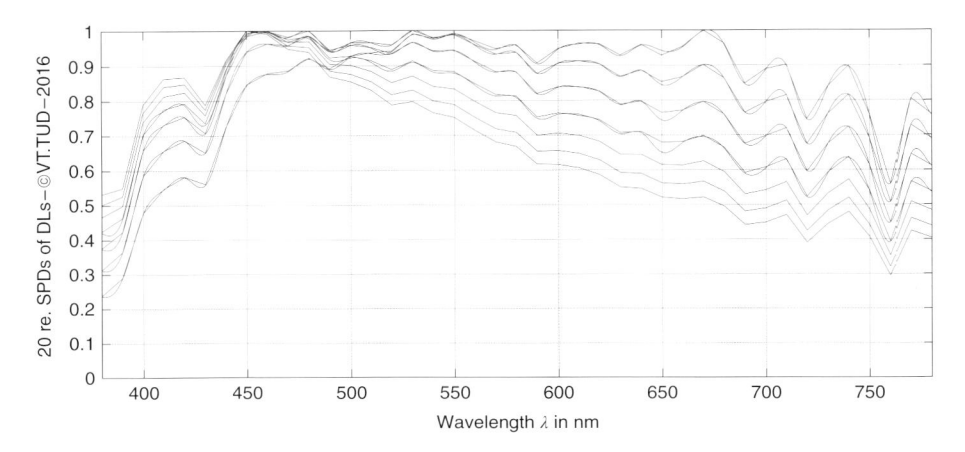

Figure 6.12 Spectral characteristics of the 29 daylight illuminants in the cold white range.

Table 6.8 Coefficients of determination (R^2) between the color quality metrics for the 272 cold white light sources.

R^2	R_a	$R_{1,14}$	GAI$_{abs.}$	GAI$_{rel.}$	Q_a	Q_f	Q_g	Q_p	CRI$_{2012}$	MCRI	FCI	R_f	R_g
R_a	1.00	0.99	0.31	0.62	0.80	0.82	0.41	0.77	0.89	0.77	0.45	0.92	0.36
$R_{1,14}$	0.99	1.00	0.35	0.66	0.80	0.82	0.44	0.79	0.92	0.79	0.47	0.93	0.40
GAI$_{abs.}$	0.31	0.35	1.00	0.52	0.33	0.27	0.58	0.51	0.30	0.41	0.23	0.28	0.66
GAI$_{rel.}$	0.62	0.66	0.52	1.00	0.47	0.43	0.84	0.65	0.54	0.78	0.75	0.57	0.61
Q_a	0.80	0.80	0.33	0.47	1.00	0.99	0.35	0.93	0.80	0.57	0.30	0.81	0.37
Q_f	0.82	0.82	0.27	0.43	0.99	1.00	0.27	0.87	0.82	0.55	0.26	0.83	0.29
Q_g	0.41	0.44	0.58	0.84	0.35	0.27	1.00	0.58	0.37	0.60	0.72	0.36	0.84
Q_p	0.77	0.79	0.51	0.65	0.93	0.87	0.58	1.00	0.78	0.65	0.44	0.76	0.61
CRI$_{2012}$	0.89	0.92	0.30	0.54	0.80	0.82	0.37	0.78	1.00	0.65	0.34	0.95	0.38
MCRI	0.77	0.79	0.41	0.78	0.57	0.55	0.60	0.65	0.65	1.00	0.75	0.62	0.51
FCI	0.45	0.47	0.23	0.75	0.30	0.26	0.72	0.44	0.34	0.75	1.00	0.32	0.51
R_f	0.92	0.93	0.28	0.57	0.81	0.83	0.36	0.76	0.95	0.62	0.32	1.00	0.33
R_g	0.36	0.40	0.66	0.61	0.37	0.29	0.84	0.61	0.38	0.51	0.51	0.33	1.00

Based on the comparison of the coefficients of determination of the warm white (see Table 6.7) and cold white spectra (see Table 6.8), the following conclusions can be drawn:

1) The conventional color-rendering indices R_a and $R_{1,14}$ correlate well with the CQS metrics Q_f, Q_a, and Q_p and also with MCRI (the memory color rendering index) for the cold white light sources. But the correlation is not good in the case of the warm white light sources.
2) The coefficients of determination between the color fidelity indices R_a, $R_{1,14}$, CIE CRI$_{2012}$, and R_f of the cold white light sources are much higher than those of the warm white light sources.
3) MCRI has a correlation with GAI and FCI that is substantially higher in the case of the cold white than in the case of the warm white light sources.
4) CQS Q_p as a color preference metric does not correlate with R_f and R_g in the case of the warm white light sources but this changes substantially in the case of the cold white light sources. Because neither the color spaces ($L^*a^*b^*$ for CQS and CAM 02-UCS for the IES metrics) nor the set of TCSs (15 CQS TCSs and 99 IES TCSs) have been changed; this can only be explained by the fact that either the referent light sources or/and the spectra of the test light sources constitute the reasons for this higher correlation in the case of the cold white light sources. Similar tendencies can also be seen in the case of the MCRI results (see Table 6.9).

The relationships between R_a and R_f, CIE CRI$_{2012}$ and R_f, MCRI and R_f, and MCRI and R_g for the cold white light sources are shown in Figures 6.13–6.16.

It can be seen from Figure 6.15 that the MCRI values remain for many spectra at the level of about 90 while the color fidelity values R_f change between 77

6 Correlations of Color Quality Metrics and a Two-Metrics Analysis

Table 6.9 Coefficients of determination between MCRI, Q_p, and the IES-metrics (TM-30).

Relationship	For warm white light sources	For cold white light sources
MCRI – R_f	0.64	0.62
MCRI – R_g	0.02	0.51
Q_p – R_f	0.50	0.76
Q_p – R_g	0.08	0.61

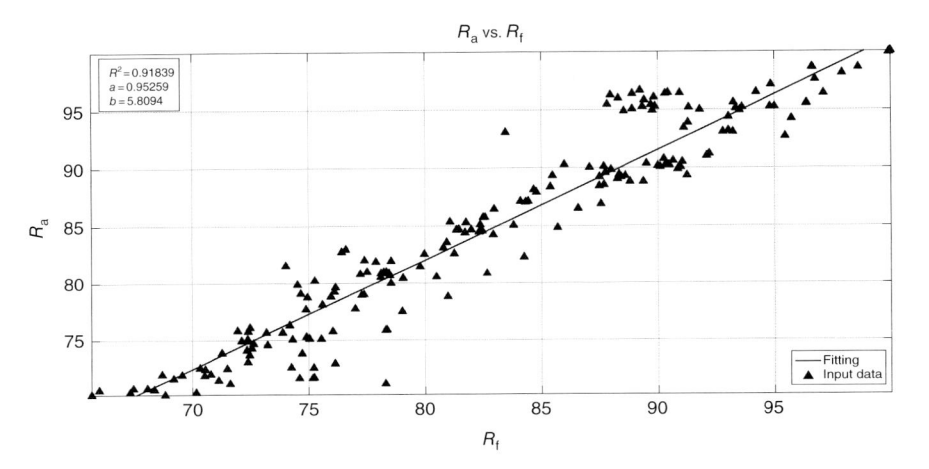

Figure 6.13 Relationship between R_a and R_f for the cold white light sources.

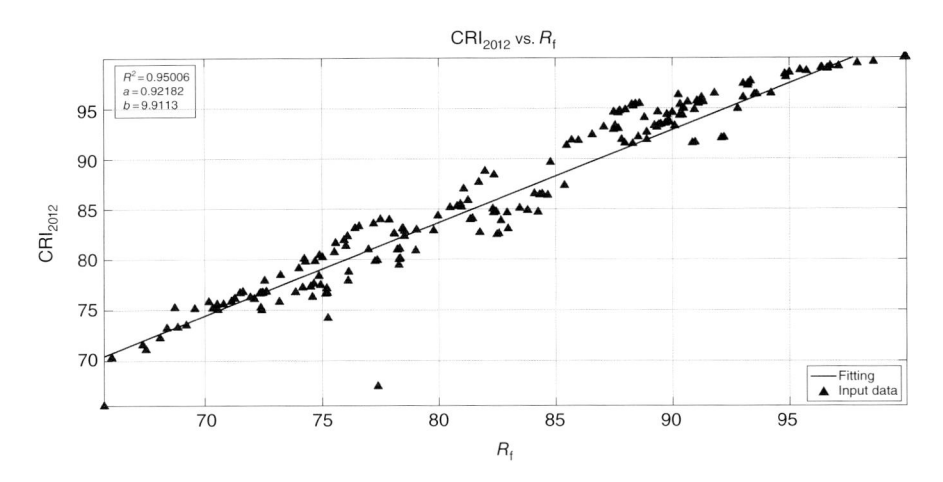

Figure 6.14 Relationship between CRI_{2012} and R_f for the cold white light sources.

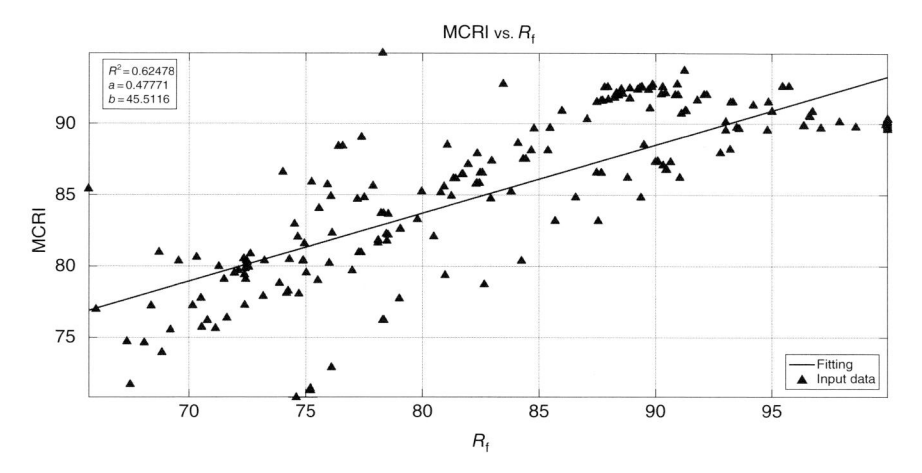

Figure 6.15 Relationship between MCRI and R_f for the cold white light sources.

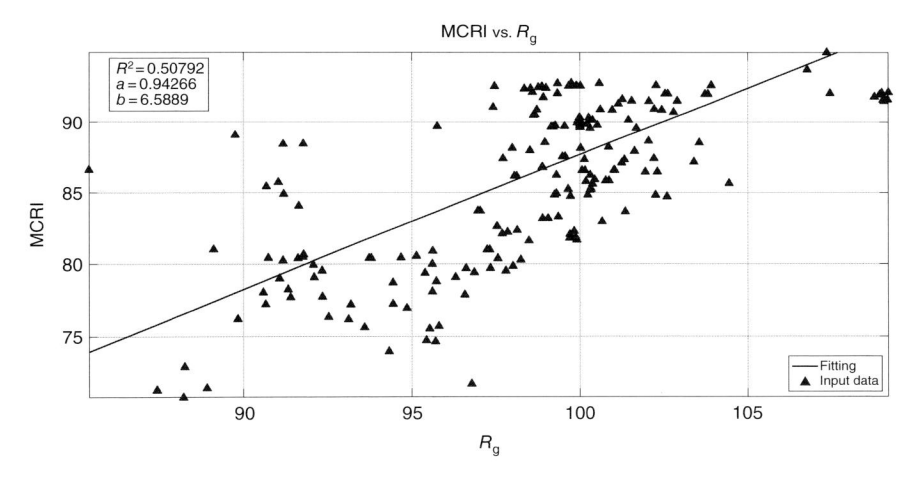

Figure 6.16 Relationship between MCRI and R_g for the cold white light sources.

and 100. In other words, the MCRI values are nearly constant around 90 while the designed light source spectra can achieve a variation in terms of R_f values between 77 and 100. In contrast, the other color preference metric CQS Q_p [3] yields a better correlation with R_f (see Figure 6.17). Specifically, if the value of R_f varies from 77 to 100 then the value of Q_p varies in the range between 82 and 100.

6.3 Color Preference and Naturalness Metrics as a Function of Two-Metrics Combinations

In order to explain the reasons for the necessity of two-metrics combinations, first some perceptual aspects of color quality [8] in practical lighting applications such

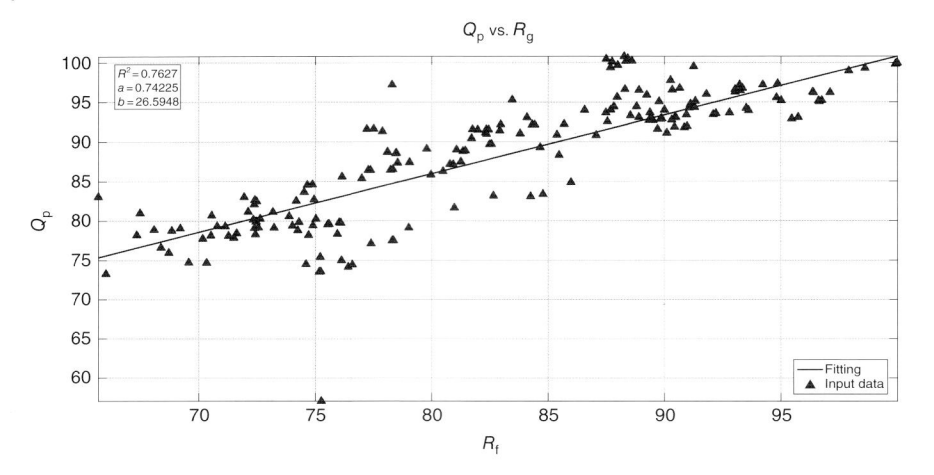

Figure 6.17 Relationship between Q_p and R_f for the cold white light sources.

as color fidelity, color preference, color vividness, or color naturalness should be considered. When the designer or the user of a general lighting system assesses the visual color quality of a lit scene, usually a color preference judgment or a color naturalness judgment is being made instead of the assessment of color fidelity, that is, the comparison with an inferred reference situation [2, 9–12]. For color fidelity assessments, the comparison between the test and reference light sources with the same correlated color temperature must be implemented either side by side, for similar colored objects, or subsequently for the same surface colors. In most lighting situations, this reference light source is not available. Therefore, to optimize the spectral power distribution of a light engine, a visually validated color preference metric or color naturalness metric is needed as a spectral optimization target with a set of boundary conditions, that is, type and quality of the illuminant's white tone, the type of the object colors (e.g., special fruits, foods, textiles, cosmetics), or the circadian stimulus level (higher for office works, lower for a more relaxing atmosphere; see Chapter 9) [13].

In this section, the color preference aspect including the memory color quality index (MCRI) [5] and the NIST CQS Q_p index [3] will be discussed in relation to two color quality indices. It was pointed out that both metrics correlate reasonably well with the observers' visual color preference judgments in the related experiments [3, 12]. It was also shown in visual experiments that neither the existing descriptors of color fidelity (color rendering indices) nor the color gamut metrics are able to describe the subjects' color preference or color naturalness judgments alone. In recent literature, it could be pointed out that the combinations of both index types (fidelity index, gamut index) can yield better visual results [2, 9–11].

The conception of the use of two-metrics combinations to describe color quality (see Figure 6.18) follows the idea that the color preference or naturalness judgments of lighting users are based on visual color experiences with daylight as light sources in the course of the evolutionary process. Under natural illumination conditions with continuous spectral contributions between 380 and

6.3 Color Preference and Naturalness Metrics as a Function of Two-Metrics Combinations

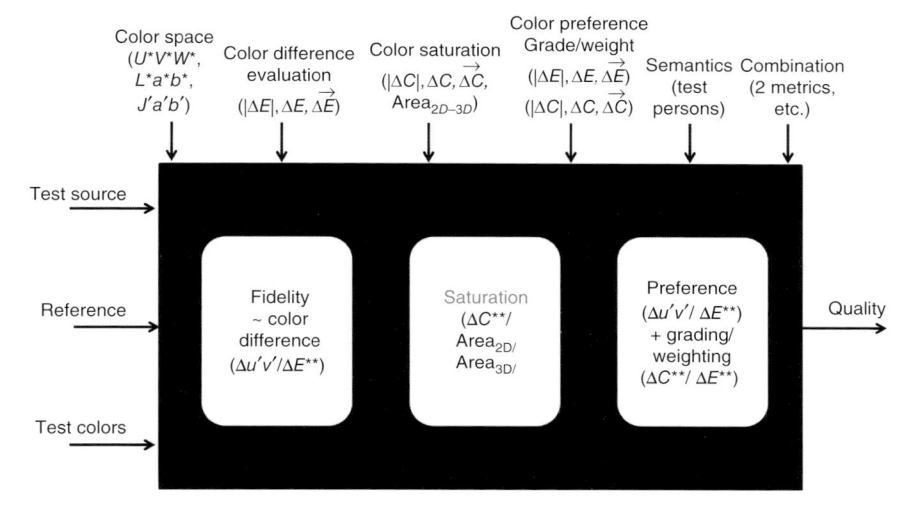

Figure 6.18 Conception for the establishment of two-metrics combinations for the description and modeling of color quality.

780 nm (see Figure 6.12), the illuminance on the colored objects is relatively high between about 30 000 and 100 000 lx so that the colored objects yield a colorful color appearance with high realness, high color contrast, and high visual acuity. For the same objects in real indoor lighting situations, in the range between about 200 and 1000 lx in the office, in the hospital, or in an industrial hall the color preference or naturalness evaluation is reduced although the light sources should have a very good color-rendering index (e.g., $R_a > 95$). In order to receive nearly similar color contrast and colorfulness, the saturation of the optical radiation falling on the color objects should be increased. The enhancement of color preference or color naturalness by a moderate enhancement of color saturation without a substantial loss of the color fidelity character for a collection of colored objects has been the subject of several research projects in the last few years. This is also one of the subjects of the present book.

In order to put this conception into lighting practice, two subsequent steps have been carried out:

First step: The present authors chose two color preference metrics, MCRI and Q_p (both of them have been validated experimentally in literature). These two metrics will be approximated by the combinations of two currently available metrics (a color fidelity metric and a color gamut metric). In this first step, the present authors assume that one or several color gamut metrics can simulate the color saturation enhancement for a certain number of colored objects under consideration. This step will be investigated in Section 6.3.

Second step: A number of visual experiments should be carried out with various spectral power distributions of different color temperatures, chromaticity, and object saturation enhancement levels of the light sources on diverse surface colors (e.g., cosmetics, foods, textiles, skin tones) with several illuminance levels (under 500 and around 1000 lx). The test subjects will judge the color preference, vividness, or naturalness of each illumination condition so that these evaluation results and the values of the color metrics for each lamp spectrum can be compared. Based on the analysis of the experimental series, appropriate two-metrics combinations of well-known metrics or new descriptors of color quality will be defined. This second step is described in Chapter 7 and it deserves further detailed studies in the future.

For a consequent analysis in the present section, three mathematical formulations will be used to approximate two state-of-the-art color preference metrics MCRI and Q_p; see Eqs. (6.2)–(6.4).

$$\text{Color Preference} = \alpha \text{ Fidelity} + (1 - \alpha) \text{ Saturation} \quad (6.2)$$

$$\text{Color Preference} = a \text{ Fidelity} + b \text{ Saturation} \quad (6.3)$$

$$\text{Color Preference} = a \text{ Saturation}^2 + b \text{ Saturation} + c \text{ Fidelity} + d \quad (6.4)$$

In Eqs. (6.2)–(6.4), it is assumed that the color fidelity metrics and color gamut descriptors (to describe saturation, that is, the saturating effect of the light source's spectrum on the colored objects) are nearly independent. In the constrained linear Eq. (6.2), the factors for color fidelity and color gamut α and $(1 - \alpha)$ must sum up to the constant value of 1.000. In the unconstrained linear Eq. (6.3), the two factors a and b can be varied and optimized independently. Generally, the two formulae of Eqs. (6.2) and (6.3) allow linear relations between the two components (fidelity and gamut) and the resulting numeric descriptor of color preference on the left side of Eqs. (6.2)–(6.4). In the other way, the formulation of Eq. (6.4) allows a quadratic relation between color preference, color saturation, and color fidelity. When the saturation of the colored objects (which is related to the color gamut of the light source) is steadily increased then the color preference judgment will also become higher. But when saturation reaches a certain limiting value then the course of color preference changes its direction and it will be reduced with increasing saturation. To implement Eqs. (6.2)–(6.4), all color fidelity and color gamut indices were calculated for the sets of test light sources described in Section 6.2 with their own specific color samples (e.g., 15 CQS color samples for Q_p, or 4 color samples in [6] for the FCI values).

6.3.1 Color Preference with the Constrained Linear Formula (Eq. (6.2))

For the warm white light sources (Section 6.2) with MCRI or Q_p as a color preference descriptor and with several different combinations of color fidelity and color gamut metrics, the coefficients of determination designated by R^2 (which, according to their definition, can also be negative) are listed in Tables 6.10 and 6.11. It can be seen that MCRI cannot be approximated well for the case of the warm white light sources with Eq. (6.2). Contrarily, Q_p can be approximated well as a two-metrics combination with Q_f and the different color gamut indices.

In case of the cold white light sources, similarly, MCRI cannot be well approximated by two-metrics combinations with Eq. (6.2); see Table 6.12. Q_p can still be considered as a color preference metric resulting from two-metrics combinations where Q_f plays the role of color fidelity and Q_g or R_g plays the role of the color gamut metric and then a higher positive coefficient of determination arises; see Table 6.13.

Table 6.10 Coefficients of determination (R^2) for MCRI as a two-metrics combination for the warm white light sources of Section 6.2.

R^2 of MCRI	R_a	$R_{1,14}$	Q_f	CRI_{2012}	R_f
$GAI_{rel.}$	−0.27	−0.13	−0.44	0.02	0.34
Q_g	−0.38	−0.03	−0.67	−0.13	0.41
FCI	−0.49	−0.84	−1.56	−0.60	0.07
R_g	−0.39	−0.02	−0.71	−0.18	0.38

Table 6.11 Coefficients of determination (R^2) for Q_p as a two-metrics combination for the warm white light sources.

R^2 of Q_p	R_a	$R_{1,14}$	Q_f	CRI_{2012}	R_f
$GAI_{rel.}$	0.40	0.39	0.91	0.46	0.54
Q_g	0.34	0.45	0.95	0.40	0.58
FCI	0.27	0.14	0.87	0.17	0.47
R_g	0.33	0.44	0.92	0.37	0.56

Table 6.12 Coefficients of determination (R^2) for MCRI as a two-metrics combination for the cold white light sources.

R^2 of MCRI	R_a	$R_{1,14}$	Q_f	CRI^{2012}	R_f
$GAI_{rel.}$	0.23	−0.08	0.15	−0.04	0.19
Q_g	0.24	0.11	0.11	−0.09	0.22
FCI	0.19	−0.39	0.01	−0.09	0.18
R_g	0.25	0.14	0.06	−0.10	0.20

Table 6.13 Coefficients of determination (R^2) for Q_p as a two-metrics combination for the cold white light sources.

R^2 of Q_p	R_a	$R_{1,14}$	Q_f	CRI_{2012}	R_f
$GAI_{rel.}$	0.76	0.65	0.89	0.81	0.77
Q_g	0.82	0.82	0.97	0.82	0.84
FCI	0.70	0.45	0.85	0.78	0.74
R_g	0.83	0.85	0.96	0.81	0.84

6.3.2 Color Preference with the Unconstrained Linear Formula (Eq. (6.3))

This formula (Eq. (6.3)) yields a better result than the constrained formula (Eq. (6.2)) for Q_p in the case of the warm white light sources (compare Tables 6.11 and 6.15). Also in this case, MCRI cannot be reconstructed well by two-metrics combinations (see Table 6.14). With this formula, the approximation for Q_p is better in the case of the cold white light sources than in the case of the warm white light sources (compare Tables 6.15 and 6.16). The three-dimensional representation of Q_p as a two-metrics combination from R_f and R_g with $R^2 = 0.85$ for the cold white light sources is illustrated in Figure 6.19.

Comparing Tables 6.11 (constrained) and 6.15 (unconstrained) for the Q_p results of the warm white light sources and comparing Tables 6.13 (constrained) and 6.16 (unconstrained) for the cold white light sources, it can be seen that the two formulae (constrained and unconstrained) yield similar results.

Table 6.14 Coefficients of determination (R^2) for MCRI as a two-metrics combination for warm white light sources.

R^2 of MCRI	R_a	R_{114}	Q_f	CRI_{2012}	R_f
GAI_{rel}	0.23	0.11	0.01	0.46	0.40
Q_g	0.31	0.47	0.20	0.62	0.61
FCI	0.32	0.08	0.07	0.32	0.40
R_g	0.29	0.47	0.16	0.54	0.58

Table 6.15 Coefficients of determination (R^2) for Q_p as a two-metrics combination for the warm white light sources.

R^2 of Q_p	R_a	$R_{1,14}$	Q_f	CRI_{2012}	R_f
$GAI_{rel.}$	0.44	0.40	0.93	0.50	0.55
Q_g	0.43	0.49	0.95	0.51	0.59
FCI	0.43	0.33	0.94	0.39	0.50
R_g	0.41	0.48	0.92	0.47	0.56

Table 6.16 Coefficients of determination (R^2) for Q_p as a two-metrics combination for the cold white light sources.

R^2 of Q_p	R_a	$R_{1,14}$	Q_f	CRI_{2012}	R_f
$GAI_{rel.}$	0.76	0.66	0.91	0.82	0.77
Q_g	0.83	0.84	0.97	0.85	0.85
FCI	0.70	0.45	0.86	0.78	0.75
R_g	0.84	0.87	0.96	0.84	0.85

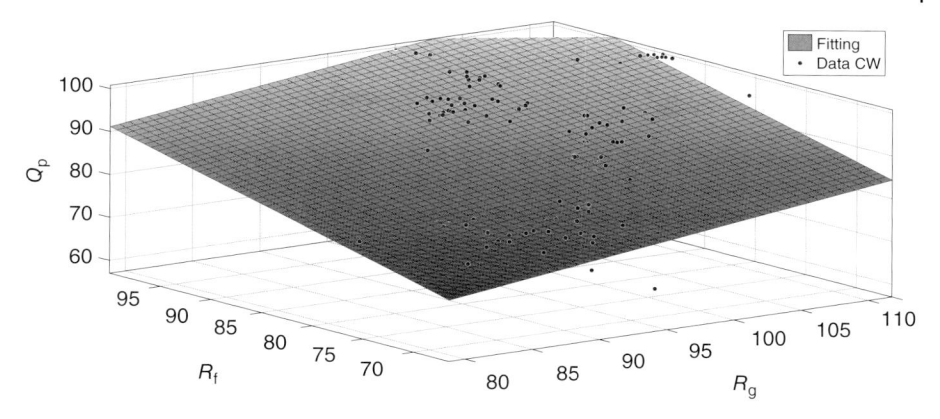

Figure 6.19 Three dimensional representation of CQS Q_p as the two-metrics combination of R_f and R_g with $R^2 = 0.85$ for the cold white light sources (the unconstrained linear 3D-description).

6.3.3 Color Preference with the Quadratic Saturation and Linear Fidelity Formula (Eq. (6.4))

Using the quadratic formula (Eq. (6.4)), the approximation for MCRI in the case of the warm white light sources is improved (see Tables 6.14 and 6.17). For the warm white light sources and using Eq. (6.4), the mathematical construction with R_f, CRI_{2012}, and $R_{1,14}$ as color fidelity components delivers better correlation for MCRI than for CQS Q_p (compare Tables 6.17 and 6.18). In the case of the cold white light sources, the correlation between the different color fidelity and color gamut metrics is on very good level for both MCRI and Q_p. The three-dimensional representation of Q_p as a two-metrics combination of R_f and R_g with the coefficient of determination $R^2 = 0.88$ in the case of the cold white light sources is shown in Figure 6.20.

In this case (see Figure 6.20), the color preference metric CQS Q_p can be expressed by Eq. (6.5).

$$Q_p = 0.008266 \cdot R_g^2 + 2.3843 \cdot R_g + 0.53381 \cdot R_f + 110.1694 \qquad (6.5)$$

Table 6.17 Coefficients of determination (R^2) for MCRI as a two-metrics combination for warm white light sources.

R^2 of MCRI	R_a	$R_{1,14}$	Q_f	CRI_{2012}	R_f
$GAI_{rel.}$	0.70	0.78	0.50	0.76	0.76
Q_g	0.56	0.72	0.46	0.72	0.73
FCI	0.65	0.78	0.60	0.79	0.78
R_g	0.50	0.67	0.50	0.71	0.73

Table 6.18 Coefficients of determination (R^2) for Q_p as a two-metrics combination for the warm white light sources.

R^2 of Q_p	R_a	$R_{1,14}$	Q_f	CRI_{2012}	R_f
$GAI_{rel.}$	0.49	0.49	0.94	0.54	0.57
Q_g	0.44	0.50	0.97	0.57	0.62
FCI	0.47	0.48	0.96	0.49	0.55
R_g	0.43	0.51	0.95	0.52	0.61

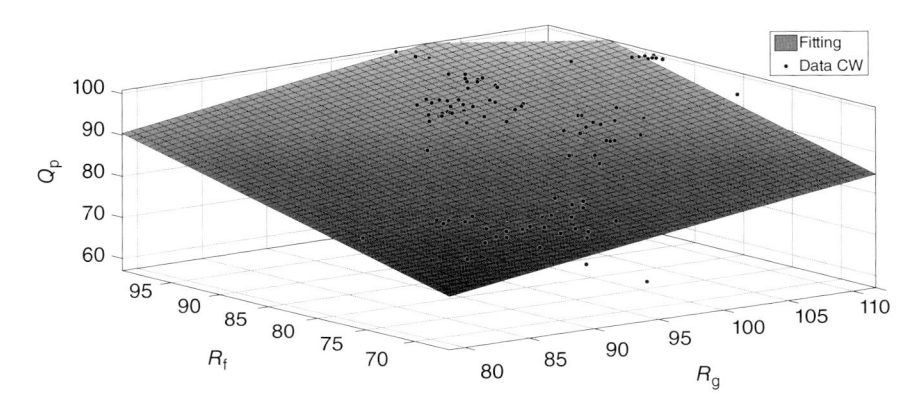

Figure 6.20 Three dimensional representation of CQS Q_p as a two-metrics combination of R_f and R_g with the coefficient of determination $R^2 = 0.88$ for the cold white light sources using the quadratic color saturation and linear color fidelity equation (Eq. (6.4)).

6.4 Conclusions and Lessons Learnt for Lighting Practice

The knowledge from this chapter can be summarized as follows.

- The three color fidelity indices $R_{1,14}$, CRI_{2012}, and R_f have a high coefficient of determination with each other for the warm white and cold white light sources. The two last mentioned metrics are defined in the uniform color space CAM 02-UCS with uniformly distributed test color samples over the hue circle in the $a' - b'$-plane and over the visible wavelength range so that these two color fidelity indices should be preferred for further international color studies and discussions.
- It appears that there are two subgroups of color gamut indices for the warm white light sources (see Table 6.7). In one of the subgroups, GAI and FCI correlate well with each other while in the other subgroup Q_g and R_g exhibit high correlation ($R^2 = 0.88$). However, FCI does not correlate well with R_g in both cases, that is, neither for the cold white nor for the warm white light sources. Based on the results in Table 6.6 for the warm white light sources, it can be

6.4 Conclusions and Lessons Learnt for Lighting Practice

recognized that the correlation between FCI − R_g and FCI − Q_g is highly dependent on the type of lamp spectrum. From the fact that GAI has been defined in a nonuniform color space with only eight not saturated surface colors and FCI has been defined with only four surface colors, it is reasonable to concentrate on the group of the color gamut metrics R_g and Q_g.

- The R_g metric is defined by the 99 IES colors and CQS Q_g is defined by the 15 saturated CQS colors. The analysis of Tables 6.17–6.20 to approximate MCRI and Q_p from a color fidelity metric and from R_g or Q_g as a color gamut metric implied that the coefficients of determination were slightly better in the case of Q_g than in the case of R_g.
- The color preference metrics MCRI and Q_p can be approximated by a quadratic combination of color gamut and color fidelity metrics with reasonable accuracy. Q_p can also be built up by the use of linear combinations of color gamut and color fidelity metrics. These approximations yield better coefficients of determination in the case of the cold white light sources than in the case of the warm white light sources. This can be explained by the choice of daylight illuminants as reference light sources in the case of the cold white lamp spectra. Additionally, Q_p is defined in the CIELAB $L^*a^*b^*$-color space with D65 illuminant in order to calculate the reference correlated color temperature factor based on the color gamut ratio (so called CCT factor) and with a chromatic adaptation formula that works better for the light sources with color temperatures higher than 4500–5000 K. Before 4500 K (or 5000 K), the reference light sources are blackbody radiators with only a small amount of spectral radiance in the wavelength range between 380 and 480 nm.

Table 6.19 Coefficients of determination (R^2) for MCRI as a two-metrics combination for the cold white light sources.

R^2 of MCRI	R_a	$R_{1,14}$	Q_f	CRI_{2012}	R_f
$GAI_{rel.}$	0.87	0.87	0.84	0.85	0.84
Q_g	0.85	0.85	0.76	0.78	0.77
FCI	0.91	0.92	0.87	0.89	0.89
R_g	0.83	0.83	0.69	0.72	0.72

Table 6.20 Coefficients of determination (R^2) for Q_p as a two-metrics combination for the cold white light sources.

R^2 of Q_p	R_a	$R_{1,14}$	Q_f	CRI_{2012}	R_f
$GAI_{rel.}$	0.83	0.85	0.94	0.87	0.86
Q_g	0.86	0.87	0.98	0.89	0.89
FCI	0.78	0.81	0.92	0.82	0.82
R_g	0.87	0.87	0.98	0.87	0.88

In this chapter, only numeric calculations were carried out to analyze color quality metrics and their mathematical relationship based on the specific definition of each color quality metric using its own test color samples. At the time of writing, the international color research community is searching for an ideal color fidelity index and an ideal color gamut index based on a suitable collection of test color samples to optimally characterize and model the visual color quality attributes *"color fidelity-color realness"* and *"color gamut, saturation of a group of colored objects."* This task has not been solved yet at the time of writing because the current metrics, if they are used as standalone values, are not able to fully describe the visual attributes color preference, naturalness, or vividness (this issue will be further dealt with in Chapter 7).

Therefore, several two-metrics solutions shall be defined and verified by conducting a number of dedicated visual color quality experiments in different research institutes with light sources having different chromaticities, color temperatures, white tones, and causing different degrees of object (over-)saturation. The visual judgment results of reliable test subjects observing different categories of colored objects (e.g., textiles, art objects, foods, and flowers) shall be modeled by reasonable two-metrics combinations and the best combination with a high correlation over different types of experiments and experimental conditions shall be chosen. A series of such experiments will be described and analyzed from the above point of view in Chapter 7.

References

1 Commission Internationale de l'Éclairage (1995) *Method of Measuring and Specifying Colour Rendering Properties of Light Sources*, Publ. CIE 13.3-1995.
2 Rea, M.S. and Freyssinier-Nova, J.P. (2008) Color rendering: a tale of two metrics. *Color Res. Appl.*, **33**, 192–202.
3 Davis, W. and Ohno, Y. (2010) Color quality scale. *Opt. Eng.*, **49**, 033602.
4 Smet, K.A.G., Schanda, J., Whitehead, L., and Luo, R.M. (2013) CRI2012: a proposal for updating the CIE color rendering index. *Light. Res. Technol.*, **45**, 689–709.
5 Smet, K.A.G., Ryckaert, W.R., Pointer, M.R., Deconinck, G., and Hanselaer, P. (2012) A memory color quality metric for white light sources. *Energy Build.*, **49**, 216–225.
6 Hashimoto, K., Yano, T., Shimizu, M., and Nayatani, Y. (2007) New method for specifying color-rendering properties of light sources based on feeling of contrast. *Color Res. Appl.*, **32**, 361–371.
7 David, A., Fini, P.T., Houser, K.W. *et al.* (2015) Development of the IES method for evaluating the color rendition of light sources. *Opt. Express*, **23**, 15888–15906.
8 Bodrogi, P., Brückner, S., Khanh, T.Q., and Winkler, H. (2013) Visual assessment of light source color quality. *Color Res. Appl.*, **38** (1), 4–13.
9 Schanda, J. (1985) A combined color preference – color rendering index. *Light. Res. Technol.*, **17**, S. 31–S. 34.

10 Islam, M., Dangol, R., Hyvärinen, M., Bhusal, P., Puolakka, M., and Halonen, L. (2013) User preferences for LED lighting in terms of light spectrum. *Light. Res. Technol.*, **45**, 641–665.
11 Dangol, R., Islam, M., Hyvarinen, M., Bhusal, P., Puolakka, M., and Halonen, L. (2013) Subjective preferences and color quality metrics of LED light sources. *Light. Res. Technol.*, **45**, 666–688.
12 Smet, K.A.G., Ryckaert, W.R., Pointer, M.R., Deconinck, G., and Hanselaer, P. (2011) Correlation between color quality metric predictions and visual appreciation of light sources. *Opt. Express*, **19**, 8151–8166.
13 Rea, M.S., Figueiro, M.G., Bierman, A., and Bullough, J.D. (2010) Circadian light. *J. Circadian Rhythms*, **8**, 2.

7

Visual Color Quality Experiments at the Technische Universität Darmstadt

In this chapter, the method and the results of the color quality experiments of the authors' laboratory will be presented and model equations for color quality will be proposed for practical use.

7.1 Motivation and Aim of the Visual Color Quality Experiments

Color quality has several different aspects that can be described by different color quality metrics and color quality indices [1]. It was shown that, in the most general case, color realness (also called *color fidelity* or *color rendering*, i.e., the similarity of the color appearance of colored test objects under the test light source to their color appearance under the reference light source) to be described by color fidelity type indices is not a suitable visual attribute to optimize color quality for general lighting.

Instead, for general indoor lighting purposes, the color quality aspects color preference, color naturalness, and color vividness turned out to be more relevant for lighting practitioners [2–8]. It was also shown that, as color preference, naturalness, and vividness are associated with more or less oversaturation of the illuminated colored objects (in comparison with the reference light source), these aspects cannot be described by a fidelity type index (alone) that describes realness (that requires no oversaturation in comparison with the reference light source).

As mentioned in Chapter 2, since the introduction of the CIE color-rendering index (CRI) [9], several other color quality metrics have been developed including gamut area index (GAI) [3], color quality scale (CQS) Q_f, Q_a, Q_p, and Q_g [10], feeling of contrast index (FCI) [11], memory color rendition index (MCRI) [12], and IES R_f and R_g [13]. Although some of these indices have been shown to correlate more or less with the visual attributes color preference, naturalness, and vividness [2, 10] (defined in Chapter 2), currently there is no internationally endorsed and experimentally validated single-value measure, which would be necessary for lighting practitioners to carry out the spectral design or evaluation of, for example, multi-LED light engines with pure semiconductor LEDs and phosphor-converted LEDs for general lighting, for general consumer use.

Moreover, it is known that the assessment of color preference depends on viewing context and lighting application [14]. For example, the immersive viewing of all colored objects in a room (e.g., office or home lighting) may give rise to a different color preference assessment than the targeted illumination of a collection of objects, for example, shop lighting. So the question is not only how to define a suitable general single color quality measure but also whether there is a dependence on lighting application. For certain applications, naturalness or vividness may be a more relevant attribute than color preference.

According to the above, the aim of the present chapter is to present the method and the results of the experiments carried out in the authors' laboratory. The presentation of the experiments follows a system that considers all relevant aspects for the lighting engineer with the motivation of understanding and modeling color quality as it is perceived and judged by the user of the light source in an interior lighting application. The final aim of this chapter is to provide a model as a usable target function for the spectral optimization of a light source (e.g., a modern multi-channel LED light engine) for the best color quality considering its different visual aspects.

In line with their long-term roadmap, authors designed and carried out the experiments to be presented in this chapter systematically, consistent with the visual brain's processing levels that are relevant to optimize color quality. Having the most basic requirements of white point characteristics and requirements (Chapter 3) in mind, Sections 7.2–7.5 (first level of color quality optimization, preparatory issues) deal with some further important basic issues (visual clarity (VC), brightness, and correlated color temperature (CCT)) to ensure the essential requirements for the subsequent spectral color quality optimization and modeling for the visual attributes color preference, naturalness, vividness, and color discrimination (second level of color quality optimization, optimization for visual color quality attributes) to be described in Sections 7.6–7.9. Finally, Section 7.10 will conclude the lessons learnt from this chapter.

At the first level of color quality optimization, in Section 7.2, VC (as defined in Chapter 2) is investigated. To achieve acceptable perceived color quality, a certain minimum level of visual acuity is required, which increases if the illuminance of the colored objects increases. In this section, the minimum illuminance level will be determined that is a prerequisite for spectral color quality optimization. It will be shown that VC, color discrimination and, as a consequence, visual comfort depend not only on illuminance level but also on the chromaticity (or CCT) of the light source so that the $V(\lambda)$-based quantity "illuminance" (in lx) alone cannot account for the impression of VC.

Section 7.3 deals with the issue of the brightness impression of the white tone (or white point) of the light source as a further step to prepare for efficient spectral color quality optimization. The brightness of white tones is important because, in a typical visual environment of interior lighting, the light of the light source is reflected from the white walls to dominate the light source user's brightness impression in the room. As was pointed out in Chapter 2, brightness impression cannot be described by the $V(\lambda)$-based quantity "luminance" (in cd/m^2) and this issue (the so-called brightness-luminance discrepancy) was examined in a brightness-matching experiment with strongly metameric light sources at two different CCT levels.

Before starting the spectral optimization procedure at the second level, a suitable CCT should be selected, according to the viewing context of the application, a room with predominantly achromatic or white objects (Section 7.4) or with red, blue, and colorful object combinations (Section 7.5). To find out the most appropriate CCT, experiments on CCT preference will be described for white objects (Section 7.4) and for red, blue, and colorful object combinations (Section 7.5). Preferred CCT ranges will be described and the Kruithof hypothesis (i.e., the preference of a lower illuminance level at lower CCTs and higher illuminance levels at higher CCTs) will be examined in these sections.

Starting the second level of color quality optimization, Section 7.6 will deal with experiments on the important color quality attributes color preference, naturalness, and vividness, which are associated with more or less oversaturation of the illuminated colored objects in comparison with the reference light source. It will be pointed out that these attributes cannot be described by a fidelity type index alone. To find a suitable model of color quality, observers scaled color preference, naturalness, and vividness visually by the use of semantic categories (as introduced in Chapter 2; e.g., "moderate," "good," "very good") in the context of office lighting (with customary office light sources excluding multi-LED light engines) in a real room (3D viewing) with colored objects on a table to be assessed by the observers. Selected color quality indices will be combined linearly to increase the correlation between the new, combined indices and the scaled visual attributes. It turned out that CCT is also a useful predicting variable according to the concept of Section 7.5. Criterion values of the most suitable combinations of indices for "good" color preference, naturalness, and vividness will be computed in order to provide user acceptance limits for spectral design.

Section 7.7 investigates color preference, naturalness, and vividness in another viewing context, that is, in the application of shop lighting, more specifically, the lighting of makeup products with a modern multi-LED light engine with flexibly variable and optimizable spectral power distributions (SPDs) at the white point of 3200 K, in a viewing booth (a different viewing condition from the real room in Section 7.6). It will be pointed out, as in Section 7.6, that spectral optimization for maximum color rendering does not guarantee a maximum visual assessment about color preference, naturalness, or vividness and also that there is a maximum degree of object oversaturation with maximum preference and naturalness. Semantics plays an important role again; observers judged the color preference, naturalness, and vividness by the aid of the semantic categories while the categories "good" or "very good" can be used in lighting practice as the limits for "good" or "very good" color quality to foster conscious spectral design. Models with combinations of color quality metrics will be shown to predict the visual attributes of color quality.

In Section 7.8, the results of the previous section dealing with reddish and skin-tone type makeup products will be compared with the results of another, similar experiment with multicolored food products by using the same questionnaire, the same multi-LED engine, the same CCT of the white point (3200 K), and the same illuminance level (500 lx) but with a different set of multi-LED spectra with different degrees of object oversaturation. It turned out that it is possible to merge the two datasets representing two lighting contexts (i.e., food and makeup).

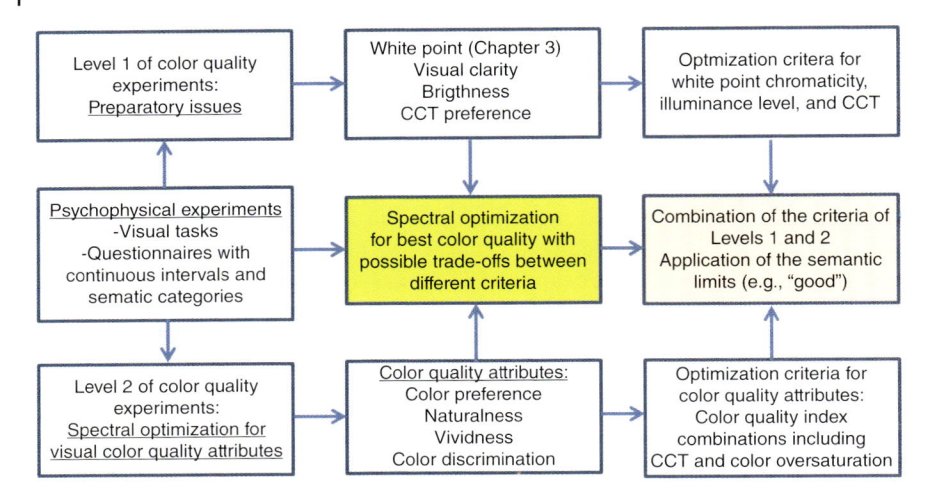

Figure 7.1 Aims and structure of this chapter.

A common model of color quality will be presented as a target function for spectral optimization. The effect of object oversaturation on color discrimination (i.e., the perception of small color differences or delicate color shadings on object surfaces as a further important aspect of color quality) will also be discussed in Section 7.8.

As seen above, semantic interpretations play a vital role in the procedure of spectral optimization as they provide acceptance criterion values for the descriptors of color quality. Accordingly, Section 7.9 summarizes all semantic interpretation aspects and provides criterion values for all types of color quality metrics. First, the semantic interpretation of color differences will be described to interpret color fidelity indices, color discrimination measures, and to provide criterion values of white tone chromaticity for the binning of white LEDs. Second, the semantic interpretation of the color appearance attributes modeled in Sections 7.6–7.8 will be presented in a synoptic view of all three visual attributes, color preference, naturalness, and vividness.

According to the above discussion, Figure 7.1 summarizes the aims and the structure of this chapter.

7.2 Experiment on Chromatic and Achromatic Visual Clarity

The concept of visual clarity was defined in Chapter 2. To achieve acceptable perceived color quality, a certain minimum level of visual acuity (both for its achromatic and chromatic aspects) is required. Visual acuity increases considerably if the illuminance of the colored objects increases. Practical experience shows that for shop lighting customers often require light levels higher than 1000 lx for acceptable VC and, as a final consequence, for good visual comfort. In the experiments of the present section, our research hypothesis was that both achromatic and chromatic assessments of VC depend not only on illuminance level but also on the chromaticity (or CCT) of the light source.

Experimental results corroborated this hypothesis and a new variable was introduced that includes S-cone contribution in order to rectify the failure of the quantity illuminance (lx) to predict visual clarity assessments. In terms of the new variable, it became possible to define a criterion value above which the subjects' visual clarity ratings saturate. The illuminance level provided by the light source should exceed this criterion value to ensure the starting point for spectral optimization for best color quality. Below the visual clarity criterion luminance level (that depends on the CCT of the light source), the impression of visual clarity breaks down, leaving no space for color quality design.

In the experiment on chromatic visual clarity a new, comprehensive experimental paradigm of several components was implemented. The experiment included two tests of color discrimination ability (the desaturated panel of the Farnsworth dichotomous test D-15 [15] and the Standard Pseudoisochromatic Plates [16]), an achromatic visual acuity test (the SKILL card [17]: bright side and dark side of the test), an achromatic test chart to scale perceived visual clarity (the so-called ARRI® test chart), and five colored objects to scale chromatic visual acuity. All test objects were illuminated by three types of light source, tungsten halogen (HAL), warm white phosphor-converted LED (LED-ww), and cool white phosphor-converted LED (LED-cw) at different horizontal illuminance levels, 40, 90, 180, 300, 470, and 1000 lx (16 light sources altogether).

7.2.1 Experimental Method

Test objects were placed into the middle of the table covered by a white cloth and illuminated homogenously by one of the test light sources at the defined horizontal illuminance level. At this position, the inhomogeneity of illuminance was less than ±5%. Only one test object was placed on the table at a time. Observers sat in front of the table and scrutinized the objects (or carried out the required visual task) from a viewing distance of 60 cm except for the SKILL card (40 cm). Figure 7.2 shows the test objects.

Figure 7.3 shows the relative spectral radiance of the three types of light source measured on the surface of a white standard with nearly diffuse reflection behavior (a Lambertian radiator).

The relative spectral radiance of the light sources (and their chromaticity coordinates and CCTs) depended on the illuminance (dimming) level. Table 7.1 shows the colorimetric properties and color quality indices of the 16 light sources used in the visual clarity experiment. Illuminance levels were logged during the experiment and their change did not exceed ±5%.

As can be seen from Table 7.1, all light sources had acceptable white tones (see Chapter 3), "good" (LEDs) to "very good" (HAL) color-rendering properties, and similar CQS Q_g values. As mentioned in Chapter 1, in order to rectify the failure of the variable illuminance (E_v in lx in the third column) to predict VC assessments (see Section 7.2.2), each one of the actual relative SPDs of the 16 light sources was multiplied by the spectral sensitivity of the S-cones (more precisely, the 10° S-cone fundamental function based on the Stiles and Burch 10° CMFs

Figure 7.2 Test objects in the visual clarity experiment. Objects were assessed (or the visual task was carried out) in the order of the numbers on the objects. 1: The book "Standard Pseudoisochromatic Plates" [16]; 2: desaturated panel of the Farnsworth dichotomous test D-15 [15]; 3: the SKILL card [17]; here its bright side is shown; 4: the ARRI® test chart; 5: "purple" painting; 6: "red" painting; 7: artificial water lily; 8: colorful pullover with well-visible colored structures; and 9: cyan textile.

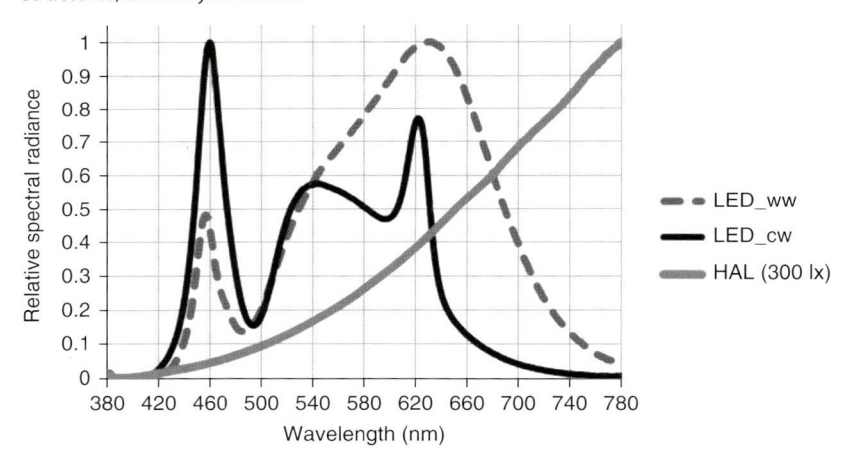

Figure 7.3 Mean relative spectral radiance of LED_ww and LED_cw used in the visual clarity experiment for the different illuminance levels. The relative spectral radiance of the HAL light source depended on the illuminance (dimming) level more strongly. In this figure, only the level 300 lx is shown. Colorimetric properties of all light sources are shown in Table 7.1.

(color matching functions); Stockman and Sharpe, 2000; see Section 2) and also with the $V(\lambda)$ function and integrated between 380 and 780 nm. The resulting quantities were divided and called S-cone/V ratios. These numbers are shown in the 10th column of Table 7.1. The last column shows the values of the new rectifying variable of VC denoted by $E_{v,S}$ for the 16 light sources. This variable was obtained by multiplying the value of illuminance (E_v in lx) by the value of the S-cone/V ratio.

First, observers had to read the numbers on the pages of the book of the Standard Pseudoisochromatic Plates [16] (see Figure 7.2) and the number of errors

Table 7.1 Illuminance levels, colorimetric properties, and color quality indices of the 16 light sources used in the visual clarity experiment.

No.	Type	E_v (lx)	CCT	Duv	R_a	R_9	Q_a	Q_g	S-cone/V	$E_{v,S}$
1	HAL	40	2166	−0.0002	98	99	98	101	0.093	3.70
2	HAL	90	2320	0.0007	98	99	98	99	0.106	9.51
3	HAL	180	2195	0.0004	99	99	99	99	0.092	16.57
4	HAL	300	2246	0.0006	100	99	98	99	0.097	29.03
5	HAL	470	2336	0.0006	100	99	98	99	0.108	50.76
6	HAL	1000	2591	0.0007	100	100	99	99	0.141	140.92
7	LED_ww	40	2792	−0.0002	89	58	88	100	0.170	6.80
8	LED_ww	90	2784	0.0005	88	56	87	99	0.164	14.78
9	LED_ww	180	2770	0.0005	88	56	88	99	0.162	29.18
10	LED_ww	300	2748	0.0002	89	58	88	99	0.161	48.36
11	LED_ww	470	2746	0.0002	89	57	88	99	0.162	75.92
12	LED_cw	40	5138	−0.0028	89	53	83	98	0.463	18.52
13	LED_cw	90	5005	−0.0018	89	49	83	98	0.444	39.95
14	LED_cw	180	4990	−0.0011	88	48	84	96	0.438	78.88
15[a]	LED_cw[a]	300	4982	−0.0009	88	48	83	96	0.436	130.87
16	LED_cw	470	5006	−0.0013	88	45	83	97	0.442	207.55

Last column: rectifying variable of visual clarity $E_{v,S}$ obtained by multiplying the value of illuminance ($E_{v,S}$ in lx) by the value of the S-cone/V ratio.
a) Reference light source.

(if any) was recorded. This number was called the first chromatic visual clarity variable ($V_{c,1}$). The number of errors in the completion of the desaturated panel of the Farnsworth dichotomous test D-15 [15] constituted the second chromatic VC variable ($V_{c,2}$). The number of the last correctly read row of the bright (dark) side of the SKILL card [17] counted from the top was the first (second) achromatic VC variable ($V_{a,1}$; $V_{a,2}$).

The VC (visibility of its achromatic structures) of the ARRI test chart constituted the third achromatic VC variable ($V_{a,3}$). This variable was assessed in the following way: first, the reference light source (No. 15, see Table 7.1, the LED_cw at 300 lx) was switched on for 30 s. Observers were taught to observe the ARRI test chart in this time interval and associate this reference light source with a VC rating of 100. Then, one of the 15 test light sources was switched on and, after 2 min of adaptation observers had to rate the VC of the test chart ($V_{a,3}$) compared to the reference light source (100).

The next task of the observer was to rate the "brilliance of the colors" ($V_{c,3}$) and the "visibility of the colored structure, colored textures, and colored shadings" ($V_{c,4}$) – according to the written instruction for the observer on the questionnaire – of the "purple" painting (see item No. 5 in Figure 7.2). The assessment was carried out for this test object (No. 5) and for all following test objects (No. 6–9) in the same way as for the ARRI test chart, that is, looking at the reference appearance for 30 s (rating = 100) and then the visual assessment of the test object under the test light source after 2 min of adaptation. The assessments of "brilliance of the colors" and the "visibility of the colored structure, colored textures, and colored shadings" of the test objects No. 6–9 constituted the chromatic VC variables $V_{c,5} - V_{c,12}$.

In the case of light sources with 40, 180, and 300 lx, observers included six men and two women between 23 and 28 years of age (all of them with normal color vision). In the case of light sources with 300, 470, and 1000 lx, observers included 17 men and 5 women between 23 and 28 years of age (all of them with normal color vision). For each observer and each VC variable, the corresponding z-score variables were computed. They were denoted by $V_{az,i}$ ($i = 1 - 3$; achromatic VC variables) and $V_{cz,i}$ ($i = 1 - 12$; chromatic VC variables). A general achromatic VC score (VC_a) was computed according to Eq. (7.1) and a general chromatic VC score (VC_c) was computed according to Eq. (7.2).

$$VC_a = V_{az,1} + V_{az,2} + V_{az,3} \tag{7.1}$$

$$VC_c = \sum_{i=3}^{12} V_{cz,i} - V_{cz,1} - V_{cz,2} \tag{7.2}$$

In Eq. (7.2), the variables $V_{cz,1}$ and $V_{cz,2}$ are subtracted from the sum of the variables $V_{cz,i}$ ($i = 3 - 12$). The reason for this subtraction is that while the values of the variables $V_{cz,i}$ ($i = 3 - 12$) increase with a more favorable response (increasing VC ratings of the test objects No. 5–9; see Figure 7.2), the values of the variables $V_{cz,1}$ and $V_{cz,2}$ (number of errors) decrease with a more favorable response. An overall VC has also been computed as the average of VC_a and VC_c; see Eq. (7.3).

$$VC = (VC_a + VC_c)/2 \tag{7.3}$$

7.2.2 Analysis and Modeling of the Visual Clarity Dataset

The visual clarity dataset to be analyzed consisted of 16 values of the VC, VC_a, and VC_c data for each one of the 16 light sources listed in Table 7.1. The aim of the analysis was to find a suitable descriptor variable for these data. First, color quality indices (including R_a, R_9, Q_a, Q_g, see Table 7.1; and others) were considered but none of them turned out to be suitable, nor the variables illuminance or scotopic illuminance. It is interesting to plot the variable VC as a function of illuminance (lx) grouped by the type of light source; see Figure 7.4.

As can be seen from Figure 7.4, although VC is a monotonically increasing function of illuminance for every type of light source, illuminance is not a suitable descriptor quantity because VC scores are significantly higher for the cool white LED than for the warm white LED or the tungsten halogen light source at the same illuminance E_v. It is important to observe that VC continues to increase beyond 500 lx, the minimum illuminance level generally required by standards of interior lighting.

To rectify the different VC scores of the different types of light source with different relative SPDs, the new variable $E_{v,S}$ was introduced by multiplying the value of illuminance (E_v in lx) by the value of the (S-cone/V) ratio for every one

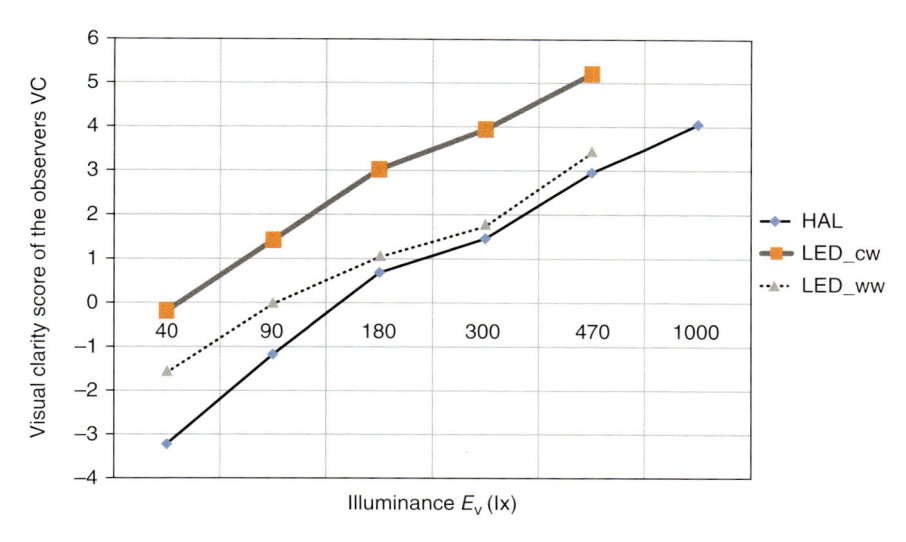

Figure 7.4 Variable VC (overall visual clarity score, Eqs. (7.1)–(7.3)) as a function of illuminance (lx) grouped by the type of light source.

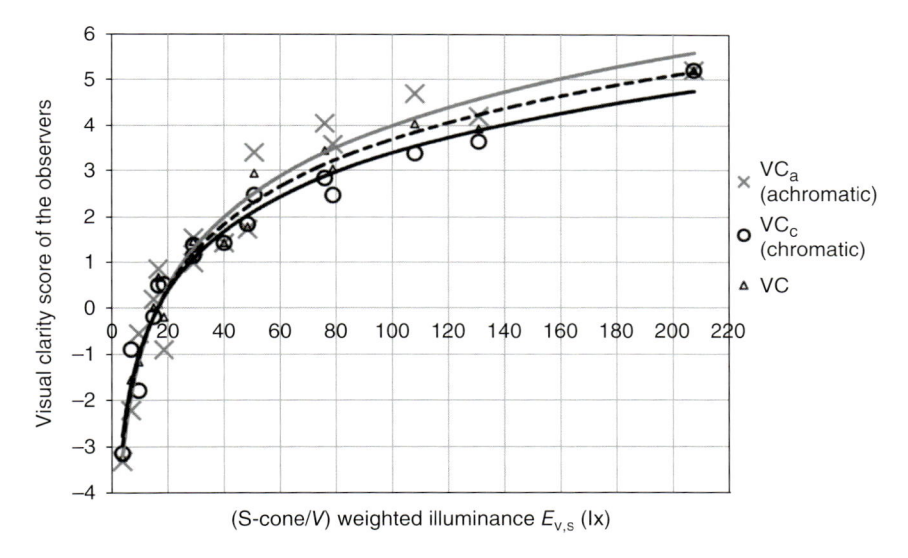

Figure 7.5 The visual clarity variables VC, VC_a and VC_c (Eqs. (7.1)–(7.3)) as a function of the new rectifying variable $E_{v.S}$ for every one of the 16 light sources.

of the 16 light sources; see the last column of Table 7.1. By the use of this variable, a monotonically increasing (logarithmic) dependence was obtained independent of the type (or CCT) of the light source; see Figure 7.5.

As can be seen from Figure 7.5, the VC scores VC_a and VC_c (Eqs. (7.1)–(7.3)) depend on the new rectifying variable $E_{v.S}$ logarithmically. The three visual clarity scores were re-scaled to have the same value (VC = VC_a = VC_c = 5.19) for

$E_{v,S} = 208$ (lx). Equations (7.4)–(7.6) contain the best fitting logarithmic equations with $r^2 \geq 0.94$.

$$VC_a = 2.1804\ \ln(E_{v,S}) - 6.0469 \quad (r^2 = 0.94) \tag{7.4}$$

$$VC = 2.0230\ \ln(E_{v,S}) - 5.6253 \quad (r^2 = 0.98) \tag{7.5}$$

$$VC_c = 1.8656\ \ln(E_{v,S}) - 5.2037 \quad (r^2 = 0.96) \tag{7.6}$$

It can also be seen from Figure 7.5 that the visual clarity curves saturate above about $E_{v,S} = 200$ (lx) and this latter value can be considered *as a criterion value for acceptable visual clarity*. If the S-cone to V ratio of the light source to be installed is known then the illuminance level required by this criterion can be estimated by dividing the critical value of $E_{v,S} = 200$ (lx) by the known S-cone to V ratio of the light source. For example, for a typical tungsten halogen light source with an S-cone to V ratio of 0.092, the necessary illuminance equals 2174 lx. In comparison, for a cool white LED light source with an S-cone to V ratio of 0.44, 455 lx is required.

Table 7.2 contains a sample computation of the illuminance values (lx) that are necessary to fulfill the criterion value of $E_{v,S} = 200$ (lx) for acceptable visual clarity for a set of typical light source SPDs used in today's interior lighting practice.

As can be seen from Table 7.2, criterion illuminance (lx) decreases with increasing CCT and the S-cone to V ratio increases with increasing CCT. These tendencies are depicted in Figure 7.6 together with two best fit curves, a sixth-order polynomial for criterion illuminance and linear in the case of the S-cone to V ratio.

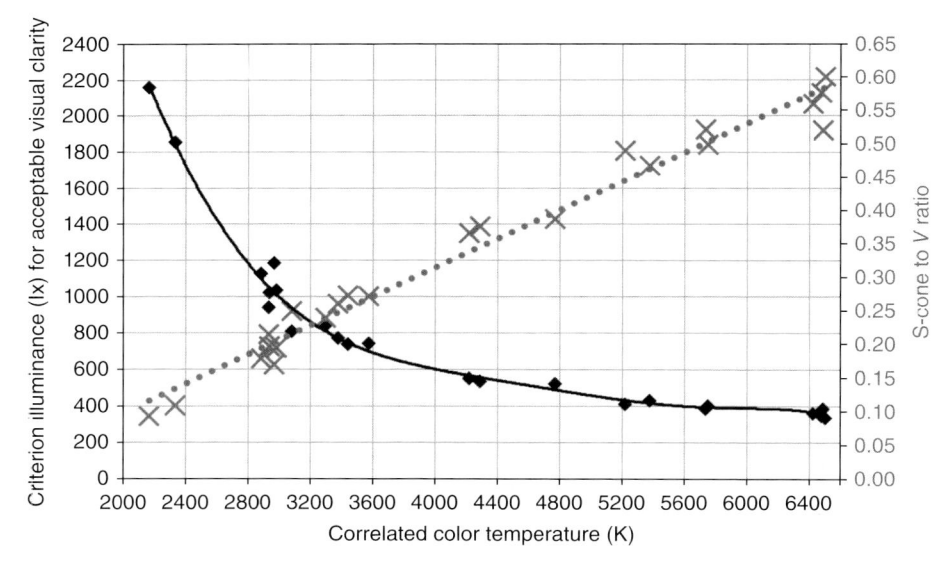

Figure 7.6 Criterion illuminance (lx) for acceptable visual clarity (left ordinate, filled black diamonds) and S-cone to V ratio (right ordinate, gray crosses) as a function of CCT for the sample set of light sources shown in Table 7.2. Best fit curves: a sixth-order polynomial for criterion illuminance (black curve) and linear (gray dots) in case of the S-cone to V ratio.

Table 7.2 Illuminance values (lx) that are necessary to fulfill the criterion value of $E_{v,S} = 200$ (lx) for acceptable visual clarity computed for a set of different light sources (examples).

Type	CCT (K)	S-cone/V	Criterion illuminance (lx)
HAL	2166	0.093	2161
HAL	2336	0.108	1852
wwLED	2887	0.178	1125
FL	2938	0.213	937
wwLED	2943	0.196	1019
wwLED	2973	0.169	1183
wwLED	2987	0.193	1034
wwLED	3083	0.248	806
wwLED	3300	0.238	840
FL	3380	0.260	770
FL	3446	0.272	736
wwLED	3578	0.271	739
FL	4224	0.365	548
FL	4290	0.375	533
cwLED	4775	0.386	518
cwLED	5223	0.489	409
cwLED	5380	0.468	428
cwLED	5739	0.520	384
cwLED	5755	0.498	401
FL	6428	0.559	358
FL	6484	0.576	347
cwLED	6496	0.520	385
cwLED	6510	0.600	333

Finally, the average number of confused color samples in the desaturated panel of the Farnsworth dichotomous test (item No. 2 in Figure 7.2) for a certain adjacent pair of color samples (at different hue angles) was computed for all observers and all 16 light sources. This average number is depicted as a function of the color difference level (the rounded value of the CAM02-UCS color difference $\Delta E'$) between the corresponding color samples and the $E_{v,S}$ level (i.e., the rounded value of $E_{v,S}/40$; see Table 7.1) in Figure 7.7.

As can be seen from Figure 7.7, the average number of confused color samples for a certain adjacent pair of color samples exhibits a generally decreasing trend as the CAM02-UCS $\Delta E'$ color difference between the corresponding color samples and the $E_{v,S}$ level increases. It can be seen that above $\Delta E' \geq 6$ and $E_{v,S} \geq 200$ (lx), the order of the color samples was not confused at all, at least in the case of the panel of color normal observers of the present experiment. Above $\Delta E' \geq 8$, there was no confusion except for $E_{v,S} < 40$ (lx) and for $\Delta E' \geq 9$, there should not be any confusion at all.

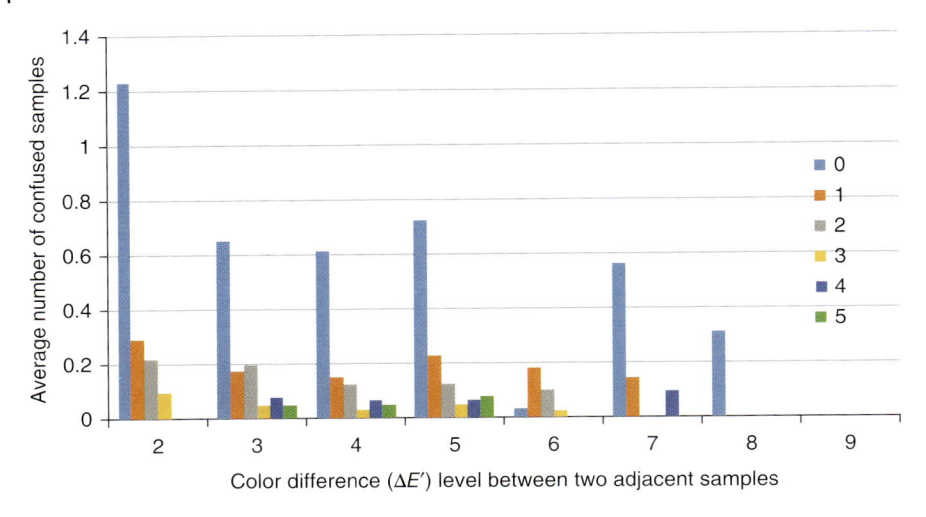

Figure 7.7 Ordinate: average number of confused color samples in the desaturated panel of the Farnsworth dichotomous test (item No. 2 in Figure 7.2) for a certain adjacent pair of color samples for all observers and all 16 light sources. Abscissa: color difference level (the rounded value of $\Delta E'$) between the corresponding color samples; legend: $E_{v,S}$ level of the light source (i.e., the rounded value of $E_{v,S}/40$; see Table 7.1).

7.3 Brightness Matching of Strongly Metameric White Light Sources

In Section 7.2, it was pointed out that the $V(\lambda)$-based quantity "illuminance" (in lx) cannot account for the impression of visual clarity subjects experience when looking at different objects with more or less visible color differences, colored structures, color transitions, and delicate color shadings. A new variable $E_{v,S}$, that is, the illuminance multiplied by the (S-cone/V) ratio of the light source, was introduced and the criterion value of $E_{v,S} = 200$ (lx) was established for acceptable visual clarity. The present section deals with the brightness impression of the white tone (or white point) of the light source. In a typical visual environment of interior lighting, the light from the light source is reflected from large white surfaces (the white walls) and these surfaces dominate brightness impression in the room.

As was pointed out in Chapter 2, brightness impression cannot be described by the quantity luminance (in cd/m^2) with reasonable accuracy. Depending on their chromaticity, two surfaces of the same luminance generally evoke different brightness perceptions. To account for this so-called brightness–luminance discrepancy, several models were developed and some selected models (Ware and Cowan conversion factor (WCCF); Berman *et al.*; Fotios and Levermore; CIE model) were introduced in Chapter 2. The common feature of these models is that instead of using luminance they derive a quantity B (brightness) from the SPD of the stimulus (e.g., a white wall illuminated by a warm white LED). Equal values of B of stimuli of different SPDs should result in equal brightness impressions and a set of different stimuli should be ordered by their B values according to their perceived brightness correctly.

The present section deals with the brightness–luminance discrepancy of highly metameric SPDs of different white tones. As described in Chapter 2, this means that although the two stimuli to be compared visually have the same chromaticity, the course of their relative SPDs is very different, for example, one of the stimuli might have a smooth, balanced SPD in the visible spectral range (so-called broadband stimulus) and the other stimulus of the same chromaticity may have a rugged SPD with high local maxima and minima (so-called narrowband stimulus).

In the present section, results of a visual experiment will be described and analyzed in which observers had to visually match the brightness of pairs of such highly (approximately) metameric white tones (a reference stimulus and a test stimulus) by changing the luminance of the test stimulus. In the last sentence, the word approximately was written because small chromaticity differences between the two white tones were allowed by the apparatus at the end of the visual brightness-matching procedure.

According to the above, our working hypothesis was that there is a significant brightness–luminance discrepancy when the two white light sources (reference and test stimuli) to be brightness matched are highly metameric and comprise small chromaticity differences between the test and the reference stimuli. The motivation of the experiment was that, in today's interior lighting practice, both types of light source (narrowband, e.g., fluorescent lamps (FLs) or broadband, e.g., phosphor-converted LEDs) are available for the lighting engineer and it is important to understand the variation of brightness impressions they evoke in the user of the light source.

The question is how this brightness–luminance discrepancy depends on CCT (warm white and cool white will be compared) and on the course of the SPD (broadband and narrowband light sources will be compared). The performance of the different brightness metrics (WCCF; Berman *et al.*; Fotios and Levermore; CIE model) and a further possible metric (the S-cone signal) to describe these highly metameric brightness-matching results will be investigated.

7.3.1 Experimental Method

Figure 7.8 shows the apparatus of the brightness-matching experiment.

As can be seen from Figure 7.8, a viewing booth with two chambers was used. The reference light sources (a tungsten halogen lamp and a compact FL) illuminated the left (reference) chamber. The test light sources (RGB-LEDs and phosphor-converted LEDs) illuminated the right (test) chamber. The walls and the bottom of both chambers had the same, spectrally aselective cover material. Thanks to the diffusers, the luminance inhomogeneity of the surfaces was less than 5%.

Observers sat in front of the viewing booth at a distance of 60 cm and adjusted the luminance of the surfaces of the right (test) chamber by the use of two buttons (up and down) in the following way. Adapting to the constant left (reference) chamber for 2 min and starting from an initial test luminance, adjustments

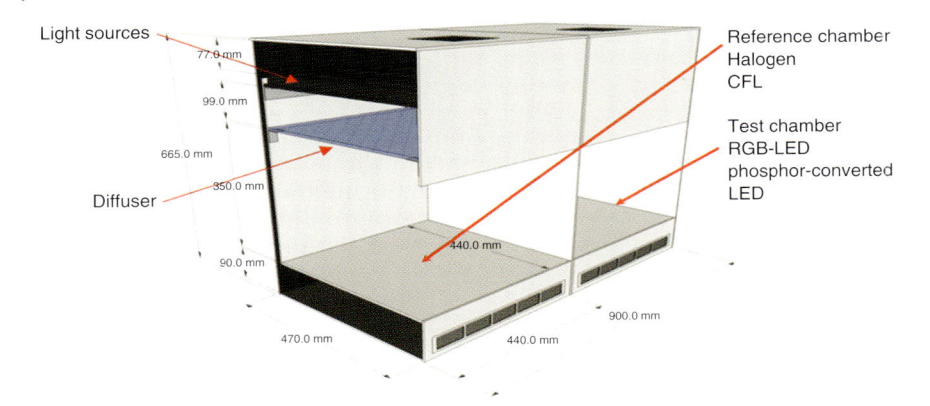

Figure 7.8 Viewing booth with two chambers (reference and test) used in the brightness matching experiment. Observers sat in front of the viewing booth at a distance of 60 cm and adjusted the luminance of the surfaces of the right (test) chamber.

were carried out by looking left and right until the perceived brightness of the two chambers matched. Observers were taught to set a higher brightness and then a lower brightness and repeat this procedure until complete brightness match despite the (slight) chromaticity difference between the two chambers. Twenty-one observers with normal color vision aged between 21 and 43 years (11 women and 10 men) took part in the experiment.

Observers indicated when they thought that the brightness impression of the two chambers was the same and then, the two matching absolute SPDs (reference and test) were measured at a well-defined point in the middle of the bottom plate of the chambers. One observer carried out only one brightness matching for each condition (see below). Two observers were excluded from the analysis owing to their inconsequent results. The brightness-matching results of the remaining 19 observers will be analyzed in Section 7.3.2.

Two CCT levels were investigated, 2500 K (relative test and reference SPDs are shown in Figure 7.9) and 5100 K (relative test and reference SPDs are shown in Figure 7.10).

As can be seen from Figure 7.9, at the 2500 K level, the reference light source R1 was a tungsten halogen lamp, the narrowband test light source T1_na was RGB LED, and the broadband test light source T1_br was a phosphor-converted LED with slight RGB LED components.

As can be seen from Figure 7.10, at the 5100 K level, the reference light source R2 was a compact FL, the narrowband test light source T2_na was RGB LED with a very slight phosphor-converted LED component, and the broadband test light source T2_br was a phosphor-converted LED with slight RGB LED components.

Table 7.3 shows the colorimetric properties of the light sources in Figures 7.9 and 7.10.

As can be seen from Table 7.3, all light sources had "good" to "very good" color-rendering properties (in terms of the general CRI CIE R_a). The narrowband light sources had a large color gamut with CQS $Q_g \geq 130$ and lower CQS Q_a values.

7.3 Brightness Matching of Strongly Metameric White Light Sources | 215

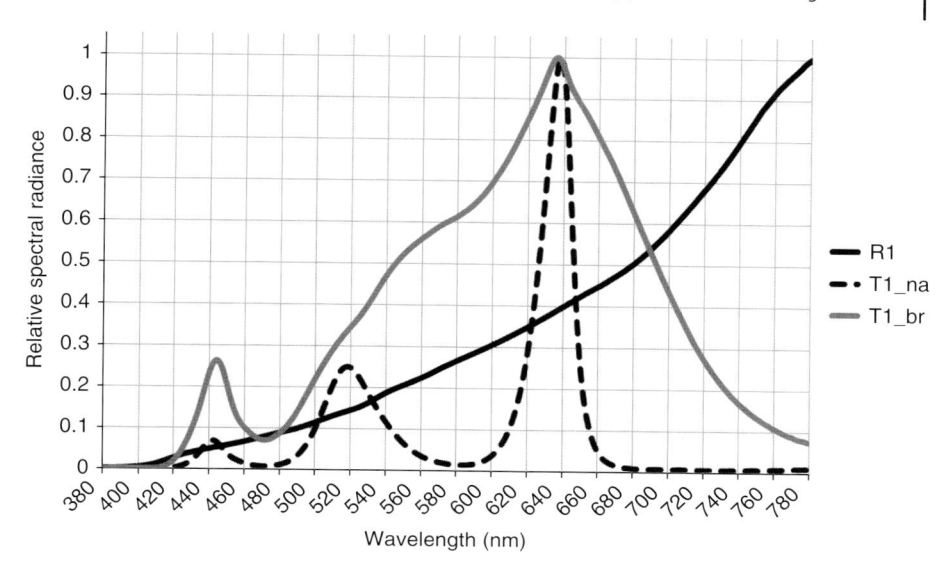

Figure 7.9 Relative test (narrowband: T1_na, broadband: T1_br) and reference (R1) SPDs at the correlated color temperature level 2500 K. The luminance of the reference R1 equaled 170 cd/m² for the brightness matching with T1_na and 157 cd/m² for T1_br (with a reproduction accuracy of ±2 cd/m²).

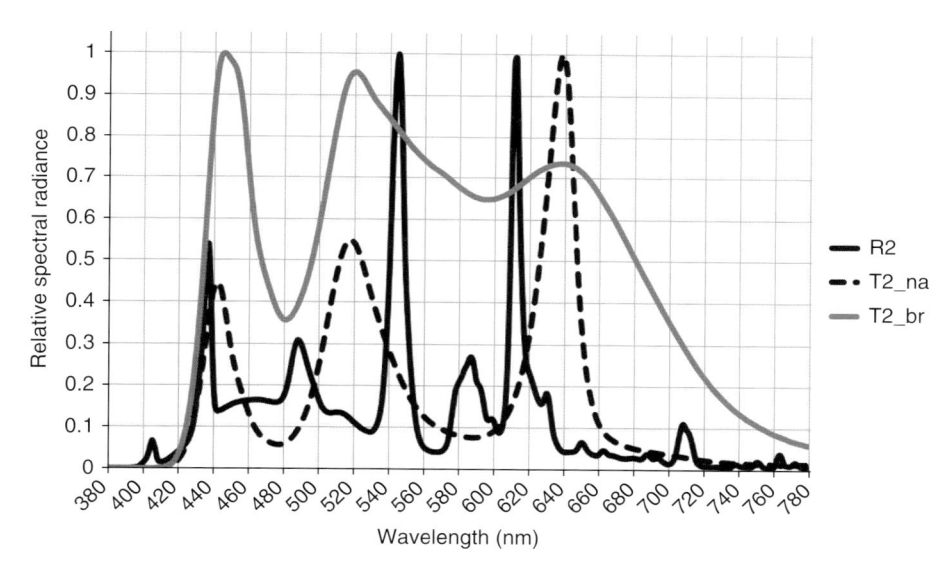

Figure 7.10 Relative test (narrowband: T2_na, broadband: T2_br) and reference (R2) SPDs at the correlated color temperature level 5100 K. The luminance of the reference R2 equaled 360 cd/m² for the brightness matching with T2_na and 375 cd/m² for T2_br (with a reproduction accuracy of ±5 cd/m²).

Table 7.3 Colorimetric properties of the light sources in Figures 7.9 and 7.10

	R1	T1_na	T1_br	R2	T2_na	T2_br
CCT (K)	2556	2334	2589	5072	5082	5279
CRI R_a	100	89	98	98	96	98
CQS Q_a	99	26	91	91	51	94
CQS Q_g	100	130	101	100	131	101

7.3.2 Results of the Brightness-Matching Experiment

As mentioned above, there was a chromaticity difference between the two white tones after the observers had completed their brightness-matching task. Figure 7.11 shows these chromaticity difference values in terms of the $\Delta u'v'$ metric.

As can be seen from Figure 7.11, the chromaticity difference ($\Delta u'v'$) between the two white tones after the observer had completed the brightness-matching task was always well above the general visibility threshold of white tone chromaticity differences ($\Delta u'v' = 0.001$). In the case of the brightness-matching condition of the tungsten halogen reference light source with the narrowband LED (R1–T1_na), the chromaticity difference between the two white tones (that were matching in brightness) was well visible, the mean $\Delta u'v'$ value being greater than 0.014. In the case of the cool white matching condition (R2–T2_br), the mean

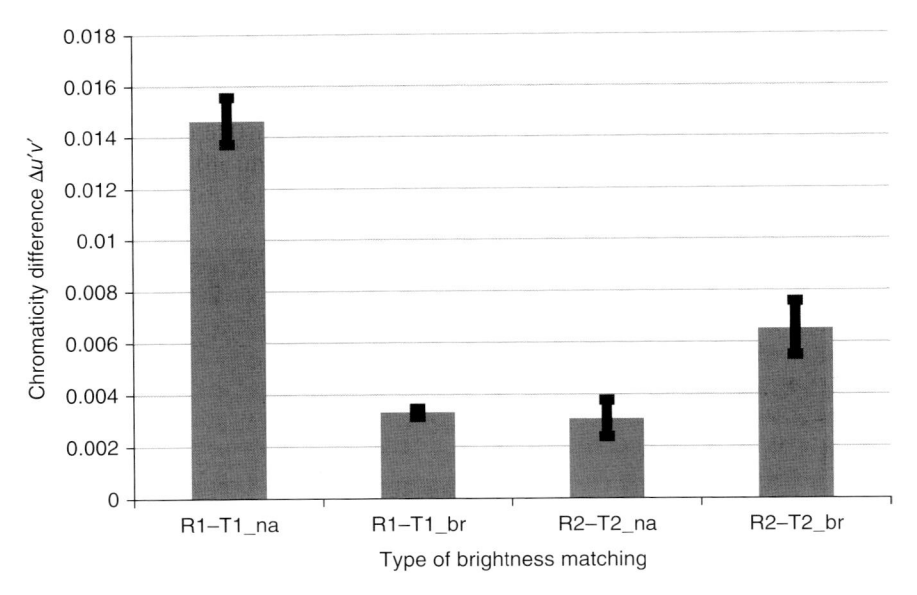

Figure 7.11 Chromaticity difference ($\Delta u'v'$) between the two white tones after the observer had completed the brightness-matching task in the case of the four types of matching conditions (see Figures 7.9 and 7.10). Mean values of 19 observers and their 95% confidence intervals.

$\Delta u'v'$ value was greater than 0.006, which is also a well-noticeable white tone chromaticity difference.

The question is, as mentioned at the beginning of the present section, whether there is a suitable brightness metric B with equal values for both of the two matching absolute SPDs. To investigate the suitability of existing brightness metrics, the ratios $R = (B_{test}/B_{reference})$ of each metric (WCCF; Berman et al.; Fotios and Levermore; CIE model; see Chapter 2) were computed for every one of the matching conditions and for every one of the 19 observers whose results were included. Values of a further possible metric, the standalone S-cone signal, were also computed from the spectrum of the light source by using the 10° S-cone fundamental based on the Stiles and Burch 10° CMFs and described by Stockman and Sharpe in 2000 (see Chapter 2). The quantity (photopic) luminance (in cd/m² units) was also included in the analysis. The ratios $R = (B_{test}/B_{reference})$ are shown in Figure 7.12.

As can be seen from Figure 7.12, the confidence intervals of the photopic luminance (cd/m²) ratios overlap with the value of 1 implying that brightness–luminance discrepancy is not significant in the case of the two "narrowband" conditions. In the case of the "broadband" conditions, test light sources (the two broadband LED spectra) exhibit higher luminance values at both CCT levels (2500 and 5100 K) than their reference light sources (the tungsten halogen and the compact fluorescent, respectively). In the case of 5100 K, the reference light source R2 was a compact FL, which itself comprises a highly structured spectrum with high local maxima (R2; see Figure 7.10). When the brightness of this reference light source is compared with its LED counterpart (T2_br; see Figure 7.10) that exhibits a more balanced spectrum in the visible wavelength range, in the case of T2_br, 7% more luminance is

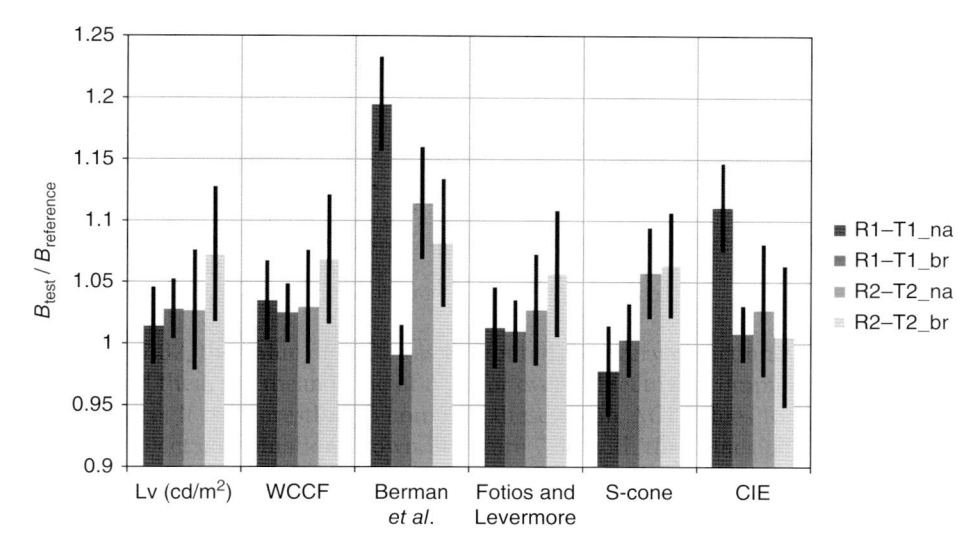

Figure 7.12 Mean ratios $R = (B_{test}/B_{reference})$ of the different brightness metrics in case of the four brightness-matching conditions R1–T1_na, R1–T1_br, R2–T2_na, and R2–T2_br. Mean values of 19 observers and their 95% confidence intervals.

necessary to obtain the match (the luminance ratio T2_br to R2 equals 1.07; see Figure 7.12).

Concerning the performance of the different brightness metrics, the WCCF metric did not perform better than luminance. The Berman *et al.* metric performs better than luminance in the case of the matching condition R1–T1_br only. The Fotios and Levermore metric has a slightly better performance than the quantity luminance but not for the R2–T2_br condition. In the latter case, this brightness metric still predicts in average 6% more brightness for the test light source despite the visual brightness match. The metric using S-cone signals can describe the R1–T1_br condition with the ideal value of $R = 1.00$ in average. The CIE metric has a discrepancy in the case of the matching condition R1–T1_na ($R = 1.11$, i.e., with 11% more predicted brightness value for the test light source T1_na).

Summarizing the above results, it can be stated that neither the quantity of photopic luminance nor any one of the above brightness metrics is able to describe the above strongly metameric brightness-matching results of white tones of more or less different chromaticity.

7.4 Correlated Color Temperature Preference for White Objects

The preference of the indoor illuminant's CCT (warm white, neutral white, cold white) is an important issue in today's light source market. CCT preference depends also on the type of colored objects in the illuminated scene, for example, a combination of multicolored objects or objects of different white shades only. As mentioned earlier, to describe visual color quality, the first step should be to specify white point or white tone characteristics, that is, the appearance of the white tone of the light source (see Section 7.3) and also the appearance of white objects first. The aim is to get to know the white preference of the users of the light source. The next step will then be the inclusion of a colorfully lit environment with different colored objects (see Sections 7.5–7.8).

To continue the discussion about the preferred white tones in Section 3.3, the present section describes the own experimental results of the authors on the CCT preference of the subjects (i.e., the light source users) about white objects placed in an experimental room at two illuminance levels (500 and 1200 lx) [28]. The motivation is that white or neutral object combinations are often encountered in office environments. Today, white furnishing also represents a trend in the private domain. In the experiment to be described in the present section, CCT was varied between 2320 and 5725 K. The preferred CCT range will be described and the Kruithof hypothesis (i.e., the preference of a lower illuminance level at lower CCTs and higher illuminance levels at higher CCTs) will be examined in this section.

7.4.1 Experimental Method

An experimental room was installed with white, diffusely reflecting walls. A thick curtain hanging at the window isolated the room from the outside. The inside

7.4 Correlated Color Temperature Preference for White Objects

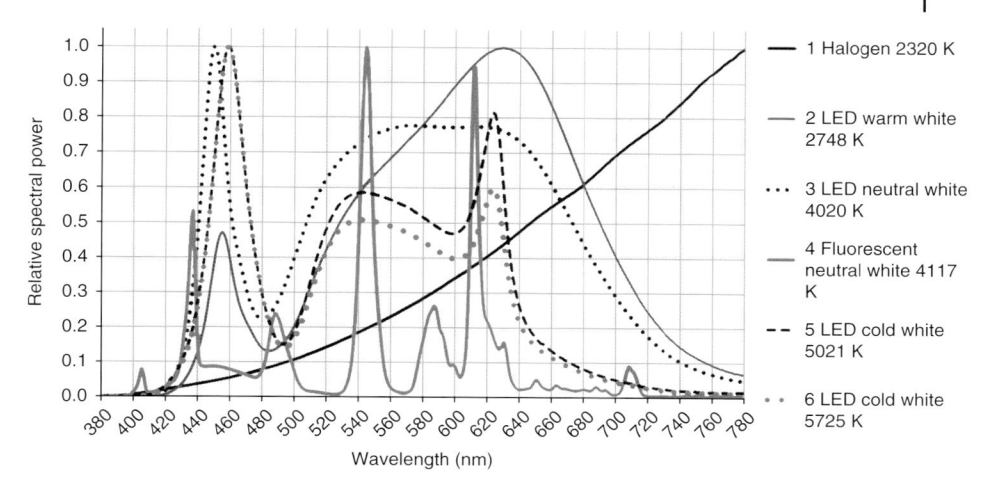

Figure 7.13 Relative spectral power distributions of the six test light sources used in the experiment. After [28].

of the curtain was also diffusely reflecting white. The light sources illuminating the room were mounted into the ceiling with spectrally aselective diffuser plates. Figure 7.13 shows the relative SPDs of the six test light sources. They could be switched on and off and dimmed by using a control panel operated by the leader of the experiment. All light sources had excellent colorimetric and photometric reproducibility and provided acceptable homogeneity of illuminance (> 85%) on the white table on which the test objects were arranged. The SPD of each light source was measured after 30 min at a white standard lying on the table.

As can be seen from Figure 7.13, the warm white and neutral white LED light sources (2748 and 4020 K) have a continuous spectrum while the two cold white LEDs (5021 and 5725 K) exhibit also an emission maximum at 625 nm. The tungsten halogen light source (including the diffuser plate) has a CCT of 2320 K (and little spectral power in the range 380–500 nm). The neutral white FL has several vacancies in its emission spectrum in the green range (500 – 530 nm), the yellow–green range (560 – 580 nm) and also from 630 nm on (red range). The colorimetric properties of the light sources in Figure 7.13 are listed in Table 7.4.

As can be seen from Table 7.4, the CCT of the six light sources ranged between 2320 and 5725 K while their general color-rendering indices (R_a) ranged between 81 and 89 except for the halogen lamp, which had $R_a = 99$. Two illuminance levels, 500 and 1200 lx were used for each one of the six light sources. The illuminance level of 500 lx represents the minimum value according to EN 12464 for most visual tasks. The value of 1200 lx is easy to implement as the higher level of a dynamic lighting installation. To achieve 1200 lx in the present experiment, simply more lamps of the same type were switched on at the ceiling without influencing relative SPDs and the homogeneity of light distribution in the room.

Two lighting situations of the test room are illustrated in Figure 7.14 with the arrangement of the white objects and the white furniture.

As can be seen from Figure 7.14, observers saw different white objects including vases, flowers, napkins, rice, sugar, flour, plates, plastic knives and forks, cups, and

Table 7.4 Colorimetric properties of the light sources in Figure 7.13. After [28].

Type	Name	CCT (K)	R_a
1	Halogen	2320	99
2	LED warm white	2748	89
3	LED neutral white	4020	88
4	FL neutral white	4117	81
5	LED cold white	5021	88
6	LED cold white	5725	87

CCT, correlated color temperature; R_a, general color rendering index; and FL, fluorescent.

Figure 7.14 Two lighting situations of the test room (illustrations) with the arrangement of the white objects and white furniture. The white standard at which the illuminance and spectral measurements were conducted can be seen in the left bottom corner. After [28].

a white table with a table cloth and white chairs. Figure 7.15 shows the spectral reflectance curves of a selected set of the objects.

As can be seen from Figure 7.15, the spectral reflectance curves of the white objects differ both in absolute value and across the visible range of wavelengths. The effect of this change (their slightly different white tones) on the object-specific color-rendering indices under the light sources of this experiment can be seen from Figure 7.16.

As can be seen from Figure 7.16, all object-specific CRI values are above 92, which corresponds to very good color rendering. This is an important point for the validity of the experimental design: according to the constantly high color-rendering level of the white objects to be assessed, the level of color rendering of the six light sources (ranging between $R_a = 81$ and 99) did not affect the assessment of CCT preference. If more colorful objects had been used then the color rendering level differences among the six light sources would have been another factor influencing preference besides CCT.

It is also interesting to look at the distribution of the different shades of white of the selected set of white objects (Figure 7.15) in the $u' - v'$ chromaticity diagram; see Figure 7.17.

7.4 Correlated Color Temperature Preference for White Objects

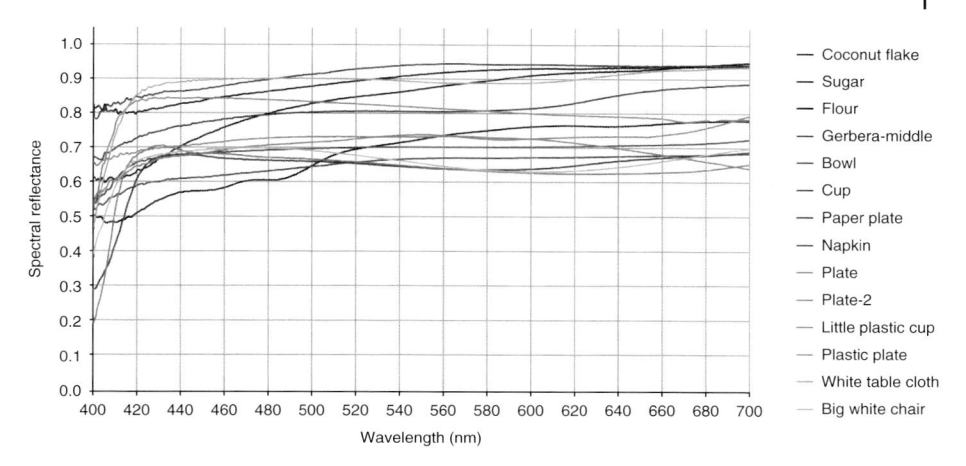

Figure 7.15 Spectral reflectance curves of a selected set of the objects.

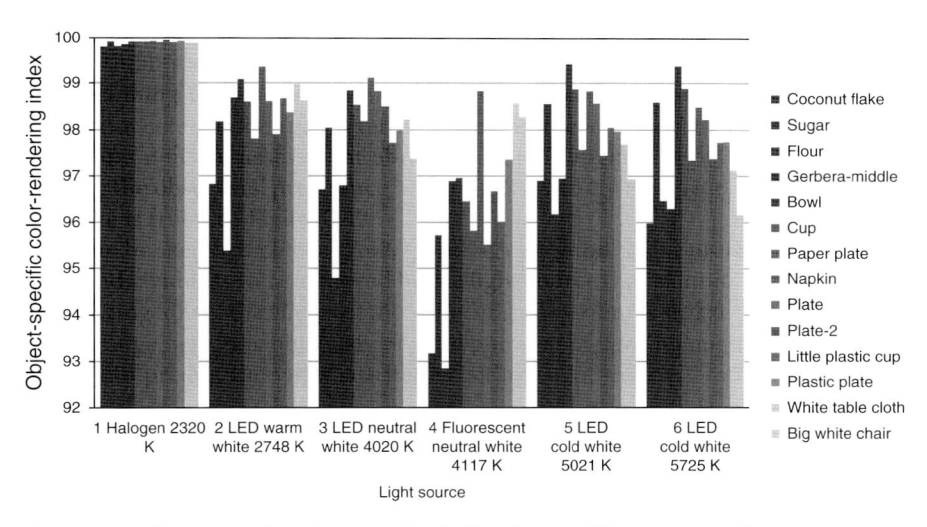

Figure 7.16 Object-specific color-rendering indices for the different white objects (see Figure 7.15) of the light sources of the experiment.

As can be seen from Figure 7.17, even in an office with predominantly white or achromatic objects, it is not only the white tone of the light source but also the whole distribution of white shades that is perceived by the observers and it is this distribution that determines the state of chromatic adaptation.

Concerning the psychophysical method, four observers were sitting around the table at a time. To specify the context of the visual assessment precisely, every subject listened to the following instructions before the session: "White furniture and white decorations are very popular today. They are often associated with a ceremony or a festive atmosphere. This room is an example for this. Imagine you are invited to a celebratory dinner (e.g., Dîner en Blanc) as a guest. You can sit down at the table, talk to each other but please do not talk about the furniture,

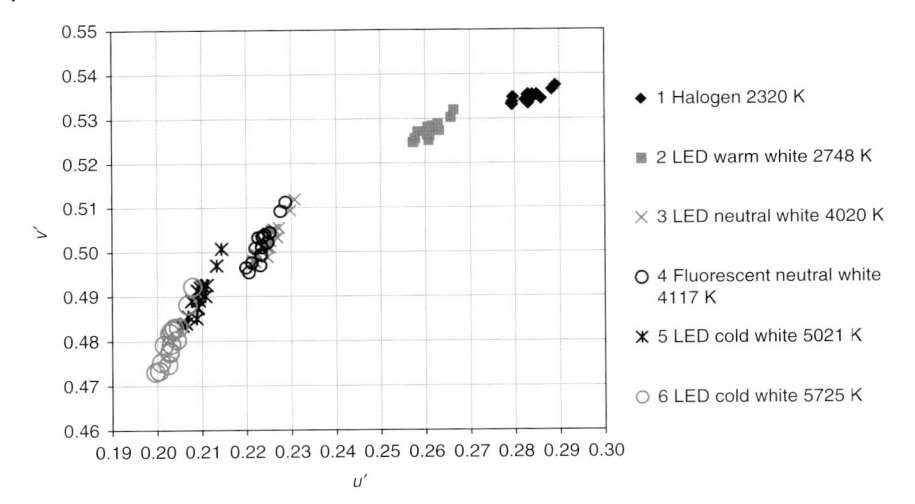

Figure 7.17 Distribution of the different shades of white of the selected set of white objects (Figure 7.15) in the $u' - v'$ chromaticity diagram.

the object or the illumination of the room. You can also stand up and go round the table to take a look at the white objects.

Your task is to assess the scene of the objects according to your preference on an interval scale between 0 and 100. The value of 100 corresponds to your maximum preference if you consider all objects including furniture in a way you would like to see them. We illuminate the objects with different light sources labeled by the numbers 1–6. Please assess your preference on the corresponding scale. The leader of the experiment tells you the number of the light source and you assess the object scene first with a pencil by a cross put along the scale. Every illumination will take 1.5 min (90 s) to be repeated three times.

You must finalize your answer about your preference at the third repetition by using a pen. There are two series with six illuminations each, a darker one and a brighter one. Please never look at the ceiling at which the light sources are located. Please wait half a minute after the switching to the next light source and then start to assess the new lighting situation. This is a 'game of patience', in fact a very easy task. Please look at the objects patiently, think about them and assess them – and do not forget, you are a guest at 'white' festive dinner."

Figure 7.18 shows the questionnaire according to the above description.

Only the third response marked by the pen was recorded and included in the statistical evaluation. Twenty European subjects of normal color vision (checked by the D-15 color vision test) living in Germany participated in that part of the experiment which is presented here (17 men, 3 women). First, these 20 observers were examined to see whether they were responsive to the question of the experiment: the standard deviation (STD) of all preference answers of every subject was computed and if the STD value of a specific subject was greater than the mean STD of all 20 subjects then this subject was taken into the cluster of "responsive" observers. Twelve subjects (11 men and 1 woman) belonged to this "responsive"

Figure 7.18 Filled questionnaire according to the description of the observer's task.

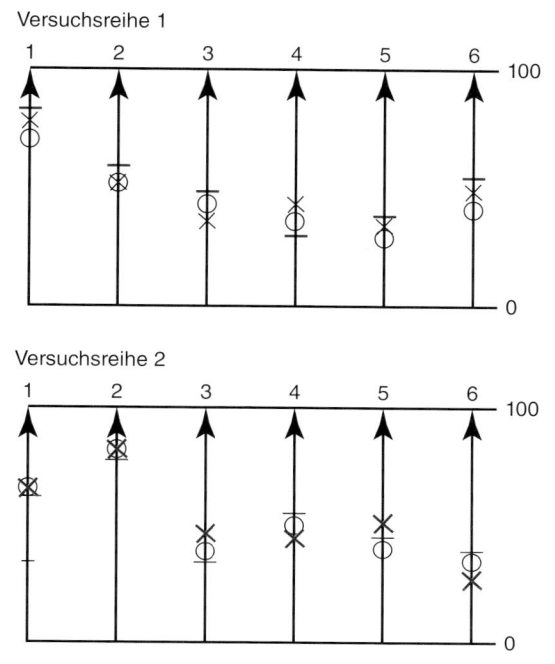

Versuchsreihe 1

Versuchsreihe 2

cluster. Only the results of this cluster will be analyzed below. The mean age of the observers in this cluster was 25.8 years (min., 22; max., 37 years).

7.4.2 Results and Discussion

Figure 7.19 shows the mean preference scale values of the 12 observers of the responsive cluster and their 95% confidence intervals grouped according to the two illuminance levels.

As can be seen from Figure 7.19, there is a general color temperature preference of neutral cool white tones peaking in the range of 4020 – 5021 K. The dislike of warm white (2320 – 2748 K) and cool white (5725 K) light sources is more accentuated at 500 lx than at 1200 lx. It can also be seen that the higher luminance level (1200 lx) is generally more preferred than the lower luminance level (500 lx) at all CCTs. Table 7.5 shows the average CCT preference values and the significance of a paired two-sided t test between the corresponding data of the 12 subjects at 500 lx and at 1200 lx.

As can be seen from Table 7.5, the mean preference is significantly higher for the higher illuminance level (1200 lx) than for the lower illuminance level (500 lx) for the two lower CCTs contradicting Kruithof's rule [18]. For neutral white and cool white, although the higher illuminance level was preferred, this difference was not significant so that Kruithof's rule could not be confirmed in the neutral and cool white range either.

7.4.3 Modeling in Terms of LMS Cone Signals and Their Combinations

Here, the aim is to find a rectifying variable to predict the CCT preference results in Figure 7.19 by using more fundamental human visual characteristics than CCT,

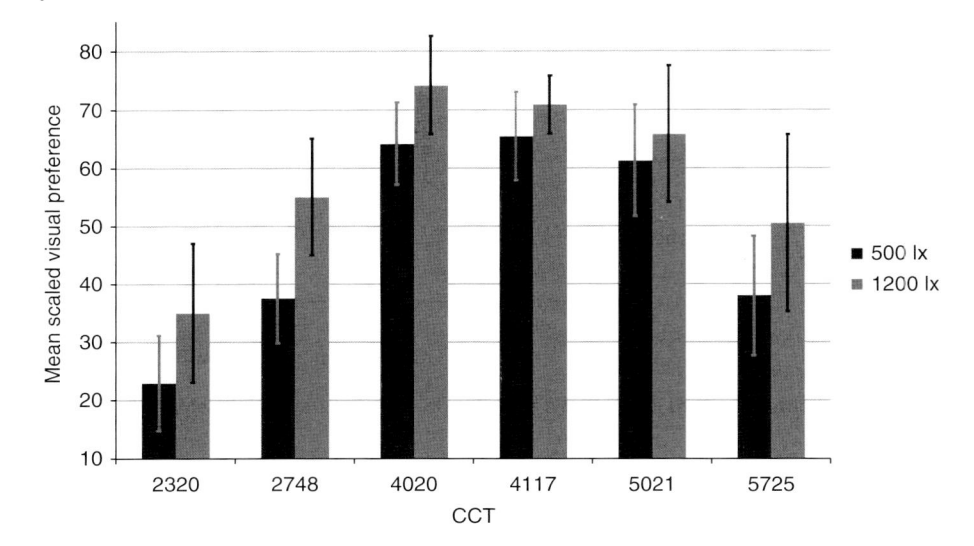

Figure 7.19 Mean preference scale values of the 12 observers of the "responsive" cluster and their 95% confidence intervals grouped according to the two illuminance levels. The light sources can be identified by their CCT and Table 7.4.

Table 7.5 Average CCT preference values and the significance of a paired two-sided t-test between the corresponding data of the 12 responsive subjects at 500 and at 1200 lx.

CCT (K)	Mean preference 500 lx	Mean preference 1200 lx	Significance p
2320	23[a]	35[a]	0.046
2748	38[a]	55[a]	0.000
4020	64	74	0.056
4117	65	71	0.084
5021	61	66	0.317
5725	38	50	0.148

a) Difference of the mean preference values is significant at the 5% level.

that is, the signals of the LMS cones. These signals are to be computed from the SPD of the light source as described in Section 2.1. By introducing the rectifying variable $r = (L - aM)/(L + aM - bS)$, that is, the ratio of the two chromatic signals (see Section 2.1) with the optimized weighting factors $a = 1.29384$; $b = 6.47002$, the mean scaled visual preference can be predicted linearly by the use of Eq. (7.7); see Figure 7.20.

$$\text{For } 1200 \text{ lx} : \text{CCT Preference} = -48.511\, r + 77.219$$
$$\text{For } 500 \text{ lx} : \text{CCT Preference} = -59.440\, r + 71.080$$
$$\text{with } r = (L - aM)/(L + aM - bS); a = 1.29384; b = 6.47002 \quad (7.7)$$

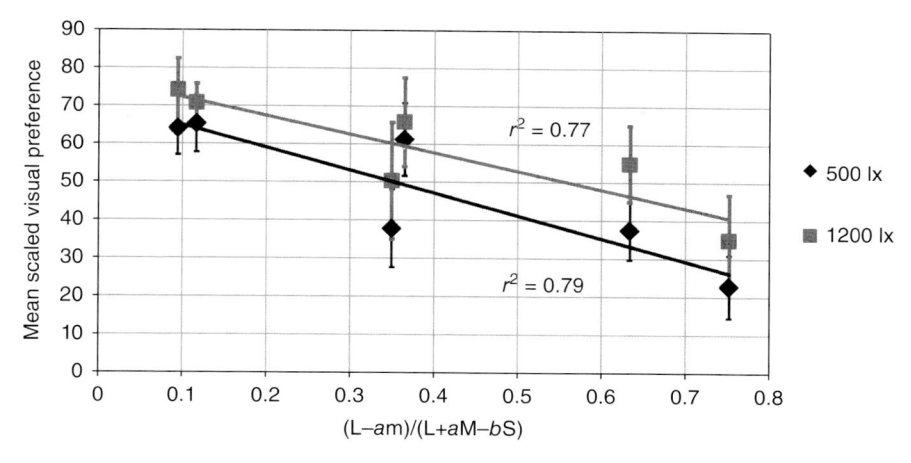

Figure 7.20 Mean preference scale values of the 12 observers of the "responsive" cluster and their 95% confidence intervals grouped according to the two illuminance levels (compare with Figure 7.19) as a function of the $r = (L - aM)/(L + aM - bS)$ value of the light sources using the optimum values $a = 1.29384; b = 6.47002$. LMS values can be computed as described in Section 2.1 if the relative spectral power distribution of the light source is known. Fit lines are based on Eq. (7.7).

As can be seen from Figure 7.20, CCT preference increases with decreasing values of the chromatic signal ratio $r = (L - aM)/(L + aM - bS)$. This ratio has its minimum at about 4000 K corresponding to the highest visual CCT preference.

7.4.4 Summary

In the experiment, European subjects living in Germany scaled their color temperature preference of a white scene with white objects depending on CCT and illuminance level. Results showed a general color temperature preference of neutral cool white. The higher luminance level was generally more preferred than the lower illuminance level. This was also true for warm white contradicting Kruithof's rule. Results are relevant to the selection of preferred color temperatures for indoor light sources to be used for household illumination. Equation (7.7) that starts from the ratio of the two chromatic signals of the human visual system can be used to predict the color temperature preference of a light source or compare two light sources according to their CCT preference.

7.5 Color Temperature Preference of Illumination with Red, Blue, and Colorful Object Combinations

In the experiment described in Section 7.4, the color temperature preference of a white scene with white objects was dealt with, with light sources of different color-rendering properties ranging between $R_a = 81$ ("moderate") and $R_a = 99$ ("excellent"). In this section, the color temperature preference of a scene with different colorful objects (red, blue, and mixed) will be investigated in a visual experiment using light sources of excellent color-rendering properties, with no or only very little oversaturation of the objects and a high illuminance level (2500 lx) representing the context of shop or retail lighting. The aims are, similar to Section 7.4, to point out the most preferred CCT range and its possible dependence on object color context (red, blue, or mixed) as well as the modeling of CCT preference under these viewing conditions.

The motivation is to provide hints for manufacturers (especially those of high-quality LED shop lighting or retail lighting) to select the CCT of lighting products in order to illuminate different types of colored objects with a high CRI ($R_a \geq 96$) at high illuminance levels ($E_v \geq 2300$ lx) to provide good user acceptance. In this section, a visual CCT preference experiment will be described that used a mechanically and thermally robust and stable, calibrated and characterized multi-LED light engine and a viewing booth containing the above-mentioned three types of arrangements of colored objects (red, blue, and mixed objects).

7.5.1 Experimental Method

Three arrangements of colored artificial objects (red–orange, blue–lilac, and mixed or "colorful," the latter containing objects of numerous different colors) were adhered on three white plastic boards [19]. Objects included the MCCC (MacBeth ColorChecker Chart®), artificial flowers, candles, pots, bowls, yarn, glove, soap, and so on. The spectral reflectance of the objects was measured (except MCCC which is known from literature) using diffuse illumination, excluding gloss and using a calibrated white standard; see Figure 7.21.

The three colored object combinations (Figure 7.21) were illuminated diffusely and homogeneously by seven different white LED light emission spectra generated by a spectrally tunable, mechanically and thermally stable and reproducible multi-LED light engine with six LED channels in the white chamber of a viewing booth at seven different CCTs: 2719, 2960, 3501, 3985, 4917, 5755, and 6428 K; see Figure 7.22. In Table 7.6, the colorimetric properties of these seven multi-LED spectra and their luminance values measured at the bottom of the viewing booth are shown. The size of the viewing chamber was 330 mm (height) × 440 mm (width) × 440 mm (length). Observers viewed the objects from a viewing distance of 50 cm. The change of the illuminance at the objects equaled 10%.

The following can be seen from Figure 7.22 and Table 7.6:

1) By fine-tuning the six LED channels of the multi-LED engine, white tones lying on the Planckian locus (for < 5000 K) or on the daylight locus (> 5000 K) with $\Delta u'v' < 0.0015$ (mean: $\Delta u'v' = 0.0008$) were achieved.
2) Every one of the seven multi-LED spectra had very good color-rendering property ($R_a \geq 96$, mean value of the special color-rendering indices $R_1 - R_{14} \geq 93$ and the special CRI of the saturated red CIE test color sample $R_9 \geq 94$). This can also be seen from the high values of their CQS parameters (Q_a, Q_f, Q_p), MCRI, and R_f values.
3) The low ΔC^* values (degree of oversaturation, see Section 2.1) indicate a low degree of oversaturation of the objects. This was important for the observers to concentrate on their white tone (CCT) preference judgment in the context of

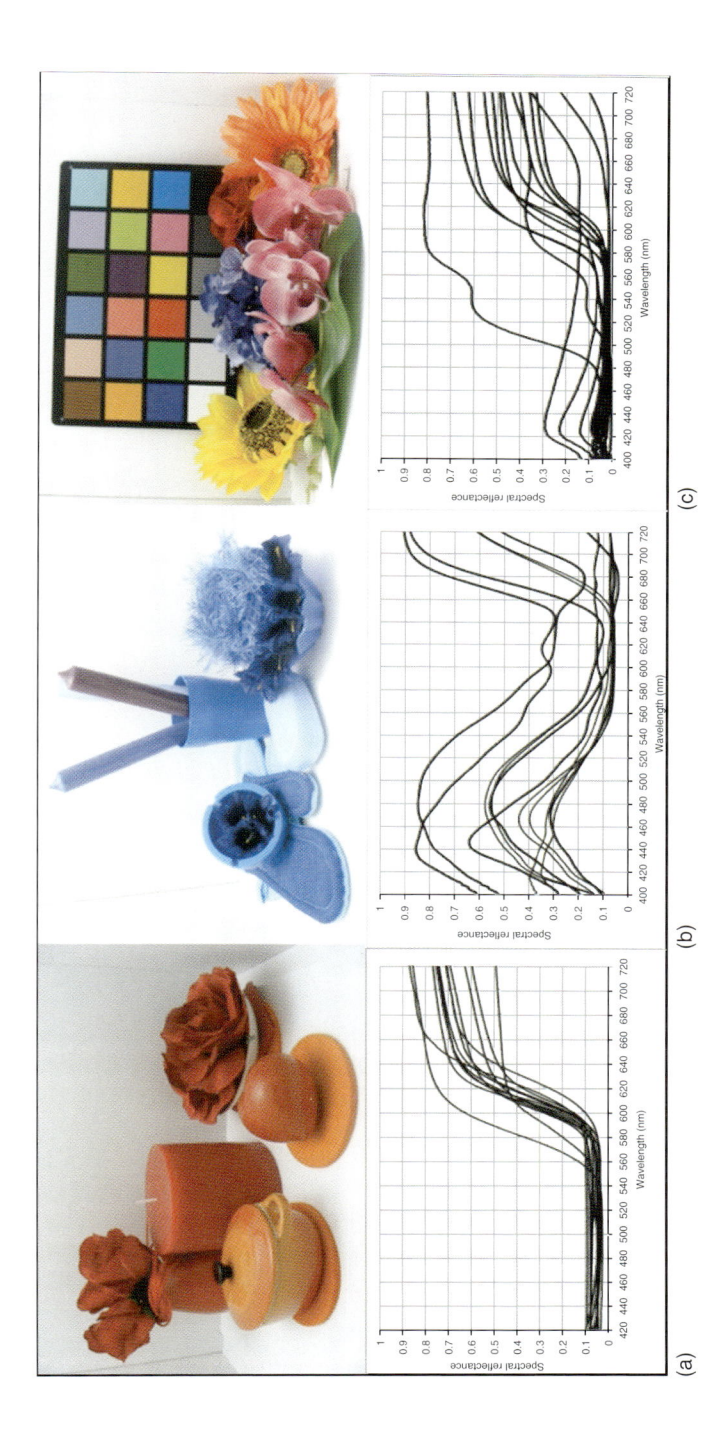

Figure 7.21 Three arrangements of real colored artificial objects. (a) "Red" with red–orange objects; (b) "blue" with blue–lilac objects; and (c) "colorful" with various colors. The "colorful" arrangement included MCCC, the MacBeth ColorChecker Chart®. Measured spectral reflectance curves of the objects (except MCCC) are shown in the lower row. (Bodrogi et al. 2015 [19]. Reproduced with permission of Sage Publications.)

Table 7.6 Colorimetric properties and color quality descriptors (see Section 2.4) of the seven multi-LED spectra and their luminance level measured at the bottom of the viewing booth. After [19].

CCT (K)	2719	2960	3501	3985	4917	5755	6428	Mean	STD
$u'\cdot v'$	0.2615·0.5276	0.2518·0.5229	0.2357·0.5112	0.2248·0.5031	0.2123·0.4859	0.2016·0.4784	0.1975·0.4700	Void	Void
$\Delta u'v'$	0.0006	0.00073	0.00007	0.00136	0.00001	0.00142	0.00109	0.0008	0.0006
CRI R_a	97	97	96	96	96	96	96	96	0.3
Mean CRI ($R_1 - R_{14}$)	95	95	94	94	93	94	93	94	0.8
GAI	101	101	101	101	100	100	101	101	0.5
FCI	132	130	127	123	117	111	110	121	8.4
CQS Q_a	96	96	95	94	93	94	94	94	1.1
CQS Q_f	94	95	94	94	92	93	93	93	0.9
CQS Q_p	95	96	97	95	94	94	95	95	0.8
CQS Q_g	100	101	102	101	101	100	101	101	0.7
MCRI	90	91	92	92	93	92	92	92	0.7
IES-R_f	93	92	91	90	88	90	89	90	1.6
IES-R_g	101	99	98	97	96	96	96	97	1.7
ΔC^* (all objects)	0.7	0.9	1.0	0.8	0.7	0.5	0.9	0.8	0.2
ΔC^* (colorful)	0.5	0.7	0.8	0.6	0.4	0.2	0.5	0.5	0.2
ΔC^* (red)	1.3	1.4	1.5	1.4	1.6	1.4	1.5	1.4	0.1
ΔC^* (blue)	0.3	0.7	0.8	0.6	0.1	0.1	0.7	0.5	0.3
Luminance (cd/m^2)	759	734	739	754	736	737	741	743	10

ΔC^*: mean values.

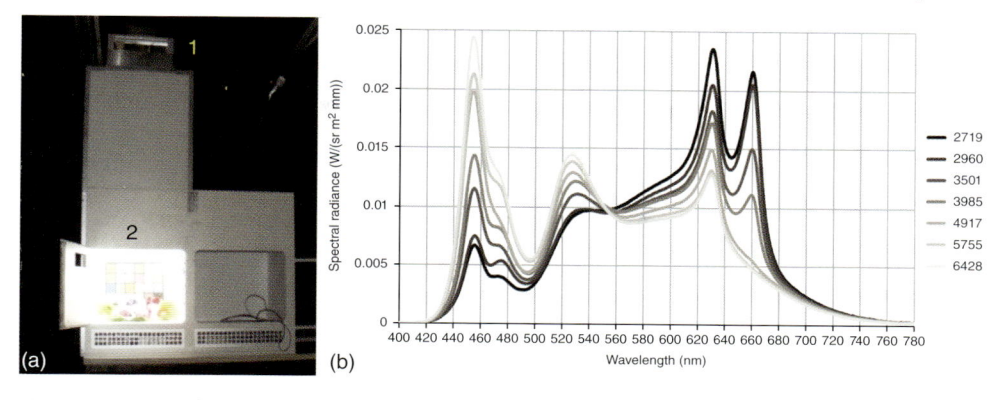

Figure 7.22 (a) Viewing booth with the multi-LED light engine at the top (No. 1). Only the white chamber No. 2 (left chamber) was used. It contained the arrangements of artificial objects (one arrangement at a time; here, the "colorful" combination is shown). Luminance at the bottom of the viewing chamber: 743 ± 10 cd/m^2 (corresponding to the horizontal illuminance of 2457 ± 27 lx at the bottom of the booth). (b) Spectral radiance of the seven multi-LED spectra used in the visual experiment (measured at a white standard at the bottom of the viewing chamber); CCT in kelvin in the legend. (Bodrogi et al. 2015 [19]. Reproduced with permission of Sage Publications.)

the colored objects and not on the varying degree of oversaturation as opposed to the high and highly variable ΔC^* values in the experiment of Section 7.7.
4) Correspondingly, the values of the color gamut measures of the seven light sources (CQS Q_g, GAI, FCI, and IES-R_g) also remain in their moderate range.
5) The luminance of a white standard positioned at the bottom of the booth equaled 743 ± 10 cd/m^2 corresponding to the horizontal illuminance of 2457 ± 27 lx at the bottom of the booth.

The multi-LED engine was very stable. Its white points could be reproduced on a long-term basis within $\Delta u'v' = 0.0004$ and $\Delta R_a < 0.08$ at every CCT. The panel of observers included 46 young subjects (18 men and 28 women) of European cultural background who all lived in Germany (average age: 25.4 ± 0.3). All subjects had good or corrected visual acuity and normal color vision. First, these 46 observers were examined to see whether they were responsive to the preference scaling task of the experiment (see below): the STD of all preference answers of every subject was computed and if the STD value of a specific subject was greater than the mean STD of all 46 subjects then this subject was taken into the cluster of "responsive" observers. Thirty-three subjects (12 men and 21 women) belonged to this "responsive" cluster. Only the results of this cluster will be analyzed below.

Subjects looked into the chamber of the viewing booth containing one of the three object combinations at a time and adapted for 1 min to one of the seven LED spectra; see Figures 7.21 and 7.22. After the adaptation to one of the seven LED spectra shown in Figure 7.22, subjects had to assess their preference about the color appearance of the current object combination under the current light source. They had the following written instruction:

"Your task is to assess the three scenes with different objects (red, blue, and colorful) after your preference on an interval scale between 0 and 100. Hundred corresponds to your maximal liking (your maximal preference) if you look at all objects as you would like to see them in a household (indoors). We illuminate the objects with different light sources numbered 1–7. Please assess your preference on the appropriate scale.

The leader of the experiment will tell you the number of the light source and you will assess the object scene first using a pencil by putting a cross along the scale. Every illuminating situation will take 1 min and will be repeated three times. You must finalize your answer at the third repetition by the aid of a pen. Please wait after the change of the light source for about 20 s and only after that time please begin the assessment of the new situation. Please look at the objects patiently, think about your task and then assess."

Figure 7.23 shows the score sheet. Note that only the third (final) response was recorded. There was a training phase before every experiment for every subject assessing one light source and one object combination.

7.5.2 Results and Discussion

Figure 7.24 shows the mean CCT preference scale values of the 33 observers of the "responsive" cluster of observers and their 95% confidence intervals grouped according to the type of object combination (red, blue, or colorful; see Figure 7.21) and also the overall mean CCT preference value as a function of CCT.

As can be seen from Figure 7.24, CCT is generally preferred in the neural white–cool white CCT range between 3985 and 6428 K. For red objects, preference peaks in the CCT range 3985–4917 K. The effect of CCT on visually scaled preference was statistically significant for every object combination and also when all results were considered (ANOVA; $\alpha = 0.05$). The effect of gender on visual preference was not significant (ANOVA; $\alpha = 0.05$). The blue and colorful object scenes were more preferred in the cool white CCT range (5755–6428 K) than the red object scene in the same range.

The red object scene was, however, more preferred in the warm white CCT range (2719–2960 K) than the blue and colorful object combination. The low preference of the blue and colorful object combinations at 2719 K is clearly visible from Figure 7.24. Table 7.7 shows the average CCT preference values for the red, blue, and colorful object combinations, the overall average preference values, and the significance of paired two-sided t test between the red–blue, red–colorful and blue–colorful data of the 33 subjects.

As can be seen from Table 7.7, the difference between the mean CCT preference values of the red and blue object combinations was significant at the 5% level for all CCT values, the red object condition being more preferred until 3985 K and the blue object condition being more preferred from 4917 K on. The mean CCT preference with the red object scene is significantly higher than the mean CCT preference with the colorful combination until 3501 K. At 4917 K, "blue" is significantly more preferred than "colorful." Finally, at 5755 K, "colorful" is significantly more preferred than "red."

7.5.3 Modeling in Terms of LMS Cone Signals and Their Combinations

In this section, the LMS signal combination model of Section 7.4 (using the rectifying variable $r = (L - aM)/(L + aM - bS)$, the ratio of the two chromatic

Training: _____

Versuchsreihe 1

Versuchsreihe 2

Versuchsreihe 3

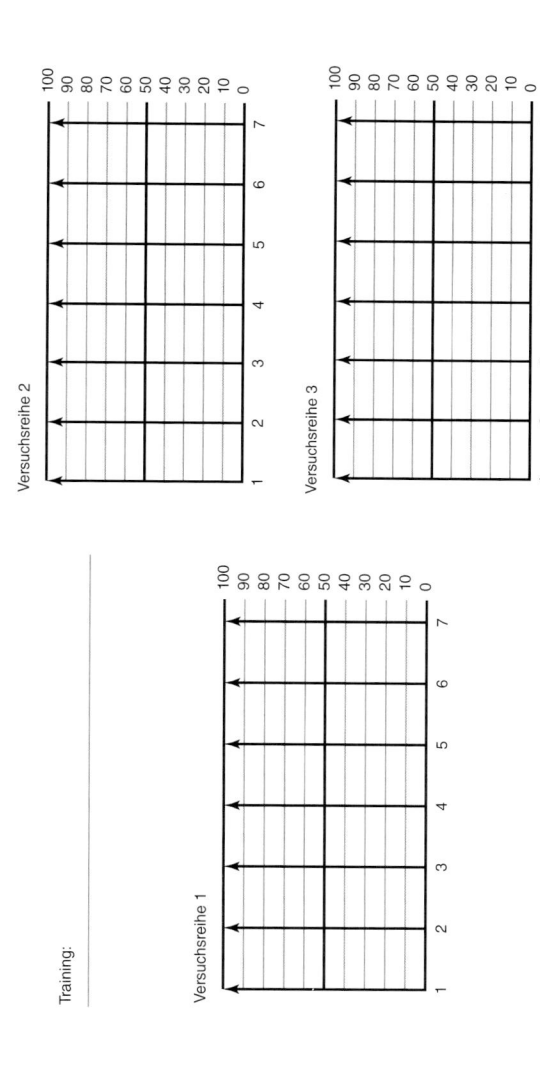

Figure 7.23 Score sheet in German language. "Versuchsreihe" = experimental series; 1: red; 2: blue; and 3: "colorful." (Bodrogi et al. 2015 [19]. Reproduced with permission of Sage Publications.)

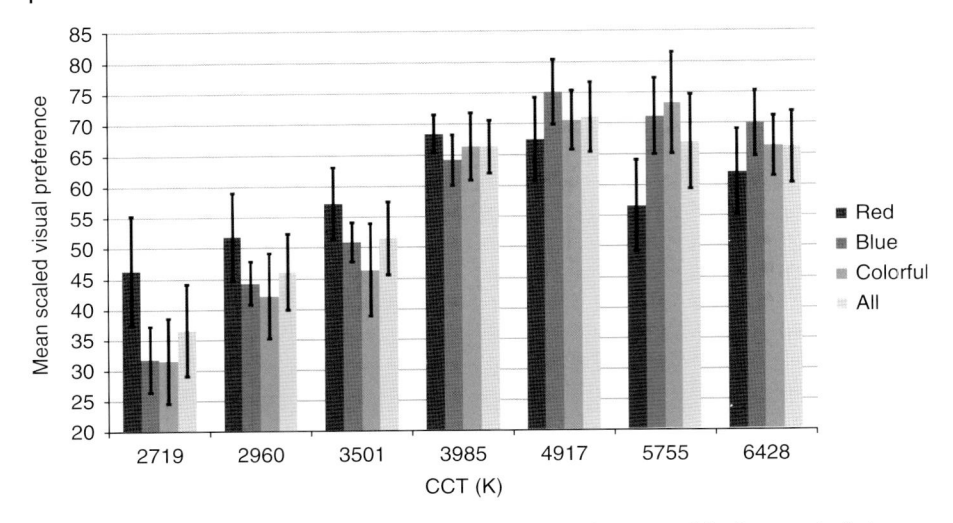

Figure 7.24 Mean CCT preference scale values of the 33 observers of the "responsive" cluster of observers and their 95% confidence intervals grouped according to the type of object combination (red, blue, or colorful; see Figure 7.21) and the overall mean CCT preference value as a function of CCT.

Table 7.7 Mean CCT (K) preference values for the red, blue, and colorful (cf.) object combinations, the overall average preference values (all), and the significance of paired two-sided t test between the red–blue (r-b), red–colorful (r-cf), and blue–colorful (b-cf) data of the 33 subjects.

CCT	Red	Blue	Cf.	All	p (b-cf)	p (b-r)	p (r-cf)
2719	46	32	32	37	0.909	0.002[a]	0.000[a]
2960	52	44	42	46	0.461	0.025[a]	0.000[a]
3501	57	51	46	52	0.149	0.009[a]	0.000[a]
3985	68	64	66	66	0.225	0.031[a]	0.269
4917	68	75	71	71	0.013[a]	0.007[a]	0.231
5755	57	71	73	67	0.465	0.000[a]	0.000[a]
6428	62	70	66	66	0.061	0.006[a]	0.054

a) Difference of the mean preference values is significant at the 5% level.

signals) will be applied to the present seven multi-LED SPDs shown in Figure 7.22. The aim is to find the optimum values of a and b for best correlation with the mean visual preference ratings for every object combination (red, blue, and colorful objects). Before doing so, the correlation coefficients between the color quality descriptors of the seven multi-LED spectra (Table 7.6) and the mean CCT preference values for the red, blue, and colorful (cf.) object combinations (Table 7.7) were also computed.

Most of these correlation coefficients were negative or very low. In the following, only the few positive, significant ($\alpha = 0.05$) correlations will be analyzed. The

Table 7.8 Parameter values (a and b) optimized to get best correlation of the variable $r = (L - aM)/(L + aM - bS)$ with the mean visual CCT preference data in comparison with the CCT preference of white objects (Figure 7.20).

Object type (illuminance level)	Blue (2500 lx)	Colorful (2500 lx)	Red (2500 lx)	All (2500 lx)	White (500 lx)	White (1200 lx)
a	1.26	1.26	1.24	1.26	1.29	1.29
b	7.07	7.16	6.45	7.04	6.47	6.7
r	−0.95	−0.93	−0.99	−0.96	−0.89	−0.88
r^2	0.91	0.87	0.99	0.93	0.79	0.77

mean "blue" ($r^2 = 0.74$) and "colorful" ($r^2 = 0.72$) CCT preference data correlated with CCT significantly (but not the "red" data, $r^2 = 0.28$). But, as mentioned above, instead of CCT, all three types of preference data (red, blue, colorful) will be modeled in terms of the more fundamental LMS cone signal quantities of the human visual system. It is worth mentioning that significant correlation exists between MCRI and the red ($r^2 = 0.76$), blue ($r^2 = 0.66$), and colorful ($r^2 = 0.84$) mean preference data, and also between the variable ΔC^* (red) shown in Table 7.6 and the "red" preference data ($r^2 = 0.60$). Possibly, the slight increase of the chroma of the red objects contributes to the tendency of CCT preference.

Turning back to the modeling in terms of LMS cone signals, Table 7.8 shows the parameter values a and b optimized to get best correlation of the variable $r = (L - aM)/(L + aM - bS)$ with the mean visual CCT preference data in comparison with the CCT preference of white objects (Figure 7.20).

As can be seen from Table 7.8, the variable $r = (L - aM)/(L + aM - bS)$ can be used to predict CCT preference for all types of object scene with good accuracy, similar to Figure 7.20 and two light sources can be compared from the point of view of their CCT preference (CCT preference decreases with increasing values of r).

7.5.4 Summary

The object scene dependence (combinations of red, blue, and mixed or "colorful" objects) of CCT preference of high illuminance ($E = 2500$ lx), high color rendering ($R_a = 96$) multi-LED illumination was investigated. A visual CCT preference experiment was conducted with a six-channel multi-LED light engine with the colored objects arranged in a viewing booth. CCT preference was significantly influenced by CCT (varied between 2700 and 6500 K) and object scene color (red, blue, or colorful). CCT was generally preferred in the neural white–cool white CCT range (4000 – 6500 K). For red objects, preference maxima occurred in the CCT range 4000 – 5000 K. CCT preference was modeled in terms of the LMS signals derived from the light source SPDs, similar to Section 7.4. Results can be used to select a high-quality lighting product that illuminates different object scenes at a high illuminance level to achieve good user acceptance especially for applications with high illuminance requirements such as shop and retail lighting.

7.6 Experiments on Color Preference, Naturalness, and Vividness in a Real Room

As mentioned in Section 7.1, the color quality attributes color preference, naturalness ,and vividness (as defined in Section 2.4) are associated with more or less oversaturation of the illuminated colored objects (in comparison with the reference light source). Therefore, these attributes cannot be described by a fidelity type index alone because the concept of color fidelity does not allow any oversaturation in comparison with the reference light source. In the experiment described in the present section, observers scaled the color preference, naturalness, and vividness visually on interval scales (0–100) labeled by semantic categories (e.g., "moderate," "good," "very good") in the context of office lighting [20].

Five customary warm white and neutral white light sources (fluorescent, halogen, phosphor-converted LED) with CCTs between 2300 and 4100 K at the Planckian locus (Duv < 0.0013) and without any object oversaturation effect illuminated a table with a collection of colored objects in a real, refurbished experimental room mimicking an office with white walls, white cupboards, and a white tablecloth on the table at a typical illuminance level of 470 lx [20]. In the present section, the assessments of the visual attributes color preference, naturalness, and vividness by the observers will be compared among each other. They will be predicted by the aid of different color quality indices.

Selected pairs of color quality indices will be combined linearly to increase the correlation between the new, combined indices and the visual results. Criterion values of the most suitable indices or linear combinations for "good" color preference, naturalness, and vividness will be computed in order to provide a usable acceptance limit for the spectral design and evaluation of light sources. A cause analysis in terms of the chroma shifts of the actual colored test objects (whose spectral reflectance curves were measured) between the test light sources used in the experiments and their reference light sources will be carried out. It will be analyzed how the visual attributes preference, naturalness, and vividness depend on the values of these chroma shifts.

In Section 7.5, observers preferred the white tone chromaticity of neutral white–cool white CCTs [19] for combinations of red, blue, and colorful objects. In the experiment of the present section, CCT turned out to be a useful independent variable to predict the dependent variable "color preference." The following, so-called "white tone preference" hypothesis was formulated: the chromaticity of the white tone that is seen together with the colored objects contributes to the color preference assessment about the colored objects. In this section, it will also be shown that according to the above hypothesis, the variable CCT can be combined with another, suitable color quality index (CQI) to describe color preference more efficiently than with a standalone index.

7.6.1 Experimental Method

Figure 7.25 shows the colored objects to be assessed visually by the subjects. These objects were arranged on a table covered by a white tablecloth. The experimental room mimicked an office environment. The room had white walls, white

Figure 7.25 Arrangement of colored objects on the table to be assessed visually by the subjects. Red–orange artificial objects (left); purple–blue–cyan artificial objects (middle); artificial colored objects of different colors (right); and various natural objects (distributed among the artificial objects on the left and on the right). Observers were allowed to walk around the table (height: 755 mm; length: 1486 mm; width: 722 mm) while assessing the colored objects. The horizontal illuminance equaled 470 ± 28 lx (among the five different light sources; see Figure 7.26) in the middle of the table. Illuminance changes on the table were within 5%. (Khanh *et al.* 2016 [20]. Reproduced with permission of Sage Publications.)

chairs, and two white cupboards. Observers were instructed to evaluate the color appearance of the colored objects on the table (Figure 7.25) by viewing all objects simultaneously. The table stood in the middle of the room and the observers were allowed to walk around the table while assessing the colored objects [20].

Figure 7.26 shows the spectral radiance of the five light sources built in the ceiling of the room and used in the visual experiment. By the use of a diffuser plate mounted at the ceiling below the light sources, the illumination was diffuse and uniform in the room. The colorimetric properties and CQI values of the five light sources are listed in Table 7.9.

As can be seen from Table 7.9, three warm white (HAL: tungsten halogen; LED-ww: phosphor-converted LED; FL-ww: fluorescent) and two neutral white (FL-nw: fluorescent; LED-nw: phosphor-converted LED) light sources were used. All of them desaturate the colored objects compared to their reference light source ($\Delta C^* < 0$). Apart from HAL, their CIE R_a values corresponded to the category of customary light sources without outstanding CRI values. Their CRI R_a values were greater than 80 as required by the standard [17]. Figure 7.27 illustrates the spectral reflectance of the colored objects on the table (Figure 7.25).

Thirty-eight observers of normal or corrected visual acuity and without any known color vision deficiency took part in the study (10 women, 28 men; between 20 and 55 years of age; mean: 35 years) in groups of four to six subjects. They were not allowed to communicate with each other during the experiment.

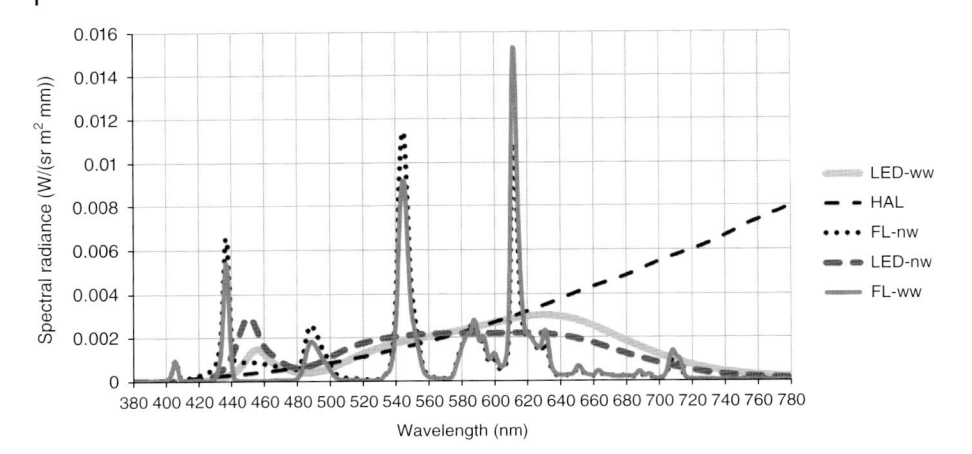

Figure 7.26 Spectral radiance of the five light sources measured at a horizontal white standard in the middle of the table. The horizontal illuminance equaled 470 ± 28 lx in the middle of the table. (Khanh et al. 2016 [20]. Reproduced with permission of Sage Publications.)

Table 7.9 Colorimetric properties of the five light sources in the experiment.

Light source	LED-ww	HAL	FL-nw	LED-nw	FL-ww
X	0.457	0.494	0.376	0.381	0.460
Y	0.410	0.416	0.374	0.380	0.408
CCT	2741	2322	4116	4013	2682
Duv	0.00001	0.00024	0.00017	0.00128	0.00092
$\Delta u'v'$	0.00001	0.0003	0.0002	0.0017	0.0013
R_a	89	100	81	88	84
GAI	50	36	83	77	52
CQS Q_a	88	98	80	88	81
CQS Q_f	87	98	79	87	79
CQS Q_g	100	100	100	99	104
CQS Q_p	92	99	87	92	87
MCRI	88	87	84	90	80
FCI	122	126	106	112	114
IES R_f	86	99	77	86	72
IES R_g	100	100	101	100	104
ΔC^*	−1.37	−0.05	−2.59	−1.40	−3.50

Source: Khanh et al. 2016 [20]. Reproduced with permission of Sage Publications.
ΔC^*: mean chroma shift for all objects (Figure 7.25) between each light source and its reference light source.

7.6 Experiments on Color Preference, Naturalness, and Vividness in a Real Room

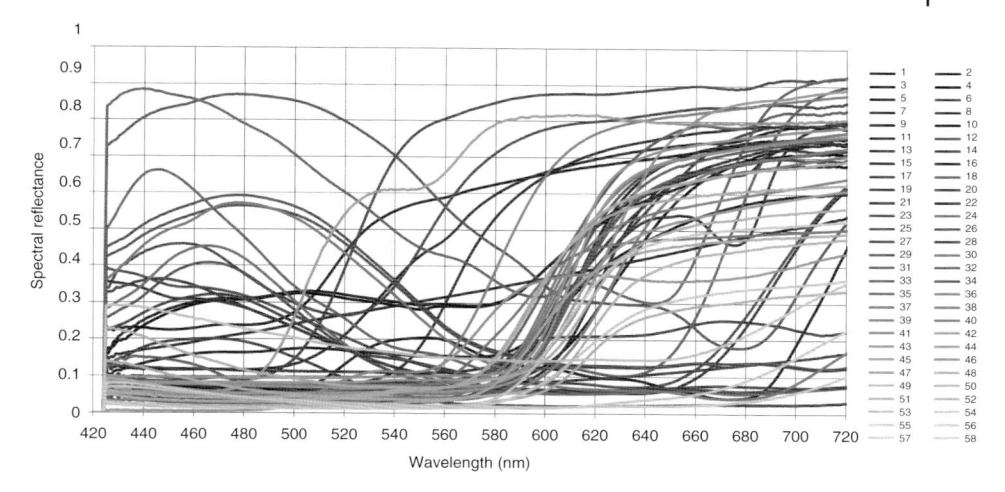

Figure 7.27 Spectral reflectance curves of the colored objects used in the experiment [20].

After a training using a randomly chosen light source, the five light sources were presented in one of three possible orders to the individual groups of observers. Figure 7.28 shows the structure of the scales on the questionnaire. There were three such scales to evaluate color preference, naturalness, and vividness, respectively. Every observer was instructed to consider all three scales simultaneously and put a cross onto the scale at the location corresponding to her/his judgment

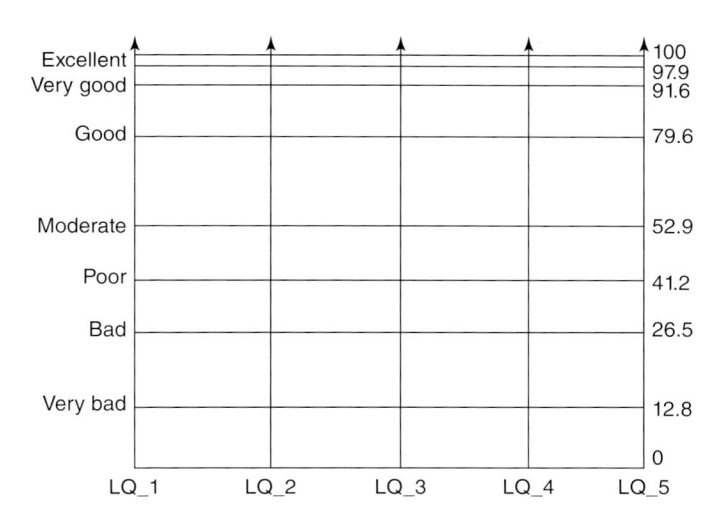

Figure 7.28 Scale on the questionnaire with one continuous interval scale (0–100) and the semantic category labels (translated from German) for every one of the five light sources (LQ_1, …, LQ_5). The nonuniform spacing of the category labels on the interval scale 0–100 is based on a previous study [22]. Every observer filled three such questionnaires (for general color preference, naturalness, and vividness) simultaneously. (Khanh *et al.* 2016 [20]. Reproduced with permission of Sage Publications.)

by considering the semantic categories. Subjects were instructed about the definitions of the three visual attributes according to Section 2.4.

The right ordinate of the questionnaire in Figure 7.28 shows the correspondence of the category labels to the interval scale values 0–100. The spacing of the categories was elaborated in a previous study [22]. Every observer carried out one judgment for every one of the five light sources. The adaptation time to any new light source switched on instead of the previous light source was 2 min and then, every light source had three further minutes for all objects on the table to be assessed. The visual dataset consisted of 38 (observers) × 5 (light sources) × 3 (visual attributes: color preference, naturalness, and vividness) = 570 numbers between 0 and 100.

7.6.2 Relationship among the Visual Interval Scale Variables Color Naturalness, Vividness, and Preference

Figure 7.29 shows the relationship among the visual interval scale variables color naturalness, vividness, and preference.

As can be seen from Figure 7.29, mean ratings or their 95% confidence intervals hardly reach the "good" level (i.e., the value of 79.6, see Figure 7.28) in the case of the five customary light sources that do not oversaturate object colors. It can also be seen from Figure 7.29a that there is strong correlation ($r = 0.73$; $r^2 = 0.54$) between scaled color preference and naturalness. The outlier point is the neutral white light source LED-nw (CCT = 4020 K) that exhibits higher color preference according to the white tone preference hypothesis described in Chapter 1. There is also strong correlation ($r = 0.71$; $r^2 = 0.54$) between scaled vividness and naturalness.

The outlier with high vividness (HAL) in Figure 7.29c can be explained by the fact that the light source HAL has the highest value of $\Delta C^*(-0.05)$ among all light sources; compare with Figure 7.33. To interpret the middle diagram of Figure 7.29 (vividness vs. color preference) it can be considered that, although the perceived vividness of the colored objects in the scene (Figure 7.25) increases for the case of LED-ww (CCT = 2757 K) and HAL (CCT = 2340 K) compared to LED-nw

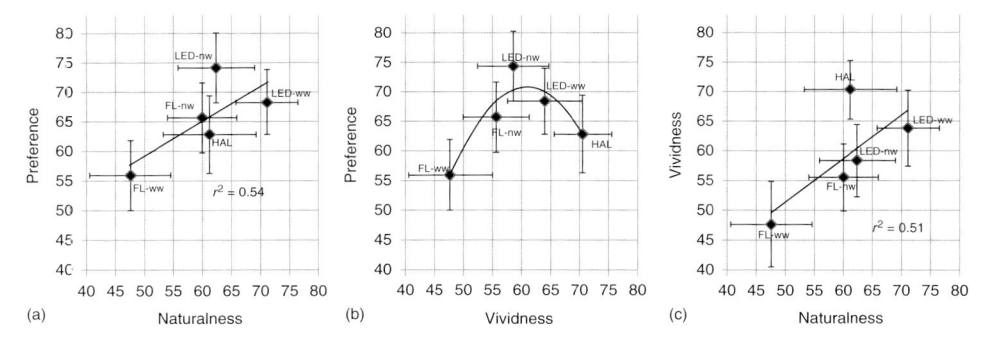

Figure 7.29 Relationship of the visually scaled attributes color preference, naturalness, and vividness. Mean values and their 95% confidence intervals for all observers (for the quadratic fit curve in the middle, $r^2 = 0.86$). (Khanh et al. 2016 [20]. Reproduced with permission of Sage Publications.)

(CCT = 4020 K), the preference of the colored objects under these two light sources is less according to the lower (warm white) CCT of the illumination. The lower CCT (with less preferred white chromaticities) of these two warm white light sources interacts with the somewhat higher perceived vividness of the colored objects. This interaction causes the decreasing branch of the vividness versus preference tendency. Concerning the right diagram of Figure 7.29, the outlier point is the tungsten halogen light source (HAL) that causes relatively high mean vividness (about 70, which corresponds to "good-moderate"; see the right ordinate of Figure 7.28) but less mean naturalness (about 62 corresponding to "moderate-good") of the illuminated colored objects.

7.6.3 Correlation of the Visual Assessments with Color Quality Indices

Table 7.10 shows the correlation coefficients r between the mean visual interval scale variables and the color quality indices plus the independent variable CCT.

As can be seen from Table 7.10, the standalone CQS metrics (Q_f, Q_a, Q_g, and Q_p), the standalone TM 30 metrics R_f and R_g, the CIE CRI R_a and the color gamut metrics FCI and GAI do not correlate well with the visual attributes naturalness and color preference. The following moderate-strong ($r > 0.5$) positive correlations can be seen from Table 7.10: color preference with MCRI and CCT (with a weak correlation between MCRI and CCT, see Table 7.9); naturalness and vividness with MCRI and R_f (with a moderate correlation between MCRI and R_f). Vividness exhibits strong correlation with the metrics R_f, R_a, Q_f, Q_a, Q_p, MCRI, and FCI while the metrics R_g and Q_g show strong negative correlation with all three perceptual attributes. The causes of these tendencies will be analyzed in Section 7.6.5 in terms of the chroma shifts and color gamut distortions.

Table 7.10 Correlation coefficients r between the mean visual interval scale variables and the color quality indices plus CCT.

	Preference	Naturalness	Vividness
IES R_f	0.32	0.54	0.95[a]
IES R_g	−0.87	−0.89[a]	−0.78
CCT (K)	0.61	0.10	−0.33
MCRI	0.89[a]	0.85	0.72
CIE CRI R_a	0.08	0.33	0.86
CQS Q_f	0.22	0.44	0.92[a]
GAI	0.49	−0.00	−0.48
CQS Q_g	−0.89[a]	−0.81	−0.68
FCI	−0.13	0.35	0.76
CQS Q_a	0.16	0.40	0.90[a]
CQS Q_p	0.17	0.38	0.88[a]

Source: Khanh et al. 2016 [20]. Reproduced with permission of Sage Publications.
a) Correlation coefficient significant at the 0.05 level.

7.6.4 Combinations of Color Quality Indices and Their Semantic Interpretation for the Set of Five Light Sources

Selected pairs of color quality indices were combined as linear combinations in order to maximize the value of the correlation coefficient between a linear combination and the mean value of a visually scaled attribute (color preference, naturalness, or vividness). The following rules were applied to select the suitable indices: (i) the correlation coefficient of both indices should be less than 0.7 in order to ensure a certain degree of independence of both indices and (ii) the resulting combined CQI should exhibit very strong correlation with an attribute ($r > 0.9$).

7.6.4.1 Prediction of Vividness

According to the above selection rules, first the indices R_f and MCRI were combined to predict visually scaled vividness; see Eq. (7.8).

$$\text{CQI}_{\text{viv}} = a\ R_f + b\ \text{MCRI} \tag{7.8}$$

The optimum parameter values equaled $a = 1.495; b = 0.506$ resulting in a correlation coefficient $r = 0.96$ between CQI_{viv} and the mean scaled vividness values. Although the correlation coefficient resulting from parameter optimization ($r = 0.96$; significant at the 0.05 level) is only slightly higher than the correlation of vividness with R_f alone ($r = 0.95$; also significant at the 0.05 level), this analysis is included as an important example for the index combination and semantic interpretation method. Figure 7.30 shows the relationship between the quantity CQI_{viv} (Eq. (7.8)) and mean scaled vividness together with the semantic interpretation of CQI_{viv} in terms of the rating category "good" associated with the interval scale value of 79.6 (see Figure 7.28).

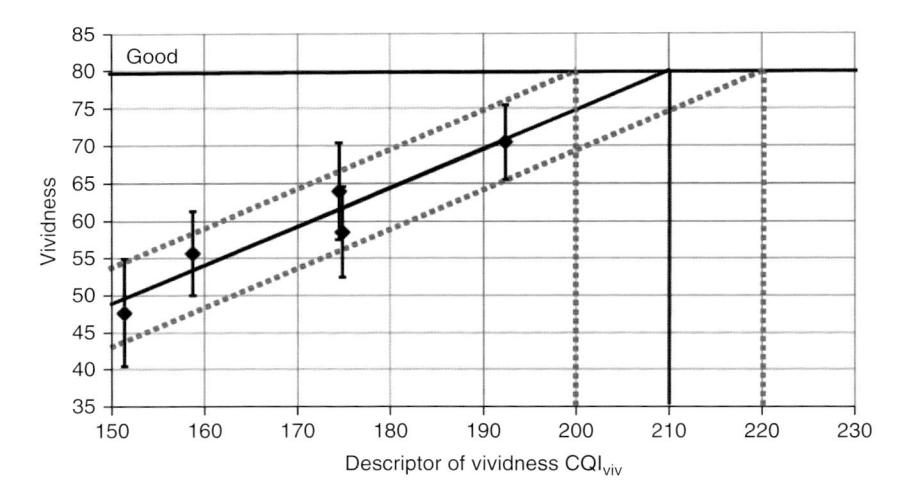

Figure 7.30 Relationship between CQI_{viv} (Eq. (7.8)) and mean scaled vividness with 95% confidence intervals (fit line: $r^2 = 0.92$). Semantic interpretation in terms of the rating category "good" (see Figure 7.28). (Khanh et al. 2016 [20]. Reproduced with permission of Sage Publications.)

As can be seen from Figure 7.30, the "good" level of color vividness could not be achieved by the five light sources of the experiment but the extrapolation of the linear tendency of vividness vs. CQI_{viv} (Figure 7.30) crosses the "good" vividness level at $CQI_{viv}("good") = 210 \pm 10$. The error indicator ($\pm 10$) was estimated by considering the two tendencies of the minimal and maximal ratings of the observers (see the gray dot lines in Figure 7.30 representing the tendencies of the lower and upper ends of the 95% confidence intervals). For spectral design, the lower gray dot line in Figure 7.30 (that represents more critical observers) can be considered as a "worst-case" color vividness acceptance criterion. According to this criterion, the SPD of the light source should be optimized to reach at least the value of $CQI_{viv} = 220$.

7.6.4.2 Prediction of Naturalness

Equation (7.8) was also used to predict the attribute naturalness but in this case, the parameter optimization procedure resulted in $a = 0$; $b = 1$ indicating that the combination of MCRI and R_f does not yield better correlation for naturalness than MCRI alone ($r = 0.85$). As the latter correlation coefficient is less than the adopted criterion value for "very strong" correlation ($r = 0.9$) the relationship between MCRI and naturalness is not shown here and no attempt was made to estimate the acceptance limit in terms of MCRI for naturalness.

7.6.4.3 Prediction of Color Preference

MCRI had the best correlation among the independent variables with color preference ($r = 0.89$, see Table 7.10). As the independent variable CCT had the second best correlation ($r = 0.61$), this variable was considered to be combined with MCRI. According to the white tone preference hypothesis, colored objects were more preferred under the two neutral white light sources (FL-nw, CCT = 4116 K, and LED-nw, CCT = 4013) than under the three warm white light sources. In Section 7.5, the mean scaled preference of European observers [19] for all colored object groups exhibited a nearly linear tendency ($r = 0.99$) as a function of CCT in the range of the present five light sources, between 2300 and 4100 K; see Figure 7.24.

Figure 7.31 reproduces the overall mean results of Figure 7.24 while the quadratic prediction curve of white chromaticity preference for European observers (WCPE) in Figure 7.31 is represented by Eq. (7.9).

$$\text{WCPE} = -4.72119343 \cdot 10^{-6} \, (\text{CCT})^2 + 0.04969531 \, (\text{CCT}) - 61.98468134 \quad (7.9)$$

As the function WCPE(CCT) in Eq. (7.9) was approximately linear in the CCT range of the five light sources (2300–4100 K), the variable (CCT/52) was used in combination with MCRI; see Eq. (7.10). The factor 52 was introduced for convenience.

$$CQI_{pref} = a \, (\text{CCT}/52) + b \, \text{MCRI} \quad (7.10)$$

In Eq. (7.10), CCT shall be given in kelvin units. The optimum parameter values to predict color preference equaled $a = 0.247$; $b = 1.767$ resulting in a correlation

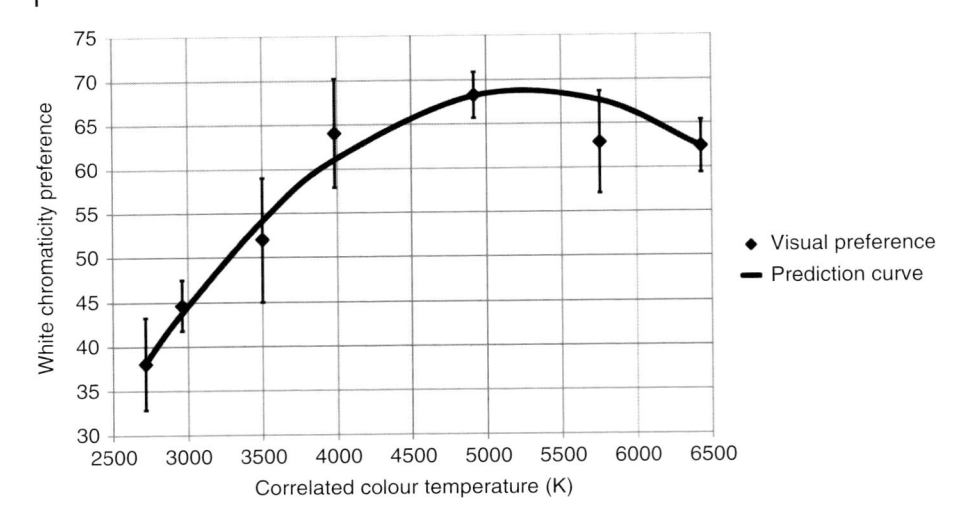

Figure 7.31 Mean scaled white chromaticity preference of European observers (WCPE) when viewing red, blue, and colorful colored objects depicted by re-analyzing the dataset [19]. Mean values and 95% confidence intervals of mean scaled preference for red, blue, and colorful objects and men and women excluding the men's results for red objects. The prediction curve ($r^2 = 0.96$) is represented by Eq. (7.9). (Khanh et al. 2016 [20]. Reproduced with permission of Sage Publications.)

coefficient of $r = 0.99$ between CQI_{pref} and the mean scaled color preference values. The correlation coefficient resulting from parameter optimization ($r = 0.99$; significant at the 0.01 level) is higher than the correlation of color preference with MCRI alone ($r = 0.89$; significant at the 0.05 level; see Table 7.10) indicating the effect of CCT. Figure 7.32 shows the relationship between the quantity CQI_{pref} (Eq. (7.10)) and mean scaled color preference together with the semantic interpretation of CQI_{pref} in terms of the rating category "good" associated with the interval scale value of 79.6 (see Figure 7.28).

As can be seen from Figure 7.32, the "good" level of color preference could not be achieved by these five light sources that do not oversaturate object colors. The effect of oversaturation will be examined in Sections 7.7 and 7.8. The extrapolation of the linear tendency of preference versus CQI_{pref} (Figure 7.32) crosses the "good" color preference level at $CQI_{pref}(\text{"good"}) = 185 \pm 8.5$. The error indicator ($\pm 8.5$) was estimated by considering the two tendencies of the minimal and maximal ratings of the observers similar to Figure 7.30. According to the "worst-case" color preference acceptance criterion, the SPD of the light source should be optimized to reach at least the value of $CQI_{pref} = 193.5$.

Finally, the combination $(R_a + GAI)/2$ [3] and its alternative version with free weighting parameters ($a \ R_a + b \ GAI$) were also applied to predict the mean values of color preference, naturalness, and vividness; see Table 7.11.

As can be seen from Table 7.11, by optimizing the coefficient values a and b, a strong correlation between ($a \ R_a + b \ GAI$) and preference ($r = 0.80$) or vividness ($r = 0.89$) can be obtained for the case of the mean visual values of the present dataset. Naturalness could not be predicted.

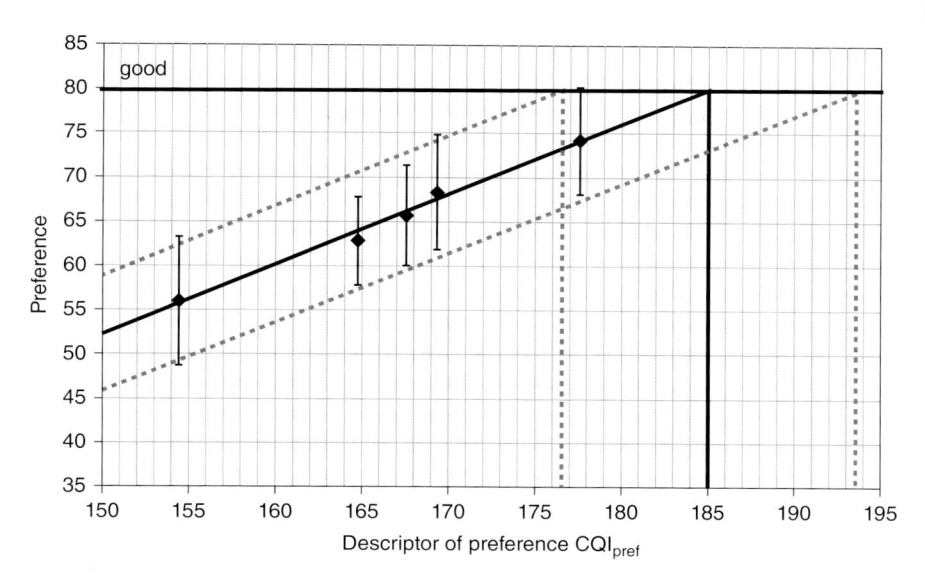

Figure 7.32 Relationship between CQI$_{pref}$ (Eq. (7.10)) and mean scaled vividness with 95% confidence intervals (fit line: $r^2 = 0.99$). Semantic interpretation in terms of the rating category "good" (see Figure 7.28). (Khanh *et al.* 2016 [20]. Reproduced with permission of Sage Publications.)

Table 7.11 Application of the combinations (R_a + GAI)/2 and ($a R_a + b$ GAI) to predict the mean values of scaled color preference, naturalness, and vividness in the case of the five light sources (Figure 7.26).

	Preference	Naturalness	Vividness
a	0.69	0.83	1.17
b	0.31	0.21	0.11
$r [a R_a + b\ \text{GAI}]$	0.80	0.47	0.89
$r[(R_a + \text{GAI})/2]$	0.67	0.14	0.22
$r[R_a]$	0.08	0.33	0.86
$r[\text{GAI}]$	0.49	−0.00	−0.48

Source: Khanh *et al.* 2016 [20]. Reproduced with permission of Sage Publications.
The parameters a and b represent the optimum parameter values for the highest correlation coefficient (r) value (none of them was not significant at the 0.05 level).

7.6.5 Cause Analysis in Terms of Chroma Shifts and Color Gamut Differences

To elucidate the reasons for the above tendencies, the CIELAB chroma shifts (ΔC^*) of the colored objects (Figure 7.27) under each one of the five light sources (in comparison to their reference light source) were computed. This calculation was carried out for all spectral reflectance curves in Figure 7.27 and also for the subset of the artificial objects of different colors (in Figure 7.27)

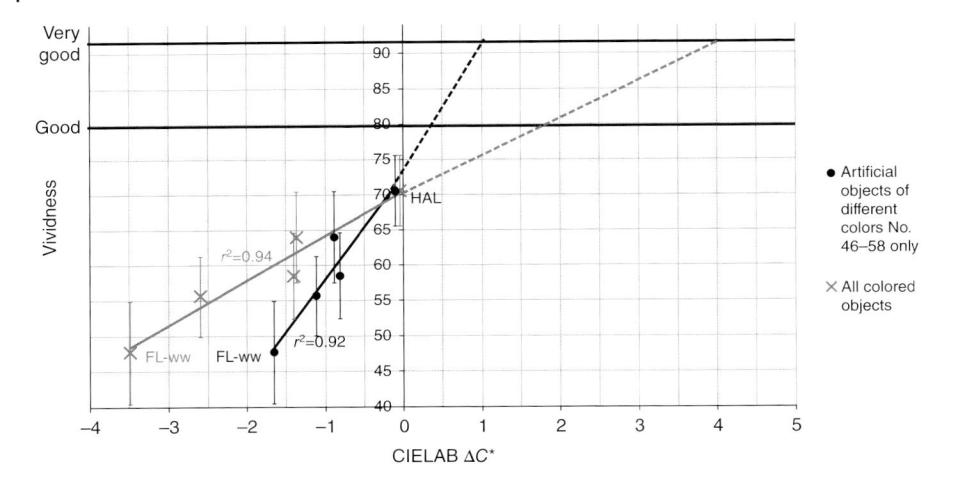

Figure 7.33 Relationship between the ΔC^* values and scaled vividness (mean of all observers with 95% confidence intervals). Extrapolation to estimate ΔC^* to be generated by the light source for "good" and "very good" vividness. (Khanh et al. 2016 [20]. Reproduced with permission of Sage Publications.)

separately. Figure 7.33 shows the relationship between the ΔC^* values and scaled vividness.

As can be seen from Figure 7.33, all five light sources desaturate the objects compared to their reference light source (i.e., the Planckian radiator of the same CCT) and the increase of perceived vividness with less object desaturation (i.e., with increasing R_a, R_f, Q_f, Q_a, and Q_p values) is apparent. Extrapolating the two trend lines (all objects; subset of artificial objects No. 46–58 only) linearly, the oversaturation to be generated by the light source (instead of de-saturation) to achieve "good" ("very good") vividness in terms of CIELAB ΔC^* can be estimated. This mean value equals $\Delta C^* = 0.4$ for "good" and $\Delta C^* = 1.0$ for "very good" in the case of subset No. 46–58 and $\Delta C^* = 1.75$ for "good" and $\Delta C^* = 4.0$ for "very good" if the mean ΔC^* value of all objects is considered. These extrapolated values shall be validated by further experiments using oversaturating light sources [20].

The correlation between the CIELAB ΔC^* values and naturalness or color preference was only low to moderate (for all objects: $r^2 = 0.43$ for naturalness and $r^2 = 0.20$ for preference). A possible reason is that the hue shifts induced by the test light source have a more accentuated effect on the attributes naturalness and color preference than on vividness. Vividness, however, seems to be influenced strongly by object chroma. Figure 7.34 elucidates by the example of the light sources FL-ww and HAL why the indices R_g and Q_g exhibit strong negative correlation with the perceptual attributes (see Table 7.10).

As can be seen from Figure 7.34, although the color gamut descriptors R_g and CQS Q_g are higher for the case of the light source FL-ww than for HAL, the color gamut of FL-ww is distorted. Although FL-ww oversaturates yellow–green (VS4–VS7) and bluish purple (VS12–VS13) objects, it desaturates the perceptually important red (VS1), purplish red (VS14–VS15), and bluish green (VS8–VS11) objects. The negative mean CIELAB ΔC^* values of FL-ww in

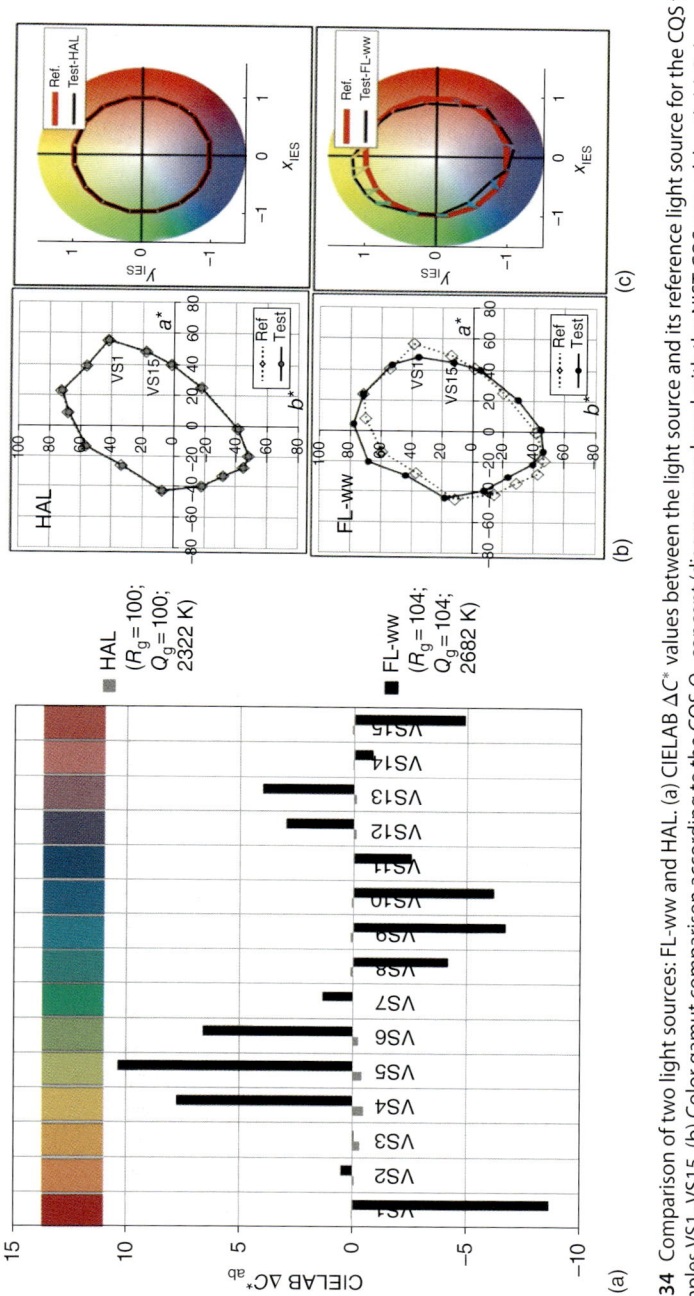

Figure 7.34 Comparison of two light sources: FL-ww and HAL. (a) CIELAB ΔC^* values between the light source and its reference light source for the CQS test color samples VS1–VS15. (b) Color gamut comparison according to the CQS Q_g concept (diagrams produced with the NIST CQS spreadsheet). (c) Color gamut comparison according to the IES concept. (Khanh et al. 2016 [20]. Reproduced with permission of Sage Publications.)

Figure 7.33 ($\Delta C^* = -3.5$ for all objects and $\Delta C^* = -1.7$ for No. 46–58) indicate that the light source FL-ww tends to desaturate the current objects in the object scene of Figure 7.25.

But this property cannot be reflected by the R_g and Q_g metrics because these metrics do not weight the test color samples according to their occurrence or perceptual relevance. However, the strong positive correlation of FCI with vividness ($r = 0.76$) can be traced back to the fact that the choice of the four test color samples of the FCI metric happens to represent more or less the test color samples in Figure 7.27 from the point of view of chroma shifts.

7.6.6 Lessons Learnt from Section 7.6

The visual attributes color preference, naturalness, and vividness were scaled with an interval scale labeled by rating categories. The advantage of the use of rating categories is the possibility of estimating an acceptance limit for the "good" level of a visual attribute in terms of a strongly correlated combined descriptor quantity. Such an acceptance limit could be derived in the case of the present visual dataset by extrapolation for two attributes, color preference and vividness, but not for naturalness [20].

The strong positive correlation of mean scaled vividness with R_f (and the other investigated color fidelity metrics) and its strong negative correlation with R_g and Q_g among these five customary light sources for the current combination of colorful objects was an important finding. For the present set of five customary light sources, R_f is able to represent vividness because vividness increases with realness (i.e., color fidelity) in the case of desaturating light sources; and secondly, care should be taken when applying a standalone color gamut metric to describe a visual attribute. If some test color samples in certain hue ranges are overrepresented or underrepresented in the color gamut index calculation algorithm then the gamut index will not correlate with the visual impression of vividness, which depends on the magnitude of actual object chroma shifts; see Figures 7.33 and 7.34 [20].

The risk of using "a relative-gamut index (R_g), that estimates the average extent to which a light source increases or decreases the saturation of surface colors" [21] (by the aid of a set of uniformly distributed color evaluation samples) as a standalone measure to describe, for example, vividness should be emphasized at this point: R_g stands for the color gamut volume only and its use as a single color quality measure cannot express the extent and form of color gamut distortions at every hue angle.

7.7 Experiments on Color Preference, Naturalness, and Vividness in a One-Chamber Viewing Booth with Makeup Products

In this section, results [24] of a visual experiment on the effect of multi-LED SPDs (with white points at 3200 K at the Planckian locus and at an illuminance level of 500 lx) with different degrees of object oversaturation (ΔC^*) on the assessment of color preference, naturalness, and vividness (see Section 2.4) of red or skin-tone-colored makeup products will be analyzed. The illumination of makeup products (those mimicking skin tones or red/reddish makeups and lipsticks for cosmetics) is very important, for example, in the retail of beauty products, in TV and cinema film production, or in theaters. In this context, the following questions arise:

1) Does the optimization of the light source spectra for maximum color rendering (in terms of a color fidelity index or CRI) guarantee a maximum visual assessment about the color appearance of the objects concerning the observers' preference, naturalness, and vividness judgments?
2) If we continue oversaturating these objects by increasing the intensity of the red channel of the multi-LED light engine, what is the maximum degree of object oversaturation (ΔC^*) with maximum preference and naturalness? Is there a breakdown of the subjects' judgments after this maximum – because subjects feel that the colored objects exhibit too much chroma?

In the experiment [24], observers judged the color preference, naturalness, and vividness by the aid of semantic categories (e.g., very good, good, low, bad; see Section 2.1) as the anchors of their visual assessment. The categories "good" or "very good" can be used in lighting practice as the limits for "good" or "very good" color quality to foster conscious spectral design. Concerning the psychophysical method, observers carried out absolute ratings about the object scene under a given light source at a time and not comparisons with a reference situation. This is an important advantage of the experimental design because in the most common general lighting situation, there is no comparison with any reference and preference judgments occur under the only light source actually illuminating the objects. During the training sessions, observers had ample time to build their own internal judgment scales to assess what color appearance corresponds to, for example, their "very good" or "low" judgment.

The aim of this section is to analyze how color quality metrics (the CQS metrics, GAI, ΔC^*, MCRI, FCI, R_a, R_f, R_g) correlate with the experimentally obtained visual color preference, naturalness, and vividness scale values of the observers. Models with combinations of color quality metrics will also be analyzed including a novel color quality (CQ) model to predict the visual attributes and to formulate the "good"–"very good" limits in terms of the values of the new, well-correlating CQ metric.

7.7.1 Experimental Method

Figure 7.35 shows the relative spectral radiance of the seven illuminating spectra constructed by optimizing the weightings of the LED channels of the six-channel multi-LED light engine used in the experiment. This was the same light engine as the one used in Section 7.5. The level of oversaturation caused by each spectrum [24] is characterized by the ΔC^* value of the deep red CQS test color sample VS1 (see Section 2.4).

As can be seen from Figure 7.35, two of the seven spectra desaturated reddish object colors ($\Delta C^* < 0$) compared to the reference illuminant (i.e., the blackbody

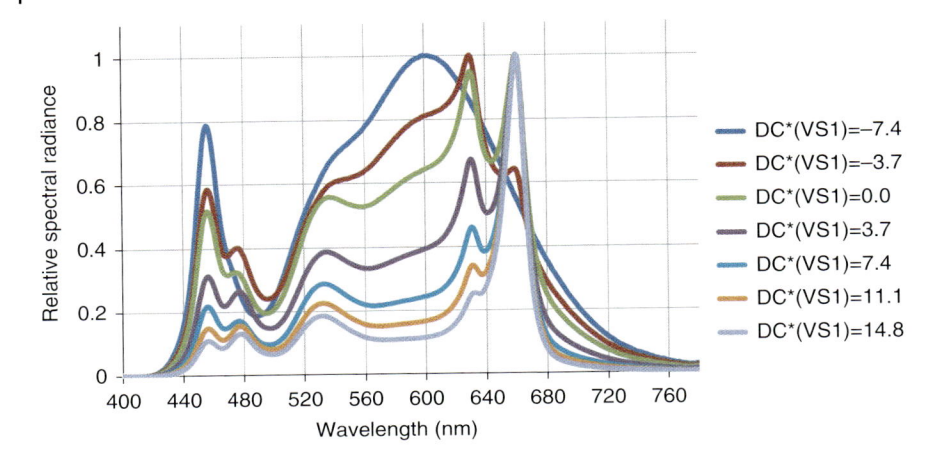

Figure 7.35 Relative spectral radiance of the seven illuminating spectra constructed from the six-channel multi-LED light engine used in the experiment. They were measured at the rear wall of the viewing chamber (Figure 7.3). The level of oversaturation caused by each spectrum is characterized by the ΔC^* value of the CQS test color sample VS1. Re-drawn after [24].

radiator at 3200 K); one of them had $\Delta C^* = 0$ and four of them caused increasing saturation ($\Delta C^* > 0$; see the increasing relative height of the deep red LED channel at 660 nm).

The same viewing booth was used as shown in Figure 7.22. The vertical luminance at the bottom wall of the viewing chamber equaled 164 ± 1 cd/m^2 corresponding to the illuminance of 548 ± 3 lx). The arrangement of colored objects included seven makeup powders and four lipsticks; see Figure 7.36. Their spectral reflectance is depicted in Figure 7.37.

As can be seen from Figure 7.37, the spectral reflectance of the objects begins to increase after 580 nm and the reflectance factor is highest in the 620 – 700 nm range implying the importance of the red LED channels (620 + 660 nm) to saturate these reddish objects. Using the actual spectral reflectance factors depicted

Figure 7.36 Arrangement of colored objects used in the experiment: seven makeup powders and four lipsticks. Reproduced from Light Research Technology [24] with permission from SAGE.

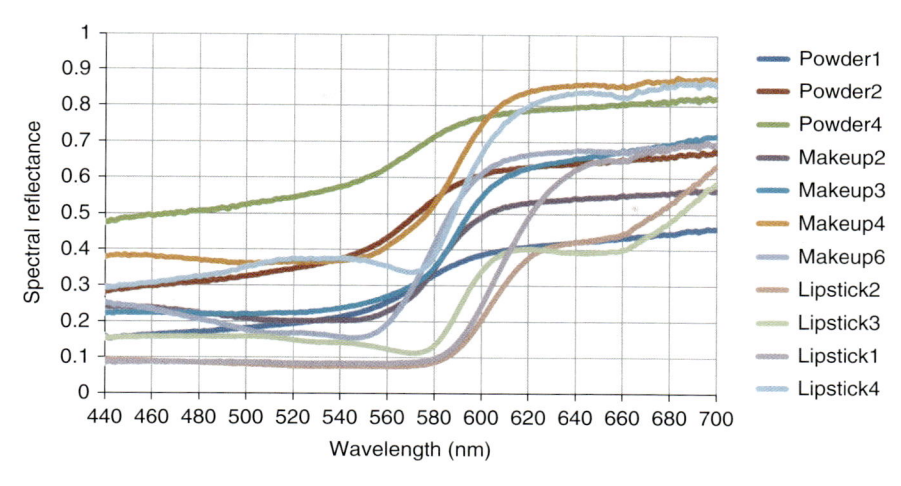

7.7 Experiments on Color Preference, Naturalness, and Vividness

Figure 7.37 Spectral reflectance of the colored objects used in the experiment. Re-drawn after [24].

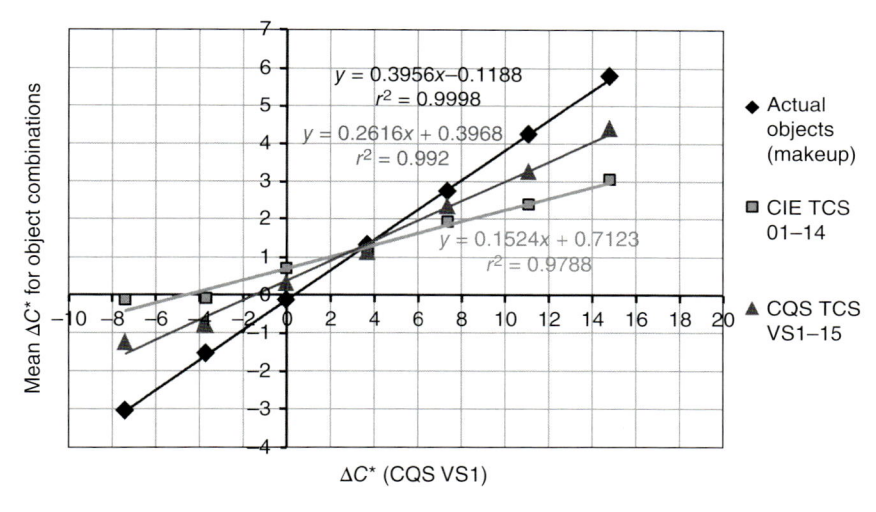

Figure 7.38 Comparison of the mean ΔC^* values of different sets of colored objects for the seven light sources (Figure 7.35). Abscissa: ΔC^* (CQS VS1), for the saturated red CQS test color only; ordinate: mean ΔC^* values for different sets of colored objects.

in Figure 7.37, the value of ΔC^* was computed for every object and every light source and then, the mean value of ΔC^* for every object was computed for each light source. It is interesting to compare these mean ΔC^* values for the actual colored object combination shown in Figure 7.36 with the values of ΔC^* (VS1) used to characterize the light source by an internationally well-known test color sample (CQS VS1). This comparison is shown in Figure 7.38.

As can be seen from Figure 7.38, there are linear relationships ($r^2 > 0.97$) between the mean ΔC^* values of the different sets of colored objects (the actual makeup objects in the present experiment; the CIE test color samples TCS01–14 and the CQS test color samples VS1–VS15) and the ΔC^* value of the standalone

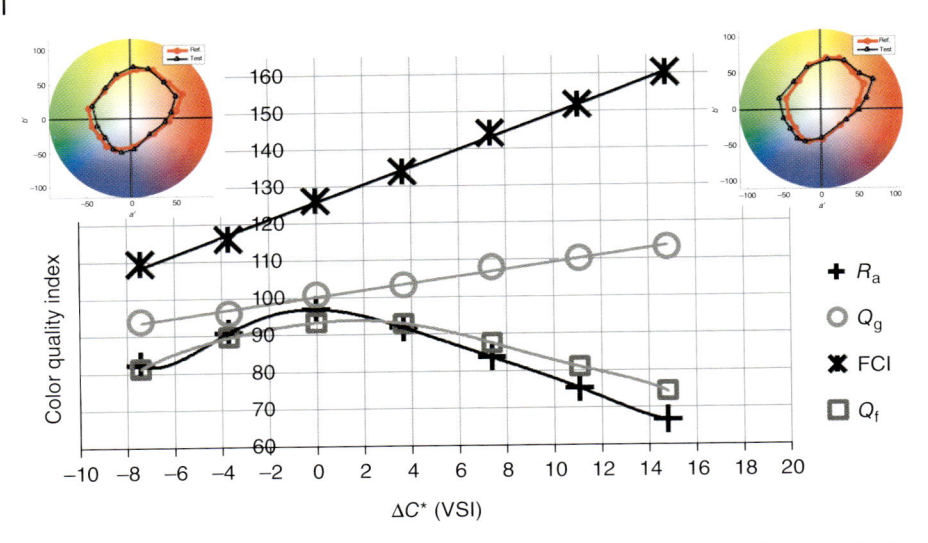

Figure 7.39 Selected color quality indices (CIE CRI R_a; CQS Q_g, Q_f, and FCI) as a function of ΔC^* (VS1). In the colored inset diagrams, black curve: a multi-LED light source from Figure 7.35; red curve: blackbody reference. Left inset diagram: color gamut in terms of the 15 CQS test color samples (VS1-VS15) for the light source with ΔC^*(VS1) = −7.4; right: with ΔC^*(VS1) = 14.8.

CQS VS1 test color sample (saturated red). For the 11 actual makeup colors of the present experiment (Figure 7.36), the mean ΔC^* value equals only about 40% of ΔC^*(VS1) for every one of the seven light sources because the test color sample VS1 is a highly saturated red one (see Section 2.4).

It is also interesting to depict some selected color quality indices (CIE CRI R_a; CQS Q_g, Q_f, and FCI) as a function of ΔC^*(VS1), see Figure 7.39.

As can be seen from Figure 7.39, the two color fidelity type indices (R_a; Q_f) peak at $\Delta C^* = 0$; otherwise, they decrease in both directions ($\Delta C^* < 0$; $\Delta C^* > 0$) because, compared to the reference light source, both desaturating and oversaturating (multi-LED) light sources cause a decrease of color fidelity (color rendering) type indices as both conditions (desaturation and oversaturation) correspond to a departure from the reference illuminant. Color gamut type indices (Q_g, FCI), however, increase with increasing ΔC^*(VS1) because the color gamut increases, especially in the red color range; see the black contours in the colored inset diagrams in Figure 7.39.

Concerning the psychophysical method, the panel of observers included 20 young subjects (7 men and 13 women; average age: 26 years; min., 22; max., 31). All subjects had good or corrected visual acuity and normal color vision. First, these 20 observers were examined to see whether they were responsive to the scaling tasks of the experiment (see below): the STD of all preference answers of every subject was computed and if the STD value of a specific subject was greater than the mean STD of all 20 subjects then this subject was taken into the cluster of "responsive" observers. Six subjects (one man and five women) belonged to this "responsive" cluster. Only the results of this cluster will be analyzed below.

Figure 7.40 Rating scale labeled by the categories very bad, bad, poor, moderate, good, very good, and excellent. Every light source and every visual attribute (color preference, naturalness, and vividness) had an individual scale. Observers were allowed to put the cross anywhere onto the scale. Reproduced from Light Research Technology [24] with permission from SAGE.

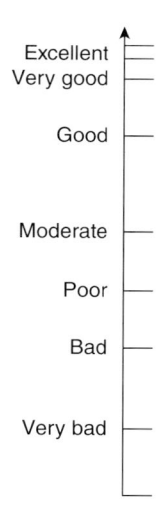

Subjects looked into the chamber of the viewing booth (see Figure 7.22) containing the objects (see Figure 7.36) and adapted for 2 min to one of the seven LED spectra; see Figure 7.35 by looking at the white rear wall of the chamber. After the adaptation period, subjects had to assess their color preference, naturalness, and vividness impression about the color appearance of all objects depicted in Figure 7.36 under the current light source. Observers were carefully instructed about the meaning and the differences of the concepts of color preference, naturalness, and vividness, according to Section 2.4. Assessments were done on a sheet of paper by putting a cross to a scale labeled by the categories; see Figure 7.40. Observers were allowed to put the cross anywhere onto the scale.

As can be seen from Figure 7.40, the rating categories are not equally spaced on the continuous interval scale. The spacing (excellent = 97.9; very good = 91.6; good = 79.6; moderate = 52.9; poor = 41.2; bad = 26.5; very bad = 12.8, see Figure 7.38) resulted from the parallel usage of the two types of scales in a previous psychophysical study [22]. First, there was a training session with all seven light sources in a randomized order and then three repetitions of the main experiment with three different randomized orders of the light sources. Only the interval scale results of the main experiment were recorded for every one of the three visual attributes (preference, naturalness, vividness). Thus, the dataset to be analyzed consisted of 3 attributes × 3 repetitions × 6 responsive observers × 7 light sources = 378 numbers between 0 and 100.

7.7.2 Color Preference, Naturalness, and Vividness and Their Modeling

First, the relationship among the visual attributes color preference, naturalness, and vividness scaled by the observers will be analyzed. This is shown in Figure 7.41.

As can be seen from Figure 7.41a, color preference correlates well with naturalness ($r^2 = 0.84$). The only outlier point is the oversaturating light source with $\Delta C^*(VS1) = 14.8$ because this light source has high vividness and high

Figure 7.41 Relationship among the mean scaled visual attributes of the observers (color preference, naturalness, and vividness, for all 6 responsive observers × all 3 repetitions). Every point corresponds to one of the seven light sources (see Figure 7.35). Intervals are 95% confidence intervals. Drawn after [24].

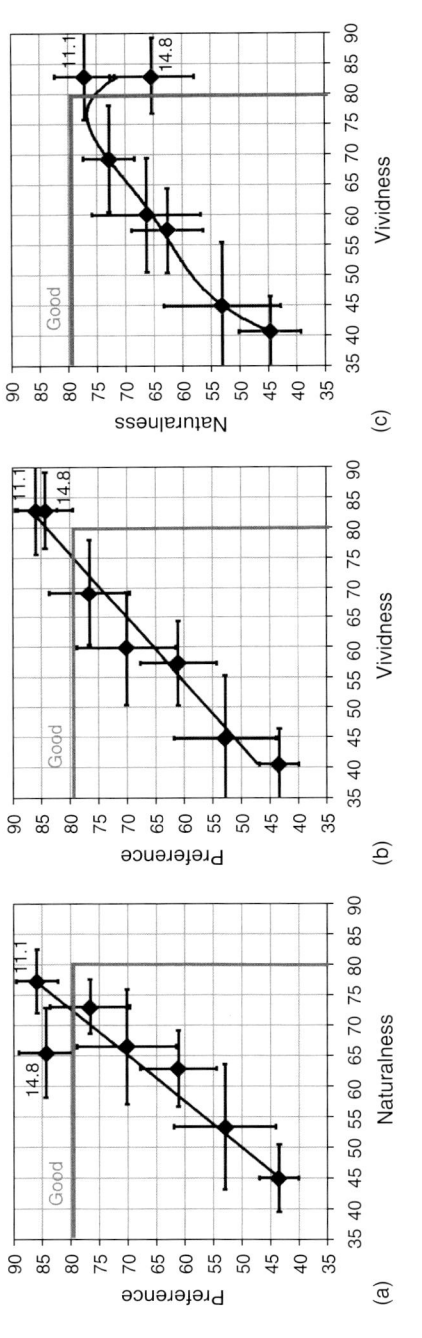

preference (Figure 7.41b) but less naturalness (Figure 7.41c). Figure 7.41b shows that preference also correlates well with vividness ($r^2 = 0.96$). However, the visual attribute naturalness breaks down if the makeup objects get oversaturated: observe in Figure 7.41c that the light source with $\Delta C^*(VS1) = 14.8$ has a mean naturalness rating of only 65, that is, "moderate-good" while the light source with $\Delta C^*(VS1) = 11.1$ exhibits a good naturalness rating (both light sources have a "good" vividness rating).

The next step is to find a suitable color quality metric to account for the visual results. To do so, the correlation of the color quality metrics CQS Q_a, Q_p, Q_g, GAI, mean $\Delta C^*(VS1 - VS15)$ MCRI, FCI, IES R_f, and R_g (see Section 2.4) with the mean visual scale values shown in Figure 7.41 will be analyzed. Table 7.12 contains the values of Pearson's correlation coefficients.

The following significant correlations ($\alpha = 5\%$) can be seen from Table 7.12: Mean ΔC^* (VS1–VS15), GAI, CQS Q_g, MCRI, FCI, and IES R_g for color preference, CQS Q_p, MCRI, and IES R_g for naturalness, and mean $\Delta C^*(VS1 - VS15)$, GAI, CQS Q_g, MCRI, FCI, and IES R_g for vividness. An important finding is that all color fidelity type indices (R_a, CQS Q_f, and IES R_f) exhibit negative correlation with all visual attributes. This finding answers the first question posed at the beginning of this section: the optimization of the light source spectra for maximum color rendering (in terms of a color fidelity index) does not guarantee a maximum visual assessment about the color appearance of the objects concerning the observers' preference, naturalness, and vividness judgments. Just the opposite is true; see Figure 7.42.

As can be seen from Figure 7.42, "good"–"very good" color preference, naturalness, and vividness were achieved by the light source that exhibits about $R_a = 75$. Optimizing for R_a, the values of the scaled visual attributes sink toward the "moderate" range, which is obviously not the purpose of lighting engineering. A better way to predict the visual attributes is the use of the ΔC^* values depicted in

Table 7.12 Values of Pearson's correlation coefficients. Reproduced from Light Research Technology [24] with permission from SAGE.

Correlation coefficient r	Preference	Naturalness	Vividness
Mean ΔC^* (VS1–VS15)	0.97[a]	0.80	0.98[a]
R_a	−0.57	−0.25	−0.66
GAI	0.98[a]	0.83	0.98[a]
CQS Q_a	0.01	0.34	−0.14
CQS Q_f	−0.35	−0.02	−0.48
CQS Q_g	0.98[a]	0.85	0.99[a]
CQS Q_p	0.82	0.94[a]	0.72
MCRI	0.97[a]	0.94[a]	0.92[a]
FCI	0.98[a]	0.83	0.99[a]
IES R_f	−0.36	−0.01	−0.48
IES R_g	0.98[a]	0.87[a]	0.98[a]

a) Correlation coefficient significant at the 5% level.

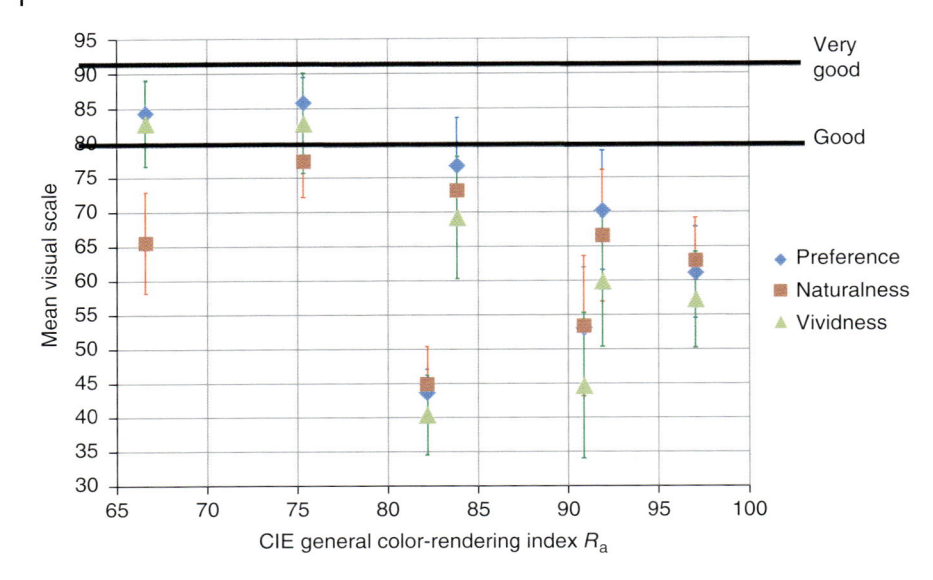

Figure 7.42 Mean values of the scaled visual attributes color preference, naturalness, and vividness as a function of the CIE general color-rendering index R_a. Intervals are 95% confidence intervals of the mean values.

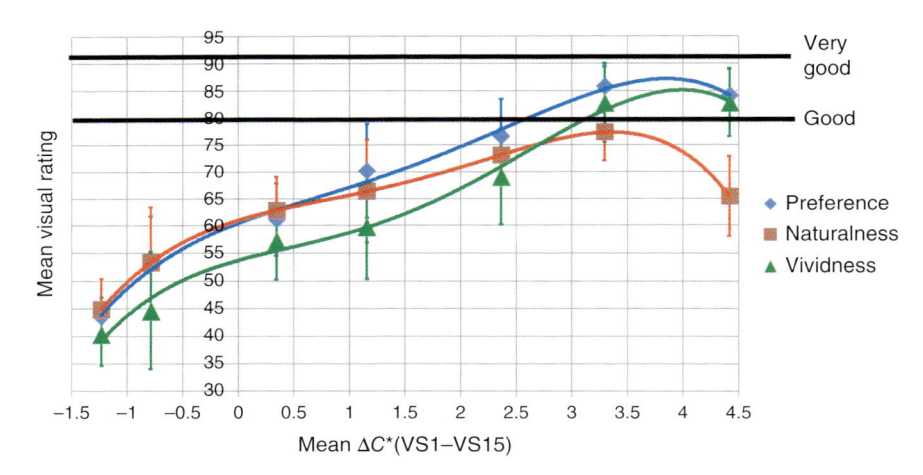

Figure 7.43 Mean color preference, naturalness, and vividness ratings of the subjects as a function of the mean $\Delta C^*(\text{VS1} - \text{VS15})$. Intervals are 95% confidence intervals of the mean values. Drawn after [24].

Figure 7.38. In this section, the mean ΔC^* value of the 15 CQS test color samples VS1–VS15 was chosen as these samples are internationally widely used and saturated enough to represent the changes of fine-tuning the spectrum of the multi-LED engine. Figure 7.43 shows the mean color preference, naturalness, and vividness ratings of the subjects as a function of the mean $\Delta C^*(\text{VS1} - \text{VS15})$.

As can be seen from Figure 7.43, in terms of the mean $\Delta C^*(\text{VS1} - \text{VS15})$, the visual attributes color preference and naturalness peak in the range of 3.3–3.5.

Naturalness drops at a fast pace if the makeup objects get oversaturated, that is, toward $\Delta C^*(\text{VS1} - \text{VS15}) > 4$. This finding answers the second question posed at the beginning of this section: if we continue oversaturating the makeup objects by increasing the intensity of the red channel of the multi-LED light engine, the maximum degree of object oversaturation (ΔC^*) with maximum preference and naturalness equals about 3.3 (in terms of the mean ΔC^* value of VS1–VS15). It is plausible to suppose that vividness ratings will continue to increase with increasing chroma (although this could not be corroborated with the present results). A value of $\Delta C^*(\text{VS1} - \text{VS15})$ of about 3.3 for the maximum of preference and naturalness corresponds to the value $\Delta C^*(\text{VS1})$ of about 11 (see Figure 7.38), which is in turn equivalent to the values of FCI = 152, $Q_g = 110$, $Q_f = 81$, and $R_a = 75$ according to Figure 7.39 for this specific experiment.

To predict the decreasing branch of the curves in Figure 7.43 after about $\Delta C^*(\text{VS1} - \text{VS15}) = 3.5$ due to oversaturation and to get a new descriptor quantity linearly related to the observer's ratings, it is plausible to add a certain degree of a color fidelity type metric (as a fidelity metric decreases with the degree of object oversaturation; see Figure 7.39) to the value of $\Delta C^*(\text{VS1} - \text{VS15})$. To do so, the CQS Q_f metric was selected for consequence as this metric is based on the same CQS test color samples. It is suggested to use Eq. (7.11) as a two-metric combination to predict the visual attributes color preference, naturalness, and vividness.

$$CQ = a + b\ \Delta C^* + c\ Q_f \quad (7.11)$$

In Eq. (7.11), ΔC^* is the mean chroma difference between the actual light source and its reference light source calculated for the 15 CQS test color samples VS1–VS15, Q_f is the CQS color fidelity index and a, b, c are weighting parameters with different values for the three visual attributes color preference, naturalness, and vividness. Table 7.13 contains the optimum values of the weighting parameters a, b, c in Eq. (7.11) resulting from the best fit to the mean visual ratings shown in Figure 7.43 and the correlation coefficients between CQ and the mean visual ratings.

As can be seen from Table 7.13, the visual color quality attribute naturalness has the highest fidelity (c) component related to the chroma increment component (b) while vividness hardly needs any fidelity component (as expected). The mean

Table 7.13 Optimum values of the weighting parameters a, b, c in Eq. (7.11) resulting from the best fit to the mean visual ratings (Figure 7.43) and correlation coefficients between the two-metric combination CQ and the mean visual ratings of the observers. After [24].

Parameter in Eq. (7.11)	Color preference	Naturalness	Vividness
Constant a	10.478	−23.000	37.174
Chroma increment b	8.331	5.896	8.224
Color fidelity c	0.536	0.913	0.164
Correlation coefficient r^2	0.977	0.880	0.972

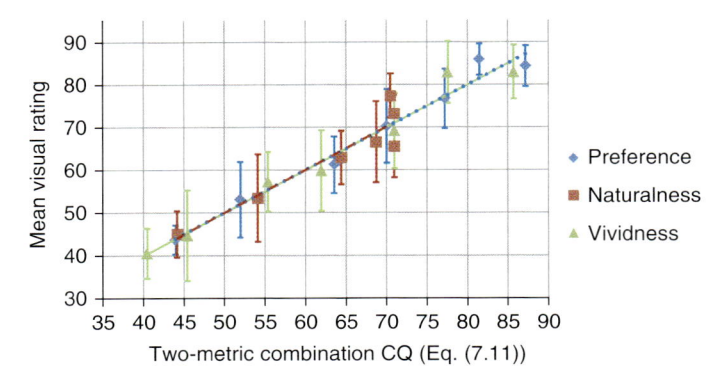

Figure 7.44 Mean visual ratings of the observers as a function of the two-metric combination CQ (Eq. (7.11)).

visual ratings of the observers as a function of the two-metric combination CQ (Eq. (7.11)) are depicted in Figure 7.44.

A comparison of Figure 7.44 with Figure 7.43 indicates that the use of the two-metric combination CQ (Eq. (7.11)) results into a good linear prediction of the visual attributes color preference, naturalness, and vividness. Two-metric combinations will be further analyzed in Section 7.9 for the case of food products in the context of office lighting. This will be compared with the visual assessment results of the present section (lighting of makeup products).

7.8 Food and Makeup Products: Comparison of Color Preference, Naturalness, and Vividness Results

In this section, the results of the visual experiment with reddish and skin-tone type makeup products (described in Section 7.7) will be compared with the results of another, similar experiment with multicolored food products. The latter experiment [23] was carried out with the same questionnaire, the same multi-LED engine, the same CCT of the white point (3200 K), and the same illuminance level (500 lx) but with a different set of multi-LED spectra with different degrees of object oversaturation (ΔC^*). The aim of this section is to compare the observers' color preference, naturalness, and vividness assessments in the two lighting contexts, food lighting and the lighting of makeup products. It turned out that it is possible to merge the two datasets and carry out an overall analysis.

The other aim is to validate the predicting Eq. (7.11), optimize the coefficients of this equation for the merged dataset, and derive a common color preference, naturalness, and vividness model in terms of the quantities ΔC^*(VS1 − VS15) and CQS Q_f (and also with IES R_f). In Section 7.8.4, the effect of object oversaturation on color discrimination (i.e., the perception of small color differences or delicate color shadings on object surfaces) will also be discussed.

7.8.1 Method of the Experiment with Food Products

Figure 7.45 shows the colored objects to be assessed by the subjects. These objects were arranged in the same viewing booth as in the experiments previously mentioned in this chapter; see Figure 7.22. Observers had to assess the color appearance of the colored objects in the chamber (Figure 7.45) by viewing all objects simultaneously. The colored objects represented typical food items. Roses were included because they are typically sold together with the food products in the groceries.

Figure 7.46 shows the spectral radiance distributions of the seven spectra used in this experiment (the same multi-LED light engine was used as in Section 7.8; compare Figure 7.45 with Figure 7.46). Table 7.14 shows their colorimetric properties.

Figure 7.47 illustrates the spectral reflectance of the colored objects (Figure 7.45) in the viewing booth.

As can be seen from Table 7.14, the seven multi-LED spectra had a mean CCT of 3221 K (STD = ±2 K). Three of them (the spectra labeled by 0, maxQg0, and rot0) were located at the Planckian locus ($\Delta u'v' \leq 0.0001$). Two of them were located below the Planckian locus (those labeled by −3 and −7) and two of them were situated above the Planckian locus (those labeled by 3 and 8). The spectrum rot0 was obtained by optimizing the weighting factors of the six LED channels of the multi-LED engine so as to maximize the ΔC^* value (mean chroma shift between the light source and its reference light source) of the deep red CQS test color sample VS1 ($\Delta C^*_{VS1} = 6.2$ was reached) by satisfying the constraints $R_a \geq 84$ (to avoid large hue shifts) and $\Delta u'v' \leq 0.001$ (to ensure the required white point chromaticity). The correlation between ΔC^*(food) and ΔC^*(VS1 − VS15) was significant at the 1% level ($r^2 = 0.93$).

Figure 7.45 Arrangement of colored food objects (typical products sold in a grocery) and four roses in the viewing booth to be assessed visually by the subjects. (Khanh *et al.* 2016 [23]. Reproduced with permission of Sage Publications.)

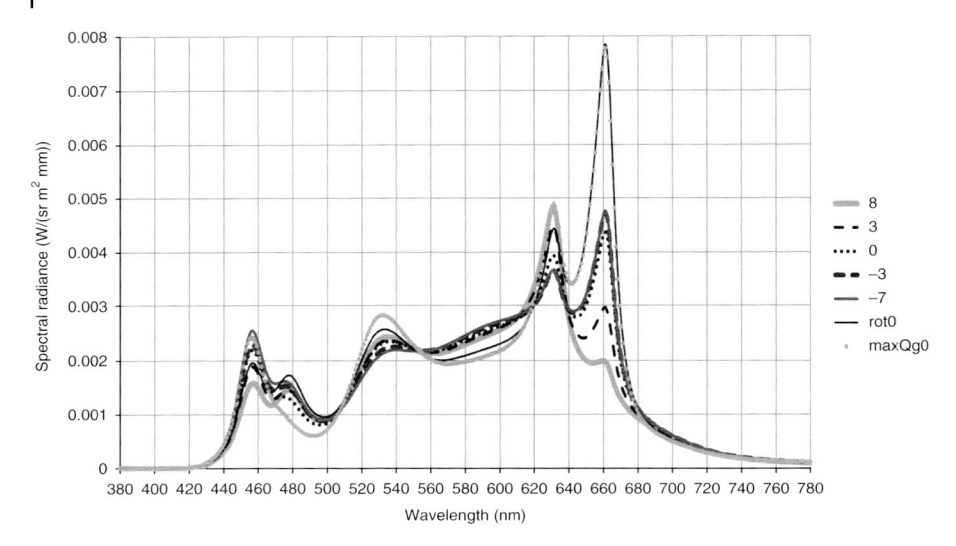

Figure 7.46 Spectral radiance of the seven multi-LED light sources used to illuminate the food objects. They were measured in the middle of the rear wall of the viewing booth. Their luminance equaled 164.0 ± 0.2 cd/m^2 (mean and standard deviation for the seven light sources); this value corresponds to a vertical illuminance level of 550 lx at the rear wall of the viewing booth. (Khanh *et al.* 2016 [23]. Reproduced with permission of Sage Publications.)

The spectrum maxQg0 was obtained in a similar way but by maximizing the CQS Q_g value of the multi-LED spectrum. Descriptor values related to the degree of object saturation (GAI, Q_g, Q_p, FCI, and R_g) are maximal for the case of the latter two light sources (rot0 and maxQg0). The other five spectra were obtained by maximizing their R_a value at the target white chromaticity, on (0), below (-3 and -7), or above (3 and 8) the Planckian locus at 3221 K. These five spectra were included to investigate the effect of the location of the white chromaticity in the chromaticity diagram on color preference, naturalness, and vividness but no significant effect was found [23]. Twenty-three responsive observers (different from the subjects in Section 7.8) took part in this experiment (1 woman, 22 men; between 27 and 55 years of age; mean, 43.7 years). The same questionnaire and the same viewing method were used as described in Section 7.8 except that there were no repetitions. After the training session, ratings were carried out only once.

7.8.2 Color Preference, Naturalness, and Vividness Assessments: Merging the Results of the Two Experiments (for Multicolored Food and Reddish and Skin-Tone Type Makeup Products)

First, it is interesting to compare the mean value of ΔC^*(VS1 − VS15) computed by using the internationally well-known set of CQS test color samples (VS1–VS15) with the mean ΔC^* values of the actual, experiment-specific colored object combination including both the makeup products (Figure 7.36) and the multicolored food objects (Figure 7.45) illuminated by the light sources

Table 7.14 Colorimetric properties of the seven multi-LED light sources in the experiment [23].

Light source label	−3	−7	maxQg0	0	3	rot0	8
x	0.419	0.416	0.422	0.422	0.425	0.422	0.429
y	0.392	0.384	0.399	0.399	0.405	0.399	0.415
CCT	3221	3221	3218	3221	3222	3220	3226
Duv	−0.0021	−0.0051	0.0000	0.0001	0.0023	0.0001	0.0055
$\Delta u'v'$	−0.0030	−0.0073	0.0000	0.0001	0.0033	0.0001	0.0078
R_a	97	97	84	97	97	86	96
GAI	99	98	110	100	101	109	101
CQS Q_a	94	93	92	94	94	93	93
CQS Q_f	93	92	87	93	94	89	92
CQS Q_g	101	103	111	101	99	106	96
CQS Q_p	95	96	99	95	94	98	92
MCRI	92	92	93	91	91	93	90
FCI	126	127	148	128	125	140	123
IES R_f	90	90	86	90	89	86	88
IES R_g	100	100	108	100	98	103	96
ΔC^*(food)	0.1	0.3	3.2	0.0	−0.5	2.7	−1.0
ΔC^*(VS1 − VS15)	0.6	1.0	2.9	0.5	−0.1	1.9	−0.8

Source: Khanh *et al.* 2016 [23]. Reproduced with permission of Sage Publications.
ΔC^*(food): Mean CIELAB chroma shift for all food objects (see Figures 7.45) between each light source and its reference light source (Planckian radiator at the same CCT) and ΔC^*(VS1 − VS15): Mean CIELAB chroma shift for the CQS test color samples VS1–VS15.

used in the corresponding experiment (7 + 7 = 14 light sources altogether). The correlation between ΔC^*(VS1 − VS15) and ΔC^*(actual) is significant at the 1% level ($r^2 = 0.97$) and Eq. (7.12) shows this relationship.

$$\Delta C^*(\text{actual}) = 1.4322 \, \Delta C^*(\text{VS1} - \text{VS15}) - 0.5515 \tag{7.12}$$

Equation (7.12) means that the variable ΔC^*(VS1 − VS15) can be used to characterize the degree of object desaturation or oversaturation for both experiments independent of the actual colored objects in a particular experiment. This finding is important to model the merged preference, naturalness, and vividness dataset in the form of Eq. (7.11); see below.

Similar to Figure 7.39, it is also interesting to depict some selected color quality indices (CIE CRI R_a; CQS Q_g, Q_f, and FCI) as a function of the value of ΔC^*(VS1 − VS15) for all 14 light sources; see Figure 7.48.

As can be seen from Figure 7.48, the color fidelity index CQS Q_f peaks at $\Delta C^* = 0$; otherwise, its value decreases in both directions ($\Delta C^* < 0$; $\Delta C^* > 0$), similar to Figure 7.39. The other color fidelity type indices (CIE R_a, IES R_f) behave similarly (this is not depicted). This finding implies that we can continue

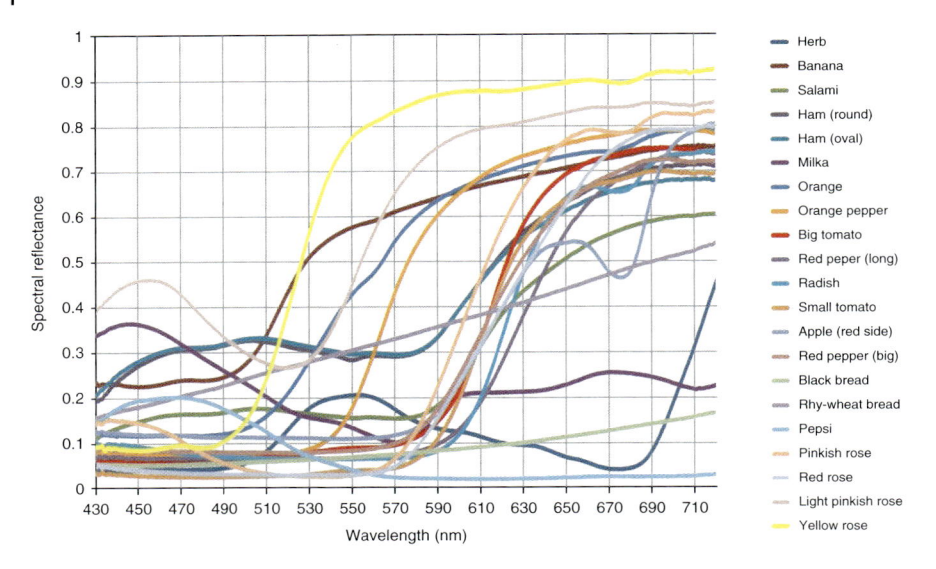

Figure 7.47 Spectral reflectance of the colored objects of food lighting (typical grocery products including four roses; see Figure 7.45.

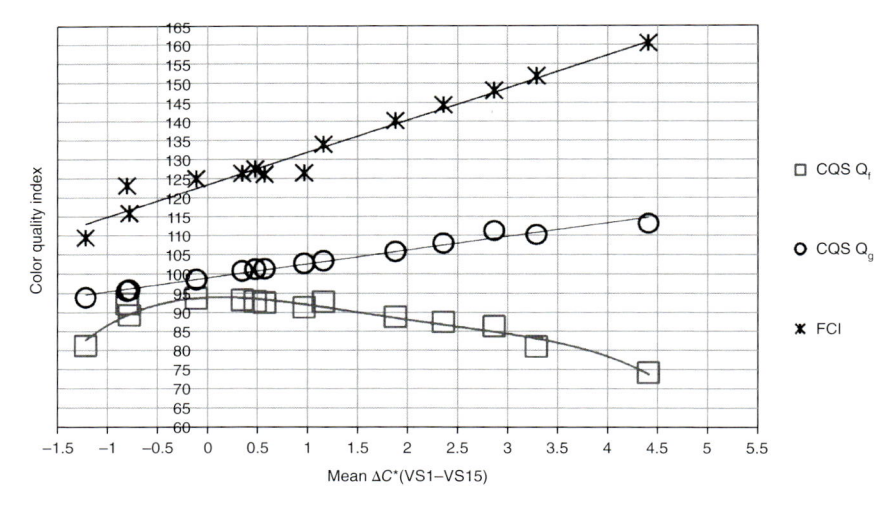

Figure 7.48 Selected color quality indices (CQS Q_g, Q_f, and FCI) as a function of the mean ΔC^*(VS1 – VS15) for all 14 light sources of the merged dataset (makeup and food objects).

to use CQS Q_f in a linear combination with ΔC^* (Eq. (7.11)) to predict the merged dataset. The reason is that the oversaturation of the objects (with, e.g., $\Delta C^* = 5$) results in a decrease of preference and naturalness (see below) according to the behavior of Q_f, which also begins to decrease in the case of $\Delta C^* > 3$ noticeably. Similar to Figure 7.39, color gamut type indices (Q_g, FCI) increase with increasing value of ΔC^*(VS1 – V14).

7.8.3 Analysis and Modeling of the Merged Results of the Two Experiments

First, the relationship among the visual attributes color preference, naturalness, and vividness of the merged dataset (makeup and food products) will be analyzed. This is shown in Figure 7.49.

As can be seen from Figure 7.49a, color preference correlates well with naturalness ($r^2 = 0.84$). The outlier point A is the oversaturating light source used in the makeup experiment (Section 7.8) with $\Delta C^*(\text{VS1} - \text{VS14}) = 4.4$. This light source has high vividness and high preference (Figure 7.49b) but less naturalness (Figure 7.49c). The light source B (also from the makeup experiment) has $\Delta C^*(\text{VS1} - \text{VS14}) = 3.3$. The light source B is not that oversaturating as the light source A; hence, B causes higher naturalness judgments than A. Figure 7.49b shows that preference correlates well with vividness ($r^2 = 0.93$). Figure 7.49c shows that the visual attribute naturalness breaks down if the objects get oversaturated. Now, the correlation of the color quality metrics CQS Q_a, Q_p, Q_g, GAI, mean $\Delta C^*(\text{VS1} - \text{VS15})$ MCRI, FCI, IES R_f, and R_g with the mean visual scale values shown in Figure 7.49 will be analyzed. Table 7.15 contains the values of Pearson's correlation coefficients.

The following significant correlations ($\alpha = 5\%$) can be seen from Table 7.15: the mean ΔC^*, GAI, CQS Q_g, MCRI, and FCI for color preference, CQS Q_p and MCRI for naturalness, and mean $\Delta C^*(\text{VS1} - \text{VS15})$, GAI, CQS Q_g, and FCI for vividness. An important finding is, again, that all color fidelity type indices (R_a, CQS Q_f, and IES R_f) exhibit negative (or positive but almost zero) correlation with all visual attributes. This finding corroborates that the optimization of the light source spectra for maximum color rendering (in terms of a color fidelity index) does not guarantee a maximum visual assessment about preference, naturalness, and vividness. Just the opposite is true; see Figure 7.50.

As can be seen from Figure 7.50, "good"–"very good" color preference was achieved by the light sources with $R_a = 65 - 87$. Another related issue is the color

Table 7.15 Values of Pearson's correlation coefficients.

Correlation coefficient r	Preference	Naturalness	Vividness
Mean ΔC^* (VS1–VS15)	0.90[a]	0.77	0.90[a]
R_a	−0.48	−0.23	−0.59
GAI	0.94[a]	0.78	0.96[a]
CQS Q_a	0.03	0.33	−0.12
CQS Q_f	−0.31	−0.03	−0.44
CQS Q_g	0.89[a]	0.81	0.89[a]
CQS Q_p	0.82	0.94[a]	0.72
MCRI	0.93[a]	0.93[a]	0.86
FCI	0.93[a]	0.80	0.94[a]
IES R_f	−0.29	0.02	−0.42
IES R_g	0.85	0.81	0.85

a) Correlation coefficient significant at the 5% level.

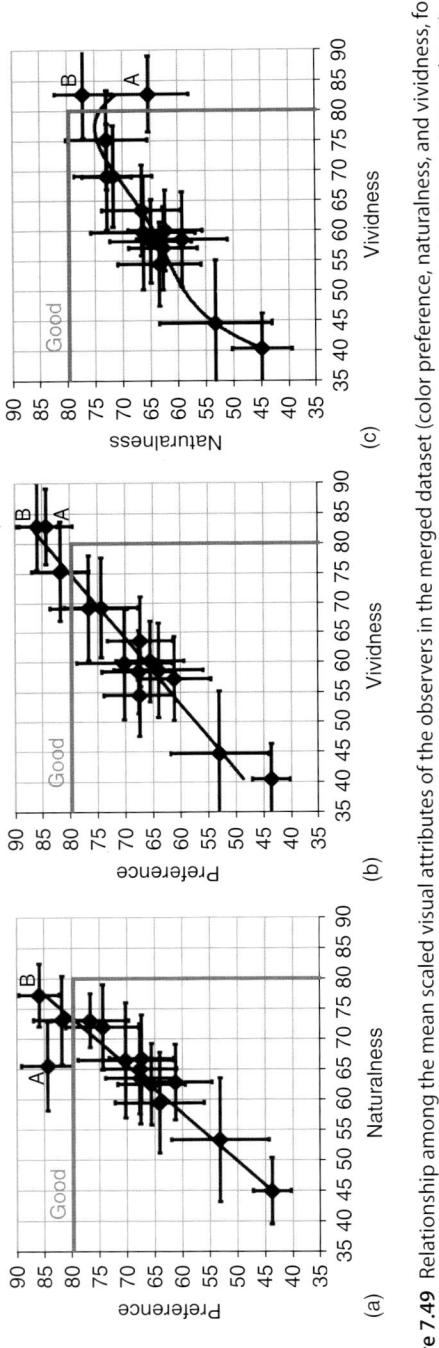

Figure 7.49 Relationship among the mean scaled visual attributes of the observers in the merged dataset (color preference, naturalness, and vividness, for all responsive observers in both experiments, makeups and food objects). Every point corresponds to one of the 14 light sources (see Figures 7.35 and 7.46). Intervals are 95% confidence intervals. The points A and B represent the two most saturating light sources (both are from the makeup experiment) with ΔC^* (VS1 − V14) = 4.4 (A) and 3.3 (B), respectively.

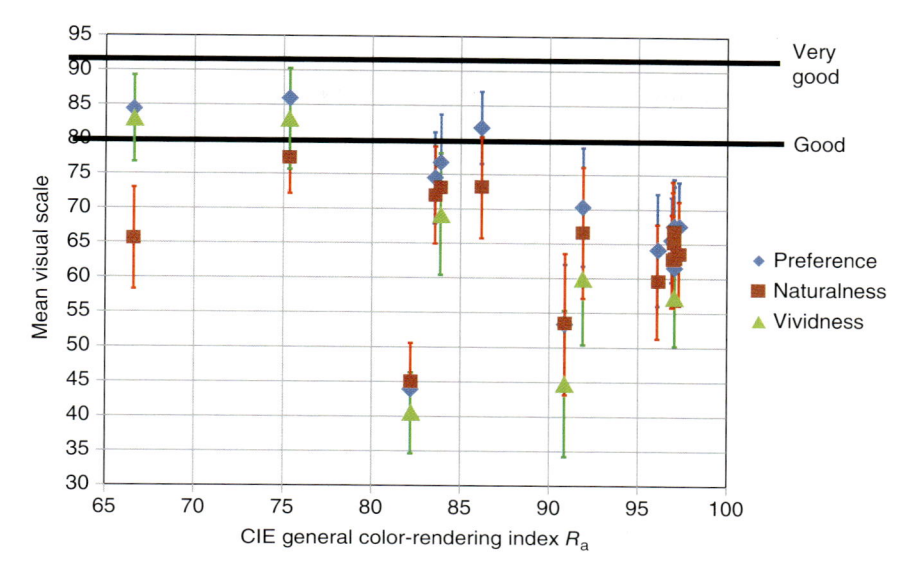

Figure 7.50 Mean values of the scaled visual attributes color preference, naturalness, and vividness as a function of the CIE general color-rendering index R_a for the merged dataset (14 light sources; makeup and food objects). Intervals are 95% confidence intervals of the mean values.

discrimination property of the light source. Subjects were not asked to assess this aspect in these experiments. Good color discrimination ability may represent a further constraint for spectral optimization. This will be discussed in Section 7.8.4. Apart from this, it can be seen from Figure 7.50 that optimizing for R_a, the values of the scaled visual attributes sink toward the range of "moderate" visual assessment. As pointed out in Section 7.7, a more usable way to predict the visual attributes is the use of the mean ΔC^*(VS1 − VS15) values in combination with a fidelity type metric as suggested by Eq. (7.11). To do so, Figure 7.51 shows the mean color preference, naturalness, and vividness ratings of the subjects as a function of the mean ΔC^*(VS1 − VS15) for all 14 light sources of the merged dataset.

As can be seen from Figure 7.51, concerning the merged dataset (14 light sources; food and makeup objects), the visual attributes color preference and naturalness peak in the range of 3.3–3.5 in terms of the mean ΔC^*(VS1 − VS15). But then, naturalness drops quickly while, in tendency, vividness continues to increase with increasing chroma. At this point, the model suggested by Eq. (7.11) including the quantity CQS Q_f (i.e., CQ = $a + b\Delta C^* + cQ_f$) will be applied to the merged dataset. The optimum coefficients arising from fitting Eq. (7.11) to the mean color preference, naturalness, and vividness data depicted in Figure 7.51 are shown in Table 7.16.

As can be seen from Table 7.16, the visual color quality attribute naturalness has the highest fidelity (c) component related to the chroma increment component (b), see the last row of Table 7.16 with the c/b ratios. For preference, a smaller (c/b) ratio arises. Vividness needs an even less fidelity component

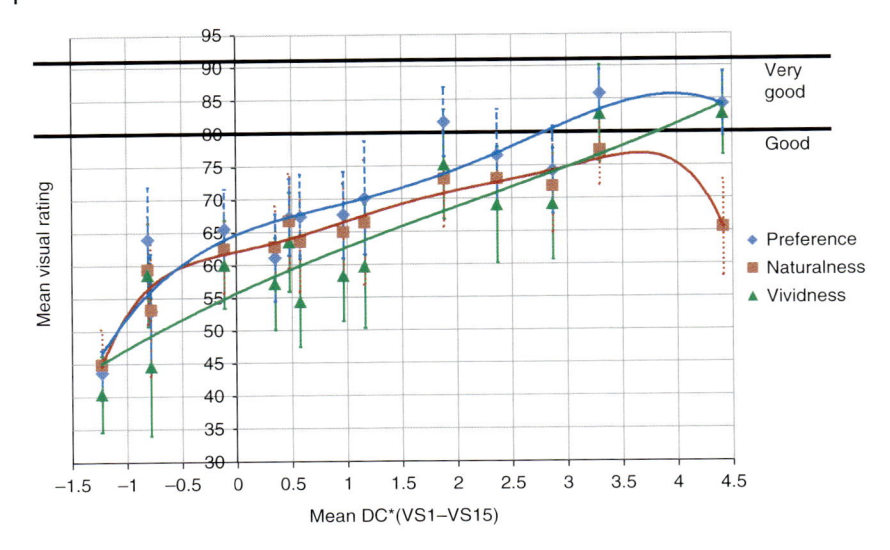

Figure 7.51 Mean color preference, naturalness, and vividness ratings of the subjects as a function of the mean ΔC^*(VS1 − VS15) for all 14 light sources of the merged dataset (food and makeup objects). Intervals are 95% confidence intervals of the mean values.

Table 7.16 Optimum values of the weighting parameters a, b, c in Eq. (7.11) (CQ = $a + b \Delta C^* + c Q_f$) resulting from the best fit to the mean visual ratings of the merged dataset (for all 14 light sources; Figure 7.51) and correlation coefficients between the two-metric combination CQ and the mean visual ratings of the observers.

Parameter	Color preference	Naturalness	Vividness
Constant a	1.154 (10.478)	−24.399 (−23.000)	27.175 (37.174)
Chroma increment b	7.624 (8.331)	5.801 (5.896)	7.376 (8.224)
Color fidelity c	0.669 (0.536)	0.934 (0.913)	0.309 (0.164)
Correlation coefficient r^2	0.875 (0.977)	0.873 (0.880)	0.830 (0.972)
c/b	0.088	0.161	0.042

In parentheses: data from Table 7.13 (makeup objects only). All correlation coefficients are significant at the 1% level.

(as expected; $c/b = 0.042$). The significant correlation coefficients between CQ and the mean visual results of the merged dataset indicate that it is reasonable to predict both lighting applications (lighting of makeup products and food products) by the same model equation (Eq. (7.11)) using the model parameters listed in Table 7.16.

Instead of the quantity CQS Q_f, the IES color fidelity index R_f can also be used; see Eq. (7.13).

$$CQ_2 = a_2 + b_2 \Delta C^* + c_2 R_f \tag{7.13}$$

In Eq. (7.13), ΔC^* is the mean ΔC^* value of the 15 CQS test color samples VS1–VS15. The optimum coefficients arising from fitting Eq. (7.13) to the mean

Table 7.17 Optimum values of the weighting parameters a_2, b_2, c_2 in Eq. (7.13) resulting from the best fit to the mean visual ratings of the merged dataset (for all 14 light sources; Figure 7.51) and correlation coefficients between the two-metric combination CQ_2 (i.e., using IES R_f instead of CQS Q_f) and the mean visual ratings of the observers.

Parameter	Color preference	Naturalness	Vividness
Constant a_2	7.390 (1.154)	−28.410 (−24.399)	37.255 (27.175)
Chroma increment b_2	7.226 (7.624)	5.475 (5.801)	7.0611 (7.376)
Color fidelity c_2	0.622 (0.669)	1.014 (0.934)	0.205 (0.309)
Correlation coefficient r^2	0.852 (0.875)	0.856 (0.873)	0.821 (0.830)
c_2/b_2	0.086 (0.088)	0.185 (0.161)	0.029 (0.042)

In parentheses: data with Eq. (7.11) (CQ) from Table 7.16 (by using Q_f to predict the merged dataset). All correlation coefficients are significant at the 1% level.

color preference, naturalness, and vividness data in Figure 7.51 are shown in Table 7.17.

As can be seen from Table 7.17, correlation coefficients between CQ_2 (using R_f, Eq. (7.13)) and the visual results are significant, similar to CQ (using Q_f, Eq. (7.11)) so that alternatively, the IES R_f metric can also be used. The (c_2/b_2) ratios are also similar to the ones in Table 7.16.

7.8.4 Effect of Object Oversaturation on Color Discrimination: a Computational Approach

In the present computational approach, by the use of each one of the relative SPDs of this set of 14 light sources, the $\Delta E'$ color differences [25] (also called *CAM02-UCS color differences*) between the adjacent color samples of the desaturated panel of the Farnsworth dichotomous test D-15 [15] were computed. The panel consists of 16 color samples numbered 0–15. Therefore, for every light source spectrum (i), 15 color differences between the pairs 0–1, 1–2, 2–3, ..., 13–14, and 14–15 were computed. They are denoted by $\Delta E'_{i,k}$ ($i = 1 - 14$; $k = 0 - 1, 1 - 2, 2 - 3, \ldots, 13 - 14, 14 - 15$).

In a previous work [22], a semantic interpretation of the similarity of the color appearance of two color stimuli in terms of categories was provided. According to the result of this series of psychophysical experiments [22], these categories were associated with limiting $\Delta E'$ [25] values; see Table 7.18.

The hypothesis is that, for a reliable and comfortable color discrimination performance of color normal observers under any test light source i, the similarity of color appearance between any pair k of the above-mentioned color samples shall be moderate ($\Delta E' = 3.98$, see Table 7.18) or worse than moderate ($\Delta E' < 3.98$).

A pilot experiment in the laboratory of the authors on color discrimination with color normal observers under different light sources indicated that the decline of reliable and comfortable color discrimination performance commences if the color difference between the adjacent colors decreases toward the category "moderate–good" ($\Delta E' = 3.03$). A visual experiment to support

Table 7.18 Semantic interpretation of the similarity of the color appearance of two color stimuli in terms of categories and their associated limiting $\Delta E'$ values [22].

Category	$\Delta E'$
Very good	0.00
Good–very good	1.05
Good	2.07
Moderate–good	3.03
Moderate	3.98
Low	5.91
Bad	8.25

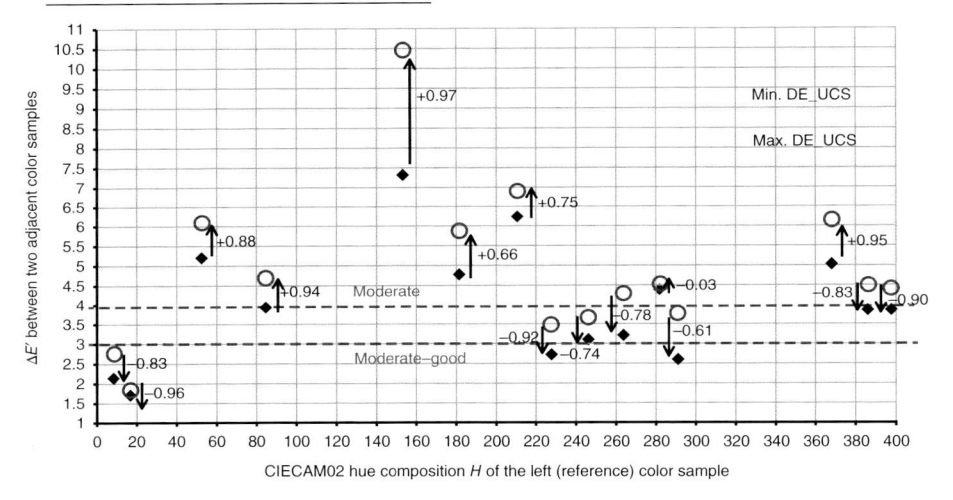

Figure 7.52 Ordinate: minimum (filled black diamonds) and maximum (gray circles) $\Delta E'$ color difference [25] values between the adjacent desaturated D-15 color samples among the 14 light sources of the two experiments (food and makeup; see Section 7.8.3). Abscissa: CIECAM02 hue composition H of the reference color sample (the reference sample is the one on the left; for example, it is No. 2 in the comparison 2–3) under the light source "maxQg0" [23]. Black arrows show the direction of increasing mean $\Delta C^*(VS1 - VS15)$ values among the 14 light sources with the corresponding Pearson's correlation coefficients r between the mean $\Delta C^*(VS1 - VS15)$ values and $\Delta E'$ for a particular pair of color samples (an example is shown in Figure 7.53).

the above hypothesis is currently underway. Figure 7.52 shows the result of the above-described computational approach.

As can be seen from Figure 7.52, if such a light source is chosen that causes an increase of chroma of the illuminated objects (i.e., a light source with a high mean $\Delta C^*(VS1 - VS15)$ value), this enhances the ability of color discrimination (by increasing the color difference between two adjacent samples) in the case of certain adjacent pairs of D-15 color samples. This is indicated by the black arrows pointing toward the top of Figure 7.52 and the positive correlation coefficient (the value next to the arrow) between the mean ΔC^*(VS1 − VS15) value of the light source and the value of $\Delta E\prime$ for this particular pair of color samples. There are seven such cases altogether. Note that all these arrows are located at or above the "moderate" level. According to the above hypothesis, it is probable that color discrimination performance does not increase significantly by increasing ΔC^* in the case of these seven pairs of adjacent color samples with upwards arrows.

Just the opposite is true for the D-15 color samples comprising an arrow that points toward the bottom of Figure 7.52 and have a negative correlation coefficient. There are eight such cases altogether. For these color samples, an increase of the ΔC^* value of the light source is associated with a decrease of color discrimination performance. The two pairs of color samples at $H = 9$ (9% yellow and 91% red) and $H = 17$ (17% yellow and 83% red) represent critical examples for which the "moderate–good" level cannot be achieved by the present set of 14 light sources.

If the viewing task requires good color discrimination performance in this reddish orange range then a different light source should be used (i.e., not from the present set of 14 light sources). Another critical example is the pair of D-15 color samples No. 0–1 with $H = 291$ (which refers to the zeroth sample containing 91% unique blue and 9% unique green). In this case, by increasing the object saturation of the light source, the similarity of the color appearance of the two adjacent color samples increases from "moderate" over "moderate–good" toward "good," which deteriorates color discrimination performance; see the descending black arrow at $H = 291$ in Figure 7.52 with $r = -0.61$. Figure 7.53 shows this tendency,

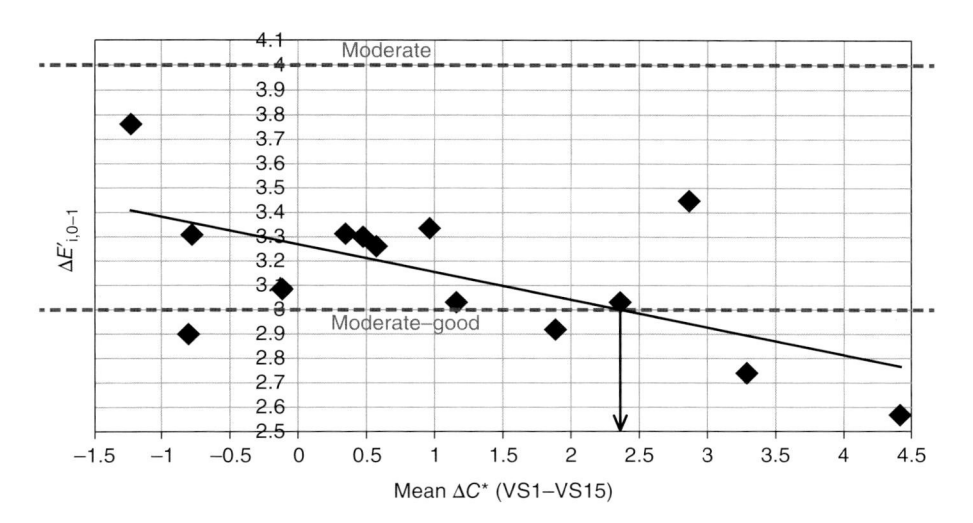

Figure 7.53 Dependence of $\Delta E'_{i,0-1}$ on the mean ΔC^*(VS1 − VS15) value of the ith light source ($i = 1 - 14$). The black arrow projects to the critical mean ΔC^*(VS1 − VS15) value, $\Delta C^*_{crit} = 2.35$. If $\Delta C^* > \Delta C^*_{crit}$ then color discrimination performance is expected to deteriorate for this particular (greenish blue) pair of test color samples (No. 0–1).

that is, the dependence of $\Delta E'_{i,0-1}$ on the mean $\Delta C^*(\text{VS1} - \text{VS15})$ value of the ith light source ($i = 1 - 14$).

As can be seen from Figure 7.53, for this particular (greenish blue) pair of test color samples (No. 0–1), the critical value of the mean $\Delta C^*(\text{VS1} - \text{VS15})$ equals $\Delta C^*_{\text{crit}} = 2.35$. If the object oversaturation caused by the light source exceeds this critical value then it can be expected that color discrimination performance begins to decline. The reason is that although an oversaturating light source spectrum is able to increase the chroma of certain saturated object colors (and extend color gamut in those regions of color space), the distance between two adjacent colors might shrink in other regions (e.g., in the greenish blue region) of color space. This is one of the reasons why the use of a color gamut metric should be avoided to characterize the color quality (or color rendition) of a light source.

7.9 Semantic Interpretation and Criterion Values of Color Quality Metrics

The concept of the semantic interpretation of color differences and color quality metrics (in terms of the categories "excellent," "good," "moderate," etc.) was introduced in Chapter 2. It was pointed out that semantic interpretation is important for lighting design in order to provide criterion values or numeric user acceptance limits in terms of suitable color quality indices. In the present section, the most important applications of semantic interpretations will be described divided into two subsections: (i) the semantic interpretation of color differences (for color fidelity indices, color discrimination, and to provide criterion values of white tone chromaticity for the binning of white LEDs) and (ii) the semantic interpretation of color appearance attributes (to provide criterion values for color preference, naturalness, and vividness, as a summary of the semantic interpretations and criterion values described in Sections 7.7–7.9).

7.9.1 Semantic Interpretation and Criterion Values of Color Differences

The semantic interpretation of color differences and their criterion values constitute the starting point to derive criterion values for color fidelity indices (Section 7.9.1.1), color discrimination (Section 7.9.1.2), and to provide criterion values of white tone chromaticity for the binning of white LEDs (Section 7.9.1.3). The semantic interpretation of color differences is based on a series of visual experiments that were carried out to rate the similarity of color appearance of two color stimuli on categorical and continuous semantic rating scales [22]. These pairs of color stimuli included two copies of the same colored real or artificial object illuminated by a test light source and a reference light source. Observers had to rate color appearance similarity on a continuous interval scale (0–6) labeled by categories; see Table 7.19

From the experimental results, a formula was developed [22] to predict the numeric value of R_U (see Table 7.19) corresponding to a category of color similarity (e.g., "moderate" or "good") or intermediate categories (e.g., "moderate–good"

7.9 Semantic Interpretation and Criterion Values of Color Quality Metrics

Table 7.19 Rating scales of color appearance similarity: continuous interval scale rating R_U (0–6) labeled by categories.

Categorical rating	Continuous rating R_U
Excellent	0
Very good	1
Good	2
Moderate	3
Low	4
Bad	5
Very bad	6

for $R_U = 2.7$) from an instrumentally measured color difference (the metric CAM02-UCS $\Delta E'$ was used). Given a numeric value of a color difference between the two members of a pair of colors, for example, $\Delta E' = 2.07$, the formula $R_{U,pred}(\Delta E')$ is able to predict a category of color similarity, for example, "good." Equation (7.14) shows the formula, a cubic polynomial. Criterion $\Delta E'$ values of the rating categories are shown in Table 7.18.

$$\text{if } \Delta E' < 11 \text{ then } R_{U,pred} = 1.03904809 + 0.4073793\,\Delta E' + 0.03377417\,\Delta E'^2 \\ - 0.00302202\,\Delta E'^3 \text{ else } R_{U,pred} = 5.585 \quad (7.14)$$

Figure 7.54 shows a plot of Eq. (7.14) together with the experimental results the formula is based on.

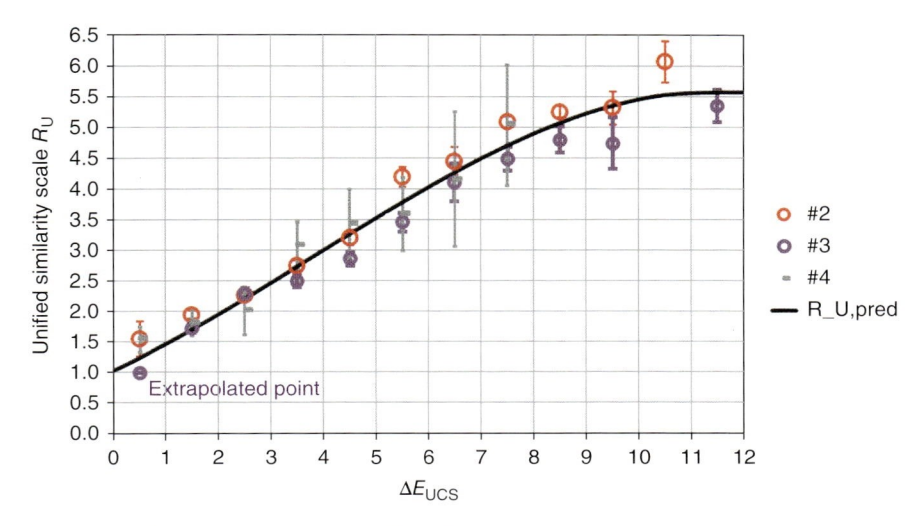

Figure 7.54 Black curve: a plot of Eq. (7.14). Symbols represent mean experimental data (observers' ratings on the so-called unified similarity scale R_U) from the three series of experiments (#2, #3, #4) the formula is based on. Intervals are 95% confidence intervals of the mean data. (Bodrogi et al. 2014 [22]. Reproduced with permission of Wiley.)

As can be seen from Figure 7.54 and Table 7.19, the categories "excellent" ($R_U = 0$) and "very bad" ($R_U = 6$) are never reached by the prediction of Eq. (7.14) because observers tend to avoid these categories.

7.9.1.1 Semantic Interpretation of Color Fidelity Indices

Because color fidelity indices are based on color differences, Eq. (7.14) was applied to interpret the values of color-rendering indices (CIE CRI R_a, CRI2012, and IES R_f) in terms of the semantic categories in Table 7.19. This semantic interpretation enables light source users to understand the CRI the light source is labeled with in terms of everyday language. For example, the user will be able to interpret a numeric value of 87 as "good color fidelity."

Figure 7.55 shows the semantic interpretation of the CIE CRI R_a, the general CRI [26] for a set of light sources using Eq. (7.14). During the computation, color differences expressed in the $\Delta E'$ metric were substituted (using a computational algorithm) for every test color sample and every test light source by the color differences computed in the outdated $U^*V^*W^*$ color space underlying the CIE CRI R_a metric.

As can be seen from Figure 7.55, although the semantic interpretation of the special color-rendering indices tend to scatter considerably, a general tendency represented by the red curve (a cubic polynomial, Eq. (7.15)) was identified to interpret the general CRI R_a.

$$\text{If } R_a > 6.7 \text{ then } R_{U,\text{pred}} = 5.6542214712829 - 0.00868708870\ R_a$$
$$- 0.00025988179\ R_a^2 - 0.00000127635\ R_a^3 \quad \text{else } R_{U,\text{pred}} = 5.585$$
(7.15)

Figure 7.56 shows the semantic interpretation of the special color-rendering indices in the CRI2012 (also called *n-CRI*) method [22].

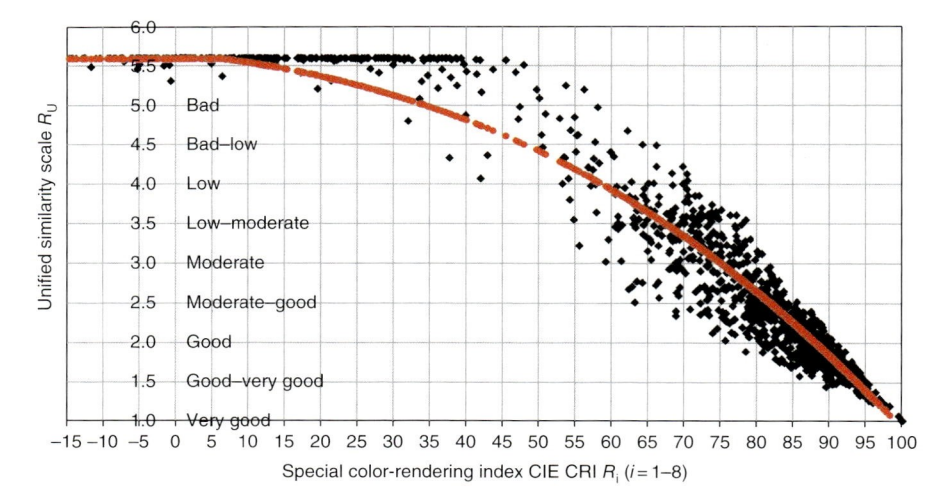

Figure 7.55 Black points: semantic interpretation on the so-called unified similarity scale R_U [22] of the special color-rendering indices CIE $R_i (i = 1 - 8)$ for a set of light sources using Eq. (7.14). Red curve: semantic interpretation of the CIE CRI R_a (mean tendency derived from the black points) [26].

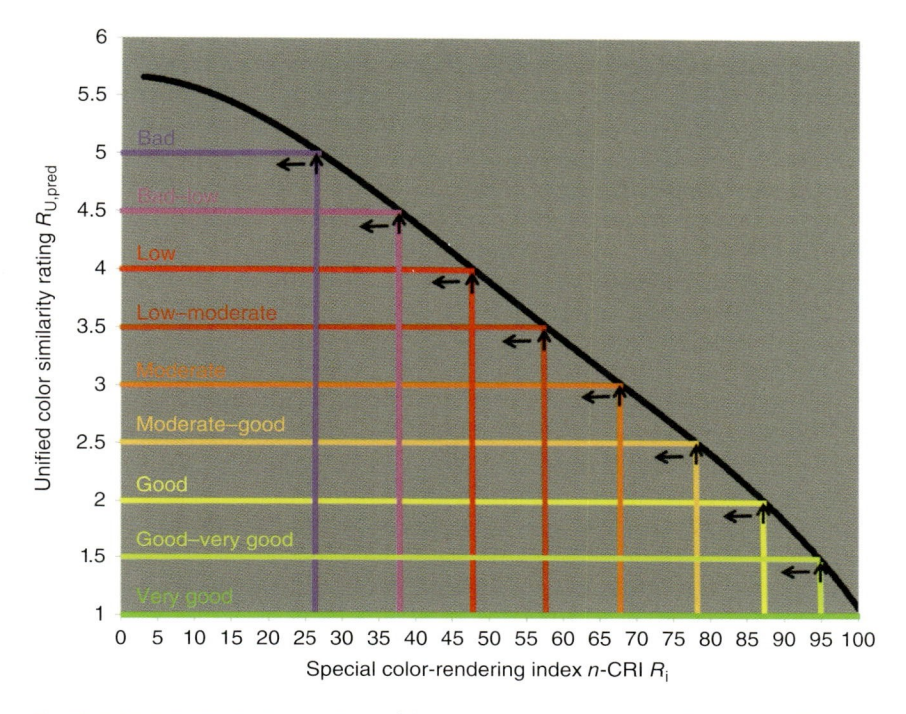

Figure 7.56 Semantic interpretation of the special color-rendering indices in the CRI2012 (also called *n-CRI*) method. Bodrogi *et al.* 2014 [22]. Reproduced with permission of Wiley.)

As can be seen from Figure 7.56, the user of the light source can interpret the values of the CRI2012 (also called *n-CRI*) color-rendering indices and their differences. The diagram can also be used to interpret the CRI2012 general CRI as these indices are based on the CAM02-UCS color difference metric $\Delta E'$. The diagram of Figure 7.56 can be used to decide whether the light source is suitable (e.g., using the criterion "good") for a given application from the point of view of color rendering. As the category "very good" corresponds to no noticeable visual color difference, this criterion should be used in the most demanding applications that require the criterion "high color fidelity," for example, high quality indoor lighting for color printout inspection.

Figure 7.57 shows the semantic interpretation of the CRI IES R_f [13].

The tendency shown in Figure 7.57 can be approximated by Eq. (7.16)

$$\text{If } R_f > 9.4 \text{ then } R_{U,\text{pred}} = 5.31566639 + 0,81498154 \ (R_f/20)$$
$$-0,6263431 \ (R_f/20)^2 + 0.0585014 \ (R_f/20)^3 \text{ else } R_{U,\text{pred}} = 5.585$$
(7.16)

Table 7.20 compares the color fidelity criterion values of the three metrics (CIE CRI R_a, CRI2012, and IES R_f) corresponding to the rating categories.

As can be seen from Table 7.20, depending on the color fidelity metric, different criterion values shall be used as decision criteria during the spectral design of the light source. The color fidelity metric in the last column of Table 7.20

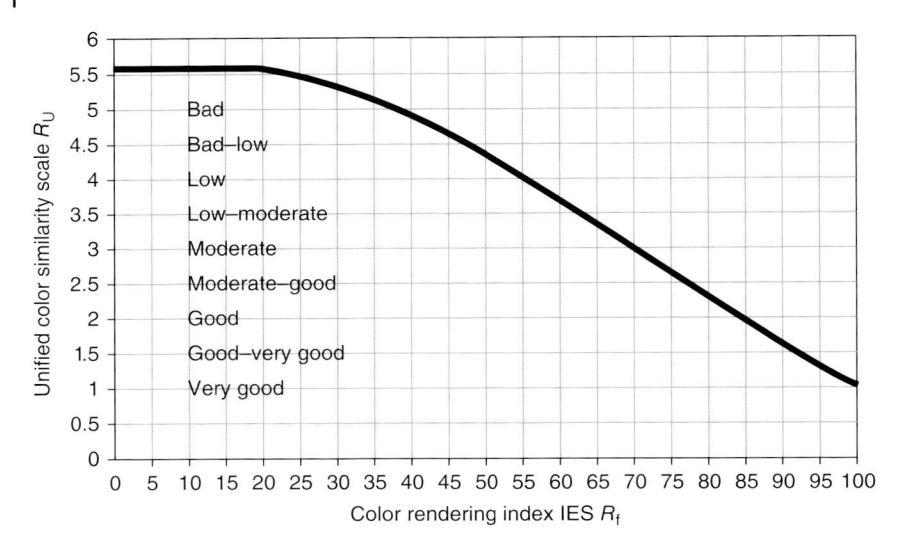

Figure 7.57 Semantic interpretation of the color rendering index IES R_f [13].

Table 7.20 Semantic interpretation and criterion values of color fidelity indices.

Category	R_a	CRI2012	IES R_f
Very good	100	100	100
Good	87	87	84
Moderate	74	67	70
Low	58	47	55
Bad	33	26	38

Comparison of the three color fidelity metrics.

(IES R_f) [13] represents the most up-to-date metric recommended for use at the time of writing.

7.9.1.2 Color Discrimination

As mentioned in Section 7.8.4, to avoid the deterioration of color discrimination ability caused by the oversaturation of the colored objects by, for example, exaggerated weightings given to the red semiconductor LED channels of a multi-LED engine, the color difference $\Delta E'$ between the color samples to be distinguished easily should not decrease below the level of the "moderate–good" category ($\Delta E' = 3.03$). In the Farnsworth–Munsell 100-hue test (a well-known test of color vision to test color vision deficiencies), the mean color difference between two adjacent color samples under CIE illuminant C has the same order of magnitude as the level of the "moderate–good" category so that observers of normal color vision are able to complete the test without difficulties. A visual experiment to support the above "moderate–good" criterion for easy color discrimination is underway in the authors' laboratory at the time of writing.

7.9.1.3 Criterion Values for White Tone Chromaticity for the Binning of White LEDs

Before the construction of a multi-LED light engine, to achieve a certain target chromaticity for its white point (see Chapter 3) and also, in order to maintain the homogeneity of the white tone of the luminaire, an efficient chromaticity binning strategy of the LED light sources is essential. Chromaticity binning means the sorting of the LED light sources into chromaticity categories (so-called bins) after production. It is required to provide criterion values in terms of a suitable chromaticity metric for the magnitude of visually acceptable chromaticity differences within a bin (i.e., the set of actual LED light sources to be installed in the luminaire) and also, between any LED light source inside the bin and a desirable target chromaticity, for example, a visually preferred white tone (see Chapter 3).

The American National Standard ANSI ANSLG C78.377-2011 specifies white light chromaticities in terms of a set of binning ranges in the CIE 1931 x, y chromaticity diagram or in the CIE 1976 u', v' chromaticity diagram. ANSI binning categories are based on MacAdam's ellipses. In this section an alternative method based on the $\Delta E'$ (CAM02-UCS) semantics shown in Figure 7.54 and in Table 7.18 will be described. The aim is to communicate the magnitude of acceptable chromaticity differences between the LED light source manufacturer and the producer of the lighting product.

Similar to Sections 7.9.1.1 and 7.9.1.2, the present computational method assigns any instrumentally measured chromaticity difference a semantic interpretation. The so-called "chromaticity center" is the desired (target) white tone chromaticity of the LED light source (e.g., a certain cool white tone). In this case, the concept of semantic interpretation means that, depending on the chromaticity center and the direction in the chromaticity diagram, every chromaticity difference is interpreted by one of the semantic categories shown in Table 7.18. For example, "very good" means a very good *agreement* of perceived white tone chromaticity between the center in the chromaticity diagram (i.e., the target chromaticity to be achieved by the LED light source) and the actual white tone chromaticity of the LED.

For the chromaticity binning of white LEDs, the present computational method [27] transforms every $\Delta E'$ value (e.g., the ones shown in Table 7.18 but also any intermediate value) into the CIE 1931 x, y chromaticity diagram around any chromaticity center in any direction. Then, the semantic contours of "very good," "good," "moderate," "low," or "bad" agreement with the chromaticity of the color center are computed in every direction around the color center. Within the areas between neighboring semantic contours in the CIE 1931 x, y chromaticity diagram, the nearest category to the considered chromaticity point shall be applied.

The above-defined semantic binning contours are shown in the CIE 1931 x, y chromaticity diagram of Figure 7.58 [27] for a warm white chromaticity center (the Planckian radiator at 2700 K) $x = 0.460; y = 0.411$. Semantic contours are compared with contours of constant chromaticity differences ($\Delta u'v' = 0.001 - \Delta u'v' = 0.007$) in the $u' - v'$ diagram measured from the chromaticity center. (These constant $\Delta u'v'$ contours approximate one-step to seven-step MacAdam ellipses; see Section 2.2.5)

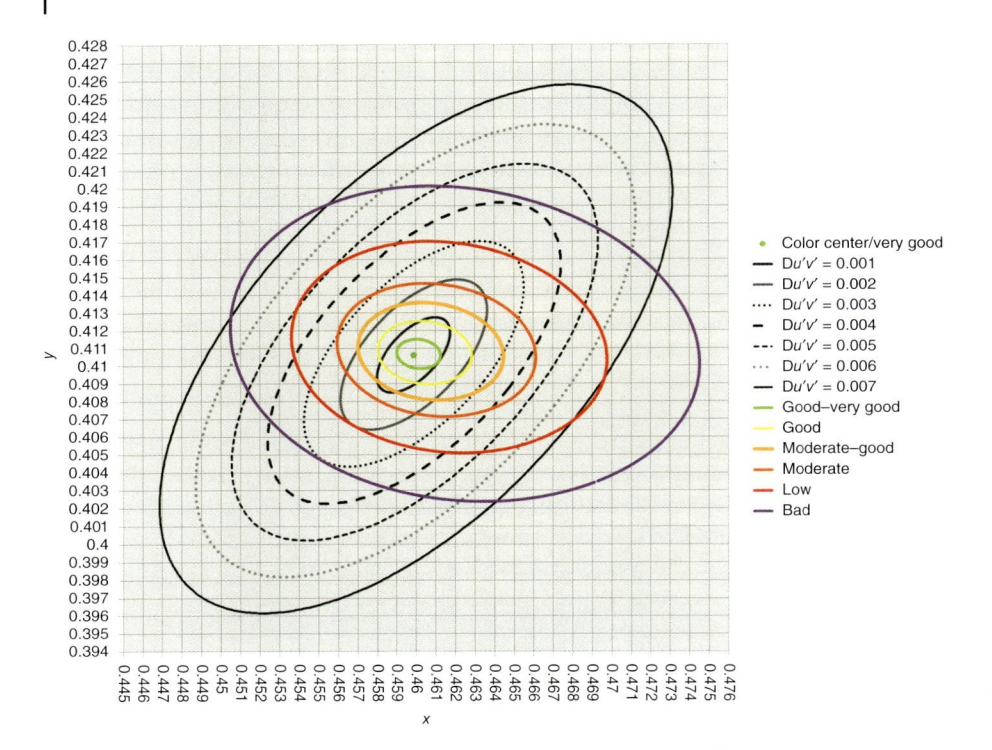

Figure 7.58 Semantic contours for chromaticity differences from the chromaticity center in the CIE x, y chromaticity diagram for a warm white chromaticity center, the Planckian radiator at 2700 K; $x = 0.460$; $y = 0.411$ (light green dot). Going off the center in any direction, contours indicate "good–very good" (green contour), "good" (yellow contour), "moderate–good" (orange contour), "low" (red contour), and "bad" (lilac contour) perceived color agreement with the center. Contours of constant chromaticity differences ($\Delta u'v' = 0.001 - \Delta u'v' = 0.007$, that is, approximations of MacAdam ellipses) measured from the chromaticity center are also shown [27].

The following can be seen from Figure 7.58: (i) the orientation of the constant $\Delta u'v' = 0.001 - \Delta u'v' = 0.007$ contours and the orientation of the semantic contours are different. (ii) At this chromaticity center ($x = 0.460$; $y = 0.411$), the different shades of white (white points) along the $\Delta u'v' = 0.001$ ellipse may correspond to "good–very good," "good," or "moderate–good" color agreement with the center (the light green point, that is, the target white point), depending on the direction; see the intersection points of the $\Delta u'v' = 0.001$ ellipse (black curve) with the green, yellow, and orange semantic contours in Figure 7.58.

Semantic contours resulting from the computation are shown in the CIE 1931 x, y chromaticity diagram of Figure 7.59 for a phase of daylight at 6500 K (D65), $x = 0.313$; $y = 0.329$. Semantic contours are compared with contours of constant chromaticity differences ($\Delta u'v' = 0.001 - \Delta u'v' = 0.007$) in the $u' - v'$ diagram measured from the chromaticity center.

The following can be seen from Figure 7.59: (i) The orientation of the $\Delta u'v' = 0.001 - \Delta u'v' = 0.007$ contours and the orientation of the semantic

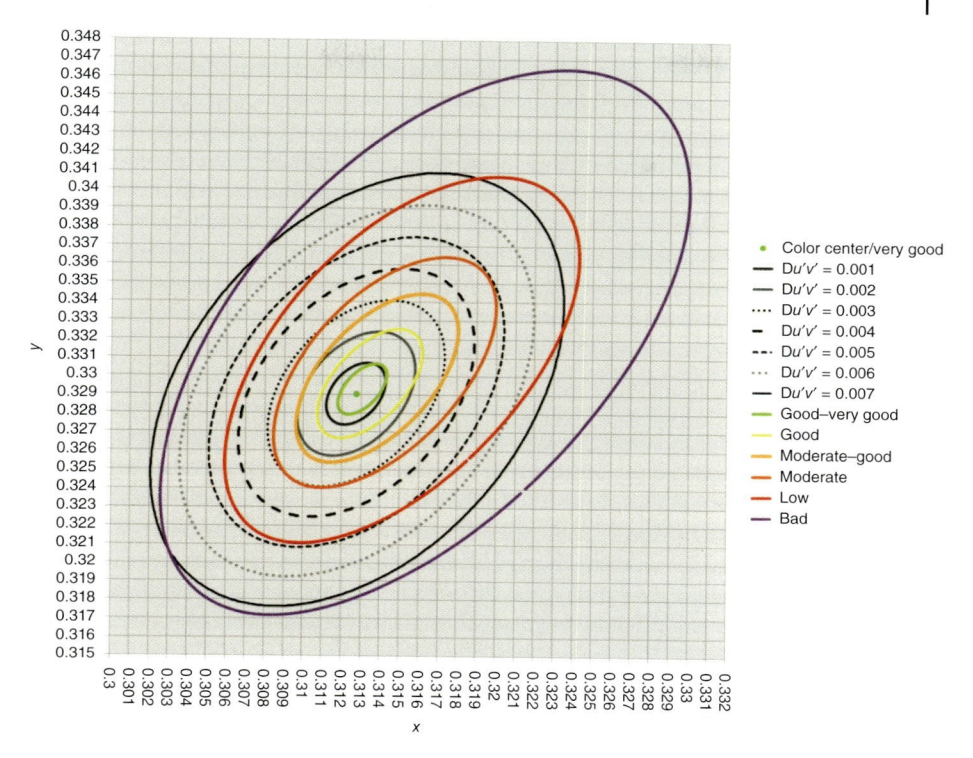

Figure 7.59 Semantic contours for chromaticity differences from the chromaticity center in the CIE x, y chromaticity diagram for a phase of daylight at 6500 K (D65), $x = 0.313$; $y = 0.329$ (light green dot). For explanations; see the caption of Figure 7.58.

contours are similar. (ii) At this chromaticity center ($x = 0.313$; $y = 0.329$), the different white tones along the $\Delta u'v' = 0.001$ ellipse (black) correspond approximately to "good–very good" agreement with the center (the green curve exhibits a similar course as the black ellipse). (iii) The different white tones along the $\Delta u'v' = 0.002$ ellipse may correspond to "good" or "moderate–good" color agreement with the center (the light green point, that is, the target white point), depending on the direction; see the intersection points of the $\Delta u'v' = 0.002$ ellipse (gray) with the yellow and orange semantic contours in Figure 7.59.

The new method to derive criterion values for white tone chromaticity for the binning of white LEDs has several advantages compared to ANSI binning. Semantic contours corresponding to "good–very good," "good," and so on, visual agreement of the considered white LED chromaticity with the chromaticity of the center (target white chromaticity) carry a straightforward meaning, even for non-expert users of light sources because these categories are formulated in terms of natural everyday language (e.g., "good") instead of using a hard-to-understand multiple of the size of an (approximated) MacAdam ellipse.

The method described in this section uses a modern color difference metric (CAM02-UCS), which is more reliable than using multiples of MacAdam's ellipses (e.g., one-step, two-step, …, seven-step MacAdam ellipses). The reason is that MacAdam's experiments were based on chromatic adaptation to a single chromaticity (daylight), on a single observer, and only on the measurement of just noticeable color differences. Upscaling these just noticeable color differences (e.g., seven-step MacAdam) might cause unpredictable visual errors. The CAM02-UCS metric, however, was designed to predict the whole range of magnitude of color differences relevant to white LED binning.

7.9.2 Semantic Interpretation and Criterion Values for the Visual Attributes of Color Appearance

Figure 7.60 shows the semantic interpretation of the two-metric combination $CQ = a + b \; \Delta C^* + c \; Q_f$ (Eq. (7.11)) for the merged dataset (food and makeup products, described in Section 7.8) with the optimum weighting parameters a, b, c taken from Table 7.16, according to Figure 7.28.

As can be seen from Figure 7.60, after computing the value of the predictor quantity CQ of the light source for the visual attribute in question (color preference, naturalness, or vividness), a semantic interpretation can be given according to Figure 7.28. Criterion values of the semantic categories are repeated in Table 7.21 for clarity. As an acceptance criterion for lighting design for general interior lighting, $CQ \geq 79.6$ ("good") can be considered. For high-quality lighting application, the category "very good" ($CQ = 91.6$) can be considered as an aim of spectral optimization.

Note that the appropriate parameter set (a, b, c) shall be selected from Table 7.16 for color preference, naturalness, or vividness, according to the intent of the lighting engineer. Concerning the choice of the visual attribute to use, naturalness might be considered for longer stays in a formal context, for example, in an office application. In contrast to naturalness, color preference represents

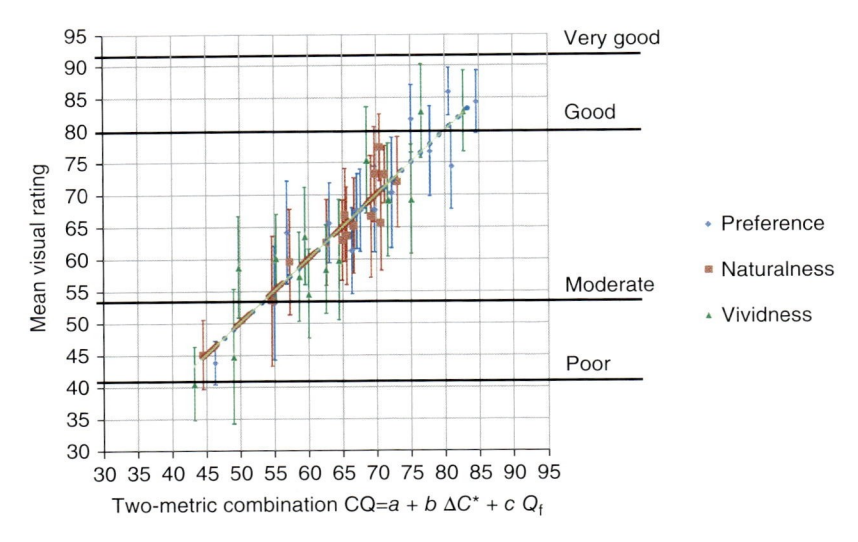

Figure 7.60 Semantic interpretation of the two-metric combination $CQ = a + b\Delta C^* + cQ_f$ (Eq. (7.11)) with the optimum weighting parameters a, b, c taken from Table 7.16, according to Figure 7.28.

Table 7.21 Criterion values of the quantity CQ (Eq. (7.11)) for the semantic categories for color preference, naturalness, and vividness.

Semantic category for color preference, naturalness, and vividness	Criterion value of the quantity CQ
Excellent	97.9
Very good	91.6
Good	79.6
Moderate	52.9
Poor	41.2
Bad	26.5
Very bad	12.8

the most general attribute for general interior lighting while vividness stands for the amazement to be evoked by the emotional effect of highly saturated object colors in special applications (e.g., for certain theater scenes as a short-term lighting effect).

In Figure 7.60, symbols represent the mean value of all (responsive) observers for a given light source and a given visual attribute (color preference, naturalness, or vividness). The criterion CQ values in Table 7.21 correspond to the linear trend of this typical (average) observer. In Figure 7.60, 95% confidence intervals are also shown representing the scatter of the observers' opinion about color quality. The trend of the lower end of these intervals can be considered as a "worst-case observer." Similarly, the trend of the upper end corresponds to a "best-case" observer. This range of opinions can be estimated at the CQ values of the indicated data points of the 14 test light sources of the merged dataset (food and makeup products) described in Section 7.8.

The semantic interpretation of the quantity CQ_2 with R_f instead of Q_f (Eq. (7.13)) can be carried out in a similar way by the use of Tables 7.21 and 7.17.

7.10 Lessons Learnt for Lighting Practice

From the considerations in this chapter, it can be concluded that a comprehensive color quality metric should have three components that take the three aspects color appearance (with its three important color quality attributes preference, naturalness, and vividness), color fidelity, and color discrimination into account. With the development of solid-state lighting products at the end of the last century and the use of white LEDs in all domains of human life (e.g., homes, hospitals, schools, offices, museums), the need for a correct description of color quality has been intensified and this requires a comprehensive and usable color quality metric that is able to describe color quality for the practice of lighting engineering. It is therefore important to systematize the state-of-the-art knowledge in order to conduct well-designed visual experiments and formulate a system of the most relevant color quality aspects for the practice of lighting engineering; see Figure 7.61.

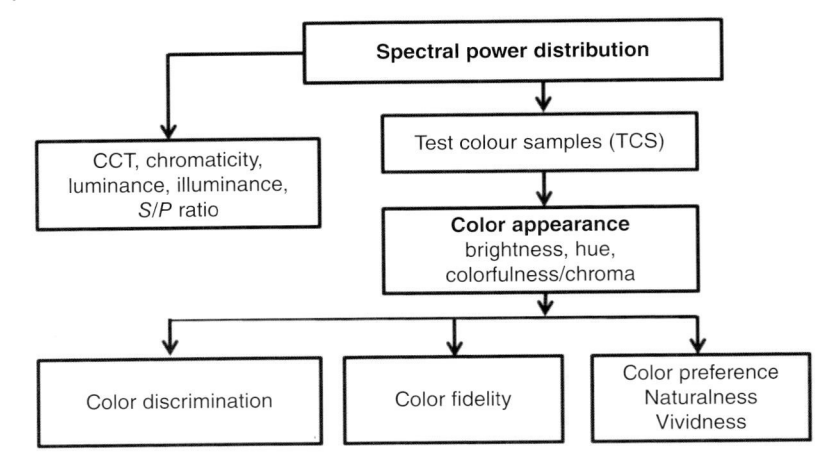

Figure 7.61 Summary of the most relevant aspects of light source color quality for the practice of lighting engineering.

From the knowledge summarized in Figure 7.61, it is important to conduct visual experiments to find out the necessary level of illuminance on the colored objects and the most suitable CCT to be used for different applications and groups of colored objects. The results in Section 7.2 showed that an illuminance level higher than 500 lx is reasonable for optimum visual acuity and optimum color discrimination. Contrary to color discrimination and color fidelity, which are based on color differences (that represent a human eye physiological issue), the color attributes color preference, vividness, and naturalness are strongly related to cognitive color processing and human color memory effects that include a comparison with natural scenarios often seen by the observer in the past in natural environments at higher illuminance levels (30 000–100 000 lx) so that visual color quality and color appearance experiments have to be designed at a level higher than 500 lx in the future.

The results in Sections 7.4 and 7.5 showed a range of preferred CCTs between 4000 and 5000 K with a specific characteristic that blue–green objects should be illuminated with spectra of higher color temperature while warm white light sources should be chosen to illuminate red–orange and skin tone objects.

From the two visual experiments in Sections 7.7 and 7.8 with two different combinations of colored objects representing two different lighting application contexts, a common multiple-metrics formula was derived to predict color quality. The proposed formulae (Eqs. (7.11) and (7.13)) show high correlation with the mean visual assessments of color preference, naturalness, and vividness. The formulae represent the idea that these visual attributes can be described as linear combinations of a color fidelity index (IES R_f or CQS Q_f) and the chroma difference ΔC^* between the test and reference appearances of suitable test color samples.

The idea behind these formulae was inspired by the concept of the CQS Q_p metric, that is, rewarding the increase of chroma ($\Delta C^* > 0$) in comparison to the color fidelity condition ($\Delta C^* = 0$). However, the authors of the present book believe that the present formulae are more advantageous than CQS Q_p because (i) the coefficients a, b, c (or a_2, b_2, c_2) were fitted to visual results on three different attributes (color preference, naturalness, vividness); (ii) CQ values per se are psychophysically relevant as these scales include a semantic interpretation with criterion values to achieve, for example, the "good" level; and (iii) by the use of the variable ΔC^*, it is straightforward to establish the link between color preference and color discrimination ability.

According to the thoughts and findings in this chapter, color quality optimization that optionally includes a color discrimination trade-off should take the following three steps into account:

1) Color fidelity (e.g., IES R_f) should be optimized. The result is a multi-LED spectrum with a high CRI so that the chroma difference ΔC^* approaches zero; see the spectrum with $\Delta C^*(VS1) = 0.0$ in Figure 7.35.
2) According to Eq. (7.13), that is, the formula $CQ_2 = a_2 + b_2\ \Delta C^* + c_2\ R_f$, the SPD of the multi-LED light engine should be optimized step by step by continuously reducing the color fidelity index and increasing the chroma difference ΔC^*. Thus, the optimum color preference rating will be achieved after a few calculation steps.
3) Developers should then compare the resulting $\Delta C^*(VS1 - VS15)$ values that provide optimum color preference, naturalness, or vividness with the values $\Delta C^*(VS1 - VS15)_{crit}$ for acceptable color discrimination. If color discrimination is important in a specific illumination project (e.g., museum lighting), then the values of $\Delta C^*(VS1 - VS15)$ that provide optimum color preference, naturalness, or vividness should be reduced. Otherwise, the values $\Delta C^*(VS1 - VS15)$ for optimum color preference, naturalness, or vividness should be adopted.

In the computational approach of this book, the critical value of $\Delta C^*(VS1 - VS15)_{crit} = 2.35$ was determined for the case of the greenish-blue pair of test color samples. This value corresponds to $R_{a,crit} = 84$). The value of $\Delta C^*(VS1 - VS15)_{crit} = 2.35$ corresponds to the visual preference rating 77 (i.e., slightly less than "good").

The maximum mean color preference assessment (87, that is, better than "good") corresponds to $R_{a,crit} = 75$ at the mean $\Delta C^*(VS1 - VS15)$ value of 3.3 corresponding to $\Delta E'_{i,0-1} = 2.9$ with deteriorated color discrimination performance. A mean color preference rating around 65 (which is rather "moderate" than "good") is associated with $R_a = 96 - 98$ and with $\Delta E'_{i,0-1} = 3.2 - 3.3$, that is, $\Delta C^* < \Delta C^*_{crit}$ in this case. These trade-off values between color preference and color discrimination in terms of the CRI (CIE R_a), the mean $\Delta C^*(VS1 - VS15)$ value, and the value of $\Delta E'_{i,0-1}$ are listed in Table 7.22. The trade-off values in Table 7.22 represent a special example for the case of the 14 light sources at CCT = 3200 K and the food and makeup objects of the present chapter.

Table 7.22 Trade-off values between color preference and color discrimination.

Mean visual color preference rating	CIE R_a	ΔC^*(VS1 − VS15)	$\Delta E'_{i,0-1}$
Maximal in the visual experiments; mean rating 87 (better than "good")	75	3.3	2.9
Color preference rating 77 (just below 80, just below "good")	84	2.35	3.03
Color preference rating around 65 (rather "moderate" than "good")	96–98	−0.2 to 0.2	3.2–3.3

References

1 Bodrogi, P., Brückner, S., Khanh, T.Q., and Winkler, H. (2013) Visual assessment of light source color quality. *Color Res. Appl.*, **38**, 4–13.
2 Smet, K.A.G., Ryckaert, W.R., Pointer, M.R., Deconinck, G., and Hanselaer, P. (2011) Correlation between color quality metric predictions and visual appreciation of light sources. *Opt. Express*, **19**, 8151–8166.
3 Rea, M.S. and Freyssinier-Nova, J.P. (2008) Color rendering: a tale of two metrics. *Color Res. Appl.*, **33**, 192–202.
4 Islam, M., Dangol, R., Hyvärinen, M., Bhusal, P., Puolakka, M., and Halonen, L. (2013) User preferences for LED lighting in terms of light spectrum. *Light. Res. Technol.*, **45**, 641–665.
5 Dangol, R., Islam, M., Hyvarinen, M., Bhusal, P., Puolakka, M., and Halonen, L. (2013) Subjective preferences and color quality metrics of LED light sources. *Light. Res. Technol.*, **45**, 666–688.
6 Dangol, R., Islam, M., Hyvarinen, M., Bhushal, P., Puolakka, M., and Halonen, L. (2013) User acceptance studies for LED office lighting: preference, naturalness and colourfulness. *Light. Res. Technol.*, **47**, 36–53.
7 Baniya, R., Dangol, R., Bhusal, P., Wilm, A., Baur, E., Puolakka, M., and Halonen, L. (2013) User-acceptance studies for simplified light-emitting diode spectra. *Light. Res. Technol.*, **47**, 177–191.
8 Jost-Boissard, S., Avouac, P., and Fontoynont, M. (2015) Assessing the color quality of LED sources: naturalness, attractiveness, colourfulness and color difference. *Light. Res. Technol.*, **47**, 769–794.
9 Commission Internationale de l'Éclairage (1995) *Method of Measuring and Specifying Color Rendering Properties of Light Sources*, CIE Publication 13.3-1995, CIE, Vienna.
10 Davis, W. and Ohno, Y. (2010) Color quality scale. *Opt. Eng.*, **49**, 033602.

11 Hashimoto, K., Yano, T., Shimizu, M., and Nayatani, Y. (2007) New method for specifying color-rendering properties of light sources based on feeling of contrast. *Color Res. Appl.*, **32**, 361–371.
12 Smet, K.A.G., Ryckaert, W.R., Pointer, M.R., Deconinck, G., and Hanselaer, P. (2012) A memory color quality metric for white light sources. *Energy Build.*, **49**, 216–225.
13 David, A., Fini, P.T., Houser, K.W., and Whitehead, L. (2015) Development of the IES method for evaluating the color rendition of light sources. *Opt. Express*, **23**, 15888–15906.
14 Lin, Y., Wei, M., Smet, K.A.G., Tsukitani, A., Bodrogi, P., and Khanh, T.Q. (2015) Color preference varies with lighting application. *Light. Res. Technol.*; Online in October 2015.
15 Lanthony, P. (1978) The desaturated panel d-15. *Doc. Ophthalmol.*, **46** (1), 185–189.
16 Ichikawa, H., Tanabe, S., and Hukami, K. (1983) *Standard Pseudoisochromatic Plates for Acquired Color Vision Defects, Part II*, Igaku-Shoin Medical Publishers.
17 Haegerstrom-Portnoy, G., Brabyn, J., Schneck, M.E., and Jampolsky, A. (1997) The SKILL card. An acuity test of reduced luminance and contrast; Smith-Kettlewell Institute Low Luminance. *Invest. Ophthalmol. Vis. Sci.*, **38** (1), 207–218.
18 Han, S. and Boyce, P.R. (2003) Illuminance, CCT, décor and the Kruithof curve, in *Proceedings of the 25th Session of the CIE, San Diego, CA, June 25–July 2, 2003*, CIE: D3, Vienna, pp. 282–285.
19 Bodrogi, P., Lin, Y., Xiao, X., Stojanovic, D., and Khanh, T.Q. (2015) Intercultural observer preference for perceived illumination chromaticity for different colored object scenes. *Light. Res. Technol.* doi: 10.1177/1477153515616435, published online before print November 24, 2015.
20 Khanh, T.Q., Bodrogi, P., Vinh, Q.T., and Stojanovic, D. (2016) Color preference, naturalness, vividness and color quality metrics-part 1: experiments in a real room. *Light. Res. Technol.* (online in April 2016).
21 Ashdown, I. et al. (2015) Correspondence: in support of the IES method of evaluating light source color rendition. *Light. Res. Technol.*, **47**, 1029–1034.
22 Bodrogi, P., Brückner, S., Krause, N., and Khanh, T.Q. (2014) Semantic interpretation of color differences and color rendering indices. *Color Res. Appl.*, **39**, 252–262.
23 Khanh, T.Q., Bodrogi, P., Vinh, Q.T., and Stojanovic, D. (2016) Color preference, naturalness, vividness and color quality metrics-part 2: experiments in a viewing booth and analysis of the combined dataset. *Light. Res. Technol.* (online in April 2016).
24 Khanh, T.Q. and Bodrogi, P. (2016) Color preference, naturalness, vividness and color quality metrics–Part 3: experiments with makeup products and analysis of the complete warm white dataset, Light Res. Technol. Online in September 2016.

25 Luo, M.R., Cui, G., and Li, C. (2006) Uniform color spaces based on CIECAM02 color appearance model. *Color Res. Appl.*, **31** (4), 320–330.
26 Khanh, T.Q., Bodrogi, P., Vinh, T.Q., and Brückner, S. (2013) *Farbwiedergabe von konventionellen und Halbleiter-Lichtquellen: Theorie, Bewertung, Praxis*, Pflaum Verlag, München.
27 Bodrogi, P. and Khanh, T.Q. (2013) Semantic interpretation of the color binning of white and colored LEDs for automotive lighting products. Proceedings of the International Symposium on Automotive Lighting, ISAL 2013, Darmstadt, Germany, 2013.
28 Bodrogi, P., Khanh, T.Q., Stojanovic, D., and Lin, Y. (2014) Intercultural Colour Temperature Preference (2200K-5800K) of Chinese and European Subjects Living in Germany and in China for Different Light Sources, SSLCHINA 2014, The 11th China International Forum on Solid State Lighting, Guangzhou, P.R. China, November 2014.

8

Optimization of LED Light Engines for High Color Quality

8.1 Overview of the Development Process of LED Luminaires

The aim of this book is to analyze the lighting and colorimetric aspects of color quality, which can be best understood and applied by the use of the visual color quality attributes such as *"color preference," "color naturalness,"* and *"color vividness."* These attributes are often used in the work of lighting architects, lighting engineers, project managers, and the users of lighting systems. In most cases, color quality as a foresight, as a philosophy, and as a constraint of technical requirements is an important issue for the design of indoor lighting. In the present section, general aspects of lighting systems for indoor lighting will be described (see also [1]).

At the first stage of thinking, the terminology *"indoor lighting"* means lighting applications inside a building so that the operation conditions are more or less independent of weather, ambient influences such as wind, rain, fog, and traffic situations related to all protection and damage aspects of the lighting systems (e.g., temperature, humidity, gas) and their users (e.g., rain or changes of daylight). At a higher stage, looking back to the history of human and industrial development, it can be concluded that indoor areas have been the main working and staying spaces since the start of the first industrial revolution at the end of the nineteenth century over the revolutions of engineering and automobile industry in the time from 1960 until 2000 and of telecommunication technology (Internet, mobile phone technology) in the last two decades. From the viewpoint of indoor lighting technology, the following applications can be taken into consideration:

- Industrial lighting (e.g., manufacturing/mounting/assembly halls, stores, humidity rooms in chemical industry)
- School lighting (e.g., primary schools, lecture halls in university buildings)
- Health area lighting (hospitals, rooms for surgical operations, and for general treatments, living houses for the elderly)
- Office lighting (e.g., in offices in industry, in governmental buildings, buildings of financial and insurance sectors)
- Hotel, restaurant, and wellness area lighting
- Museum and gallery lighting

- Shop lighting, lighting of areas and spaces for representative purposes such as trading fairs, automobile, or fashion shows
- Theater, film, and TV studio lighting.

These applications are very different concerning their philosophy of lighting design and their aims and requirements; hence, dissimilar demands arise for the development of their light sources and luminaires. Currently, most indoor lighting designs are based on international and national standards in which lighting parameters are mostly defined on the basis of the $V(\lambda)$-function of the CIE (1924). Consequently, it is necessary to satisfy the following two primary requirements of lighting:

1) The aim of the desired activities and work performance under the lighting system to be designed shall be achieved (e.g., reading texts, mounting engineering tools, or building machines).
2) The errors and possible accidents during the desired activities under the lighting system under design shall be minimized (e.g., error-free handling of machines or the avoidance of traffic accidents during the movements inside industrial working halls).

The important lighting parameters are glare (in terms of UGR), horizontal illuminance E_m in lx, and homogeneity U_o. Since the introduction of LED luminaire technology, vertical illuminance E_v has also been considered in order to avoid the dark areas inside the lit space. In Table 8.1, a set of important lighting parameters and their values for the different working areas [2] are listed.

The illuminance values in Table 8.1 are the minimal values. They can be increased depending on the application and context of illumination. In Section 7.2, experimental results implied that an illuminance level above 500 lx improves the color appearance and color difference ability of the test subjects. In the last two decades of lighting and color research and with the introduction of LED technology including both colored and white pc-LED components, a rethinking process has taken place with much new fundamental knowledge, summarized as follows:

- *About lighting quantities*: Lighting quantities should not be fixed, that is, not to be used with similar values for all applications and all light source users working in the same area. Luminaire design should allow various setups and different lighting levels with both direct and indirect illuminating elements and diverse correlated color temperatures (CCT) individually according to the appropriate consideration of the differences among the age, gender, cultural background, and working habits of the different users of the lighting system.
- *About the dynamic properties of lighting:* The CCT or/and the illuminance should not only be preprogrammed for several fixed settings depending on weather or/and daytime but should also be flexibly adjustable depending on daytime, season, and the available daylight contribution during the day to improve the concentration of the light source users during their working time. The lighting atmosphere shall be changed to allow for relaxation after the working hours (see also Chapter 9).

Table 8.1 Important lighting parameters and their required minimal values [2].

Number	Application	Working area and its specific activities	Minimal allowed average illuminance E_m (lx)	Glare (UGR)	Homogeneity U_o	R_a
	Office	Writing, reading, processing data	500	19	0.6	80
		Technical design, drawing	750	16	0.7	80
		Conference, meeting rooms	500	19	0.6	80
	School	Class rooms in elementary schools	300	19	0.6	80
		Class rooms in evening schools, class rooms for adults	500	19	0.6	80
	Art school	Class rooms for drawing	750 (5000 K < CCT < 6500 K)	19	0.7	90
	Hospitals	Waiting rooms	200	22	0.4	80
		Analysis and treatment	1000	19	0.7	90
		Operation rooms	1000	19	0.6	90

- *About color quality and quantity*: The chromatic content of optical radiation should be analyzed as it plays an important role in modern lighting concepts, especially in shop lighting, hotel lighting, museum lighting, and in film and TV studio lighting to improve the perceived quality of the illuminated areas and colored objects (e.g., car show lighting, food lighting).
- *Other aspects:* Besides the well-known color-rendering index in the context of color fidelity, color temperature preference, which depends on cultural and regional individuality and on the context of the illuminated colored objects (see Chapter 7), is also important for the light source users' acceptance. Some recent studies showed that a color temperature around 3300–3500 K is preferred in food lighting and museum lighting. In addition, the investigations in Sections 7.4 and 7.5 also reveal that the color temperature range between 4000 and 5000 K is preferred for general lighting systems.

In the last few years, intensive discussions on the concept of Human Centric Lighting (HCL; see also Chapter 9) have taken place. The concept of HCL places the needs of lighting users (as human subjects) in the focus of the design of light sources, luminaires, and lighting installations. The HCL concept is associated with the visual and nonvisual aspects summarized in Figure 8.1.

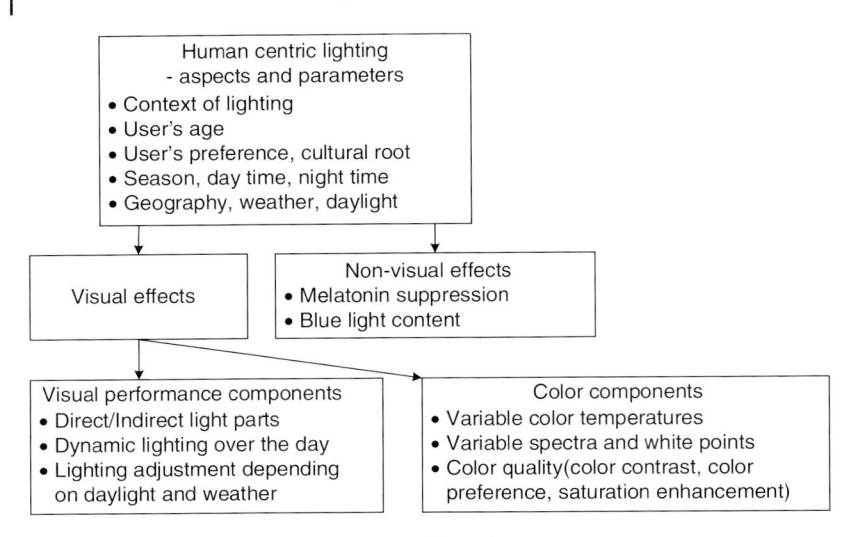

```
┌─────────────────────────────────────┐
│     Human centric lighting          │
│     - aspects and parameters        │
│  • Context of lighting              │
│  • User's age                       │
│  • User's preference, cultural root │
│  • Season, day time, night time     │
│  • Geography, weather, daylight     │
└─────────────────────────────────────┘
          │                  │
          ▼                  ▼
   ┌──────────────┐  ┌──────────────────────┐
   │Visual effects│  │  Non-visual effects  │
   │              │  │ • Melatonin suppression│
   │              │  │ • Blue light content │
   └──────────────┘  └──────────────────────┘
          │
          ▼
┌──────────────────────────────┐  ┌──────────────────────────────────┐
│Visual performance components │  │       Color components           │
│ • Direct/Indirect light parts│  │ • Variable color temperatures    │
│ • Dynamic lighting over the day│ │ • Variable spectra and white points│
│ • Lighting adjustment depending│ │ • Color quality(color contrast, color│
│   on daylight and weather    │  │   preference, saturation enhancement)│
└──────────────────────────────┘  └──────────────────────────────────┘
```

Figure 8.1 Visual and nonvisual aspects of the HCL concept.

Besides the nonvisual and visual components shown in Figure 8.1, current color temperature depends on the available daylight or on the current weather conditions. Note that color quality aspects such as color contrasts, color preference, or color vividness are also important psychological and physiological factors to be considered to improve human activity and work performance, the light source user's well-being, user acceptance, and health.

Regarding color quality in the context of color fidelity, the general color-rendering index R_a for office, industrial, and school lighting has to be above 80. For more complicated applications in medical treatment and operation rooms, the CIE CRI R_a value should be better than 90. Based on the analysis in Section 7.9, it can be recognized that a general color-rendering index of $R_a = 80$ means a visual semantic interpretation between "*moderate*" and "*good*" only and the one of 90 is associated with a semantic interpretation that is slightly better than "*good*." An LED lighting system can only deliver a color-rendering quality of "*just good*" if the general color-rendering index equals at least 86. In another study at the Technische Universität Darmstadt with 34 white LED types, a general CIE CRI R_a value (the general color-rendering index calculated by the use of the first eight CIE test color samples; $TCS_1 - TCS_8$) of 86 or 90 corresponds to an average special color-rendering index value of the CIE CRI R_9 (the saturated red test color sample) of 40 or 60. In Figure 8.2, the relationship between the general CRI R_a and the average special CRI value of all 14 CIE test color samples (also R_9) is illustrated computed for the set of 34 white LEDs.

In the American product regulation Energy Star Eligibility Criteria (version 2.1) [3] that is also being used in many countries in the field of solid state lighting (SSL), the following requirement is formulated: "*The luminaire (directional luminaires) or replaceable LED light engine or GU24 based integrated lamp (non-directional luminaires) shall meet or exceed $R_a \geq 80$.*" In Table 8.2, the

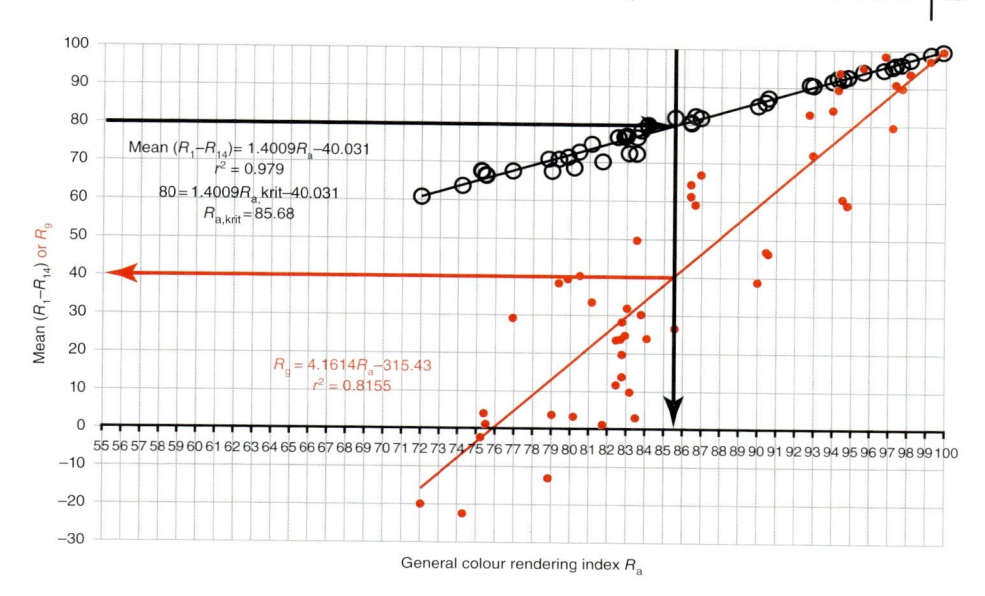

Figure 8.2 Relationship between the general CRI (R_a), mean CRI for all 14 CIE test color samples and R_9 for a sample set of 34 white LEDs.

Table 8.2 Requirements of two American authorities regarding color fidelity specification.

Parameter	Energy star [3]	California Quality LED Lamp Specification [4]
R_a	≥ 80	≥ 90
R_9	> 0	> 50

requirements of two American authorities regarding color fidelity specification are listed.

Table 8.2 implies that, in the case of LED luminaires in indoor lighting applications with just "good" semantic interpretation of color fidelity quality of the colored objects, a minimal general color-rendering index CRI R_a of 86 has to be ensured. According to the fact that LED-based indoor luminaires are technical products for various lighting applications, the following approaches should be considered in the design and development process of these luminaires:

- An electrotechnical physical method structuring the whole luminaire as a system of different functional groups (optics, electronics, as well as thermal and mechanical group)
- A lighting method analyzing the luminaire from the viewpoint of lighting and color aspects.

The above two approaches are necessary and should be considered separately before they become the object of analysis in the common context. The

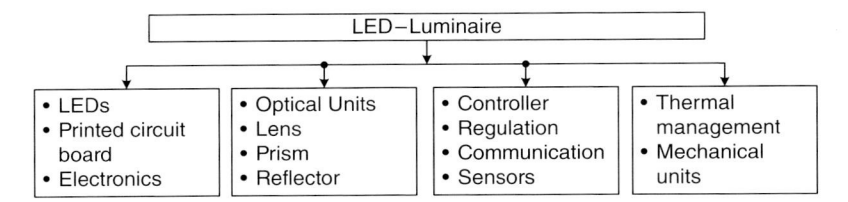

```
┌─────────────────────────────────────┐
│           LED–Luminaire             │
└─────────────────────────────────────┘
      │          │          │          │
┌──────────┐ ┌──────────┐ ┌──────────────┐ ┌──────────────┐
│• LEDs    │ │• Optical │ │• Controller  │ │• Thermal     │
│• Printed │ │  Units   │ │• Regulation  │ │  management  │
│  circuit │ │• Lens    │ │• Communication│ │• Mechanical │
│  board   │ │• Prism   │ │• Sensors     │ │  units       │
│• Electronics│ │• Reflector│ │           │ │              │
└──────────┘ └──────────┘ └──────────────┘ └──────────────┘
```

Figure 8.3 Principal structure of an LED luminaire.

LED indoor luminaire consists, from a technical point of view, generally of the following functional units with different tasks and parameters shown in Figure 8.3.

The structure of the LED luminaire shown in Figure 8.3 will be analyzed below.

- *LED, printed circuit board, electronics:* The LEDs have to be selected and arranged in parallel and in series strings. The number of the LEDs shall be chosen according to the required luminous flux of the LED luminaire and by the calculation of the luminaire's optical efficiency. Three conventional LED packaging systems can be considered as follows:
  – White and/or color LEDs are SMD components soldered on the circuit board with a predefined pitch. The luminous flux of the board is the sum of the luminous fluxes of all LEDs. This arrangement is preferred for large-sized indoor luminaires in office, hospital, school lighting applications and, as soft light with diffusely emitted light, also for gallery and photographic studio lighting. The secondary optics is micro-optics (a microprisms plate), a lens for each SMD LED, or a set of diffuse plastic plates (PMMA, i.e., polycarbonate; see Figure 8.4). The micro-optics and the diffuse plates have the additional task of mixing the optical radiation from different LED groups of white and RGB colors.
  – LED unit as a chip-on-board (COB) molded phosphor mixture with a high luminous flux from a relatively small light-emitting surface. This concept is typically used for downlight luminaires with reflector optics for lighting systems in hotels, shops, restaurants, and in the high LED power class between 50 and 300 W for industrial lighting inside big factory halls.
  – LED unit with remote phosphor plate and, most frequently, reflector optics for downlight luminaires.

The printed circuit board can be made of different materials such as FR4 as standard material, FR4 with thermal vias to improve thermal conduction, metal core board (MCB), or a ceramic material. In most applications, MCB is often used if the pitch of the LEDs on the board is not too small. For COB configurations of very high thermal density, the ceramic material AlN with its high thermal conductivity of about 170 W/(m·K) is recommended.

The driving electronics is selected according to the number of the LEDs for a series string and their current. At 350 mA, most high-power blue and white LEDs have a forward voltage of about 2.7–2.9 V and, at 700 mA the voltage can be between 3.0 and 3.2 V so that the driving electronics can be chosen based on the allowed maximal voltage and electrical power. The selection of

Figure 8.4 LED-luminaire Lunexo. (Photo Source: ® Company Trilux/Ansberg – Germany.)

the suitable electronics is also a question of the lifetime of the electronic components under the ambient and system operation conditions (maximal load, maximal temperature, humidity). Thinking about the electronics of the LEDs, it should also be considered how the LEDs can be dimmed – either by DC dimming or by pulse width modulation (PWM) dimming. The correct use of these two dimming methods can help minimize the negative effect of the color shifts of white and color LEDs. A comprehensive study of this aspect will follow in detail in Section 8.3.

- *Optical systems:* In the time of conventional incandescent and discharge lamps, optical radiation has been emitted into all solid angle directions (4π) and hence reflector optics has often been used. With the introduction of white and color LEDs that emit optical radiation rather into the half-sphere solid angle (2π), lens optics is used in many different applications. This optical technology has the advantages of scalability, low cost, and better manufacturing control. However, there are also some serious disadvantages:
  - The lens of the optics-LED arrangement leads to a periodic system of the light-emitting area with dark bright patterns that can intensify glare effects in indoor and outdoor lighting applications under certain circumstances (see Figure 8.5).
  - The plastic lenses (PMMA, PC) undergo an aging and yellowing process depending on the ultraviolet and blue radiation energy content of the illumination, on humidity, and on the temperature inside the material.

  Based on the above two aspects, reflector optics can be meaningfully used for LED indoor luminaires and for COB LED configurations.
- *Controller–regulation electronics:* Control and regulation electronics have different tasks that are important for current and future smart lighting systems:

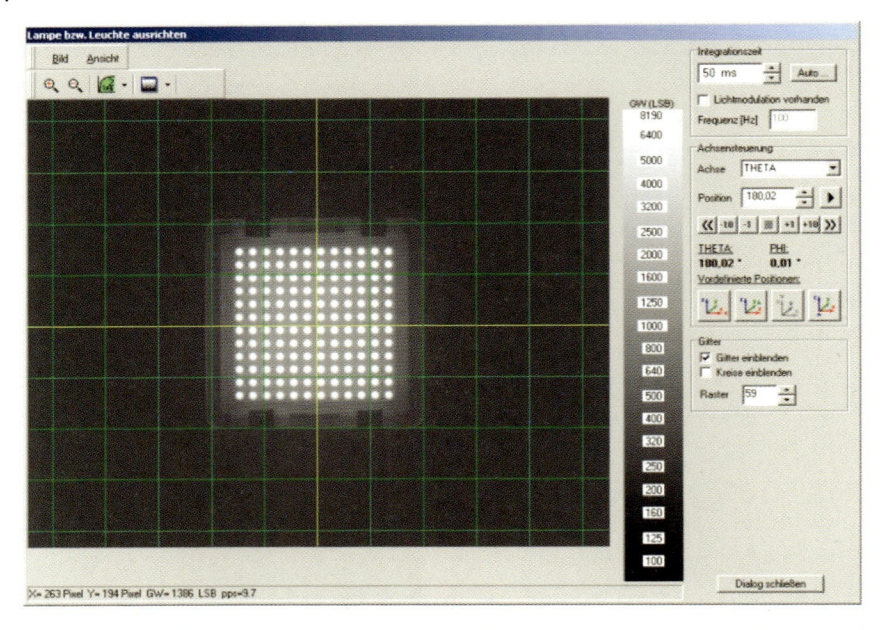

Figure 8.5 An LED luminaire exhibiting a dark–light pattern. (Photo Source: Company ILEXA, Ilmenau/Germany.)

- With temperature sensors on the board or in the heat sinks, control electronics dims the LED current or reduces the PWM duty cycle in order to avoid high temperatures at the LEDs.
- With optical sensors (such as CCD (constant current dimming) or CMOS cameras or infrared sensors) or ultrasonic sensors, the movement or presence of the lighting users can be detected so that the LEDs can be dimmed accordingly.
- With microcontrollers (μC-hardware), the spectra, color temperature, luminous flux, and the ratio of the light for the direct and indirect part of the LED luminaire or also the luminous intensity distribution (LID-curve) can be selected or changed either individually or preprogrammed as a function of season, time of the day, or the orientation of the building to sunlight direction.
- Communication with the building's management and with other building service systems (heating, warm water), with other luminaire systems in the building via DALI or DMX systems, with Internet or other telecommunication platforms (e.g., Bluetooth, KNX).
- With time counters or status data loggers built on the board, the lifetime, electrical current, and temperature profile of the LEDs (loading scheme) can be recorded over a long time (even 50 000–100 000 h) and can contribute to the decision whether to replace the LED module.

- *Thermal management:* To conceptualize thermal management, the following steps should be taken into account:

- Measurement of the LEDs to be used at the predefined currents so that the electrical power, optical power, luminous flux, color coordinates, and the thermal power can be calculated.
- Measurement or estimation of the thermal resistance for the whole chain from the pn-junction to the heat sink including the luminaire housing.
- Based on the thermal power of the LEDs, the thermal resistance of the system at the considered maximal ambient temperature of the luminaire and the maximal junction temperature shall be calculated. The maximal junction temperature at the predefined current strongly influences the LEDs lifetime. Therefore, the following approaches should be considered in order to reduce the negative effects of higher temperatures in all applications:
  Reduction of the predefined current; if doing so then more LEDs have to be used on board.
  Selection of appropriate materials for the circuit board and the adhesive layers with better thermal conductivity.
  Selection of better heat sinks (materials, forms, geometry) with lower thermal resistance.
  Application of active forced cooling systems (fan, water cooling, heat pipe, or Peltier cooling).

Generally, all steps of the optimization of the thermal management can be simulated with modern thermal simulation software being available on the market. Based on the correct input data about the geometry and material characteristics, luminaire designers can change and then select the best configuration of the luminaire system from the viewpoint of thermal management. The process of lighting design, simulation, and optimization can be structured according to the flowchart shown in Figure 8.6.

The definition of the luminaire's luminous flux is either customer specific or it can be oriented to the luminous flux of conventional lamp types and mirror-reflector luminaires for T5 fluorescent lamps. Generally, a downlight luminaire for shop lighting should deliver 3000 lm and an office luminaire illuminated by tubular fluorescent lamps with the optical efficiency of about 80% and the luminous flux of 5800 lm (58W-T5 lamps) should have a luminous flux of about 4100 lm. All above-mentioned design principles and requirements are valid for LED luminaires that are constructed by the use of white phosphor-converted LEDs (pc-LEDs) with one single CCT. In order to achieve excellent color quality with a reasonable saturation enhancement of the illuminated colored objects at different color temperatures and at a well-defined chromaticity, color LEDs (e.g., RGB LEDs) and white pc-LEDs should be combined on a common electronic board or on different boards inside an LED luminaire. The optical radiation of each LED group has to be carefully mixed (an example is shown in the chromaticity diagram of Figure 8.7) by special software and the LED channels shall be calibrated to yield white light with the required white point at a reference temperature (e.g., 25 °C in the calibration laboratory of the manufacturer). Owing to the different current and temperature behavior of dissimilar LEDs from various semiconductor materials, the spectra of the luminaire's white light will be shifted because of the spectral change of each

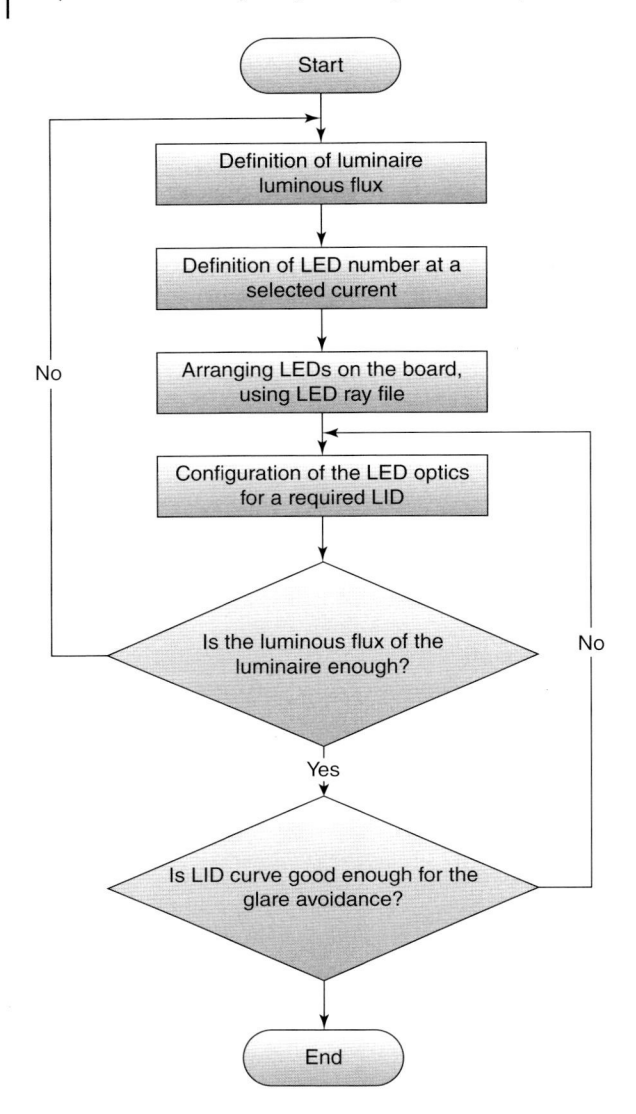

Figure 8.6 Flowchart of the optical and lighting design process.

LED color channel when using the luminaire at different climate regions and under different environmental conditions. As a consequence, the color quality, white point, and color temperature will not remain inside the desired tolerance limits. In order to avoid this artifact, an appropriate color regulation shall be designed and implemented. This is shown in the schematic operation diagram of Figure 8.8.

In Figure 8.8, the power path includes an electronic driver that converts alternating current (AC-110/230 V, 50/60 Hz) into direct current (DC) and the LED luminaire supplied by a desired current from this electronic driver. The signal path includes a color sensor with its RGB signals processed by a slave

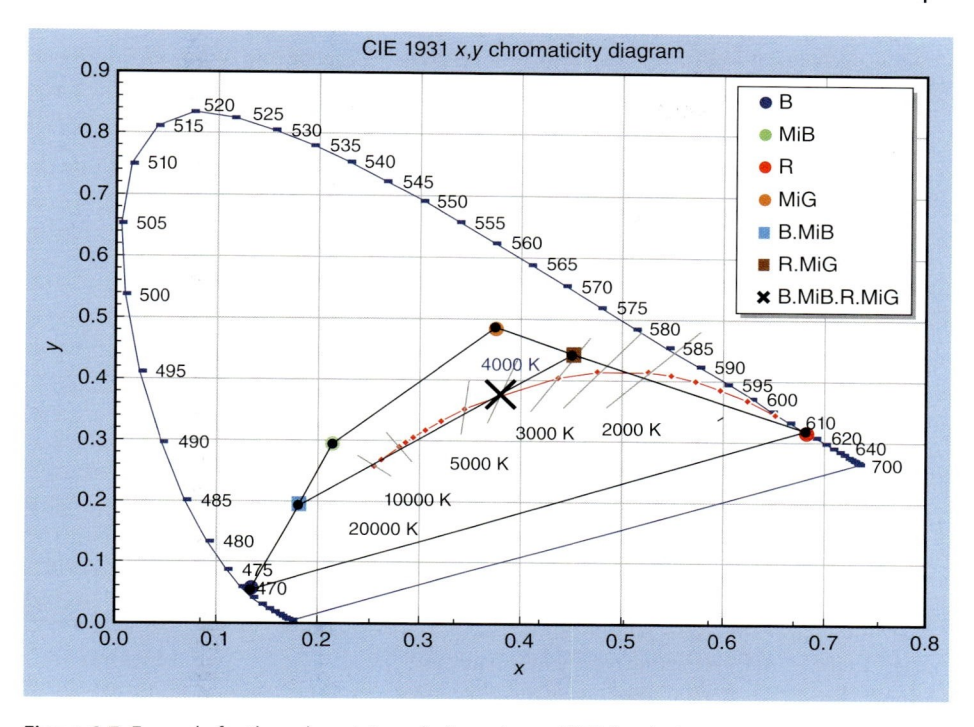

8.1 Overview of the Development Process of LED Luminaires | 293

Figure 8.7 Example for the color mixing of a four-channel LED luminaire.

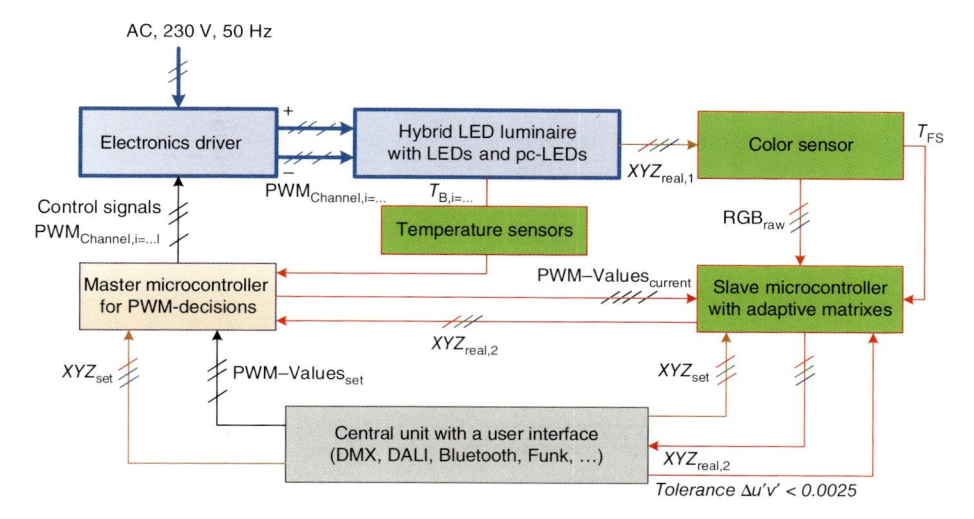

Figure 8.8 General structure of an LED luminaire with both color and temperature sensors.

microcontroller in order to achieve the accurate tristimulus values $XYZ_{\mathrm{real,2}}$. Temperature sensors have also to be applied to supply the information about the temperatures $T_{\mathrm{B,i}}$ of all channels. The control path includes a master microcontroller that processes the input signals such as the tristimulus values and board temperatures and then it compares the actual color state of the luminaire's radiation with the reference color values XYZ_{set} (the users had adjusted or defined them at a previous stage). If the color difference between the reference color values and the luminaire's color values is greater than a permitted tolerance value (e.g., $\Delta u'v' = 0.0025$ or 0.002) then a correction and regulation algorithm will be activated to generate new output values for the color regulation. Based on these new regulation values, the driver electronics produces corresponding electrical signals (currents in the case of CCD dimming or PWM signals in the case of PWM dimming) for each LED color channel so that the required reference color values XYZ_{set} can be achieved for the LED luminaire.

For practical implementations, the scheme of the color regulation is depicted in Figure 8.9.

In Figure 8.9, the continuous connecting lines stand for the transmission of signals while dashed lines indicate the processes that take place inside the master microcontroller. The role of the slave microcontroller is similar to that of a translator. The role of the translator is to ensure understandable information exchange between the source and the target so that the translated signals or information should be as accurate as possible. The source is the actual spectral output and the target is the corresponding chromaticity and brightness. The accurateness of this conversion depends on the quality of the spectral sensitivity as the intrinsic nature of the applied color sensor (i.e., on the optical filters and silicon/germanium or selenium photodiodes) and the spectral form of the received optical radiation. In fact, spectral sensitive changes with the change of the sensor's operational conditions such as its operation temperature (T_{FS}). The information on the actual form of the spectrum being detected can only be guessed based

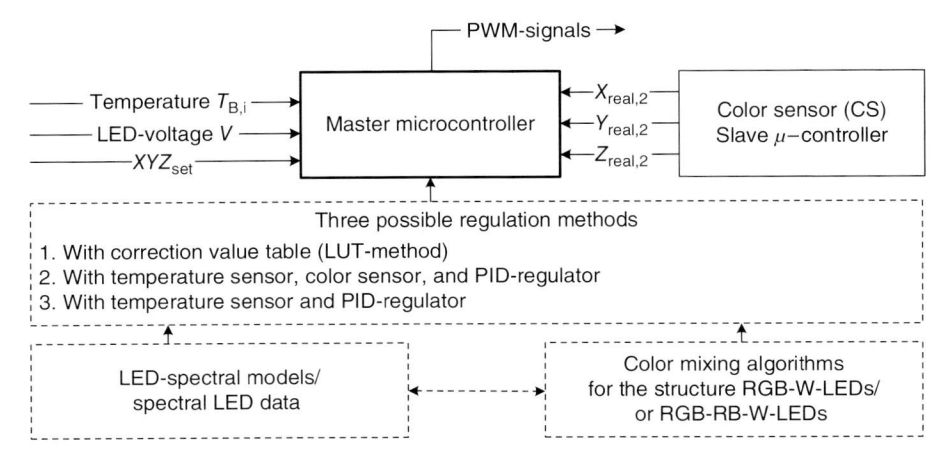

Figure 8.9 General structure of a color regulation unit with both color sensors and temperature sensors.

on the actual values of PWMs. However, the core component of the system in Figures 8.8 and 8.9 is the master microcontroller. During real operation, the LEDs voltage can also be necessary to supply for this microcontroller as an input. Based on this input, the board temperature ($T_{B,i}$) and also some premeasured thermal properties such as the LEDs junction temperature can be determined to apply an LED model constructed for a given temperature range. The central component of the master microcontroller is the color regulation in Figure 8.9. The aim of one of the three possible regulation methods is to generate the necessary new PWM signals adjusted following the captured information such as the actually measured board temperature ($T_{B,i}$), LED voltages, and chromaticity ($XYZ_{real,2}$). Based on the PWM signals, the adapted or adjusted output spectrum is issued according to the desired settings. These color regulations are supported by spectral LED models or spectral LED data and color mixing algorithms.

Based on the above descriptions, it can be concluded that there are five basic problems that should be solved for the development of an intelligent LED-luminaire:

1) Temperature–current behavior and their estimation/characterization
2) LED spectral models
3) Dimming methods
4) Color mixing method for a well-defined color quality criterion from the users' side
5) Color regulation methods.

In the next sections, the first four problems will be dealt with. The analysis of the color regulation method is not a subject of this book. The topic "*color mixing method for a well-defined color quality criterion from the users' side*" or "*color quality optimization*" for an LED engine will be the focus of the present chapter.

8.2 Thermal and Electric Behavior of Typical LEDs

8.2.1 Temperature and Current Dependence of Warm White LED Spectra

The consequences of the LED's spectral change as a function of the operating temperature and the forward current are important. Herewith, the variations of all other optical and colorimetric parameters can be explained qualitatively and determined quantitatively. Therefore, in this section six typical LED types that can be the representatives for the spectral behavior of other warm white pc-LEDs are selected to illustrate the spectral changes as a function of the operating temperature in Figure 8.10 and as a function of forward current in Figure 8.11. More details can be taken from another work [1] of the same authors.

8.2.1.1 Temperature Dependence of Warm White pc-LED Spectra

In Figure 8.10, the temperature dependence of warm white pc-LED spectra is shown in the operating range from 40 to 80 °C at 350 mA. Here, the indices "1" and "2" (attached to the name of each LED in Figure 8.10) are used to indicate the spectra at 40 and 80 °C, respectively. Fundamentally, in warm white pc-LEDs a luminescent material system with two different phosphor types (such as the

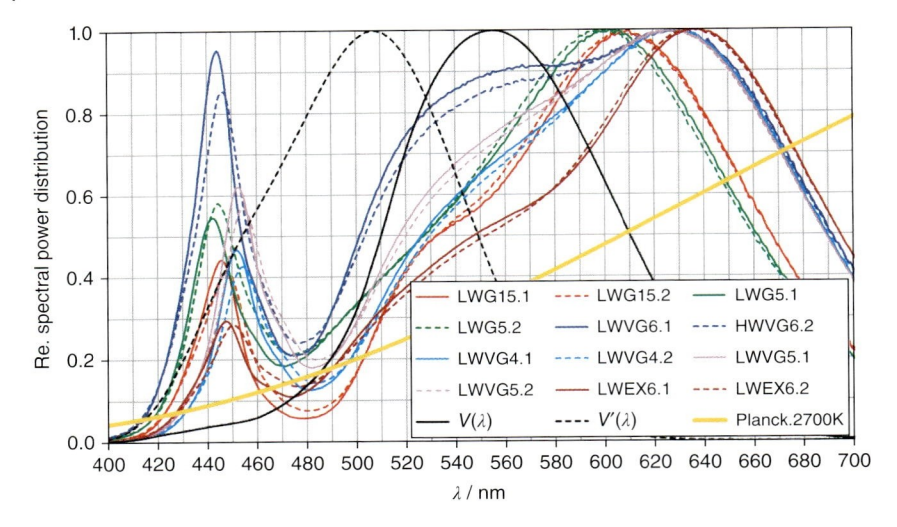

Figure 8.10 Temperature dependence of six typical warm white LEDs (2700–3500 K) at 350 mA. Solid lines correspond to 40 °C and dashed lines to 80 °C.

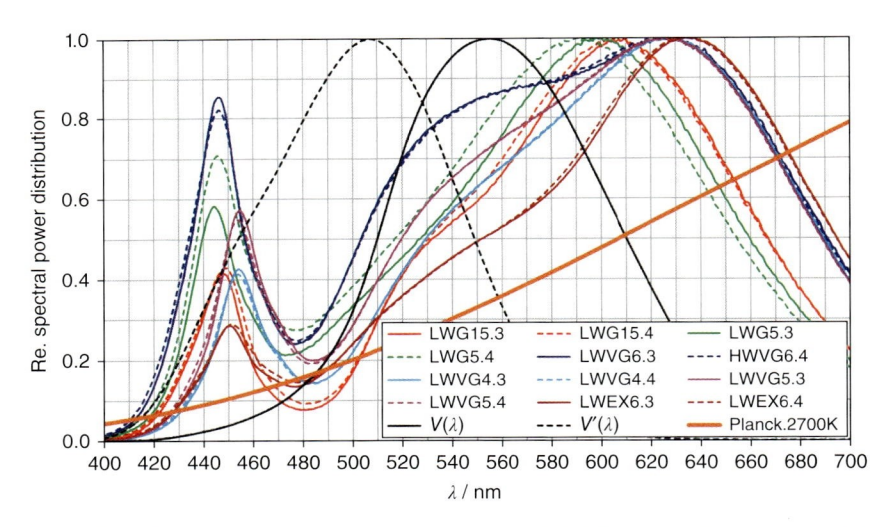

Figure 8.11 Current dependence of six typical warm white LEDs (2700–3500 K) at 80 °C. Solid lines correspond to 350 mA and dashed lines to 700 mA.

mixture of a green phosphor and a red phosphor or of a yellow phosphor and a red phosphor) is usually built on a blue LED chip. Therefore, in the investigation of warm white pc-LED spectra, three spectral components (the spectral component of a blue chip, a green/yellow phosphor and a red phosphor) shall be considered. Based on the measured spectra of the warm white pc-LEDs in Figure 8.10, it can be considered that the peak wavelength of the red phosphor component of the LEDs with high CRI values (such as LWVG4, HWVG6, and LWEX6) are always situated at longer wavelengths, at about 630–640 nm. One of the white LEDs with lower CRI has a similar peak at about 600 nm only (LWG5) or at about 610 nm (LWG15). In addition, the comparison between the solid curves (designating the spectra at 40 °C) and the dashed curves (designating the spectra at 80 °C) in Figure 8.10 reveals that the temperature stability of the spectral components belonging to the red phosphors is better than that of the spectral components belonging to green, yellow, or orange phosphors. Similarly, the temperature stability of the middle spectral components is better than that of the short spectral components belonging to the blue chip.

Indeed, the four LED types (HWVG6, LWG5, LWVG5, and LWEX6) can be regarded as the examples of non-optimal temperature stability of warm white pc-LEDs. In these cases, the short and the middle spectral components are shifted very strongly compared with the long spectral components. Therefore, their white point is shifted much into the red direction (called *red shift*) causing a strong decrease of CCT. In contrast to this fact, in the case of LWG5, there is nearly no change for the middle spectral component while the short spectral component strongly increases and the long spectral component shifts into the shorter wavelength range. As a result, its white point shifts into the blue direction (called blue *shift*) causing a remarkable increase of CCT.

8.2.1.2 Current Dependence of Warm White pc-LED Spectra

The current dependence of the warm white pc-LED spectra is shown in Figure 8.11 in the operating range from 350 to 700 mA at 80 °C. Herewith, the indices "3" and "4" are used to denote the spectra at 350 and 700 mA, respectively. In most cases of the investigated warm white pc-LEDs, the increase of the spectral components is proportional to the forward current with the exception of LWG5 and LWG15. Therefore, the relative spectral change is small when the forward current increases from 350 mA (denoted by solid curves) to 700 mA (denoted by dashed curves) as shown in Figure 8.11. In the worst case of LWG5, the visible spectral change takes place with all spectral components (the short, middle, and long spectral components). Thus, this is an example for a low quality warm white LED. In this case, both the blue chip and the phosphor system have neither good temperature stability nor current stability.

Based on the spectral change, when changing the temperatures at a constant current, the temperature dependence of the color difference for the warm white LEDs can be calculated. This is shown in Figure 8.12 at the temperature of 80 °C as a reference point. In addition, based on the analysis in Chapter 3 and Section 7.9, it can be summarized that a white point shift of $\Delta u'v' = 0.003$ should be acceptable in most lighting applications of indoor lighting. This leads to the conclusion that the operating temperature must be always higher than 45 °C in case of LWEX6, LWG5, and HWVG6. In the case of the other LED types, there is an acceptable color difference when the operating temperature is varied between 40 and 80 °C.

8.2.1.3 Current Dependence of the Color Difference of Warm White pc-LEDs

In Figure 8.13, the current dependence of the color difference of the warm white LEDs is shown in the operating range from 100 to 700 mA at 80 °C. It can be seen that the operating current range should be chosen from 100 to 700 mA

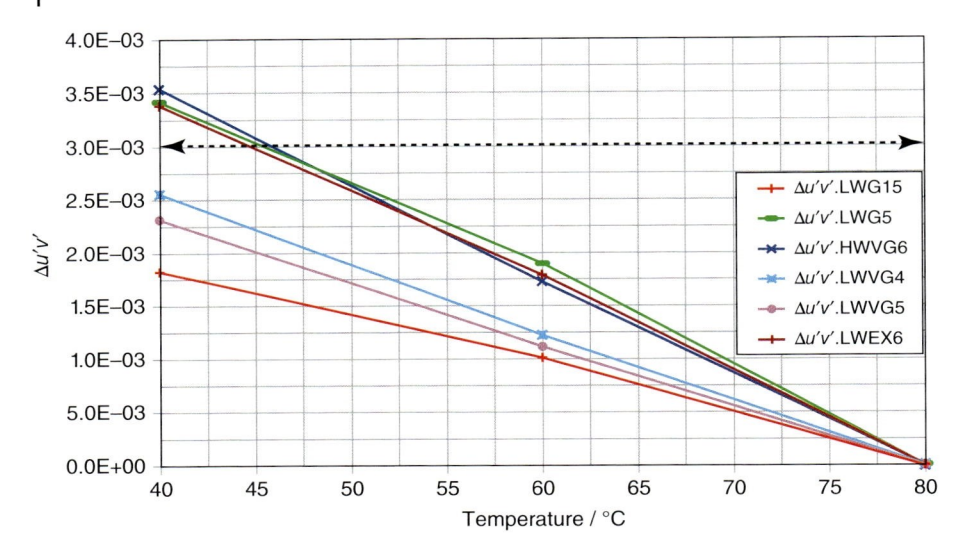

Figure 8.12 Temperature dependence of the chromaticity difference $\Delta u'v'$ of six typical warm white pc-LEDs at 350 mA.

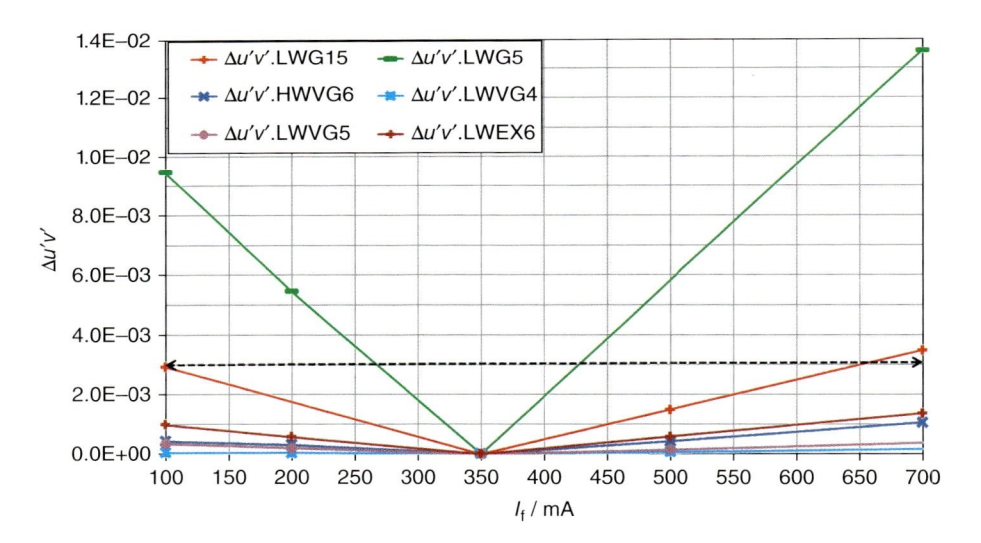

Figure 8.13 Current dependence of chromaticity differences $\Delta u'v'$ of six typical warm white pc-LEDs at 80 °C.

for most warm white pc-LEDs in order to keep the chromaticity difference less than $\Delta u'v' = 0.003$. In other words, when the forward current increases from 100 to 700 mA, there is an acceptable chromaticity difference for most warm white pc-LEDs. The LED types LWG5 and LWG15 are hereby the exception with the permitted forward current range extending only from about 100 to about 650 mA for LWG15 and from about 275 to about 425 mA for LWG5.

8.2.2 Temperature and Current Dependence of Color LED Spectra

In order to analyze the temperature and current dependence of modern color (i.e., pure semiconductor without phosphor conversion) high-power LEDs, blue, green, and red LEDs manufactured by three LED-manufacturers were measured spectroradiometrically.

In Figure 8.14, the temperature dependence of blue LEDs with a peak wavelength shorter than 450 nm at 350 mA is shown when varying the board temperature between 25 and 90 °C. The optical efficiency is slightly reduced. With the temperature of 80 °C as the binning point (the reference point in this case) almost all blue LEDs can be driven to operate between 60 and about 90–95 °C because their chromaticity difference shift shall not exceed the value of $\Delta u'v' = 0.003$.

The temperature dependence of the green LEDs with the peak wavelength around 520–530 nm is presented in Figure 8.14 at 350 mA and 80 °C as the reference point. These green LEDs are very stable concerning the aspect of their optical power behavior and this allows for an operating range between about 50 °C and more than 90 °C if the chromaticity shift of $\Delta u'v' = 0.003$ is considered as a visual white point quality criterion. The temperature range of 50–90 °C is herewith acceptable for both blue and green LEDs so that these two LED types can be soldered on the same electronic board. The chromaticity difference characteristics of the red LEDs with a peak wavelength around 630 nm shows a similar tendency, in comparison to green and blue LEDs, but their optical power is more rapidly reduced if the temperature is varied between 25 and 90 °C (see Figure 8.15). This fact has the consequence that a hybrid LED luminaire consisting of RGB LEDs and white LEDs calibrated at 25 °C in

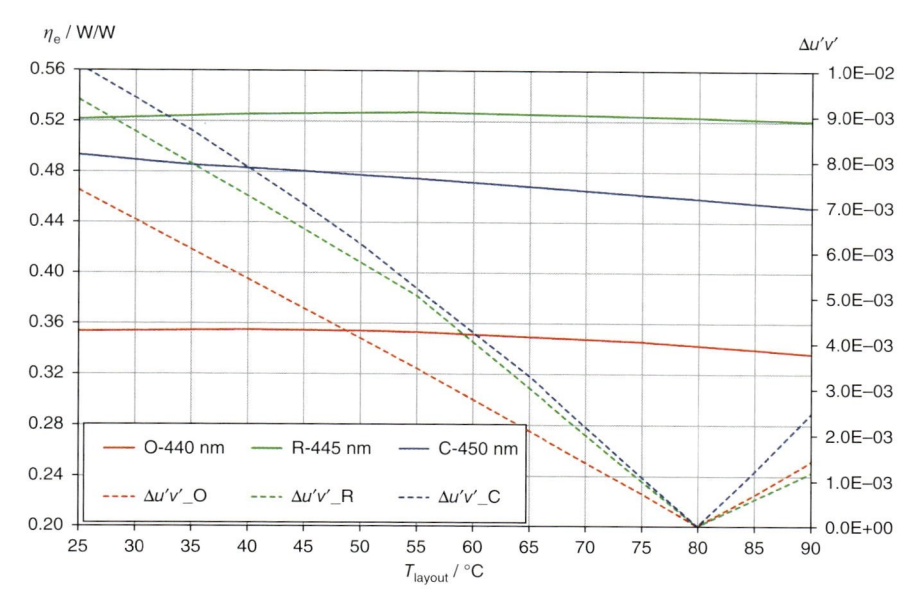

Figure 8.14 Temperature dependence of blue LEDs at 350 mA and with 80 °C as the reference point.

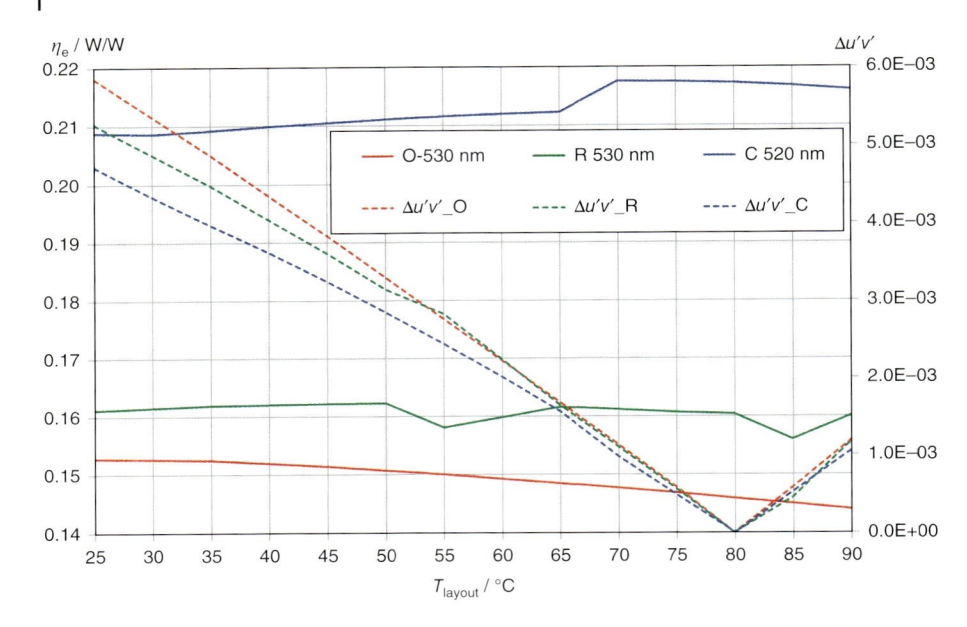

Figure 8.15 Temperature dependence of green LEDs at 350 mA and with 80 °C as the reference point.

the laboratory of the manufacturers for a certain white point (e.g., for 4000 or 3200 K) will change this white point if the luminaire is used in a TV studio or in a shop with higher ambient temperature (e.g., 38 °C) because the optical radiation of the red LEDs will be reduced causing a relative loss of the red spectral part in the spectral power distribution (SPD) of the whole LED luminaire (Figure 8.16).

Based on the discussions in Section 8.2, it can be recognized that the temperature and current dependence of white and color LEDs is not negligible and it should be compensated for by a fully programmed color regulator (see Figure 8.9) to mix the LED channels appropriately in case of any new temperature so that the targeted white point and the targeted spectra of the whole luminaire can be achieved. Then, the complicated changes of the spectral outputs of the LED channels in the LED luminaire can be adjusted adaptively according to the continuous variation of the operation temperature by the appropriate generation of new PWM values corresponding to new duty cycles. Consequently, it is necessary to discuss LED dimming methods in the next section.

8.3 Colorimetric Behavior of LEDs under PWM and CCD Dimming

A dimming method for color and white LEDs has the aim to change the absolute brightness and radiant flux of the light source without changing its color temperature, spectrum, or chromaticity. In practical use, each dimming method changes more or less the temperature and thermal state of the LEDs. The question

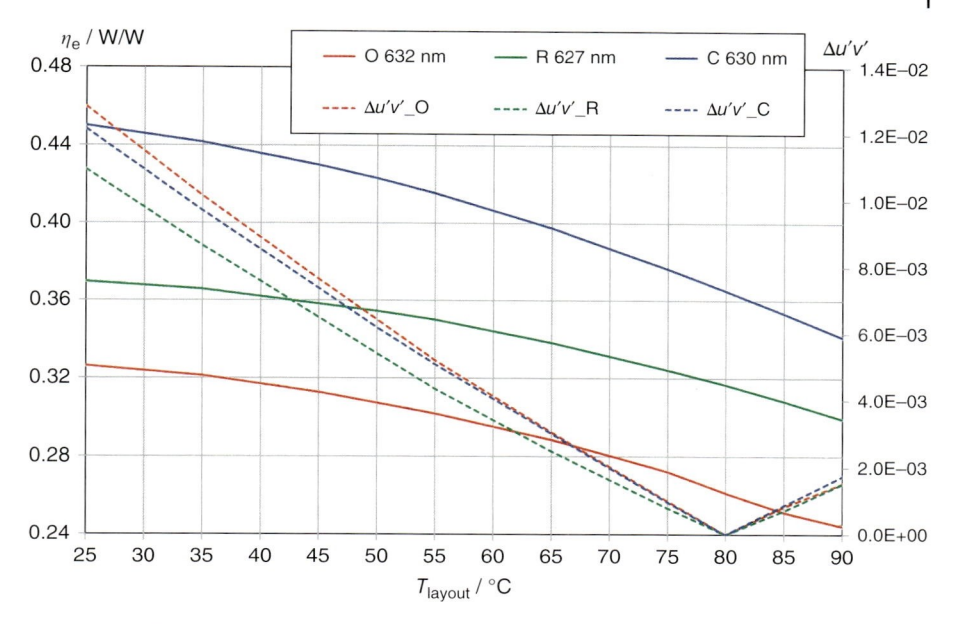

Figure 8.16 Temperature dependence of red LEDs at 350 mA and with 80 °C as a reference point.

is which method is useable for a specific application or luminaire type. There are two conventional dimming methods used in current LED technology (see [1]):

1) *The method of pulse width modulation (PWM-method)*: LED components are operated with a series of nearly rectangular pulses and this enables switching on and off in a very fast manner. With a switching frequency higher than 400 Hz the human visual system's processing apparatus cannot resolve the intensity variation in the temporal domain. Consequently, brightness will be detected as an average intensity value. In the PWM method, the amplitude of the pulse remains constant and the perceived brightness can be varied by changing the ratio of the pulse width to the period of the whole pulse (see Figure 8.17).

2) *Constant current dimming (CCD-method)*: The radiant flux or luminous flux and the luminance of the LED components can also be dimmed or increased by the variation of the LEDs current. The variation of LED current defines the number of recombinations between electrons and positive charges and consequently, photon energy is generated. The electronic circuit of the CCD method is more complicated than the one of the PWM method and, therefore, it is more expensive. The advantages of the CCD method are that this method is free from visual flicker perception and the visual stroboscopic effect.

In the Lighting Laboratory of the present authors, several measurements were carried out to characterize the chromaticity shifts and the luminous flux behavior of white and color LEDs under different dimming methods [1]. In a recent study, the spectral irradiance of warm white LEDs on an illuminated area was absolutely measured if the current of these LEDs or the duty cycles were varied while the

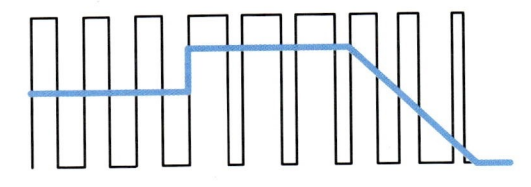

Figure 8.17 Principle of pulse width modulation (PWM).

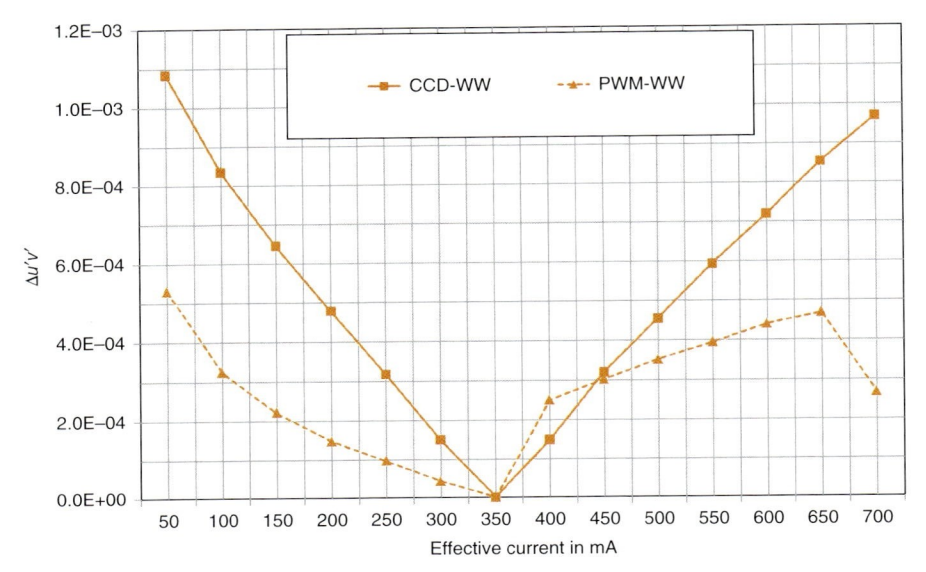

Figure 8.18 Chromaticity difference of a warm white LED unit operated by the two different dimming methods, CCD and PWM.

temperature of the LED unit was kept at a temperature with a tolerance of ±0.2 K. The chromaticity difference $\Delta u'v'$ was determined for the two dimming methods with 350 mA as the reference point for the chromaticity difference calculation. Based on the results shown in Figure 8.18, it can be recognized that both methods do not remarkably change the chromaticity difference of the white pc-LEDs if the operating current varies between 50 and 700 mA. In most cases, both methods can be used if the LED luminaires are constructed with white pc-LEDs only.

In Figure 8.19, the chromaticity differences of the color LEDs (450, 520, and 660 nm) dimmed by the two different dimming methods are shown. It can be recognized that the PWM method is able to dim the color LEDs with much smaller chromaticity differences in comparison with the current dimming method. Therefore, for highly qualitative LED systems, to achieve good accuracy of the LED spectra, the PWM method should be preferred.

8.4 Spectral Models of Color LEDs and White pc-LEDs

The SPD of LEDs is the most important characteristic to determinate luminous flux, chromaticity, color quality, and many other optical parameters. If the

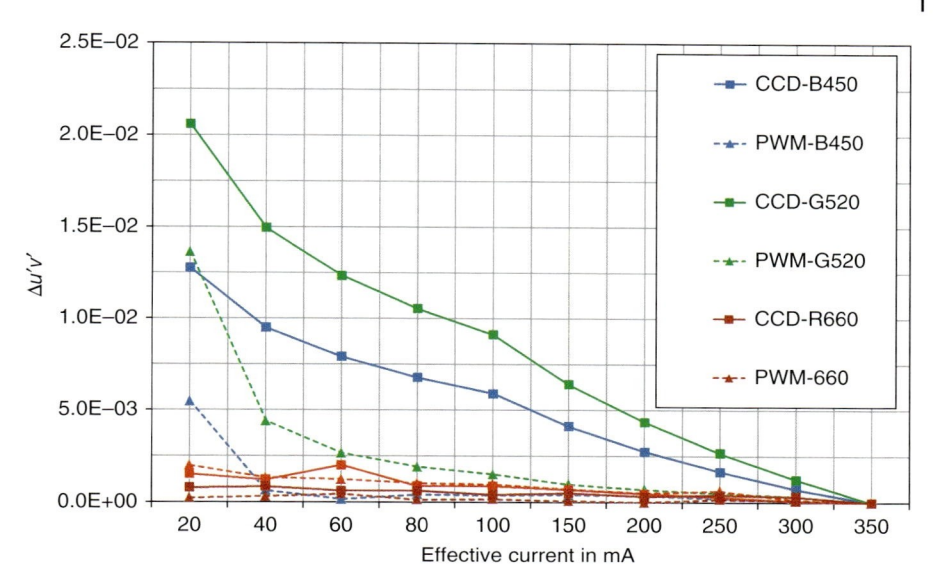

Figure 8.19 Chromaticity difference of color LEDs (450, 520, and 660 nm) operated by the two different dimming methods, PWM and CCD.

Figure 8.20 A multi-input multi-output (MIMO) system of LED spectral models.

thermal and electrical data of the actual color LEDs and white LEDs are well known or measurable then the absolute SPDs of the LEDs and LED channels can be simulated accurately by suitable spectral models. Subsequently, a color mixing procedure can be carried out to generate the spectrum of the whole LED lighting system with a defined color quality. Since about 2005, there have been a series of publications studying this subject. Until now, there have been two main approaches including a purely mathematical approach and a combination between mathematical descriptions and physical LED properties. In the present section, the LED's spectral model is assumed to be a multiple-input–multiple-output system (MIMO system). The input of the MIMO system is the junction temperature (T_j/°C) and the forward current (I_f/mA). The output is the LED spectrum that can be described by the peak wavelength (λ_p/nm), full width at half maximum (λ_{FWHM}/nm), and the spectral radiant power (S_p/W/nm) (Figure 8.20).

In the following sections, some aspects and proposals on LED spectral models will be described. The spectral models were analyzed in a previous comprehensive study [1].

Gaussian function: At the beginning of LED simulation with purely mathematical methods, the authors in [5, 6] assumed that the first order Gaussian function can describe LED spectra. This function models the spectrum by peak wavelength (λ_p/nm) and full width at half maximum (λ_{FWHM}/nm) as described by Eq. (8.1).

$$S(\lambda_p, \lambda_{FWHM}, \lambda) = \exp\left\{\frac{-2.7725(\lambda - \lambda_p)^2}{\lambda_{FWHM}^2}\right\} \quad (8.1)$$

This Gaussian function (Eq. (8.1)) has been used until now in several documents characterizing LED spectra, although its simulation quality is not good enough for practical applications. In addition, the current dependence and temperature dependence of the parameters of the LED spectral models are not reflected in this approach.

Developing further models from the Gaussian function: The theoretical background of the LED's SPD is described in [1] where it is explained qualitatively that the LED's SPD is asymmetric. Indeed, there is a difference between the left side and the right side of the LED's emission spectrum. While the left-hand side distribution has the form of a square root function ($S_{left}(E) \sim \sqrt{E - E_g}$), the right-hand side distribution has an exponential form ($S_{right}(E) \sim e^{-E/(kT_j \alpha)}$). Thus, the rising edge of the left distribution is faster while the falling edge of the right distribution is slower. In contrast to this asymmetry, the previously used Gaussian functions are always symmetric mathematical forms. Therefore, the authors in [7–9] attempted to vary the mathematical formulations from the Gaussian function into the new form with a more realistic character.

Particularly, Chien and Tien [9] proposed a double Gaussian function with two sets of parameters including the first set of the first spectral power (P/W), peak wavelength (λ_p/nm), and full width at half maximum (λ_{FWHM}/nm), and the second set of the second spectral power (P'/W), peak wavelength (λ_p'/nm) and full width at half maximum (λ_{FWHM}'/nm) for the spectral model of both pc-LEDs and color semiconductor LEDs. All the parameters are functions of both the junction temperature and forward current. The estimated $M \times N$ spectral matrix $\tilde{S}$ for color semiconductor ("semiconductor" in the sense that they are purely semiconductor LEDs without phosphor conversion) LEDs was introduced according to Eq. (8.2).

$$\tilde{S} = G + G' \quad (8.2)$$

In Eq. (8.2), $G = (gg_1, gg_2, \ldots, gg_M)^T$ and $G' = (gg_1', gg_2', \ldots, gg_M')^T$; they are Gaussian spectral matrices. The matrix G has M spectral vectors gg with N sampling wavelengths. From the nth point of mth row vector gg_m (denoted by g_{mn}), its value can be determined according to Eq. (8.3).

$$gg_{mn} = p_m \exp\left\{\frac{[\lambda_n - (\lambda_p)_m]^2}{(\lambda_{FWHM})_m^2}\right\} \quad (8.3)$$

In Eq. (8.3), the parameters p_m, $(\lambda_p)_m$, and $(\lambda_{FWHM})_m$ correspond to the mth power, the mth peak wavelength, and the mth full width at half maximum whose values can be obtained by satisfying the minimization of Eq. (8.4).

$$\arg\min\left[|S_m - \tilde{S}_m|^2, \{p_m, (\lambda_p)_m, (\lambda_{FWHM})_m, p_m', (\lambda_p')_m, (\lambda_{FWHM}')_m\}\right] \quad (8.4)$$

In Eq. (8.4), S_m and $\tilde{S}_m = gg_m + gg'_m$ are the mth row vectors of S and $\tilde{S}$, respectively. In order to evaluate the accuracy of this approach, the simulated and measured spectra of a green semiconductor LED (a) and a pc-LED (b) are compared in Figure 8.21.

Other mathematical forms in Table 8.3 are listed by Reifegerste and Lienig in [7]. These authors conclude that Eq. (8.5) is the most suitable one for the approximation of color semiconductor LED spectra. Particularly, Eq. (8.5) can be rewritten with its extension according to Eqs. (8.6)–(8.9).

$$f(\lambda) = \frac{A}{S}\left(1 + e^{\frac{\lambda - C + Wln(S)}{W}}\right)^{\frac{-S-1}{S}} \left\{e^{\frac{\lambda - C + Wln(S)}{W}}\right\} (S+1)^{\frac{S+1}{S}} \tag{8.5}$$

$$A(T_j, I_f) = a_0 T_j^{a_T} I_f^{a_I} \tag{8.6}$$

$$C(T_j, I_f) = c_0 + c_T T_j + c_I \log(I_f) \tag{8.7}$$

$$S(T_j, I_f) = s_0 + s_T T_j + s_I \log(I_f) \tag{8.8}$$

$$W(T_j, I_f) = w_0 + w_T T_j + w_I I_f \tag{8.9}$$

The accuracy of Eq. (8.5) was checked in an actual measurement comparing simulated and measured spectra at 10 mA and 76.8 °C and 20 mA and 104.7 °C; see Figure 8.22. All potential spectral models with physical background can be studied in more detail in [1].

The potential spectral models for color and white LEDs can be programmed and regarded as part of an LED color mixing tool. If the temperature and the voltage of the LED channels as input parameters of these models are varied (e.g., by change of the environmental condition of the luminaire) the signals of the temperature sensor and color sensor automatically initiate a new calculation with the spectral model. Herewith, the new SPDs of each LED channel can be generated so that a new color-mixing procedure is possible. In the next sections the color-mixing process or optimization of the luminaire spectra will be presented in order to achieve an optical radiation characteristic and an illumination of the objects under consideration with a well-defined color quality.

8.5 General Aspects of Color Quality Optimization

Color quality optimization or the color-mixing procedure is a process where the absolute spectral power of each LED channel (white and color LEDs) has to be combined for a defined application. These well-known applications are shop and museum lighting, event and entertainment lighting, and illumination of TV and film productions. Recently, the concept of HCL (see Chapter 9) has arisen and this concept can only be put into practice by implementing the principle of varying the LED light source's spectra, color temperature, and lighting direction. Based on the methodological and colorimetric approaches, the following aspects should be considered:

1) What colorimetric and lighting criteria should be taken for the optimization of the light source spectra and what color quality metrics should be taken into consideration?

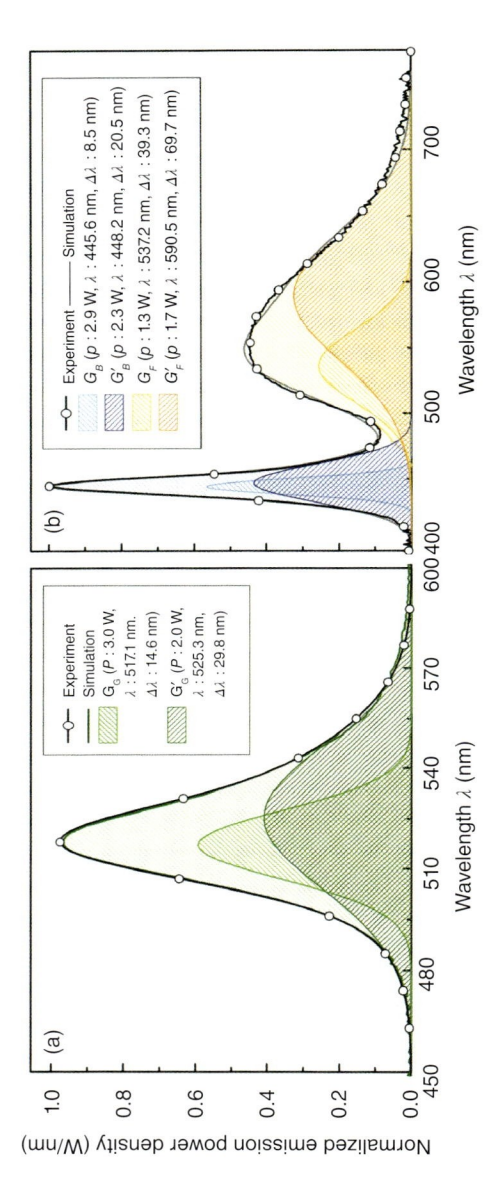

Figure 8.21 Double Gaussian model for the green semiconductor LED (a) and the white pc-LED (b) at $T_j = 25\,°C$ and $I_f = 350$ mA. For the pc-LED, the blue and phosphor spectral components are individually considered by two double functions GB and $G'B$ and GF and $G'F$, respectively. (Chien and Tien 2012 [9] https://www.osapublishing.org/oe/abstract.cfm?uri=oe-20-S2-A245.)

8.5 General Aspects of Color Quality Optimization

Table 8.3 Functions to approximate the LEDs spectral power distribution [7].

Order	Name	Function $f(\lambda)$
1	Gaussian	$f(\lambda) = Ae^{-\left(\frac{\lambda-C}{W}\right)^2}$
2	Split Gaussian	$f(\lambda) = Ae^{-\left(\frac{\lambda-C}{W}\right)^2}$ with $W = W_1$, for $\lambda < C$ and $W = W_2$ otherwise
3	Sum of Gaussian	$f(\lambda) = A_1 e^{-\left(\frac{\lambda-C_1}{W_1}\right)^2} + A_2 e^{-\left(\frac{\lambda-C_2}{W_2}\right)^2}$
4	Second order Lorentz	$f(\lambda) = \dfrac{A}{\left(1+\left(\frac{\lambda-C}{W}\right)^2\right)^2}$
5	Logistic power peak	$f(\lambda) = \dfrac{A}{S}\left(1 + e^{-\frac{\lambda-C+Wln(S)}{W}}\right)^{-S-1}\left\{e^{-\frac{\lambda-C+Wln(S)}{W}}\right\}(S+1)^{\frac{S+1}{S}}$
6	Asymmetric logistic peak	$f(\lambda) = A\left(1 + e^{-\frac{\lambda-C+Wln(S)}{W}}\right)^{-S-1}\left\{e^{-\frac{\lambda-C+Wln(S)}{W}}\right\}S^{-S}(S+1)^{S+1}$
7	Pearson VII	$f(\lambda) = \dfrac{A}{\left(1+\left(\frac{\lambda-C}{W}\right)^2 \left(2^{\frac{1}{S}}-1\right)\right)^S}$
8	Split Pearson VII	$f(\lambda) = \dfrac{A}{\left(1+\left(\frac{\lambda-C}{W}\right)^2 \left(2^{\frac{1}{S}}-1\right)\right)^S}$ with $W = W_1$, $S = S_1$ for $\lambda < C$ and $W = W_2$, $S = S_2$ otherwise
9	Asymmetric double sigmoidal	$f(\lambda) = \dfrac{A}{1+e^{-\frac{\lambda-C+\frac{W}{2}}{S_1}}}\left(1 - \dfrac{1}{1+e^{-\frac{\lambda-C+\frac{W}{2}}{S_2}}}\right)$
10	Piecewise third order polynomial (spline)	Piecewise: $f(\lambda) = a_3 x^3 + a_2 x^2 + a_1 x^1 + a_0$, piecewise defined for n range $x_{k-1} \leq x < x_k$, $k = 1, \ldots, n$

Source: F. Reifegerste, private communication, 2014, copyright ©2014 with permission from Dr. Frank Reifegerste

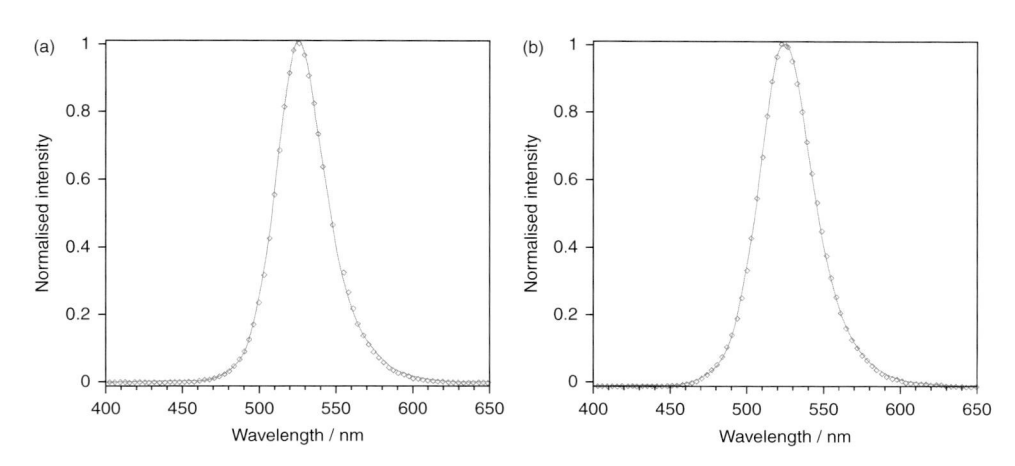

Figure 8.22 Comparison between the simulated and measured spectral results of the color semiconductor LED at 10 mA and 76.8 °C (a) and at 20 mA and 104.7 °C (b). Reproduced from F. Reifegerste, private communication, 2014, copyright ©2014 with permission from Dr. Frank Reifegerste.)

2) What quality or, more specifically, what semantic category (in terms of a color quality metric) shall be required in order to fulfill the color quality expectation of the lighting users?
3) What wavelength of the color LEDs shall be used in order to optimize the color quality metrics to achieve best human physiological and perceptual effects?
4) What test color sample collection shall be applied as the set of representative colors for a context-specific or general illumination purpose?

In order to answer the first question, visual experiments and numeric analyses of international color research and in the Lighting Laboratory of the authors were studied. In [10], Smet *et al.* analyzed a number of visual experiments in international publications and concluded that the color quality metrics CQS Q_p, MCRI (memory color rendering index), and $(GAI + R_a)/2$ correspond well with visual assessments (concerning the attribute color preference) and the attribute "naturalness" could be well described by the color metrics CQS Q_p and $(GAI + R_a)/2$.

In Sections 7.7 and 7.8, results of visual experiments for the lighting context of cosmetics, food, and colorful colored objects in a test room were analyzed [11, 12]. In Table 8.4, the correlation coefficients r between the mean visual interval scale variables and the color quality indices plus CCT and ΔC^* resulting from these experiments are listed.

The following aspects can be seen from Table 8.4:

- The color fidelity indices CRI R_a, CQS Q_f, and IES R_f do not correspond to the visual ratings on color preference, naturalness, and vividness.

Table 8.4 Correlation coefficients r between the mean visual interval scale variables and the color quality indices plus CCT and ΔC^* resulting from the experiments described in Chapter 7.

No.	Parameter	Preference	Naturalness	Vividness
1	CCT	0.38	0.08	−0.23
2	CRI R_a	−0.23	0.02	0.07
3	CQS Q_a	0.24	0.45	0.60
4	CQS Q_g	0.41	0.37	0.35
5	CQS Q_f	0.09	0.32	0.47
6	CQS Q_p	0.51	0.64	**0.76**[a]
7	GAI	0.53	0.42	0.15
8	MCRI	**0.72**[a]	**0.80**[a]	0.57
9	FCI	0.55	0.68	**0.72**[a]
10	IES R_f	0.21	0.46	0.58
11	IES R_g	0.28	0.21	0.23
12	ΔC^*	**0.74**[a]	**0.79**[a]	**0.79**[a]

a) Correlation significant at the level $p < 0.05$.

8.5 General Aspects of Color Quality Optimization

Table 8.5 Selected linear combinations of the independent variables with parameter values that maximize the correlation coefficient (r) with naturalness.

Order	Combination	a	b	c	r
1	a MCRI + $b \Delta C^*$	1.000	1.745	—	0.83[a),b)]
2	a MCRI + b FCI	1.000	0.106	—	0.81[a),c)]
3	a MCRI + b Q_p	1.000	0.086	—	0.80[c)]
4	a R_f + b R_g	1.000	1.600	—	0.61
5	$(R_a + GAI)/2$	—	—	—	0.41
6	a MCRI + b (CCT/52)	1.000	0.000		0.80[c)]
7	a MCRI + b (CCT/52) + c ΔC^*	1.000	0.057	2.117	0.83[a),b)]
8	MCRI + CCT/52 + Q_p	1.000	0.001	0.087	0.80[c)]
9	MCRI + CCT/52 + FCI	1.000	0.059	0.162	0.82[a),b)]

Source: Khanh et al. 2016 [12]. Reproduced with permission of Sage Publications.
a) The maximal correlation coefficient (r) obtained by parameter optimization was greater than the values realized by MCRI as a standalone variable for preference ($r = 0.80$); see Table 8.4.
b) Correlation significant at $p < 0.01$ level.
c) Correlation significant at the $p < 0.05$ level.

- The color gamut metrics GAI, Q_g, and IES R_g do not correspond to the visual ratings on color preference, naturalness, and vividness, either.
- The color metrics Q_p, MCRI, and the chroma difference ΔC^* reflect the characteristics and the level of these visual attributes (color preference, naturalness, and vividness) in a reasonable manner.

In order to achieve better correlations between the visual attributes and the color quality metrics, the single color metrics were combined into two metrics or multiple metrics in order to reflect the complex visual perception and evaluation of the light source users (see Chapter 7). The analysis for the attribute "naturalness" in [12] is presented in Table 8.5.

As can be seen from Table 8.5, the correlation coefficient r for naturalness can be improved if two or three metrics are combined. From the point of view of the moment of writing this book, a combination of MCRI, ΔC^*, and CCT should be the best solution. The combined color quality metric $(GAI + R_a)/2$ could not describe the visual results in the experiments of the present authors (Chapter 7) well.

Based on the above analysis, the following criteria can be defined for color quality optimization:

1) *Color fidelity*: The current philosophy and practical implementation of color quality aspects in most countries, guidelines, international standards, and regulations are color-rendering indices in the context of color fidelity. Therefore, the first criterion to be considered shall be color fidelity. Based on the analysis in Chapter 6 and in [13], it can be recognized that the most often used color fidelity metrics IES R_f, CIE CRI_{2012}, and CIE $R_{1,14}$ do have a high correlation with each other. Consequently, the optimization of color quality can be

implemented by the use of any of these color metrics and their own test color sample collections. Based on the conclusions of Chapter 4, it can be stated that the color collection of the 99 IES test color samples is reasonable and usable for the optimization of color fidelity. After the optimization, the values of other color fidelity metrics can be checked and possibly co-analyzed.

2) The criteria for color quality in the approach of color preference and color naturalness should be MCRI, CQS Q_p, and chroma difference ΔC^* or a combination of the latter parameters. Based on the analysis of lighting quality for red cosmetics in Section 7.7, the maximum of naturalness and color preference will be achieved for the light source spectra giving rise to a chroma difference for CQS VS_1 (the saturated red test color sample) in the range of 11–12 that corresponds to the average saturation enhancement for all 15 CQS test color samples of about 3.2 and for all 14 CIE CRI test color samples (1995) of about 2.5 (see Figure 8.23). It is an important hint for luminaire developers to optimize the spectral output for their application with their specific illuminated colored objects.

3) Besides the color collections that are defined in the most well-known color quality metrics, some specific color collections must be taken into account if the application is well defined and it should play an important role in the development of the LED luminaire. Some examples for specific colors are the reddish cosmetic colors analyzed in Section 7.7 or the spectral radiance coefficients of foods and fruits with red hue tones (see Figure 8.24) if the system developers have to illuminate shopping centers for foods and fruits.

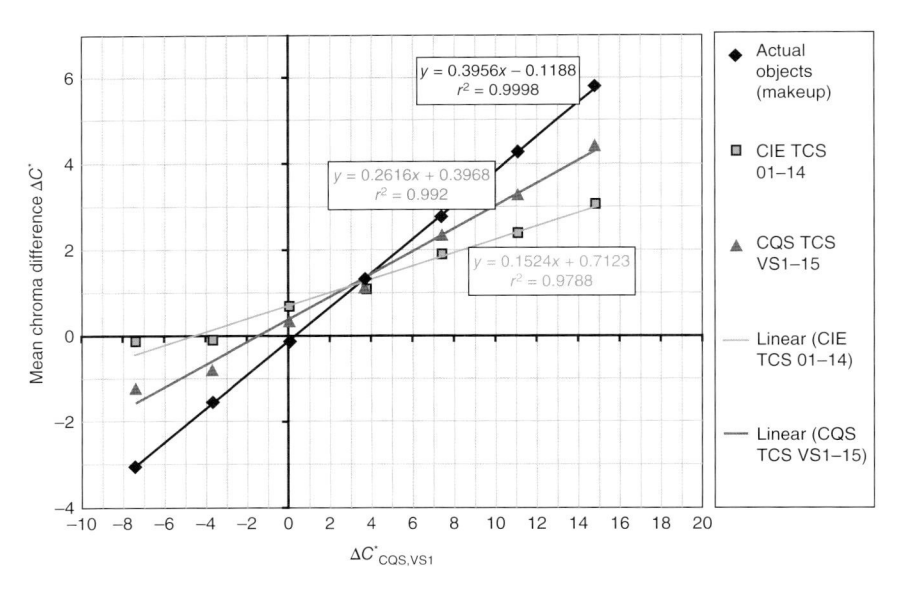

Figure 8.23 Relationship between the chroma difference of CQS TCS VS_1 ($\Delta C^*_{CQS,VS1}$) and the mean chroma difference (ΔC^*) of 15 CQS TCS VS_1 – VS_{15}, 14 CIE CRI TCS_1 – TCS_{14} and the specific makeup TCSs.

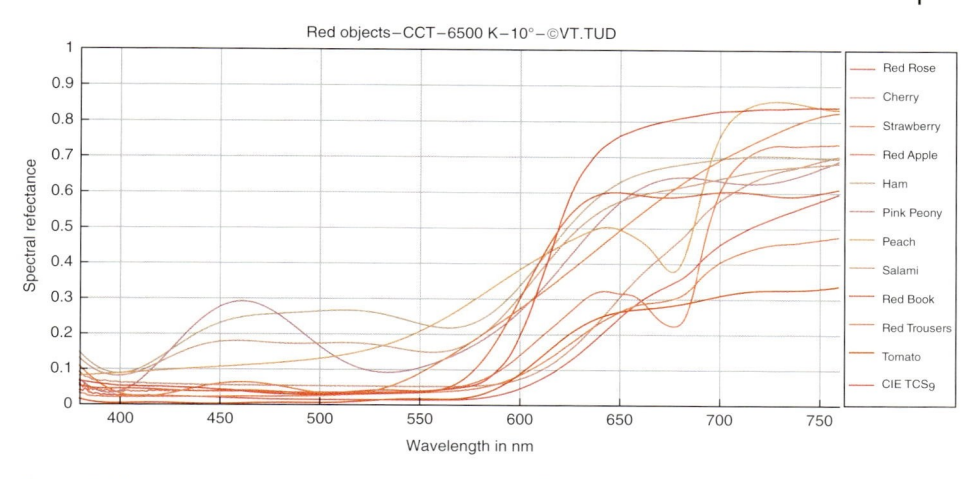

Figure 8.24 Spectral reflectance (spectral radiance coefficients) of some reddish objects (flowers, fruits, foods, book). (Copyright TU Darmstadt.)

8.6 Appropriate Wavelengths of the LEDs to Apply and a System of Color Quality Optimization for LED Luminaires

8.6.1 Appropriate Wavelengths of the LEDs to Apply

Beside the consideration of the illuminated colored objects and their spectral reflectance (see Figure 8.24), the selection of the appropriate wavelength of the applied LEDs plays an important role in the optimization of color quality and the development of LED luminaires. The aim of the wavelength selection of the LEDs to apply is to meet those wavelengths and wavelength regions of the highest spectral sensitivity regarding the human visual brain's channels of brightness, color opponency, and the regulation of the circadian rhythm, the LMS cone photoreceptors and the rod photoreceptors. In Figure 8.25, the spectral sensitivity of the retinal receptors, ipRGC-cells, the opponent channels, and the $V(\lambda)$-function for human photopic vision, which is a combination of the L and M channels, are shown for 2° (foveal) viewing (see also Chapter 2).

Based on the curves in Figure 8.25, the following wavelength ranges can be taken into account for a reasonable optimization:

- The blue LED chips should have the peak wavelength in the range of 440–460 nm to overlap with the S cones' spectral sensitivity. An additional blue LED-chip with a peak wavelength at 465 nm in cooperation with a green LED with a peak wavelength at 510 or 520 nm or with a green phosphor at 510 nm (cyan-green phosphor) can meet the channel of the ipRGC cells (see Chapter 9) at a high level of sensitivity. This green color semiconductor LED and the cyan–green phosphor can also cover rod sensitivity with a maximum at 507 nm.

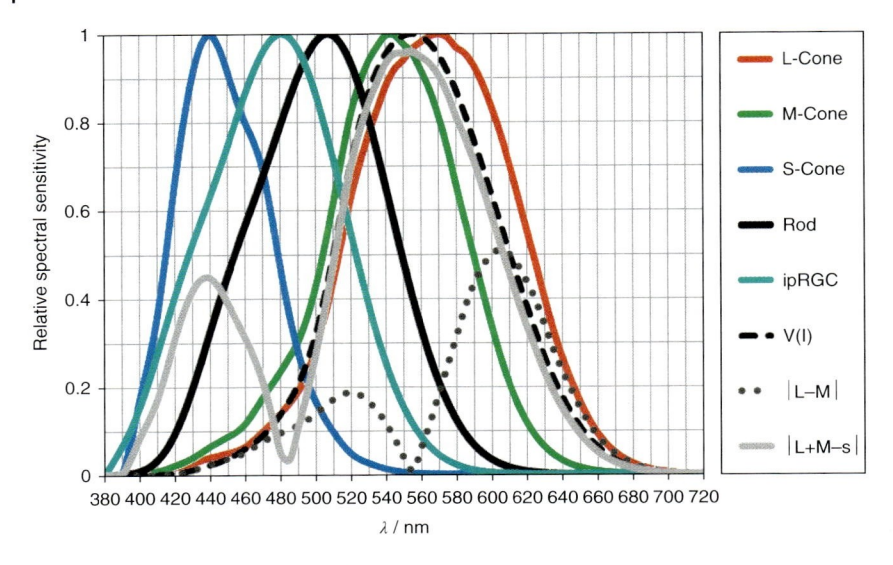

Figure 8.25 Spectral sensitivity of the retinal photoreceptors, ipRGC-cells, the opponent channels, and the $V(\lambda)$-function.

- Because many objects in the nature, in household and in industry are green and green-yellow products with a high reflectance in the range between 520 and 580 nm (see Figure 8.26) and the spectral sensitivity of the photoreceptors of L- and M-type have their maximum in the range between 540 and 570 nm, a color semiconductor LED type or phosphor systems with a high emission in this range should be optimal. Also, because the green LEDs in the range 540–570 nm are commercially not available, a phosphor system is the best option.
- Owing to the phenomenon that red objects have a high reflectance in the range from 600 until 700 nm (see Figure 8.24), the L cones' spectral responsibility is low from 660 nm and the opponent channel L − M has its maximal sensitivity at 615 nm, the optimal LED wavelength to render or emphasize red objects shall be in the range between 610 and 660 nm.

In order to prove the assumption that a longer wavelength in the red range can support color quality optimization, a calculation was done for a white LED consisting of a blue chip, a phosphor system, and a red semiconductor LED. In the calculation, the peak wavelength of the red color LED was shifted virtually and the specific color rendering index R_9 according to the CIE definition 1995 and the chromatic lightness L^{**} according to the formula of Fairchild and Pirrotta [14] (see Chapter 2) were calculated. As can be seen from Figure 8.27, the spectral reflectance of the red objects increases with wavelength from 600 nm and with a shift of the wavelength of the red LED, the signal product of the spectral reflectance and the LED emission will become higher. Consequently, the spectral reflectance capacity of the red objects can be supported by the radiation of the red semiconductor LED.

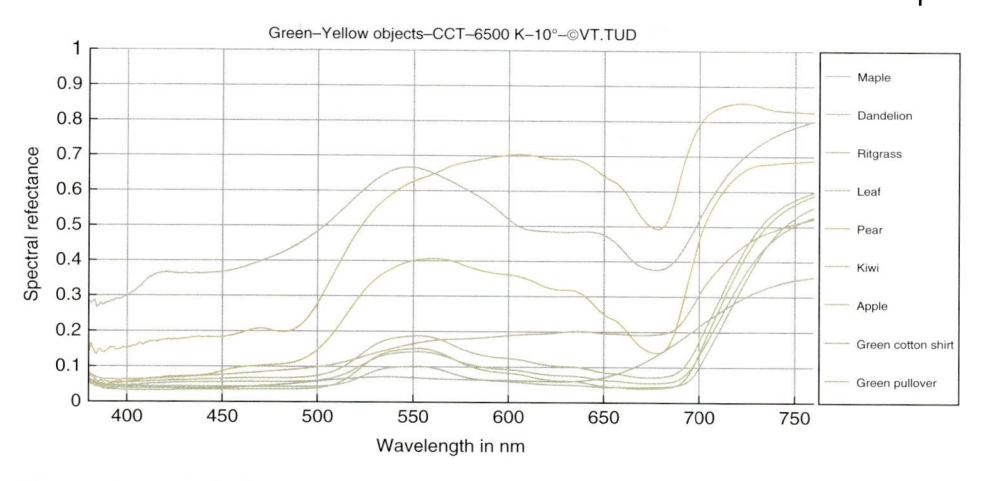

Figure 8.26 Spectral reflectance of some green–yellow objects (fruits, tree leaf, and textiles). (Copyright TU Darmstadt.)

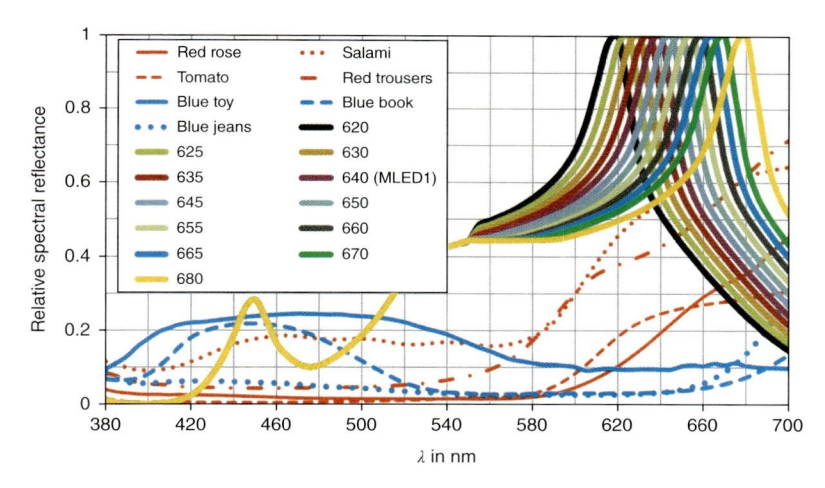

Figure 8.27 Spectral reflectance of several typical red objects together with the spectral shift of the red semiconductor LED's peak.

In Figure 8.28, both the object-specific color-rendering index of a deep red rose and its chromatic lightness L^{**} increase continuously when the peak wavelength of the applied red semiconductor LED varies from 620 to 650 nm; and, finally, the value of R_9 reaches its maximum at 650 nm. The value of chromatic lightness L^{**} approaches a high level as the wavelength of the LEDs touches the limit of 650 nm. This analysis shows the necessity to have red semiconductor LEDs with a peak wavelength in the range of 650–660 nm for spectral optimization with high color quality.

From the beginning of lamp technology with the introduction of the tungsten incandescent lamp in the nineteenth century to date, the basis of light source

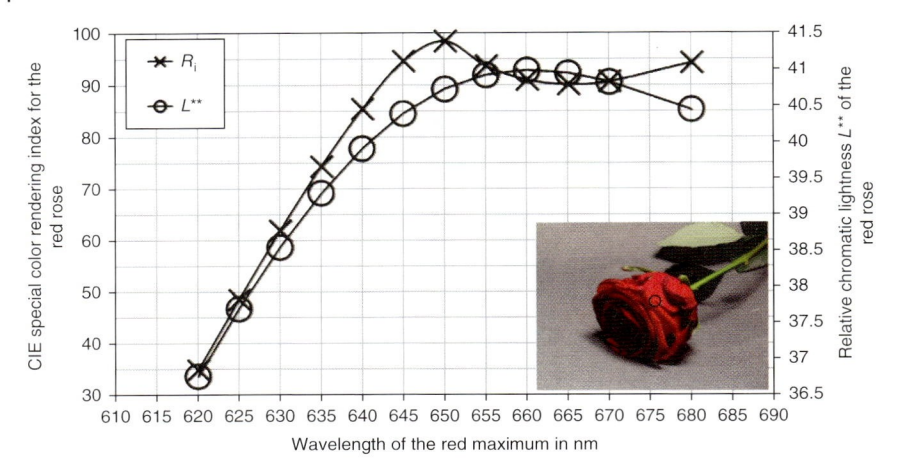

Figure 8.28 Enhancement of color quality (R_9, L^{**}) for a red rose by shifting the peak wavelength of the red semiconductor LED.

evaluation has been luminous efficacy defined by Eq. (8.10).

$$\eta = \frac{K_m \int_{380}^{780} \Phi_{e\lambda} \cdot V(\lambda) \cdot d\lambda}{P_{el}} = \frac{\Phi_v}{P_{el}} \quad (8.10)$$

The quantity luminous flux Φ_v in Eq. (8.10) is calculated by the integration of the spectral radiant power in the visible wavelength range between 380 and 780 nm weighted by the $V(\lambda)$ function. The latter function is, from a human-physiological point of view, the spectral sensitivity of the L + M channel, responsible for the achromatic aspects of visual perceptions in the photopic range at high spatial and temporal frequencies (see Figure 8.25). The main idea of most lamp developers in the last century has been to "*compress*" the optical power into the wavelength range between about 500 and 600 nm in order to have high luminous efficacy. The consequence was that the chromatic component (|L − M| and |L + M − S| in Figure 8.25) of the optical radiation reaching the eye and responsible for color quality was ignored. In most applications and under most viewing conditions (e.g., working in the office, meeting in a conference, observation of a scene in nature, or visiting a museum), both the chromatic and the achromatic parts of the radiation establish the total brightness impression of the viewers (i.e., the light source users) at the same time. In Figure 8.29 the ratios of the 2° and 10° spectral brightness functions of the CIE (1988), $V_{b2}(\lambda)$ and $V_{b10}(\lambda)$, to the $V(\lambda)$ function and the $V_{10}(\lambda)$ function are illustrated to demonstrate that the contribution to brightness from the wavelength range 600–670 nm (red range) and from the wavelengths in the blue range (until 480 nm) is very high and it cannot be ignored in the development process of light sources.

Figure 8.29 suggests that light source and luminaire developers, scientists, and lighting architects should change their lighting philosophy. They should rethink that the blue and red wavelength regions are also very important for perceptual efficiency and color quality, and hence for the light source users' acceptance. In

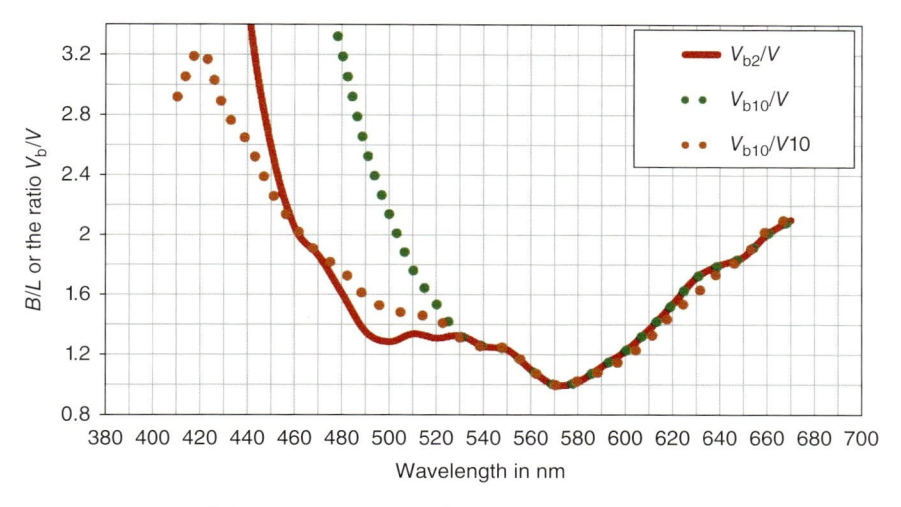

Figure 8.29 Ratio of the spectral brightness functions $V_{b2}(\lambda)$ and $V_{b10}(\lambda)$ of the CIE (1988) to the $V(\lambda)$ and $V_{10}(\lambda)$ functions.

the context of energy saving and light source efficacy, the criterion of evaluating light sources and luminaires shall be the ratio of a suitable descriptor quantity of the users' acceptance level (e.g., VC in Eq. (7.5)) to electric power, or the ratio of a color quality metric to electric power. The concept of luminous efficacy used in the last century should be integrated into the new concept as a part of it but it is not to be used as a holistic criterion.

8.6.2 Systematization for the Color Quality Optimization of LED Luminaires

8.6.2.1 Conventional Structures of LED Luminaires in Real Applications

Figures 8.30–8.32 show three LED combinations in a systematic way. These combinations are considered as conventional LED structures.

As can be seen from Figure 8.30, the three color LEDs include a blue LED chip (with the peak wavelength λ_p of 465 nm), a green LED chip ,(λ_p = 525 nm) and a red LED (λ_p = 635 nm). The warm white LED has its peak wavelength at around 600 nm. Thus, Figure 8.30 is a demonstration of a four-channel LED engine used frequently as an LED light source in entertainment industry. The six-channel LED light engine (Figure 8.31) comprises two blue LED chips (λ_p = 450 and 470 nm), two red LED chips (λ_p = 640 and 670 nm), one green LED chip (λ_p = 525 nm), and a warm white pc-LED with the peak wavelength of 605 nm. This configuration is part of a concept for a shop lighting system. An LED combination consisting of a green LED at 505 nm, a red LED at 640 nm, and two pc-LEDs can be seen in Figure 8.32. This can be an interesting proposal for today's shop lighting applications and as an LED luminaire for medical applications.

8.6.2.2 Schematic Description of the Color Quality Optimization of LED Luminaires

In the present approach of systematization, the elements of a hybrid LED system should be understood. In detail, the control system structure of hybrid LED luminaires can be assumed as a MIMO system; see the scheme of Figure 8.33.

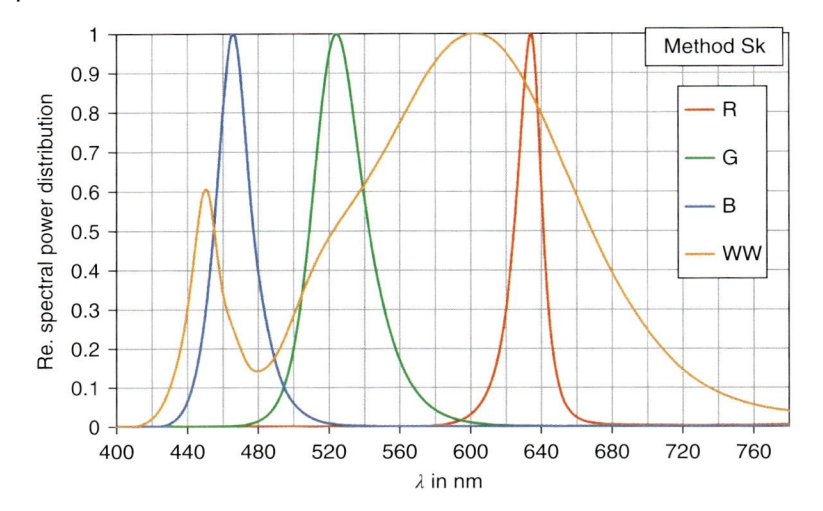

Figure 8.30 LED combination with RGB-LEDs and a warm white LED with a phosphor peak wavelength of 600 nm.

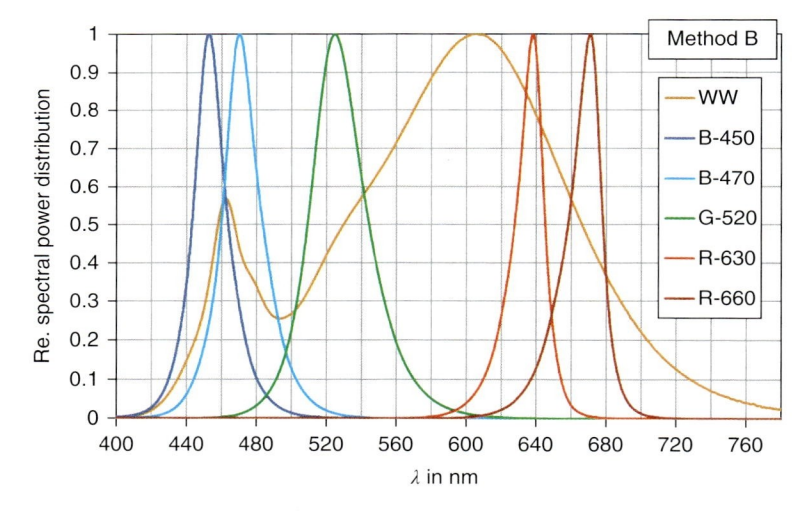

Figure 8.31 LED combination with six LED channels.

As can be seen from Figure 8.33, the setting values and the feedback values can be classified into three categories including the background values, a main value, and subvalues or subconditions. Firstly, the background values consisting of the CCT in kelvin and the chromaticity distance from the Planckian locus ($\Delta u'v'_{CCT}$) are the basic values to be necessarily achieved in a hybrid LED system in order to have acceptable white light sources with a desired CCT. Secondly, the main value of the hybrid LED luminaire is always the most important unique value of a desired color quality metric such as a color fidelity parameter (CIE CRI R_a, CIE CRI $R_{1,14}$, CIE CRI$_{2012}$, CQS Q_f, or IES TM30-15 R_f), which can also be expressed in terms of a color difference (ΔE or $\Delta E'$ in the color spaces $U^*V^*W^*$, CIELAB

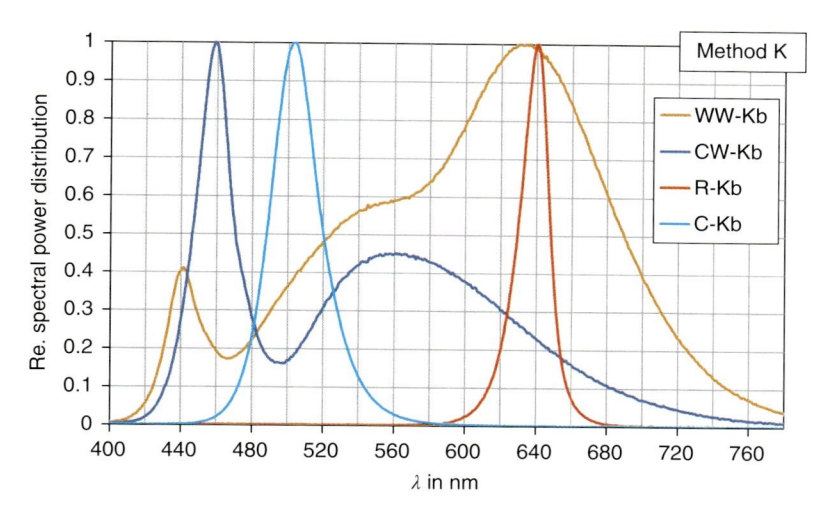

Figure 8.32 LED combination with a green LED at 505 nm, a red LED at 640 nm, and two pc-LEDs (warm white and cold white).

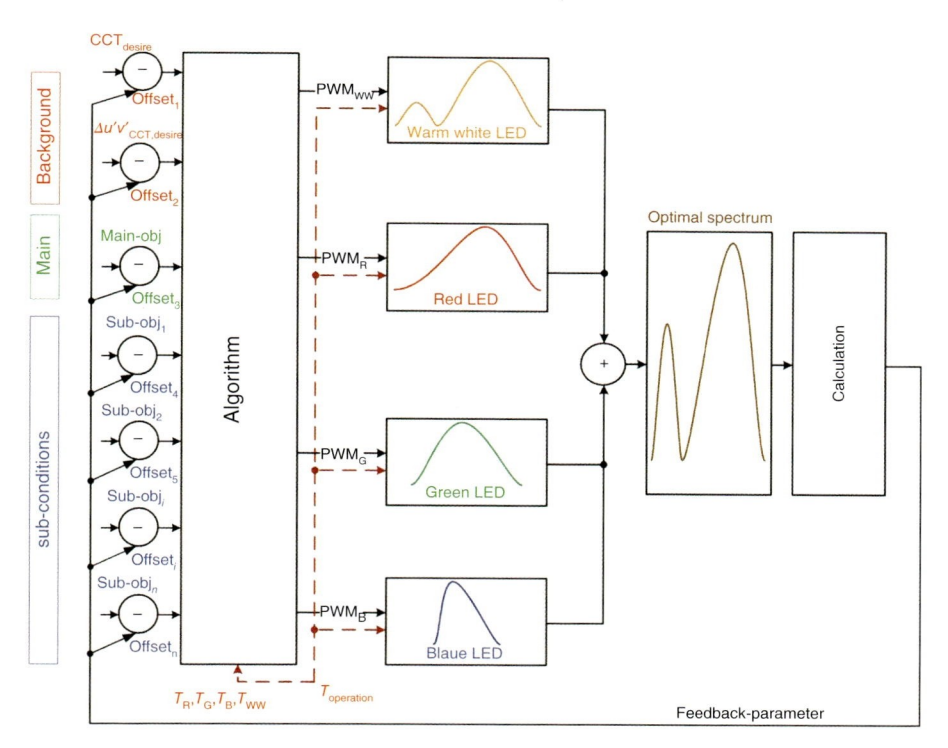

Figure 8.33 Schematic view of an LED luminaire system in the context of color quality optimization. (Khanh *et al*. 2014 [1]. Reproduced with permission of Wiley-VCH.)

$L^*a^*b^*$ or CAM0m2-UCS $J'a'b'$), or a color gamut/color saturation parameter (GAI, FCI, CQS Q_g, or TM30-15 IES R_g) or only simply chroma difference (ΔC^* in the color spaces $U^*V^*W^*$, CIELAB $L^*a^*b^*$, or CAM02-UCS $J'a'b'$), or their appropriate combinations to obtain a descriptor of color preference (MCRI, CQS Q_p, or others). Every metric should be calculated based on its own TCS set or an object-specific TCS set for a desired special scene of colored objects.

Beside the main value or the main condition, although the subconditions are the necessary subvalues of the desired color quality metric of the hybrid LED luminaire, subconditions also play an important role depending on the demand of a specific lighting context. While in optimization theory, the main condition is represented uniquely by only one value, the subconditions or subvalues shall play the role of subconstraints so that the optimization shall run in permitted or desired directions. This can be recognized most clearly in the optimization procedure to illuminate skin tones. The best color quality can be achieved when the main color fidelity parameter is already "good" or "very good" and then the subcondition of chroma difference can be adjusted gradually to reach the value of about $\Delta C^*_{\text{CQS VS1-VS15}} = 2$–$3$. In other words, in this case, a compromise between the total color difference (color fidelity) and chroma difference (chroma enhancement) takes place so that the total color difference is small enough for human color quality assessment and the chroma difference is big enough so that the beauty of skin tones is best exposed visually. Generally, the subconditions can be also other color fidelity parameters, other color gamut/color saturation parameters, or their appropriate combinations.

The central objects in the scheme of Figure 8.33 are the LEDs characterized by appropriate LED models. In Figure 8.33 the example of a warm white pc-LED, a red, a deep red, a green, a royal blue, and a blue LED is shown. The inputs for these LEDs are PWM signals and operation temperatures. The output is their spectra to be mixed together into the total spectrum of the hybrid LED luminaire. Then, the spectrum is calculated to issue the feedback values for the calculation of color quality offsets to be supplied for the optimization algorithm. The word "*object*" in this situation should be understood as the main parameter and the subparameters to be satisfied during the optimization process.

8.6.2.3 Algorithmic Description of Color Quality Optimization in the Development of LED Luminaries

The algorithm or its decision-making part is the core component of the system in the schematic view of Figure 8.33. In Figure 8.34, the corresponding algorithmic description is presented.

The concept of "*optimization*" in Figure 8.34 should not only be understood simply as a computing algorithm or software. Instead, it should be a complex progress including both a computing algorithm and an appropriate selection of the LEDs. It can be recognized from experience that if the desired CCT range should vary between 2500 and 6500 K while the built LED structure has only a cold white LED as a pc-LED then the desired CCT range will become unavailable. A similar situation might happen with an unsuitable basis pc-LED during the development of LED luminaires. Therefore, optimization should consider both aspects at the same time, hardware and software. At the beginning of the

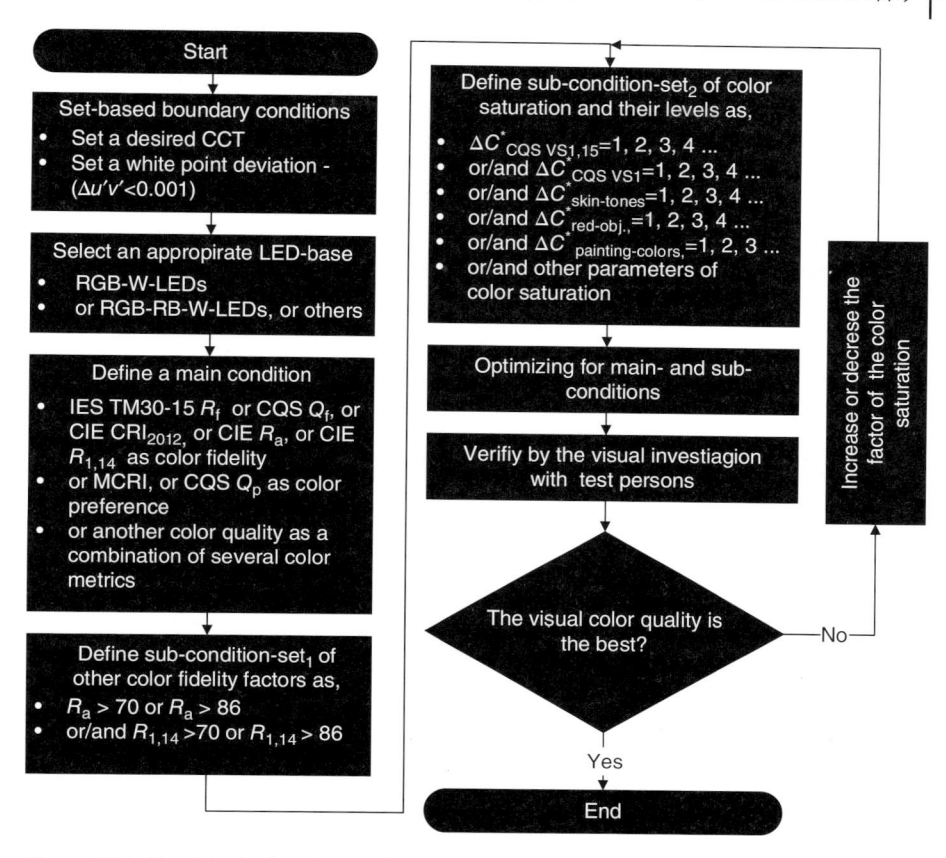

Figure 8.34 Algorithmic description of color quality optimization for the development of LED luminaries.

process shown in Figure 8.34, the desired CCT and the chromaticity distance from the Planckian or the daylight locus shall be set as background parameters to be absolutely fulfilled. Then, the choice of the basis pc-LED shall be considered to yield appropriate combinations such as RGB-W, RGB-RB-W, or R-G-CW-WW. Then, the main condition must be defined such as color fidelity, color preference, or color saturation or a combination of several color quality parameters. Sub-conditions can be classified as subcondition-Set$_1$ (with the conventional color fidelity parameters) and subcondition-Set$_2$ (with color saturation). Their appropriate value ranges shall also be defined by the use of the desired test color samples. Finally, the optimal outcome should be verified with an appropriate visual check of the color quality by test subjects.

8.6.2.4 Optimization Solutions

One boundary condition should be that the general color-rendering index R_a is greater than 70, 80, or 86 if the optimization for maximal MCRI or Q_p should be implemented as the main condition. A general color-rendering index of 80 only means a visual semantic interpretation between "*moderate*" and "*good*" and the one of 86 just has a similar semantic interpretation to "*good*." In most international and national regulations and standards for indoor lighting, the general color rendering index of 80 is recommended (see Table 8.1). In the case of $R_a >$ 70 the optimizer or other lighting developers should balance between the color fidelity aspect and the maximum of MCRI or Q_p. This decision often depends on the specific lighting application under consideration.

For academic research and also for product development, the answers to the following questions are interesting and relevant:

- How can the light source developer balance between high color fidelity (with small color difference ΔE and small chroma difference ΔC^*) and high color preference/naturalness with higher ΔC^* and slightly reduced color fidelity?
- Is there any remarkable difference for the achieved values of the various color quality metrics if the test color samples CQS TCS_{1-15} or CIE CRI TCS_{1-14} are applied in the optimization?
- Are there differences for the achieved values of color fidelity (R_a, $R_{1,14}$, R_f, CRI_{2012}), color preference (MCRI, Q_p), or color saturation (FCI, GAI, R_g) if the optimized LED engine is constructed from RGB + warm white LED (four-channel LED engine) or from R_1-R_2-B_1-B_2-G + warm white LED (six-channel LED – engine with two blue LEDs, two red LEDs, a green LED, and a warm white LED)?

8.7 Optimization of LED Light Engines on Color Fidelity and Chroma Enhancement in the Case of Skin Tones

In Chapter 4, skin tones were analyzed with the illuminants D65 and a thermal radiator with a color temperature of 3200 K (see Figures 4.8 and 4.9). Concerning the hybrid LED engine consisting of four LED components shown in Figure 8.30 (nowadays frequently used in entertainment industry), the questions are as listed below:

- How do the LED engine's spectra look in case of desaturation, high saturation, or best color fidelity if the 15 representative skin tones in Figure 4.7 are the desired target objects to be illuminated?
- How do the color metrics color fidelity, color gamut, and color preference vary if the chroma difference $\Delta C^*_{\text{skin-tone-set}}$ is varied from -2 (desaturation) up to $+8$ (high saturation) in comparison to the reference thermal radiator of 3100 K? In each case, light source spectra must always satisfy the same white point of the reference light source ($\Delta u'v' < 0.001$).

For this purpose, 15 spectral reflectance curves of 15 skin tones measured in the Lighting Laboratory of the authors were used together with the LED combination shown in Figure 8.30. From the 11 chroma difference levels (from -2 up to $+8$) computed, four SPDs and the corresponding color gamut diagrams for the levels $\Delta C^*_{\text{skin-tone-set}} = -2, 0, 4,$ and $+8$ are illustrated in Figure 8.35.

As can be seen from Figure 8.35, if the SPD is referred to 635 nm as the wavelength of maximum power, the SPD for the LED combination in case of color

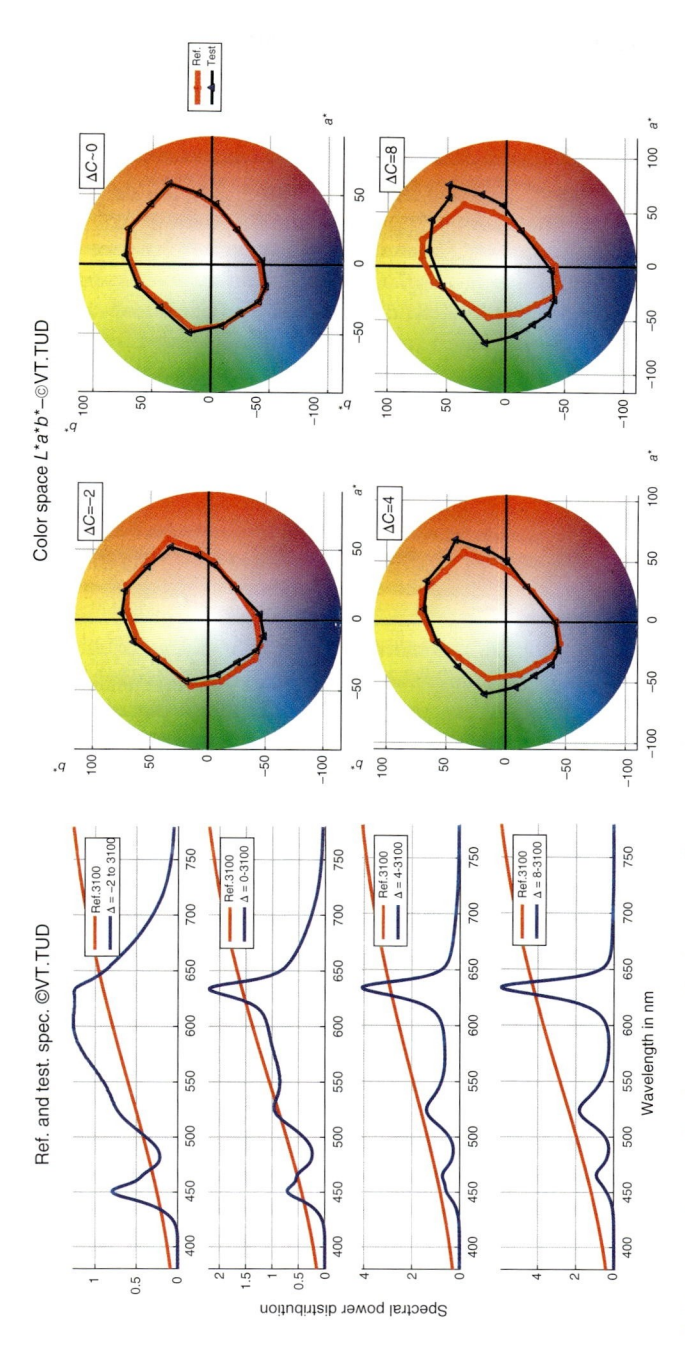

Figure 8.35 Spectral power distributions and color gamut diagrams in the cases $\Delta C^*_{\text{skin-tone-set}}$ = −2, 0, +4, and +8.

desaturation ($\Delta C^*_{\text{skin-tone-set}} = -2$) will have a remarkable amount of blue light at 450 nm and green and yellow light from 550 up to 600 nm. In this case, the part of the phosphor-converted warm white LED is high. If this spectrum is changed to the next spectrum allowing for zero chroma difference ($\Delta C^*_{\text{skin-tone-set}} = 0$; small color difference = high color fidelity) then the contribution of the warm white pc-LED is strongly reduced and the dominance of the red LED is visible in the spectral distribution. Generally, the optical radiation of the blue, green, and green-yellow range is then reduced relative to the red radiation at 635 nm. In the case of the chroma difference $\Delta C^*_{\text{skin-tone-set}} = +8$ (a strong chroma increase), the contribution of the warm white LED is minimized and the three peaks of the three color LEDs in the blue, green, and red spectral ranges appear more obviously. The radiation gap between 550 and 610 nm is distinctive indicating the dominance of the red LED radiation. In the color gamut diagram for the case of $\Delta C^*_{\text{skin-tone-set}} = +8$, it can be seen that the gamut of the LED light source is much bigger than the gamut of the reference thermal radiator at the same color temperature of 3100 K in the red and green-cyan hue tone region. In the case of the light source with the best color fidelity or $\Delta C^*_{\text{skin-tone-set}} = 0$, the two gamut areas of the reference and LED light source are nearly identical (see Figure 8.35).

Table 8.6 shows the values of the color quality metrics color fidelity, color gamut, color preference, and MCRI for 11 LED spectra at 3100 K and with 11 different chroma differences.

Table 8.6 Values of the color quality metrics color fidelity, color gamut, color preference, and memory color rendering index (MCRI) for 11 LED spectra at 3100 K and with 11 different chroma differences $\Delta C^*_{\text{skin-tone-set}}$.

ΔC^*	−2	−1	0	1	2	3	4	5	6	7	8
CCT	3100	3101	3100	3101	3100	3101	3100	3100	3100	3100	3100
$\Delta u'v'$	9.2E−5	1.0E−4	8.7E−5	1.1E−4	9.2E−5	1.0E−4	9.2E−5	9.55E−5	8.56E−5	1.05E−4	1.11E−4
R_{13}	84.60	93.47	97.79	89.12	80.44	71.81	63.16	54.51	45.82	37.10	28.28
R_a	84.69	92.35	95.75	88.85	80.91	73.00	65.12	57.29	49.49	41.72	34.00
$R_{1,14}$	79.26	88.85	93.97	86.97	78.03	68.84	59.57	50.37	41.23	32.17	23.16
GAI	89.39	95.57	101.53	107.40	113.15	118.80	124.37	129.85	135.28	140.67	145.91
Q_f	84.25	90.35	93.38	90.48	84.90	78.78	72.52	66.26	60.00	53.78	47.66
Q_g	97.05	101.11	104.76	108.04	111.04	113.71	116.12	118.24	120.14	121.77	123.39
Q_p	75.40	85.50	94.10	99.11	99.97	98.17	94.68	90.21	85.88	81.55	77.71
CRI_{2012}	88.10	92.23	94.11	93.51	90.85	86.90	82.15	76.92	71.41	65.76	60.17
MCRI	87.51	89.80	91.33	92.23	92.73	92.84	92.58	91.93	90.86	89.21	86.94
FCI	112.34	121.57	130.22	138.40	146.14	153.44	160.39	166.92	173.15	179.02	184.70
R_f	83.46	89.07	92.43	90.94	86.78	81.86	76.68	71.48	66.33	61.25	56.29
R_g	98.67	101.57	104.02	106.08	107.81	109.21	110.34	111.18	111.79	112.16	112.40

8.8 Optimization of LED Light Engines on Color Quality with the Workflow

The following tendencies can be seen from Table 8.6:

- It is the aim of the optimizer to change the spectra for various chroma differences at the same CCT and constant white point ($\Delta u'v' < 0.0001$). It means that the spectra do have different chroma for the skin tones but the chromaticity distance from the Planckian locus is nearly zero.
- For the condition of nearly zero chroma difference ($\Delta C^*_{\text{skin-tone-set}} = 0$), the corresponding spectrum enables all color fidelity metrics (R_a, $R_{1,14}$, Q_f, R_f, CRI_{2012}, and R_{13} of CIE 1995 for skin tone) to be maximal. In this case, the general color-rendering index achieves the value of $R_a = 95.75$.
- If the average chroma difference values for the 15 skin tones increases from -2 to $+8$, the color gamut value for Q_g, R_g, GAI, and FCI also continuously increases.
- The color preference values for CQS Q_p rapidly increase if the chroma difference values $\Delta C^*_{\text{skin-tone-set}}$ scale up from -2 to $+1$ and reach the maximum at $\Delta C^*_{\text{skin-tone-set}} = +2$ before they scale down, while the values of $\Delta C^*_{\text{skin-tone-set}}$ increase further.
- A similar tendency can be observed for MCRI but the speed for scaling up and down is much lower in comparison to CQS Q_p. MCRI has a maximum at $\Delta C^*_{\text{skin-tone-set}} = +3$.

In order to analyze the behavior of the chroma enhancement of the 15 different skin tones, various levels of specific chroma differences $\Delta C^*_{\text{skin-tone-TCS},i}$ and two-dimensional hue circles in the color space $L^*a^*b^*$ are shown in Figure 8.36.

As can be seen from Figure 8.36, the values $\Delta C^*_{\text{skin-tone-TCS},i}$ are not similar for all skin tones. This means that a light source with a predefined average chroma difference does not have equal chroma difference levels for all skin tones under consideration. Also, based on the hue circles on the right side it can be observed that the skin tones' locations under the reference light source and the LED light source with maximal color fidelity have nearly the same area. In the case of the high chroma difference, the two clouds are situated at very different locations. The higher the chroma differences are, the smaller the overlapping of the two clouds is or the clearer the distinguishing distance is. Herewith, the points belonging to the LED light sources always have higher chroma values.

8.8 Optimization of LED Light Engines on Color Quality with the Workflow

In this section, two LED combinations are optimized using the method of Section 8.6 with two different test color sample collections (14 TCSs; CIE 1995, and 15 CQS TCSs; NIST 2010). Then the achieved results are compared with each other.

8.8.1 Optimization of the LED Light Engine on Color Quality Using the RGB-W-LED Configuration

The first LED combination shown in Figure 8.30 includes a blue LED-chip (the peak wavelength λ_p of 465 nm), a green LED-chip ($\lambda_p = 525$ nm), a red LED

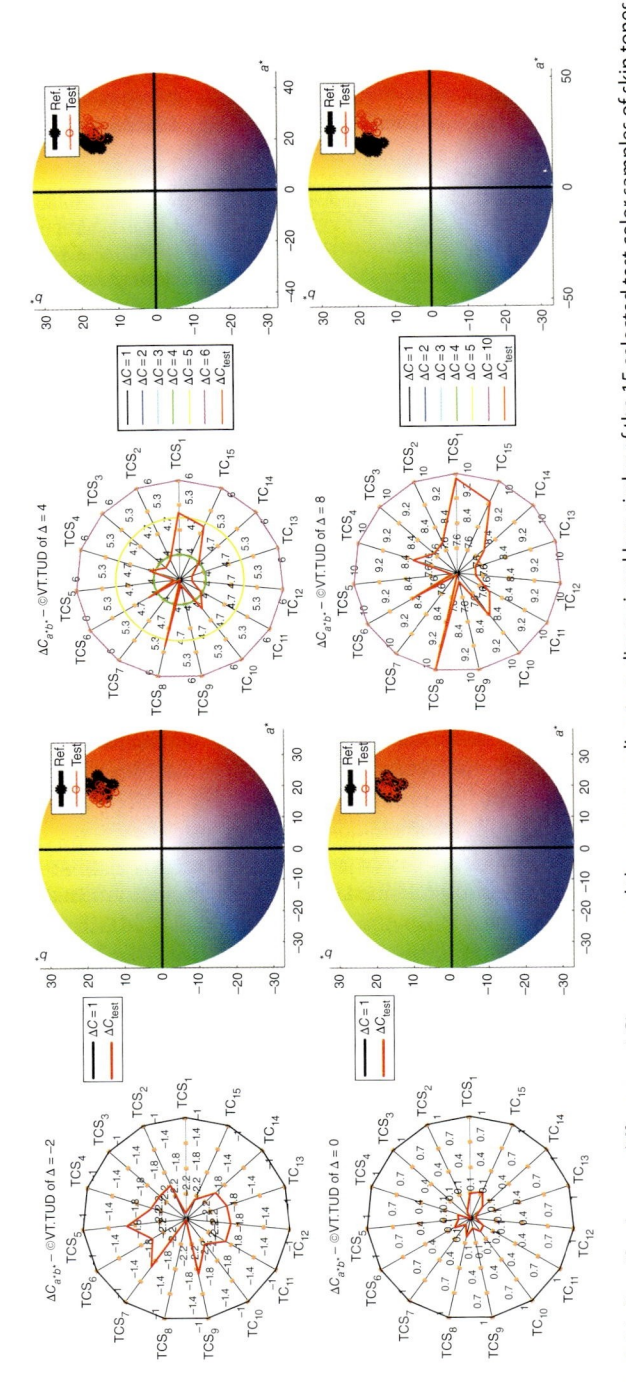

Figure 8.36 Specific chroma differences $\Delta C^*_{\text{skin-tone-TCS},i}$ and the corresponding two-dimensional hue circles of the 15 selected test color samples of skin tones under different LED test light sources in the four cases of average chroma differences ($\Delta C^*_{\text{skin-tone-set}} = -2, 0, 4, 8$). The black cloud is the chromaticity of the skin tones under reference light source 3100 K and the red cloud is the one under the LED test light sources.

($\lambda_p = 635$ nm), and a warm white LED with the peak wavelength of about 600 nm. The color temperatures investigated in this section are 3200 K (warm white) and 5700 K (cold white). In the first step, the data for the optimization in the case of 3200 K will be analyzed. The following properties were achieved:

- All spectra had a CCT of 3200 K with a very small Planckian distance $\Delta u'v'_{CCT} = 0.0001$.
- In the best case of the color fidelity (named 3200 K – $R_{1,14}$) with $R_f = 92$, $R_a = 97$, and $R_{1,14} = 95$, the average chroma difference $\Delta C^*_{CQS\,VS1-VS15}$ for all 15 CQS TCSs was only 0.85.
- In the best case of the color preference parameter Q_p (named 3200 K – Q_p – $R_a 86$), this spectrum had a value of $Q_p = 100.74$ at $\Delta C^*_{CQS\,VS1-VS15} = 2.33$. It was optimized with the boundary condition of $R_a > 86$.

Table 8.7 summarizes the properties of the optimum spectra at 3200 K.

In Table 8.7, three additional spectra are also shown, one spectrum with maximal MCRI (3200 K–MCRI–unconstrained) with the maximal value of MCRI of 93.08 at a very high value of $\Delta C^*_{CQS\,VS1-VS15}$ of 3.69 and $R_a = 75$. For this spectrum, the value for FCI reached a very high value of 151. FCI is often used as the color quality metric for the evaluation of vividness. Another spectrum named 3200 K – MCRI – R_a 86 had the boundary condition of $R_a > 86$. It could achieve the value of $\Delta C^*_{CQS\,VS1-VS15} = 2.33$. The third spectrum named 3200 K – MCRI – $R_{1,14}$ – 86 had to satisfy the boundary condition $R_{1,14} > 86$. It means that only one step into the direction of more color fidelity can cause a slightly smaller value of $\Delta C^*_{CQS\,VS1-VS15} = 2.06$.

The above-mentioned spectra are shown in Figure 8.37.

As can be seen from Figure 8.37 the greatest difference between the spectrum with the best color fidelity (3200 K – $R_{1,14}$) and the spectrum with the maximal MCRI (3200 K –MCRI – unconstrained) is the spectral power contribution in the range between 520 and 620 nm. Generally, this spectral power amount has to be drastically reduced to achieve high chroma difference. The three other spectra for the optimization of CQS Q_p and MCRI with the requirement of color fidelity $R_a > 86$ or $R_{1,14} > 86$ establish two stages between these two extreme spectra.

Because the value of $\Delta C^*_{CQS\,VS1-VS15}$ used for the optimization is an average value for all 15 test color samples $VS_1 - VS_{15}$ in the CQS system, there is an interest to know the specific chroma differences $\Delta C^*_{CQS\,VSi}$ for each one of the test color samples of this color collection. In order to analyze this, Figure 8.38 shows these chroma differences.

As can be seen from Figure 8.38, the spectrum exhibiting maximal MCRI (3200 K – MCRI – unconstrained) causes the maximal chroma differences in the range of $\Delta C^* = 9 - 10$ at the red TCS VS_1 and green TCS VS_7, which correspond to the SPD in Figure 8.37.

In Figure 8.39, the relative SPDs for the different optimization criteria in the case of CCT = 5700 K are illustrated.

As can be seen from Figure 8.39, the SPD in the blue and green ranges up to 540 nm is nearly identical for all analyzed spectra. From 540 nm on, the peak part of the red LED at 635 nm is conspicuous for those spectra that cause a higher

Table 8.7 Values of the color quality metrics color fidelity, color gamut, color preference, and MCRI for the five optimum LED spectra at 3200 K on the basis of the four-channel LED engine

Name	3200 K − $R_{1,14}$	3200 K − Q_p − R_a 86	3200 K − MCRI − unconstrained	3200 K − MCRI − R_a − 86	3200 K − MCRI − $R_{1,14}$ − 86
CCT	3200	3200	3200	3200	3200
$\Delta u'v'_{CCT}$	1.00E−04	1.00E−04	1.00E−04	1.00E−04	1.00E−04
R_9	95	53	10	53	62
R_a	97	**86**	75	**86**	88
$R_{1,14}$	**95**	84	71	84	**86**
GAI	101	110	118	110	108
Q_f	94	89	80	89	90
Q_g	104	109	113	109	108
Q_p	93.22	**100.74**	99.21	100.74	100.15
CRI_{2012}	94	93	88	93	93
MCRI	91.34	92.70	**93.08**	**92.70**	**92.53**
FCI	128	141	151	141	138
R_f	92	89	83	89	90
R_g	103	106	109	106	106
ΔC^*_{CQS} VS1–VS15	**0.85**	2.33	3.69	2.33	2.06

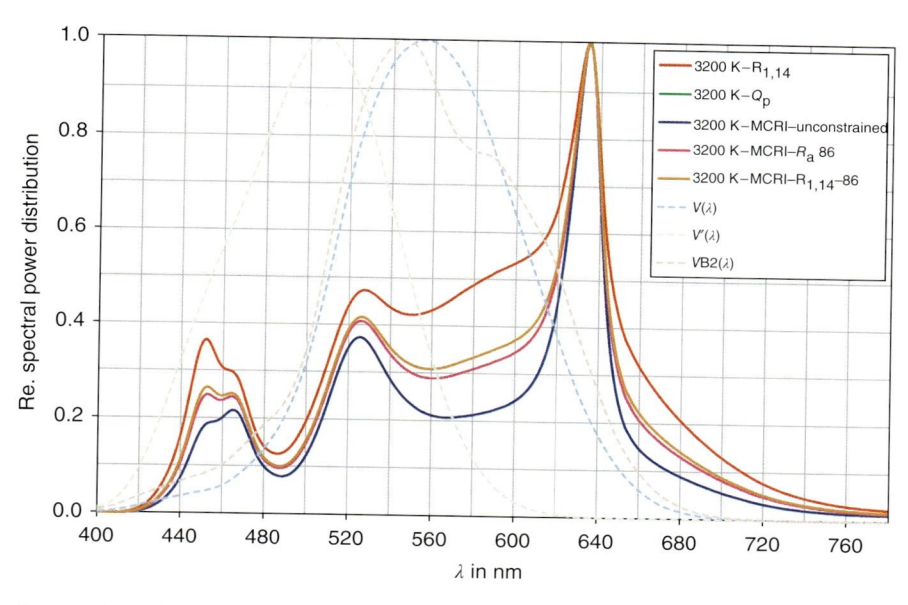

Figure 8.37 Relative spectral power distributions of the RGB-W-LED combination delivering the maximal values of color fidelity, color preference (MCRI and Q_p), and constrained MCRIs in the case of 3200 K.

amount of both chroma differences and CQS Q_p. However, with the exception of the spectrum for the best color fidelity (5700 K – $R_{1,14}$), all other spectra have similar relative SPDs. Therefore, they cause nearly equal colorimetric values, which are listed in Table 8.8.

As can be seen from Table 8.8, the maximally achievable value of the average chroma difference $\Delta C^*_{\text{CQS VS1–VS15}}$ is about 1.319 in this case. A more detailed analysis is shown in Figure 8.40.

As can be seen from Figure 8.40, the highest chroma differences occur in case of the red TCS VS_1, green TCS VS_7, green–blue TCS VS_{10}, and dark blue–green TCS VS_{11}. The violet TCSs VS_{13} and VS_{14} exhibit a very small chroma difference due to the selection of the blue chip for the LED combination with the peak wavelength at 465 nm. In order to have a higher chroma difference in the hue range blue and violet, a blue chip at 440 nm is surely the best choice.

8.8.2 Optimization of the LED Light Engine on Color Quality with the R_1-R_2-G-B_1-B_2-W-LED-configuration

As an example for a reasonable LED combination with six channels, the relative SPDs of the selected color LEDs and white LEDs were shown in Figure 8.31 including two blue LED chips (λ_p = 450 and 470 nm), two red LED chips (λ_p = 640 and 670 nm), one green LED chip (λ_p = 525 nm) and one warm white LED with the peak wavelength of 605 nm. For the case of the color temperature of 3200 K, the five optimal LED spectra are illustrated in Figure 8.41.

As can be seen from Figure 8.41, for every one of the five optimal spectra, the contribution of the blue chip at 450 nm is not really visible. From the spectrum

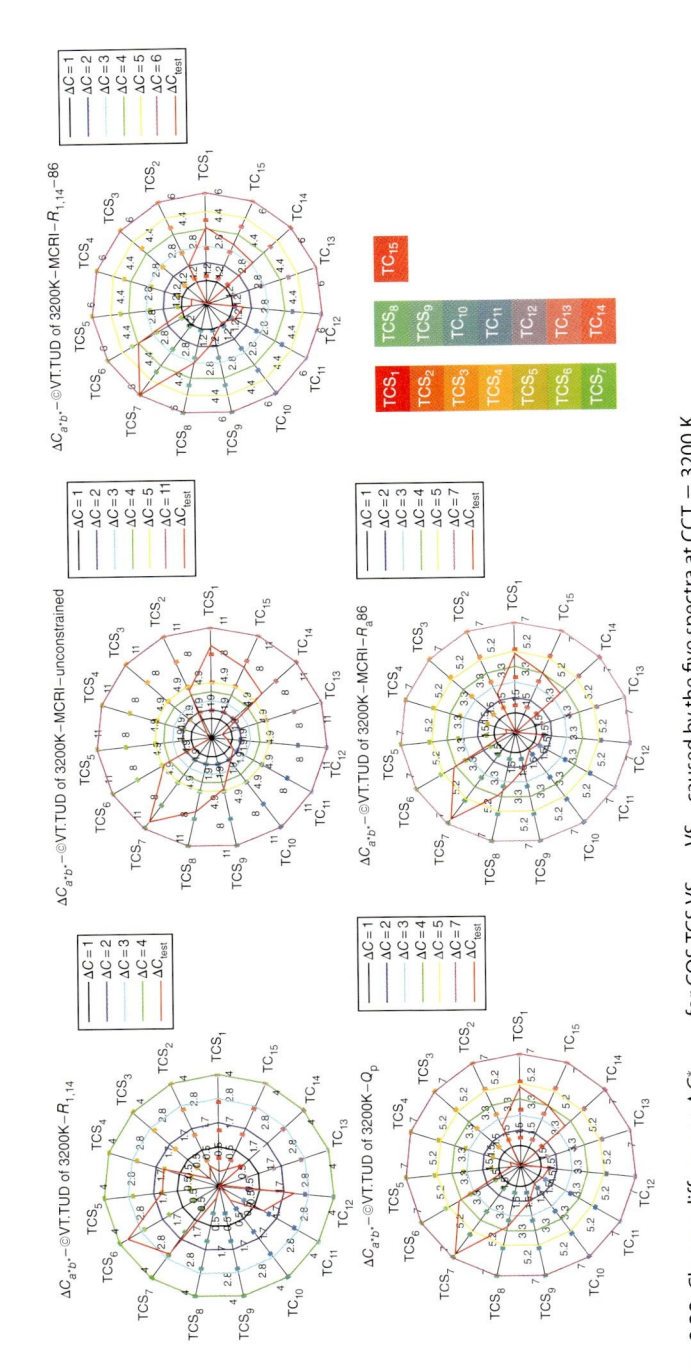

Figure 8.38 Chroma differences $\Delta C^*_{CQS\ VSi}$ for CQS TCS $VS_1 - VS_{15}$ caused by the five spectra at CCT = 3200 K.

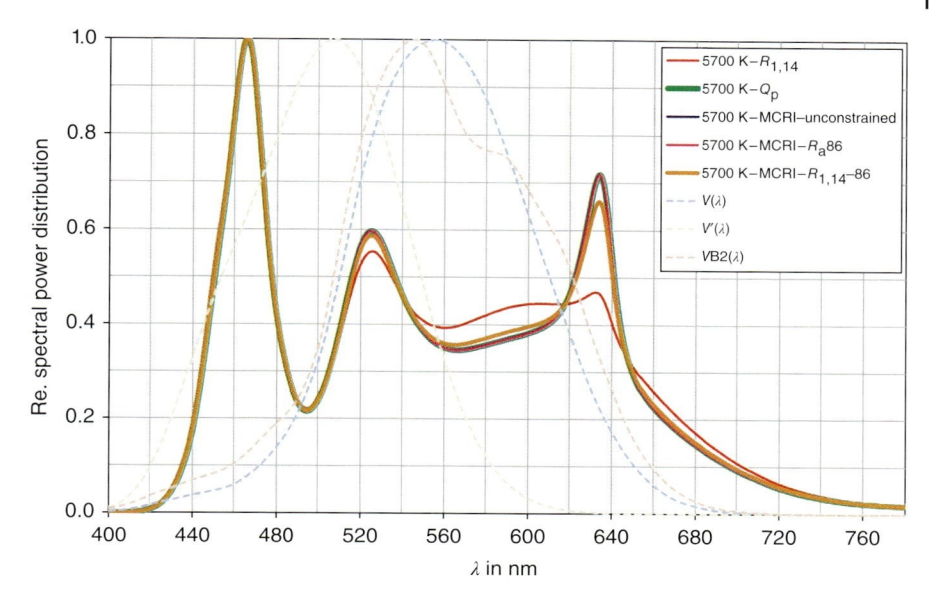

Figure 8.39 Relative spectral power distributions of the LED combination delivering the maximal values of color fidelity, color preference (MCRI and Q_p), and constrained MCRIs in the case of 5700 K.

with best color fidelity ($3K2 - R_{1,14}$) to the other spectra with higher values of both Q_p and $\Delta C^*_{CQS\,VS1-VS15}$, the spectral power from 420 until 660 nm is strongly reduced.

If the colorimetric data of the four-LED channels in the case of 3200 K listed in Table 8.7 are compared with the values of Table 8.9 for the six-LED channels solution with the selected LED wavelengths, the following can be recognized:

- Although the color fidelity values (R_f, Q_f, R_a, CRI_{2012}, or $R_{1,14}$) of the two LED-combinations for the case of the best color fidelity are nearly identical, the color gamut and color preference values Q_g, GAI, FCI, R_g, and Q_p in the case of the four-channel LED engine are always higher than those of the six-channel LED unit.
- In the case of the four-channel LED unit and for the spectrum optimized for $Q_p(3200\text{ K} - Q_p - R_a 86)$, the Q_p value equals 100.74 and in the case of the six-channel LED unit this value equals only 93.385 ($3K2 - Q_p$) although the values of color fidelity do have much smaller differences. The average chroma difference of 15 CQS TCS $VS_1 - VS_{15}$ for the four-channel LED-unit ($\Delta C^*_{CQS\,VS1-VS15} = 2.33$) is higher than the value of $\Delta C^*_{CQS\,VS1-VS15}$ of 1.41 in the case of the six-channel LED engine.

The reason why the developers of intelligent LED engines can have a higher chroma difference and higher color preference value Q_p at nearly similar high values of color fidelity with the four-channel LED engine is of great interest and it has to be kept in mind. Based on the above descriptions, it can be recognized that a chroma enhancement can be realized if the part of

Table 8.8 Values of the color quality metrics for color fidelity, color gamut, color preference, and MCRI for the five LED spectra at 5700 K on the basis of the four-channel LED engine.

Name	5700 K – $R_{1,14}$	5700 K – Q_p – R_a86	5700 K – MCRI – unconstrained	5700 K – MCRI – R_a86	5700 K – MCRI – $R_{1,14}$ – 86
CCT	5700	5700	5700	5702	5700
$\Delta u'v'$	9.99E–05	0.0001	0.0001	9.92E–05	9.47E–05
R_{12}	68	70	70	70	70
R_a	93	88	88	88	90
$R_{1,14}$	92	84	84	84	86
GAI	99	108	108	108	106
Q_f	92	90	90	90	91
Q_g	99	104	104	104	103
Q_p	85	95	95	95	93
CRI_{2012}	89	87	87	87	88
MCRI	92.1	93.4	93.4	93.4	93.2
FCI	109	121	121	121	118
R_f	86	84	84	84	85
R_g	96	100	100	100	99
ΔC^*_{CQS} VS1–VS15	–0.513	1.319	1.319	1.319	0.978

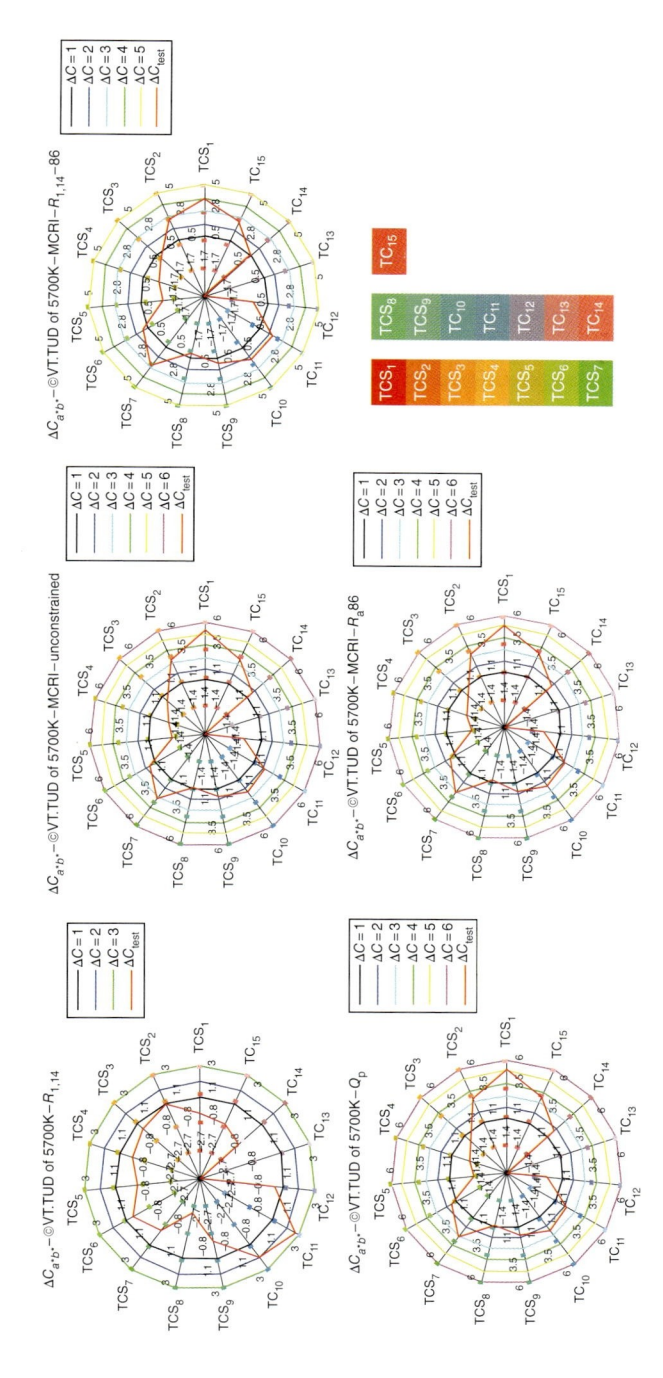

Figure 8.40 Chroma differences $\Delta C^*_{CQS\,VSi}$ for CQS TCS $VS_1 - VS_{15}$ caused by the five spectra at CCT = 5700 K.

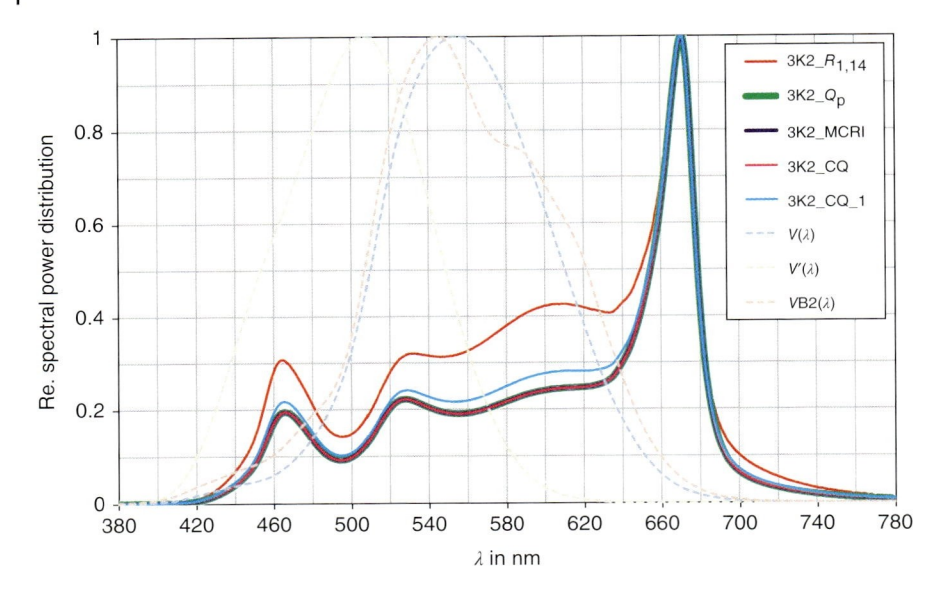

Figure 8.41 Relative spectral power distributions of the R1-R2-G-B1-B2-W LED combination delivering the maximal values of color fidelity, color preference (MCRI and Q_p), and constrained MCRI in the case of 3200 K.

optical power between 520 and 600 nm is relatively small. The ratio of optical power at 560 nm to the optical power at 610 nm is about 1 : 1.3 in the case of the six-channel LED unit in Figure 8.41. This ratio increases up to 1 : 1.7 in the case of the four-channel LED engine in Figure 8.37. The relative SPD has a higher slope in the four-channel configuration; or, in other words, the optical power in the range of 610–630 nm shall be increased more strongly compared to the one in the range of 550–600 nm. This is an explanation from the mathematical point of view. To explain this phenomenon from the human eye physiological point of view, it should be considered that the $V(\lambda)$ function or the (L + M) channel is responsible for achromatic vision and the opponent channels |L − M| and |L + M − S| are responsible for chromatic perception. Consequently, the chroma difference enhancement in the hue areas of red and green (see Figure 8.38) can only be explained by the |L − M| channel. A chroma difference enhancement shall be related with the increase of the ratio of the chromatic channel to the achromatic channel. It means that an SPD shall be optimized to increase the |L − M| signal and reduce the $V(\lambda)$-weighted signal at the same time. In Figure 8.25, the course of the spectral sensitivity curves implies that the spectral sensitivity of the |L − M| channel has a high slope from 560 until 615–620 nm. If the spectral optical power can be relatively increased in the range around 620 nm then chroma enhancement will become realizable.

Finally, the values of the color quality metrics for color fidelity, color gamut, color preference, and MCRI for the five LED-spectra at 3200 K on the basis of the six-channel LED engine are listed in Table 8.9.

Table 8.9 Values of the color quality metrics for color fidelity, color gamut, color preference, and MCRI for the five LED spectra at 3200 K on the basis of the six-channel LED engine.

Name	3K2 – $R_{1,14}$	3K2 – Q_p	3K2 – MCRI	3K2 – CQ_{max}	3K2 – Q_1
CCT	3200	3200	3200	3200	3200
$\Delta u'v'$	1.0E−04	1.0E−04	1.0E−04	1.0E−04	1.0E−04
R_9	99	42	42	42	58
R_{12}	79	82	82	82	81
R_a	97	89	89	89	93
$R_{1,14}$	**96**	86	86	86	90
GAI	99	107	107	107	105
Q_f	94	91	91	91	93
Q_g	99	104	104	104	103
Q_p	85.125	**93.385**	93.385	93.385	91.735
CRI_{2012}	94	89	89	89	91
MCRI	91.1803	92.6333	**92.6333**	92.6333	92.2979
FCI	124	136	136	136	133
R_f	90	86	86	86	87
R_g	97	101	101	101	100
$\Delta C^*_{CQS\ VS1-VS15}$	−0.0076	1.4176	1.4176	1.4176	1.0000

8.9 Conclusions: Lessons Learnt for Lighting Practice

The optimization of the LED light engine for variable color temperature with high color fidelity and/or high color preference property (in terms of Q_p or MCRI) requires knowledge on the temperature and current behavior of the LEDs and on the control of the spectral models for LEDs in order to stabilize and regulate the mixed optical radiation of the LED light engine. For the optimization, the selection of suitable peak wavelengths for colored (pure) semiconductor LEDs and appropriate phosphor-converted white LEDs is essential. If the color fidelity is not a unique optimization criterion then another metric, that is, a color preference metric, shall be used. For example, CQS Q_p or MCRI can be chosen from the current state of the art of color research at the time of writing of this book. The optimization of MCRI and/or CQS Q_p is always related to the chroma enhancement of the colored objects. This means that the value of ΔC^* (the difference between the chroma of the colored object under the LED light source and its reference light source, a Planckian radiator, or a phase of daylight) is positive; the LED light source should enhance the chroma of the object. With the boundary condition of keeping the color fidelity property of the LED light source at a moderate or high level of $R_a > 80$ or $R_a > 86$, the maximum of Q_p can be achieved with a chroma difference $\Delta C^*_{CQS\ VS1-VS15}$ of about 2.2–2.4 for the mean value of the 15 CQS test color samples $VS_1 - VS_{15}$. First visual experiments (see Section 7.7) resulted in a "good"–"very good" visual evaluation regarding the perceived color quality attribute *"color preference"* if the average $\Delta C^*_{\text{CQS VS1-VS15}}$ equaled about 3.0–3.5. This chroma enhancement criterion requires a reduction of the CIE CRI R_a values down to the range of about 75.

References

1. Khanh, T.Q., Bodrogi, P., Vinh, Q.T., and Winkler, H. (2015) *LED Lighting-Technology and Perception*, 1st, Englisch edn, Wiley-VCH Verlag GmbH & Co. KGaA. ISBN: 10: 3527412123, ISBN 13: 978-3527412129.
2. DIN EN 12464-1 (2011-08) Licht und Beleuchtung – Beleuchtung von Arbeitsstätten – Teil 1: Arbeitsstätten in Innen-räumen, Beuth Verlag GmbH, Berlin.
3. Energy Star (2012) Energy Star Program Requirements for Luminaires, Partner Commitments, Eligibility Criteria Version 1.2.
4. McGaraghan, M. (2013) Energy Solutions, on behalf of PG&E, November 14, 2013, California, Quality LED Lamp Specification.
5. Khanh, T.Q. (2004) Ist die LED – Photometrie mit V(λ) – Funktion wahnehmungsgerecht. Licht 06/2004, www.lichtnet.de (accessed 01 September 2016).
6. Ohno, Y. (1999) LED Phometric Standards, Lecture 1999.
7. Reifegerste, F. and Lienig, J. (2014) Private Communication.
8. Ohno, Y. (2005) Spectral design considerations for white LED color rendering. *Opt. Eng.*, **44** (11), 111302.
9. Chien, M.C. and Tien, C.H. (2012) Multispectral mixing scheme for LED clusters with extended operational temperature window. *Opt. Express*, **20** (S2), 245–254.
10. Smet, K.A.G. and Hanselaer, P. (2015) Memory and preferred colours and the color rendition of white light sources. *Light. Res. Technol.*, **48** (4), 393–411.
11. Khanh, T.Q., Bodrogi, P., Vinh, Q.T., and Stojanovic, D. (2016) Colour preference, naturalness, vividness and color quality metrics – Part 1: experiments in a real room. *Light. Res. Technol.* (online in April 2016).
12. Khanh, T.Q., Bodrogi, P., Vinh, Q.T., and Stojanovic, D. (2016) Colour preference, naturalness, vividness and colour quality metrics-part 2: experiments in a viewing booth and analysis of the combined dataset. *Light. Res. Technol.* (online in April 2016).
13. Khanh, T.Q., Vinh, Q.T., and Bodrogi, P. (2016) A numerical analysis of recent color rendition metrics. *Light. Res. Technol.* doi: 10.1177/1477153516632568
14. Fairchild, M.D. and Pirrotta, E. (1991) Predicting the lightness of chromatic object colours using CIELAB. *Color Res. Appl.*, **16** (6), 385–393.

9

Human Centric Lighting and Color Quality

In this chapter, the concept of Human Centric Lighting (HCL) is defined with its visual and nonvisual components (Section 9.1). One of the nonvisual components, the so-called circadian stimulus (CS), will be described in detail (Section 9.2). An example of a possible technological implementation of spectral HCL design will be shown that combines color quality and circadian optimization (Section 9.3). Considering the issue of recent demographic developments, one selected aspect of spectral HCL design, the effect of the change of spectral transmission of the eye media with age, will be dealt with (Section 9.4). Finally, the most important lessons learnt from this chapter will be summarized (Section 9.5).

9.1 Principles of Color Quality Optimization for Human Centric Lighting

Since the beginning of the evolutionary development of mankind, lighting and light sources have always accompanied all activities of human beings. Until the end of the nineteenth century, daylight was most generally used during the daytime exhibiting dynamically changing spectral power distributions, color temperatures, and illuminance levels depending on geographical position, weather, the time of the day, and the time of the year. In those days, society was reliant on the use of oil lamps, gas lamps, firewood, and candles in the evening. Visual perceptions of light sources and illuminated colored objects were memorized by individuals in the context of these light sources and aesthetic requirements about light source color and object colors arose, which were incorporated in culture [1].

Since about 1880, industrial electrification and the broad dissemination of incandescent lamps have resulted in a reorganization of working schedules and a more and more dramatic change of lifestyles. This tendency has been accelerated in the twentieth century by the emergence of new light source technologies: mercury vapor lamps, sodium discharge lamps, fluorescent lamps, and metal halide lamps. But the possibility of variation in the color properties of these light sources, that is, their spectral power distribution, color temperature, or chromaticity was limited (by changing and optimizing their metallic and phosphor components) and classic lighting engineering has primarily concentrated on (achromatic) visual performance parameters such as luminance, illuminance, glare, and illuminance distributions often ignoring the perceived quality of the color appearance of the illuminated colored objects in the lit environment [1].

Since about 2002, a new tendency of lighting design and light source design has appeared: HCL that exploits the flexibly variable spectral power distributions of LED-based light engines. The concept of HCL can be defined as follows. HCL concentrates on the human user of the lighting system including all relevant subjective properties, that is, age, gender, health, culture, region of living, region of origin, profession, individual needs and social conditions; visual comfort, emotional well-being, work performance, productivity, and motivation; individual visual behavior, and user acceptance of lighting installations. HCL combines the above subjective properties with the objective conditions of lighting, that is, the type of lighting application, for example, shop, retail, office, home, industry (shift work), hospital, retirement home, hotel, wellness, vehicle, or school lighting; and also the time of the day, season, month, weather, and the availability of daylight as well as its combination with artificial lighting. The different aspects of the HCL concept are depicted in Figure 9.1.

For a successful HCL design, it is not only the illuminance level, spatial distribution (direct and indirect light components), color temperature, or white tone chromaticity that should be varied and optimized. More importantly, the relative spectral power distribution of the light source should be flexibly variable by using recent LED light engines with multiple, separately driven LED color channels (multi-LED light engines) [1]. Color quality aspects of the light source (color fidelity; color preference, naturalness, vividness; color discrimination) represent an important optimization target for HCL: color quality aspects should be co-optimized with other visual and nonvisual HCL aspects (e.g., the circadian aspect, see Sections 9.2 and 9.3). To implement the HCL concept in lighting engineering, it is worth keeping in mind that there exists only one single human brain that reacts to electromagnetic radiation as a result of complex interactions of its visual and nonvisual pathways [2] including the subsystem of color quality

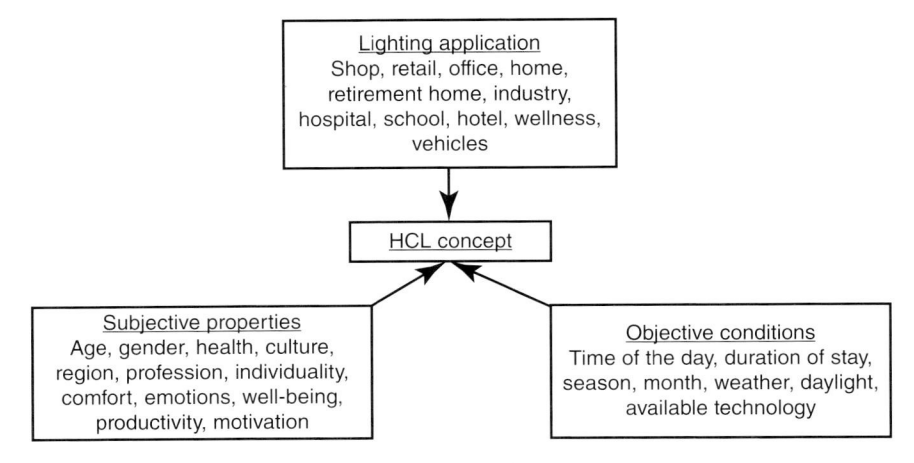

Figure 9.1 Aspects of the HCL concept.

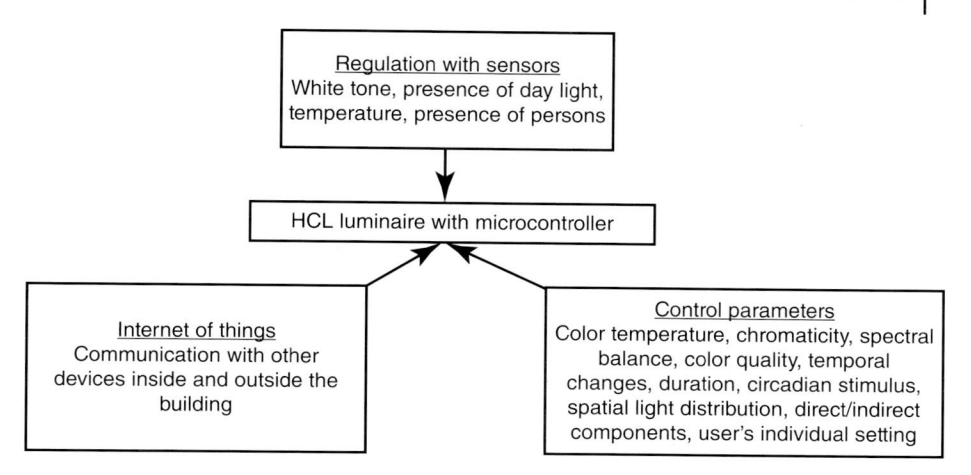

```
                    ┌─────────────────────────────────┐
                    │      Regulation with sensors    │
                    │ White tone, presence of day light,│
                    │ temperature, presence of persons │
                    └─────────────────┬───────────────┘
                                      ▼
                    ┌─────────────────────────────────┐
                    │ HCL luminaire with microcontroller│
                    └──▲───────────────────────────▲──┘
                       │                           │
┌──────────────────────┴──┐         ┌──────────────┴──────────────────────┐
│    Internet of things   │         │          Control parameters         │
│   Communication with other│        │ Color temperature, chromaticity, spectral│
│ devices inside and outside the│    │   balance, color quality, temporal  │
│         building        │         │  changes, duration, circadian stimulus,│
└─────────────────────────┘         │ spatial light distribution, direct/indirect│
                                    │ components, user's individual setting│
                                    └─────────────────────────────────────┘
```

Figure 9.2 Design concept for a modern, intelligent HCL multi-LED luminaire [3].

assessments. According to the above discussions, Figure 9.2 shows the design concept for a modern, intelligent HCL multi-LED luminaire [3].

As can be seen from Figures 9.1 and 9.2 for an effective HCL design, different aspects of lighting quality (including, e.g., the avoidance of glare) or color quality (e.g., color fidelity, color preference, or color discrimination) should be considered [4]. As described in Chapters 5 and 7, when the user of a general indoor lighting system assesses the visual color quality of the lit scene in a general lighting situation, usually a color preference judgment is being made instead of the assessment of color fidelity, that is, the comparison with an inferred reference situation. Therefore, to optimize the spectral power distribution of, for example, a multi-LED light engine, a visually validated color preference metric is needed as a spectral optimization target with a set of boundary conditions such as the type and quality of the illuminant's white tone (Chapter 3) or the level of the CS [5] in the late afternoon hours, which should be higher for office work and lower for a more relaxing atmosphere (see Section 9.3).

In the above sense, the optimization of the spectral power distribution of the light source as a part of the HCL optimization framework should consider the constraints of including a high-quality white point, excellent color quality, and an appropriate correlated color temperature (CCT) range associated with the intended amount of CS [5] of the spectrum (a high circadian component corresponds to a high CCT; this issue will be discussed in Section 9.2). It should be noted, however, that the use of the concept of LER (luminous efficacy of radiation) should be avoided in HCL design as the concept of LER considers only the luminance mechanism represented by the $V(\lambda)$ function, which is just one of the numerous other retinal mechanisms that are relevant for HCL (see Chapter 2).

Generally, the use of $V(\lambda)$ weighted quantities (e.g., LER, luminous efficiency, luminance, illuminance) results in a loss of important light rays in the red and blue spectral ranges, which are important for HCL, both for high color quality and for the accentuation of emotional effects (for which the red-orange spectral range is of vital importance) and also to be able to set the level of the CS

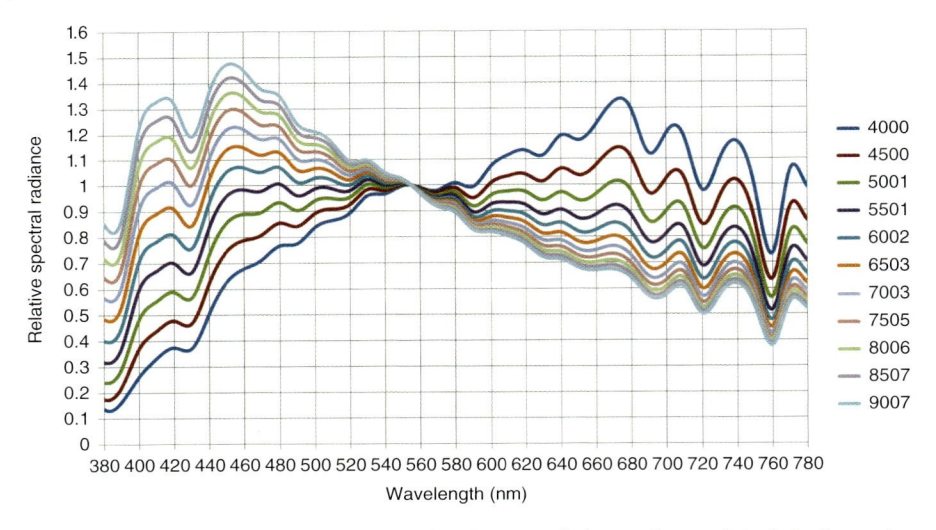

Figure 9.3 Typical relative spectral power distributions of phases of natural daylight. Legend: correlated color temperature in kelvin.

(by the aid of the bluish rays of the HCL luminaire) [1]. A spectrally balanced light source such as a phase of typical natural daylight (see Figure 9.3) comprises all important wavelengths in the visible range stimulating all types of receptoral and post-receptoral mechanisms on the retina; compare the spectral radiance distributions of Figure 9.3 with the spectral sensitivities of the mechanisms depicted in Figure 9.6. Therefore, daylight evokes all types of visual and nonvisual effects and its typical spectra suggest inspiring examples for the spectral design of future multi-LED engines. It should also be noted that natural daylight also delivers spectral components in the ultraviolet and infrared range. A restricted UV component is advantageous to generate vitamin D, and restricted amounts of infrared components foster blood flow in the human body.

9.2 The Circadian Stimulus in the Rea *et al.* Model

The development of a modern, intelligent HCL luminaire (Figure 9.2) requires knowledge of all mechanisms of light and color perception and also the mechanisms of nonvisual brain signal processing including their effect on biorhythms. As was pointed out in Section 9.1, visual and nonvisual effects should be considered simultaneously. In Section 9.3, an actual example will be shown for the spectral co-optimization of a multi-LED engine for a selected nonvisual aspect (the control of the CS) and a selected visual aspect (the chroma enhancement of red objects for better color preference). To prepare for Section 9.3, this section is devoted to the description of the circadian effect and its modeling by the quantity CS in the Rea *et al.* model.

Turning to the explanation of the circadian effect, electromagnetic radiation affects not only the human visual system but also the so-called *circadian clock*,

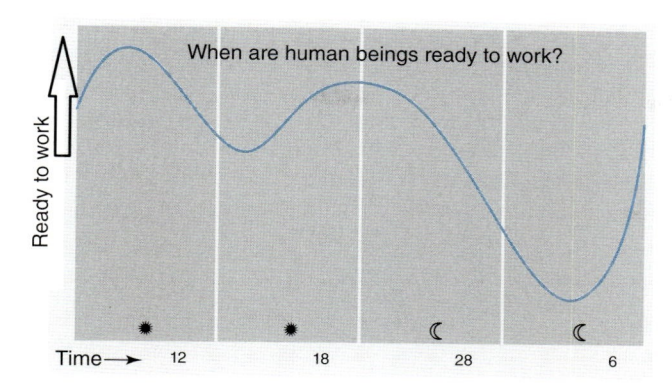

Figure 9.4 Human activity (readiness to work) in the 24 h cycle [8]. (Reproduced with permission from Licht.de.)

which is responsible for the timing of all biological functions according to daily rhythm [6]. This effect depends on the intensity, the spatial and spectral power distribution, and the timing (e.g., morning, evening, or night) and duration of visible electromagnetic radiation reaching the human eye [7]. Life has a cyclic rhythm: chronobiologists distinguish between ultradian (a couple of hours like hunger or short phases of sleep by babies), circadian (24 h, lat. circa = about, dies = day), and infradian (> 24 h like seasonal changes) rhythms. For all aspects of human life, the most important rhythm is the circadian one. Figure 9.4 shows this 24 h cycle of human activity in terms of readiness to work.

As can be seen from Figure 9.4, human activity exhibits a 24 h cycle that is controlled by light. In other words, light is an important *zeitgeber* (literally from German: "time giver"). But how does this timing mechanism work in the human neural system? In 2002, the discovery of a new photoreceptor in the eye, the so-called intrinsically photoreceptive retinal ganglion cell (ipRGC) was published [9]. It turned out that 2% of all ganglion cells contain the photosensitive protein melanopsin, which has a spectral sensitivity that is different from the sensitivity of the rods and the cones (see Chapter 2). The ipRGCs are distributed nonuniformly across the retina: in the lower retinal area (where light sources in the upper part of the viewing field are imaged) ipRGC density is much larger than in the upper area.

Light sources emitting from homogeneous large areas in the upper half of the viewing field cause an increased circadian effect (i.e., timing effect of optical radiation) compared to punctual light sources and those in the lower half of the viewing field [7]. Optical radiation reaches the ipRGCs that send neural signals toward the circadian clock located at the suprachiasmatic nucleus (SCN) which, in turn, regulates hormone (melatonin) release. Melatonin is a sleep hormone that fosters sleep-in: it is an important component of the circadian clock. Exposure to light suppresses the production of melatonin and hence alertness increases [7]. Figure 9.5 shows the cyclic change of melatonin level together with cortisol level, which is a stress hormone.

As can be seen from Figure 9.5 melatonin peaks at about 3 o'clock in the night (in the sleeping period) while cortisol (and alertness) is maximal at about 9 o'clock

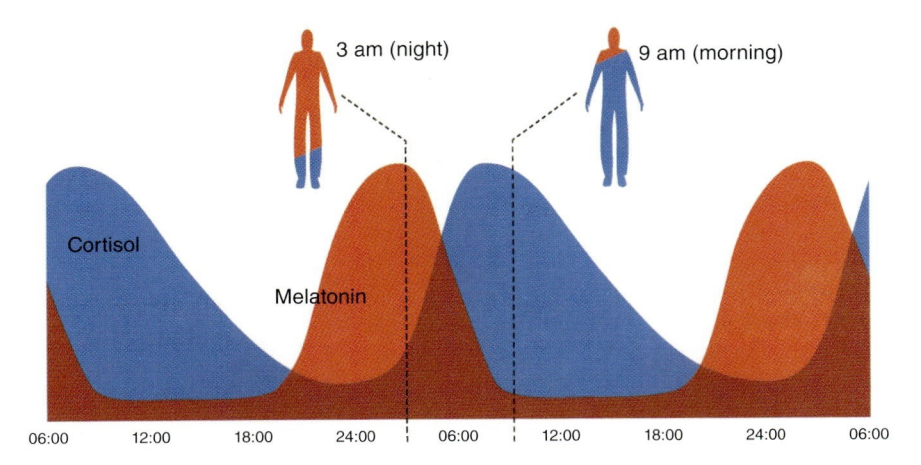

Figure 9.5 Cyclic change of melatonin (sleep hormone) level and cortisol (stress hormone) level [8]. (Reproduced with permission from Licht.de.)

in the morning. In the daytime, melatonin is suppressed and this can be hardly influenced by any artificial light source. But during the dark hours, it is possible to suppress melatonin in a HCL design concept (e.g., for shift work) and increase evening-time or night-time work performance by using an appropriate spectral power distribution. But possible long-term adverse health effects of such an optimization are currently unknown. For example, if an office worker would like to continue working during the late hours, will she or he be able to fall asleep afterwards and work the next day at all? Alternatively, the aim of another HCL spectral optimization concept can be to create a relaxing atmosphere after work in order not to suppress melatonin artificially. To do so, it is essential to know the effect of light source spectral power distribution on nocturnal melatonin suppression.

Experimental data on the extent of nocturnal melatonin suppression as a function of the peak wavelength of narrowband and broadband stimuli (abscissa) [6, 10–12] showed that the maximum of nocturnal melatonin suppression occurs at 460 nm and not at 480 nm, which is the peak spectral sensitivity of melanopsin, the ipRGC photopigment. This indicates a possible interaction of ipRGC signal with S cone signals at the postreceptoral signal processing stage. Moreover, the local maximum for narrowband spectra at 510 nm disappeared for more broadband spectra, indicating that there was a spectrally opponent retinal mechanism working with cone signal differences and contributing to nonvisual information processing.

Rea *et al.* concluded [6] that multiple retinal mechanisms contribute to the light-induced control of nocturnal melatonin suppression. Hence the use of a single spectral weighting function (corresponding to a hypothetical single retinal mechanism) to model the spectral dependence of the circadian effect of a light source is not appropriate and a more complex modeling (the so-called Rea *et al.* model [6]) is needed. The Rea *et al.* model [6] uses the signals of more visual retinal mechanisms in addition to the (nonvisual) ipRGC signal to compute an output quantity called circadian stimulus (CS). Experimental melatonin

9.2 The Circadian Stimulus in the Rea et al. Model

suppression data (responsible for the extent of alertness) were found to correlate well with the logarithm of CS [6], which can be computed if the absolute spectral irradiance distribution from a light source at the human eye is known. The spectral sensitivity of the retinal mechanisms contributing to the computation of CS is depicted in Figure 9.6.

As can be seen from Figure 9.6 the signals of the S cones, the rods, the ipRGCs, as well as the L and M cones (see Chapter 2) represented by $V_{10}(\lambda)$ contribute to the CS. $V_{10}(\lambda)$ is the large-field spectral luminous efficiency function [13] shown in comparison with $V(\lambda)$ in Figure 9.6 while $V(\lambda)$ itself is not used in the Rea *et al.* model. Figure 9.7 illustrates the workflow of the Rea *et al.* model [6].

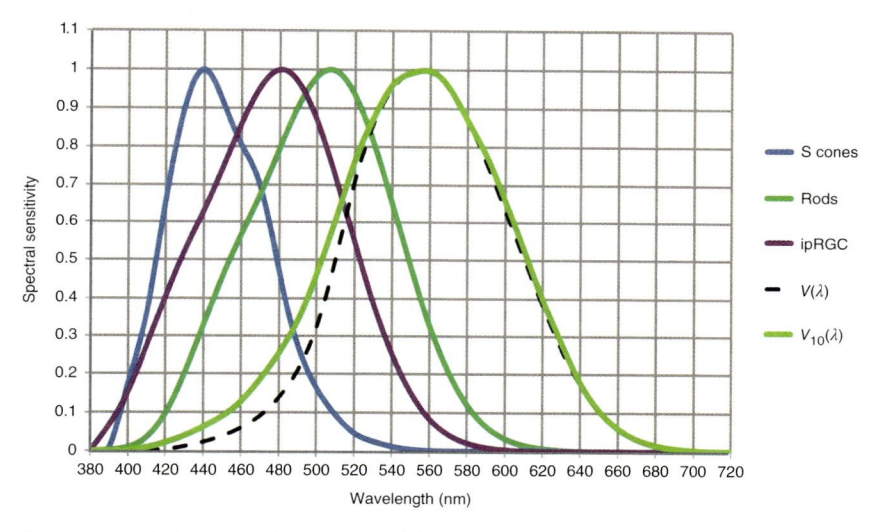

Figure 9.6 Relative spectral sensitivity of the retinal mechanisms contributing to the circadian stimulus (CS), the output of the Rea *et al.* model [6]. The additive combination of the L and M cone mechanisms is represented by $V_{10}(\lambda)$, the large-field spectral luminous efficiency function shown in comparison with $V(\lambda)$. $V(\lambda)$ itself is not used in the Rea *et al.* model. (Khanh *et al.* 2015 [7]. Reproduced with permission of Wiley-VCH.)

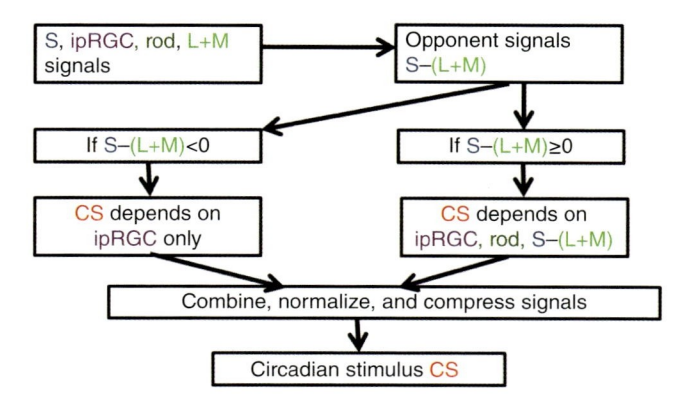

Figure 9.7 Workflow of the Rea *et al.* model [6]. (Khanh *et al.* 2015 [7]. Reproduced with permission of Wiley.)

As can be seen from Figure 9.7, the Rea et al. model [6] consists of the following steps:

Step 1: S-cone, ipRGC, rod, and (L + M) signals are computed by multiplying the relative spectral radiance $L_{e,\text{rel}}(\lambda)$ of the stimulus (e.g., a large wall surface in the office) by the spectral sensitivity of the corresponding mechanism (see Figure 9.6, $V'(\lambda)$ stands for rods) and integrating between 380 and 780 nm (using $V_{10}(\lambda)$ to represent L + M; see Eqs. (9.1)–(9.5). Equation (9.5) represents the connection to the absolute value of the vertical illuminance at the eye (in lx).

$$S = k \int_{380\text{nm}}^{780\text{nm}} L_{e,\text{rel}}(\lambda) S(\lambda) d\lambda \tag{9.1}$$

$$\text{ipRGC} = k \int_{380\text{nm}}^{780\text{nm}} L_{e,\text{rel}}(\lambda) \, \text{ipRGC}(\lambda) d\lambda \tag{9.2}$$

$$V' = k \int_{380\text{nm}}^{780\text{nm}} L_{e,\text{rel}}(\lambda) V'(\lambda) d\lambda \tag{9.3}$$

$$(L + M) = 0.31 \, k \int_{380\text{nm}}^{780\text{nm}} L_{e,\text{rel}}(\lambda) V_{10}(\lambda) d\lambda \tag{9.4}$$

with

$$k = \frac{E_v}{\int_{380\text{nm}}^{780\text{nm}} L_{e,\text{rel}}(\lambda) V(\lambda) d\lambda} \tag{9.5}$$

Step 2: The opponent signal S − (L + M) is computed from the S and (L + M) signals by the aid of Eqs. (9.1), (9.4), and (9.5).

Step 3: If S − (L + M) is negative then the CS depends only on the ipRGC signal and the intermediate quantity circadian light CL is computed according to Eq. (9.6).

$$\text{CL} = 0.285 \, \text{ipRGC} - 0.01 \, \text{lx} \tag{9.6}$$

Step 4: If S − (L + M) is non-negative then CS depends on the ipRGC, the S − (L + M), and the rod signals, and CL is computed by using Eqs. (9.7) and (9.8). For high irradiance levels, the model predicts rod saturation, that is, no rod contribution.

$$R_{\text{saturated}} = 1 - e^{-(V'/6.5 \, \text{lx})} \cdot (\text{lx}) \tag{9.7}$$

$$\text{CL} = [(0.285 \, \text{ipRGC} - 0.01) + (0.2 \, \{S - (L + M)\} - 0.001)] \\ - 0.72 \, R_{\text{saturated}} \tag{9.8}$$

Step 5: The resulting CL value is normalized to CIE illuminant A to obtain the quantity CL_A according to Eq. (9.9). The factor 5831 in Eq. (9.9) corresponds to the criterion that a color stimulus with the chromaticity of the

CIE standard illuminant A and with an illuminance of 1000 lx at the eye should have $CL_A = 1000$ lx.

$$CL_A = 5831\,CL \tag{9.9}$$

Step 6: Signal compression is carried out to fit experimental nocturnal human melatonin suppression data results in the CS value; see Eq. (9.10)

$$CS = 0.75 - \frac{0.75}{1 + \left(\frac{CL_A}{215.75\,lx}\right)^{0.864}} \tag{9.10}$$

By the use of Eqs. (9.1)–(9.10), the CS was computed for a set of different light sources and illuminants (commercially available white LEDs, fluorescent lamps, phases of daylight; see Figure 9.3; and Planckian radiators) at different CCTs (in kelvin). The illuminance at the eye in Eq. (9.5) as constant, $E_v = 700$ lx. Figure 9.8 shows CS as a function of CCT for these light sources.

As can be seen from Figure 9.8, the function CS(CCT) is nonlinear at the fixed illuminance of $E_v = 700$ lx at the eye: at least in the case of this representative set of light sources, CS becomes saturated for CCT > 6500 K. The minimum of CS equaled 0.48 at about 2500 K and the maximum of CS (0.66) occurred at CCT > 6500 K. The resulting HCL concept for spectral circadian optimization is shown in Figure 9.9.

According to Figure 9.9, using the spectral model of the light source, the relative spectral radiance distribution of the light source is being varied during the procedure of spectral optimization and the CS [6] is being computed to optimize the circadian effect. As mentioned above, illumination in the evening or during the night might have the purpose of relaxation (to do so, melatonin level should be increased by using the spectral optimization criterion CS = min.) or "continue working" (with CS = max. corresponding to higher melatonin suppression).

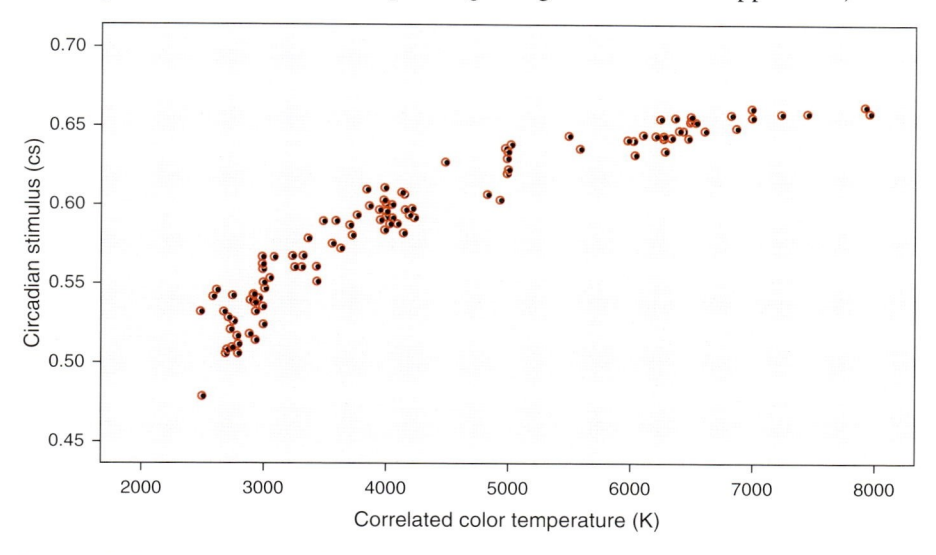

Figure 9.8 Circadian stimulus (CS) after the Rea *et al.* model (Eqs. (9.1)–(9.10)) for different light sources [14] and illuminants (commercially available white LEDs, fluorescent lamps, phases of daylight, see Figure 9.3; and Planckian radiators) at different correlated color temperatures (CCT in kelvin). The illuminance at the eye in Eq. (9.5) has constant $E_v = 700$ lx.

344 | *9 Human Centric Lighting and Color Quality*

```
┌─────────────────────┐          ┌─────────────────────┐
│ Vary relative spectral │          │ Mathematical model  │
│ radiance             │          │ fot the circadian   │
│ systematically       │          │ stimulus            │
└──────────┬──────────┘          └──────────┬──────────┘
           │                                │
           ▼                                ▼
┌─────────────┐     ┌─────────────────────┐     ┌─────────────┐
│ 1. Relaxing │     │                     │     │ 2. (Continue)│
│ (foster     │◄────│ Circadian optimization│────►│ working     │
│ melatonin   │     │ of the light source │     │ (with melatonin│
│ release)    │     │                     │     │ suppression) │
└─────────────┘     └─────────────────────┘     └─────────────┘
```

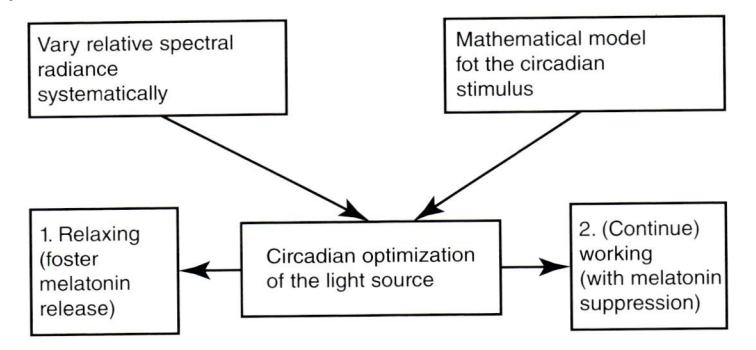

Figure 9.9 HCL concept for spectral circadian optimization.

9.3 Spectral Design for HCL: Co-optimizing Circadian Aspects and Color Quality

In this section, an actual computational example [15] will be shown for a possible technological implementation of spectral HCL design: the spectral co-optimization of a simulated multi-LED engine for the nonvisual aspect CS and a selected visual aspect, the chroma enhancement of red objects. The chroma enhancement of reddish objects or surfaces (of large spatial extent) fosters emotional well-being affecting the residents of elderly peoples' homes [16]. Figure 9.10 shows the relative spectral radiance of the six LED channels of the simulated multi-LED engine in the example together with the spectral sensitivity of the visual and the ipRGC mechanisms (see also Chapter 2) and also the spectral reflectance of the deep red CIE test color sample TCS09 used to model reddish objects.

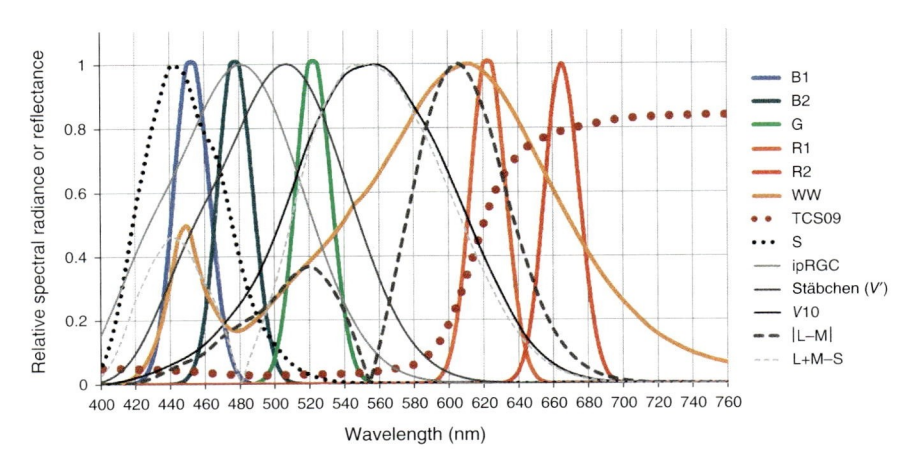

Figure 9.10 Relative spectral radiance of the six LED channels of the simulated multi-LED engine in the sample HCL optimization of Section 9.3, spectral sensitivity of the visual and the ipRGC mechanisms, and spectral reflectance of the deep red CIE test color sample TCS09 used to model reddish objects [15].

As a basic LED channel to ensure high efficiency, a warm white LED (WW in Figure 9.10) was selected supplemented by five colored LED channels (B1, B2, G, R1, R2) with overlapping peak wavelengths with the spectral sensitivity curves of the individual mechanisms. For example, the channel B1 (maximum at 450 nm) corresponds to the spectral sensitivity of the S cones, B2 to the ipRGC mechanism, G to the shorter wavelength branch of the $|L - M|$ mechanism. The LED channels R1 and R2 stimulate the long wavelength branch of the $|L - M|$ mechanism strongly because the R1 and R2 LED emission curves overlap with the spectral reflectance of the deep red CIE test color sample TCS09 used to represent reddish objects [15].

Equation (9.11) describes the additive mixture of the six LED channels with six weighting factors simulating the intensity of the individual LED channels.

$$L_e(\lambda)_{\text{Multi-LED}} = a_{\text{WW}} \text{ WW}(\lambda) + a_{\text{B1}} \text{ B1}(\lambda) + a_{\text{B2}} \text{ B2}(\lambda) \\ + a_{\text{G}} \text{ G}(\lambda) + a_{\text{R1}} \text{ R1}(\lambda) + a_{\text{R2}} \text{ R2}(\lambda) \quad (9.11)$$

During the spectral optimization, the five parameters a_{B1}, a_{B2}, a_{G}, a_{R1}, and a_{R2} of Eq. (9.11) were varied systematically to fulfill eight different optimization criteria (see the second row of Table 9.1 [15]) while the weighting of the warm white LED was constant, $a_{\text{WW}} = 1$. The boundary condition $\Delta u'v' < 0.002$ was used for the distance of the chromaticity of the white tone of the multi-LED spectrum $L_e(\lambda)_{\text{Multi-LED}}$ in Eq. (9.11) from the blackbody or daylight loci, in order to ensure a white tone of high perceived quality.

In Table 9.1, the optimization criteria No. 1 and No. 2 represent the aim of either minimizing (No. 1) or maximizing (No. 2) the CS by fulfilling the boundary conditions $R_a > 90$ and $T_{cp} \leq 7800$ K. The criteria No. 3–5 aim at excellent color rendering by using the boundary conditions $R_i > 82$ (for all 14 CIE test color samples TCS01–TCS14) and $R_a > 92$ while CS was either minimized (No. 3), or the

Table 9.1 Colorimetric properties and color quality descriptors of the eight optimized multi-LED spectra (Eq. (9.11)) for HCL design with different optimization criteria.

Number	1	2	3	4	5	6	7	8
Crit.	CS min.; $R_a > 90$	CS max.; $R_a > 90$	CS min.; $R_i > 82$; $R_a > 92$	CS > 0.6; $R_i > 82$; $R_a > 92$	CS max.; $R_i > 82$; $R_a > 92$	$C^*_{ab,09}$ max.; CS = min.	$C^*_{ab,09}$ max.	$C^*_{ab,09}$ max.; CS ≥ 0.6
CCT(K)	2670	7800	3179	3854	4271	2683	2826	3679
CS	0.541	0.660	0.577	0.601	0.619	0.543	0.553	0.600
$\Delta u'v'$	0.0018	0.0018	0.0018	0.0018	0.0018	0.0018	0.0018	0.0018
CIE R_a	91	91	93	93	90	86	86	86
CIE R_9	80	74	92	97	88	60	37	42
CIELAB $C^*_{ab,09}$	74	65	74	72	72	79	81	78
L^{**}_{09}	66	59	65	63	63	68	68	66
$R_{i,\min}$	69	74	82	84	80	60	37	42

Crit.: optimization criterion (see text) [15].

condition CS > 0.6 had to be fulfilled (No. 4) or CS was maximized (No. 5); see Table 9.1.

Concerning the criteria No. 6–8, the optimization target was to maximize the CIELAB chroma $C^*_{ab,09}$ of the saturated red CIE test color sample TCS09 (see Figure 9.10). At the same time, a high chromatic lightness L^{**} was achieved for TCS09 (denoted by L^{**}_{09} in Table 9.1) while the quantity CS was either minimized (No. 6) or maximized (No. 7), or the criterion CS ≥ 0.6 had to be fulfilled (No. 8); see Table 9.1. Concerning the colorimetric properties and color quality descriptors of the eight optimum multi-LED spectra, the following can be seen from Table 9.1.

1) Different optimization criteria could be achieved at different CCTs (CCT > 7800 K was not allowed).
2) CS values are inside the interval 0.541–0.660. The CS interval of 0.48 (CCT = 2500 K) to 0.66 (CCT > 6500 K) can be seen from Figure 9.8.
3) Using the criterion of oversaturating the red test color sample TCS09 (No. 6–8), the CIE CRI of $R_a = 86$ ("good") could be achieved while the value of R_9 is between 37 and 60 (the lower R_9 values result from the fact that the red test color sample is more saturated under the optimum multi-LED spectrum than under the reference light source and this causes a color difference penalized by the CRI definition).
4) In the case of criterion No. 5, the criterion $R_a > 92$ could not be fulfilled.
5) The descriptors CIELAB chroma $C^*_{ab,09}$ and chromatic lightness L^{**}_{09} were increased by red oversaturation (No. 6–8). According to the findings described in Chapters 5 and 7, this improves the color preference of the users of the multi-LED light source.

Figures 9.11–9.13 show the relative spectral radiance distributions of the eight optimum multi-LED light sources.

As can be seen from Figure 9.11, the criterion CS = max. implies higher B1-, B2-, and G-radiation components (and higher CCT) in comparison to CS = min.

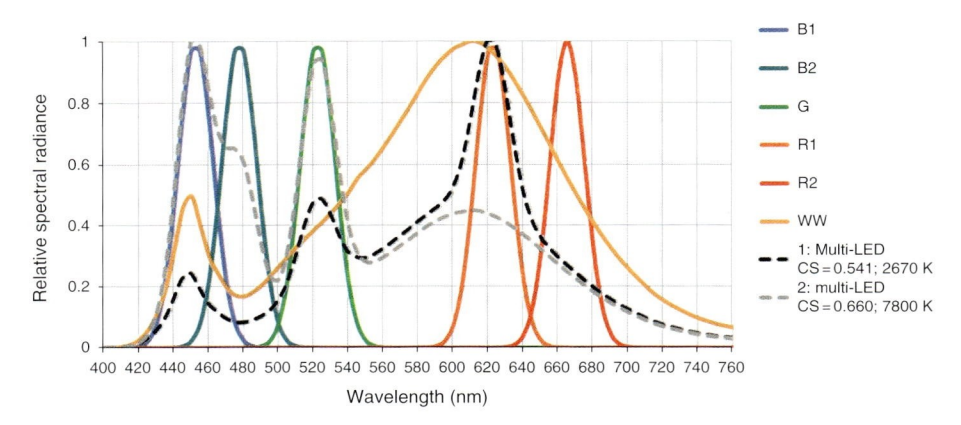

Figure 9.11 Optimized emission spectra No. 1 and 2 of the simulated multi-LED engine. Optimization criteria: $R_a > 90$; CS = min. (No. 1), and CS = max. (No. 2) [15].

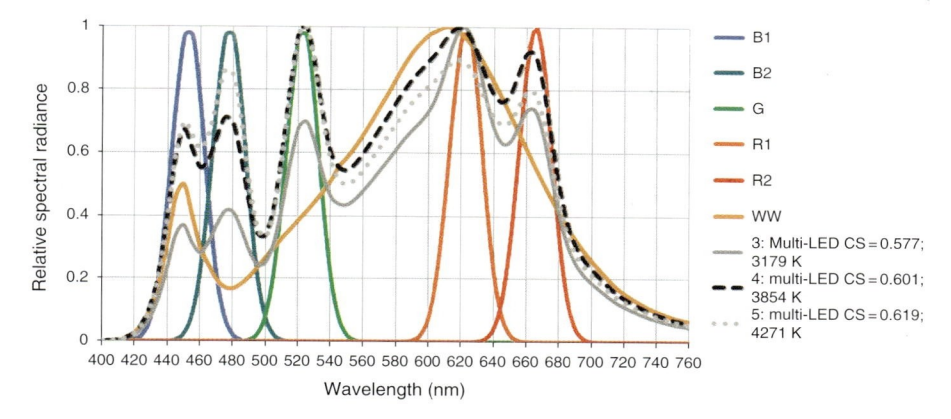

9.3 Spectral Design for HCL: Co-optimizing Circadian Aspects and Color Quality

Figure 9.12 Optimized emission spectra No. 3–5 of the simulated multi-LED engine. Optimization criteria: $R_a > 92$; all CIE special CRI $R_i > 82 (i = 1–14)$; CS = min. (No. 3); CS > 0.6 (No. 4) and CS = max. (No. 5) [15].

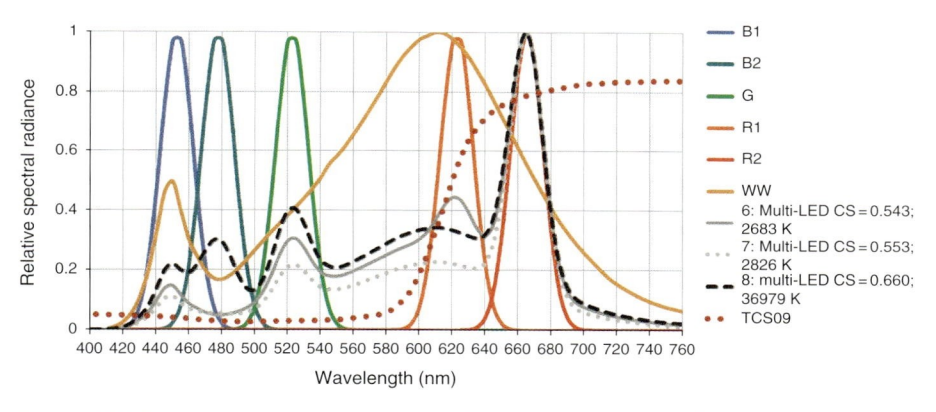

Figure 9.13 Optimum emission spectra No. 6–8 of the simulated multi-LED engine. Optimization criteria: maximum CIELAB chroma C^*_{ab} for the red CIE test color sample TCS09; No. 6: CS = min.; No. 7: no target value for CS; No. 8: CS ≥ 0.6 [15].

These higher green and blue components stimulate the S- and the ipRGC mechanisms, respectively (see Figure 9.11) to give rise to a strong CS, for example, on the wall of the office illuminated by the multi-LED.

As can be seen from Figure 9.12, to fulfill the "good" to "very good" color-rendering criteria (in case of every CS criterion), the resulting optimum multi-LED spectra are spectrally balanced (i.e., there are no deep and/or broad spectral minima in the visible range; compare with Figure 9.3. To achieve this, the deep red channel R2 plays an important role.

As can be seen from Figure 9.13, to increase the chroma and the chromatic lightness of the deep red test color sample TCS09, it is necessary to use the deep red LED channel R2.

The eight optimum multi-LED spectra are represented in the three-dimensional diagram of the HCL aspects $\{L^*_{09}; R_a; CS\}$ in Figure 9.14 while

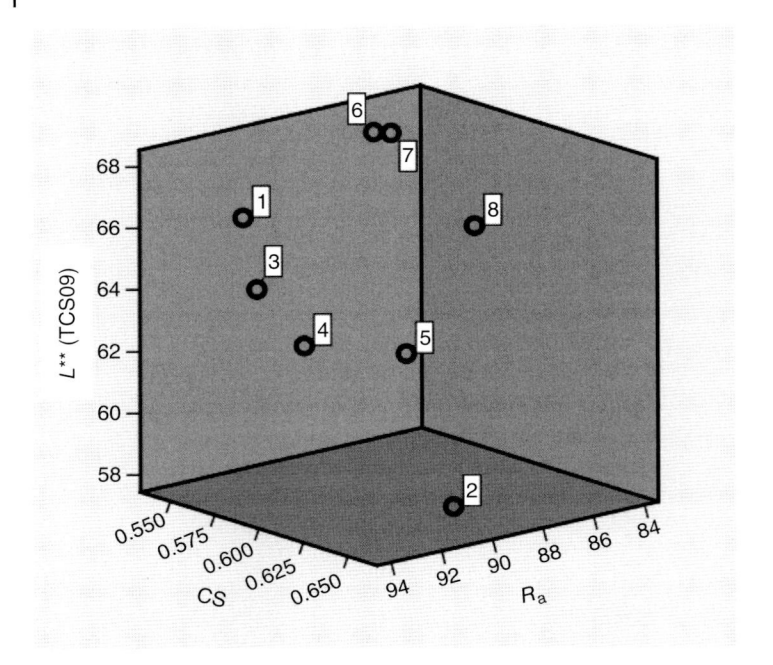

Figure 9.14 The eight optimum multi-LED spectra (Table 9.1) in the three-dimensional diagram of the HCL aspects $\{L_{09}^{**}; R_a; CS\}$. The aspects L_{09}^* and R_a are visual aspects and the dimension CS is a nonvisual aspect in the sample computation of Section 9.3 [15].

the aspects L_{09}^{**} and R_a are visual aspects and the dimension CS is a nonvisual aspect in this sample computation.

As can be seen from Figure 9.14, by the aid of the six LED channels of the simulated multi-LED engine in the present example, the criteria $\{R_a = \text{max.} + CS = \text{max}\}$ (i.e., "to continue working with good color rendering in the evening") and $\{L_{09}^{**} = \text{max.} + CS = \text{min.}\}$ ("ensure an atmosphere with emotionally stimulating red tones and fostering relaxation in the evening") could be reached. The multi-LED spectrum could not be optimized to both "high circadian stimulus" and "strong red oversaturation" owing to the contradicting spectral sensitivity ranges (greenish blue vs. red) of the mechanisms that are responsible for these two effects.

9.4 Spectral Design for HCL: Change of Spectral Transmittance of the Eye Lens with Age

Recent demographic developments lead to aging societies all around the world. An example of age distribution development (forecast) in the People's Republic of China is shown in Figure 9.15. In most other countries, a similar development is expected.

In this section, results of a sample HCL computation will be shown to point out the importance of the age of the light source user from the point of view

9.4 Spectral Design for HCL: Change of Spectral Transmittance of the Eye Lens with Age | 349

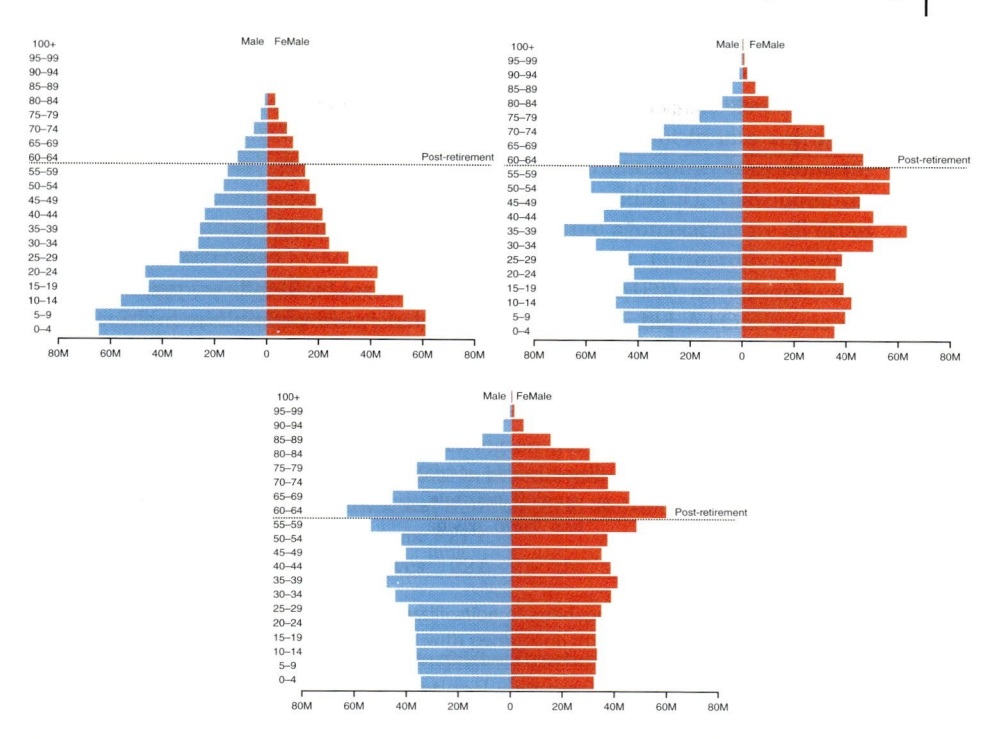

Figure 9.15 Age distribution development (forecast) in the People's Republic of China. Source: United Nations.

of HCL design by considering the change of the spectral transmittance of the human eye lens with age [17], shown in Figure 9.16. In this sample computation, only this factor (i.e., the changing spectral transmittance of the eye lens) will be considered. It should be noted that other factors such as the deterioration of other eye media, the retina, and post-retinal processing of visual signals as well as the general weakening of the nervous system contribute to age-related losses that will not be considered in this example.

As can be seen from Figure 9.16, the most accentuated deterioration of the spectral transmittance of the human eye lens occurs in the blue–bluish green spectral range. In the first example, the relative spectral power distributions of the phases of natural daylight shown in Figure 9.3 were weighted, that is, multiplied at every wavelength by the spectral transmittance of the human eye lens for every age shown in the legend of Figure 9.16. Then, four quantities (luminous flux Φ_v, CS, CQS Q_p, and the S-cone signals) were computed for these weighted spectral power distributions for every age, related to the age of 20(= 1.000).

Figure 9.17 shows the dependence of weighted relative luminous flux on age.

As can be seen from Figure 9.17, relative weighted luminous flux (= 1.000 for the age of 20) decreases as a function of age. There is a steep decline after 60 years of age down to about 60% of the value found in 20-year-old observers according to the computation based on the human lens spectral transmittance data [17]. It can also be seen that the decline of luminous flux is more accentuated in the case

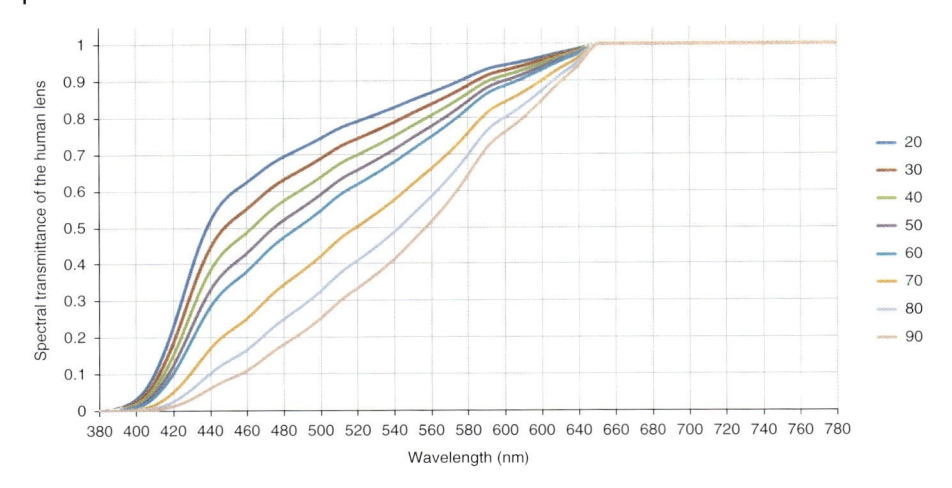

Figure 9.16 Change of the spectral transmittance of the human eye lens with age [17]. The curves are based on the optical density data of the aging human lens taken from Table I in [17]. Legend: age of the observer in years.

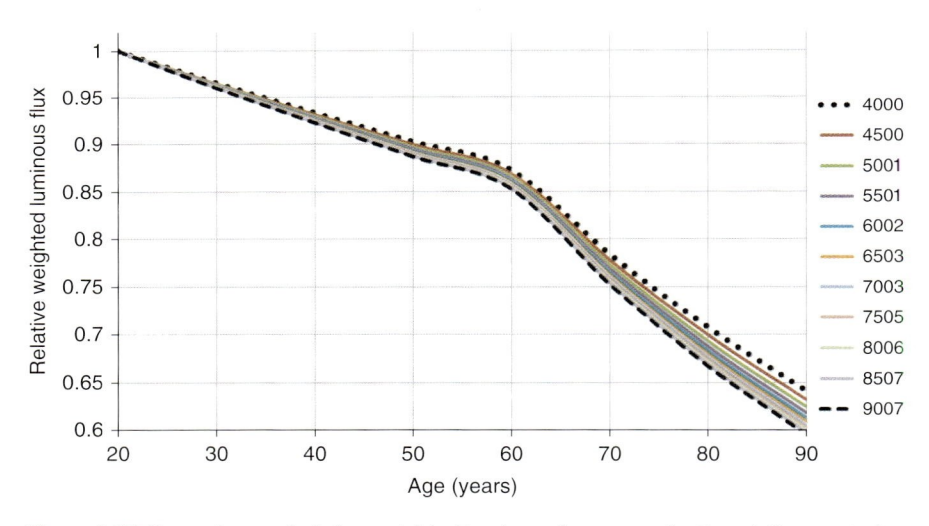

Figure 9.17 Dependence of relative weighted luminous flux on age for the relative spectral power distributions of the phases of natural daylight (Figure 9.3; CCTs are shown in the legend) weighted by the spectral transmittance of the human eye lens (Figure 9.16).

of higher CCTs; compare the dotted curve (a phase of daylight at 4000 K) with the dashed curve (a phase of daylight at 9007 K).

Figure 9.18 shows the dependence of the relative CS according to Eq. (9.10) ($CS_{rel} = 1.000$ for the age of 20) on age for the same weighted phases of natural daylight as in Figure 9.17.

As can be seen from Figure 9.18, the value of the relative CS (= 1.000 for the age of 20) also decreases as a function of age, similar to the relative luminous flux data shown in Figure 9.17. According to the human lens spectral transmittance

9.4 Spectral Design for HCL: Change of Spectral Transmittance of the Eye Lens with Age

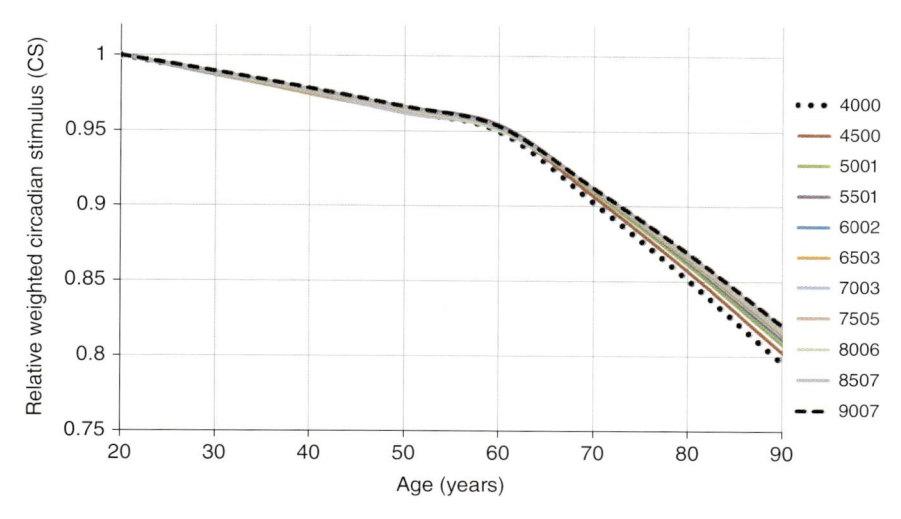

Figure 9.18 Dependence of relative circadian stimulus CS according to Eq. (9.10) on age for the relative spectral power distributions of the phases of natural daylight (Figure 9.3; CCTs are shown in the legend) weighted by the spectral transmittance of the human eye lens (Figure 9.16).

data [17], there is a steep decline after 60 years, again down to about 80% of the value found in 20-year-old observers. This means that, according to this sample computation for daylight phases (as ideally balanced natural light source spectra), an elderly user of the light source needs, depending on her/his age, 10–20% more CS to achieve the same circadian HCL effect. It can also be seen that the decline of the CS is slightly more accentuated in case of lower daylight CCTs; compare the dotted curve (a phase of daylight at 4000 K) with the dashed curve (a phase of daylight at 9007 K).

As pointed out in Sections 9.1 and 9.3, color quality is an important component of HCL. Therefore, the dependence of a descriptor of color preference, CQS Q_p was also computed as a function of age by using the same weighted spectral power distributions of daylight phases as in the case of Figures 9.17 and 9.18. Figure 9.19 shows the dependence of the relative CQS Q_p values ($Q_{p,rel}$ = 1.000 for the age of 20) on age.

As can be seen from Figure 9.19, the relative value of CQS Q_p (= 1.000 for the age of 20) decreases as a function of age, similar to the earlier parameters (Figures 9.17 and 9.18), down to about 77% of the value found in 20-year-old observers in the worst case of 4000 K and about 88% in case of 9007 K. This finding implies that HCL design for good color preference should consider the observer's age, and special spectral power distributions should be used for elderly users to ensure a good color preference.

As was mentioned in Section 7.2, visual clarity represents an important requirement to carry out visual tasks effectively and to assess the color quality of the lit environment. Visual clarity was modeled by the aid of the S-cone signals that have a spectral sensitivity in the bluish spectral range. It is the bluish spectral range in which the spectral transmittance of the human eye lens

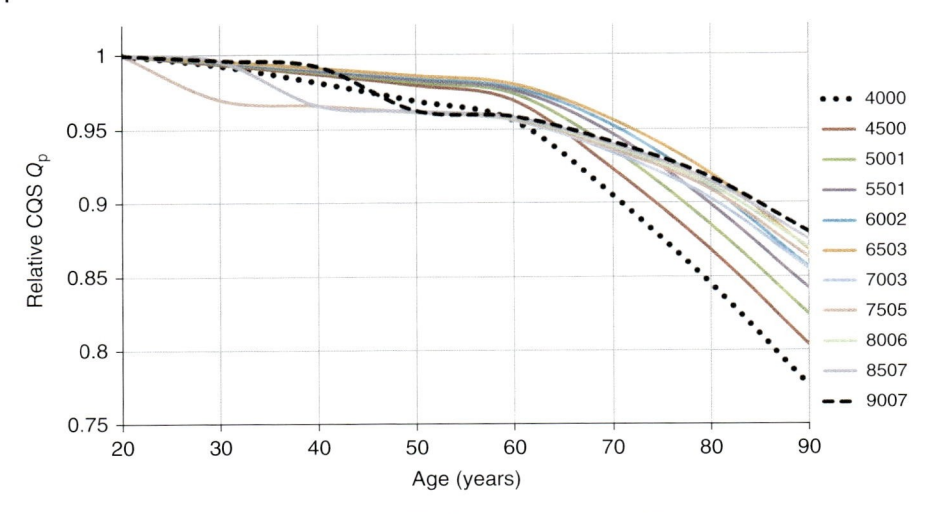

Figure 9.19 Dependence of the relative CQS Q_p values ($Q_{p,\text{rel}} = 1.000$ for the age of 20) on age for the relative spectral power distributions of the phases of natural daylight (Figure 9.3; CCTs are shown in the legend) weighted by the spectral transmittance of the human eye lens (Figure 9.16).

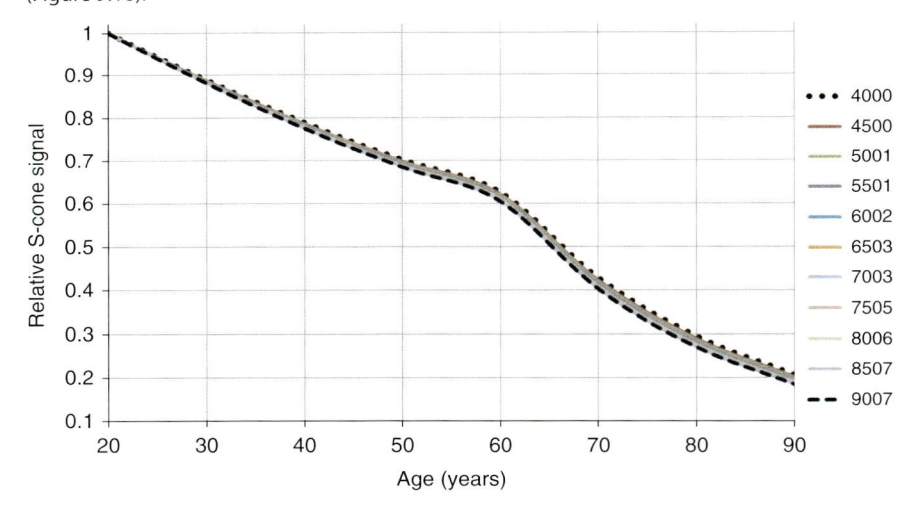

Figure 9.20 Dependence of the relative S-cone signal values (1.000 for the age of 20) on age for the relative spectral power distributions of the phases of natural daylight (Figure 9.3; CCTs are shown in the legend) weighted by the spectral transmittance of the human eye lens (Figure 9.16).

decreases with age most rapidly (Figure 9.16). This is why it is important to examine the dependence of the relative S-cone signal values (1.000 for the age of 20) on age for the present example of the daylight spectra weighted by the spectral transmittance of the human eye lens; see Figure 9.20.

As can be seen from Figure 9.20, S-cone signals decrease rapidly with the HCL luminaire user's age, especially after 60 years of age, down to about 20% (at the age of 90) of the S-cone signal of a 20–year-old-observer. As visual clarity was found to be closely related to S-cone signals (see Figure 7.5), this finding means that visual clarity should deteriorate quickly with age to be compensated for by HCL design. Note that Figures 9.17–9.20 show the results of a sample HCL computation that considers the decline of the spectral transmittance of the aging human lens only. Other factors such as neural factors of later visual or nonvisual information processing have not been taken into account. For a reliable spectral HCL design, further visual experiments including color preference experiments with elderly observers are necessary. The aim of the present section is just to draw the attention of the reader to the possible consequences of aging society on HCL product design to elaborate more conscious design principles in the future.

In a second example, a similar computation was carried out but with a set of 11 other light sources different from the phases of daylight including typical light sources of exterior lighting (car headlamps and typical street lamps) and interior lighting (three typical fluorescent lamps and two multi-LEDs). Table 9.2 shows the relative weighted luminous flux, circadian stimulus, CQS Q_p, and S-cone signal (compare with Figure 7.5) values of this second computation for two ages, 60- and 90-year-old observers, respectively, compared to a 20-year-old observer (1.000). Light sources of exterior lighting were included because their color quality and the CS they evoke during the late hours (lighting of pedestrians, colored cars, traffic signs, and building facades) are important for HCL.

Table 9.2 Relative weighted luminous flux (Φ_v), relative circadian stimulus CS, relative CQS Q_p, and S-cone signal values of the second HCL computation with different light sources for two ages, 60 and 90 years of observers, respectively, compared to a 20-year-old observer (1.000) for the relative spectral power distributions of typical light sources of exterior and interior lighting weighted by the spectral transmittance of the human eye lens (Figure 9.16).

Quantity	R_a	Q_p	Rel. Φ_v	Rel. Φ_v	Rel. CS	Rel. CS	Rel. Q_p	Rel. Q_p	S-cone	S-cone
Age	—	—	60	90	60	90	60	90	60	90
Car headlamp H7	98	96	0.88	0.66	0.94	0.78	0.92	0.68	0.63	0.22
Car headlamp LED	82	76	0.87	0.62	0.94	0.77	1.00	0.96	0.60	0.17
Car headlamp Xe	70	67	0.88	0.65	0.94	0.78	0.95	0.70	0.61	0.20
HPS[a]	20	31	0.91	0.73	0.90	0.66	0.83	0.19	0.63	0.21
Mercury vapor[a]	43	52	0.88	0.66	0.90	0.68	0.91	0.48	0.54	0.12
Metal halide[a]	64	60	0.88	0.64	0.94	0.78	0.94	0.68	0.60	0.19
MLED-WW[b]	97	95	0.89	0.68	0.94	0.75	0.93	0.64	0.64	0.22
MLED-CW[b]	96	95	0.86	0.61	0.95	0.81	1.00	0.93	0.62	0.20
triphosphor FL[b]	82	83	0.88	0.65	0.93	0.75	0.94	0.64	0.59	0.17
Cool white FL[b]	63	62	0.88	0.64	0.94	0.76	0.96	0.76	0.59	0.17
Daylight FL[b]	77	75	0.86	0.61	0.95	0.80	0.97	0.85	0.60	0.18

a) Typical street lamps.
b) Light sources of interior lighting.
S-cone, signal of the S-cones relevant for visual clarity (see Figure 7.5); HPS, high-pressure sodium; MLED-WW (CW), warm white (cool white) multi-LED; and FL, fluorescent lamp. R_a, Q_p, original values of the unweighted spectra.
Rel.: Relative

As can be seen from Table 9.2, the values of the four HCL descriptor quantities of this second sample computation decline at the age of 60 and even more explicitly at the age of 90. Concerning relative weighted luminous flux, the deterioration of the light sources MLED-CW (the cool white multi-LED light source for interior lighting) and the daylight fluorescent lamp is most obvious. Both light sources are subject to losing 39% of their luminous output at the age of 90 (Note that, as pointed out in the previous sections, as the quantity luminous flux represents only the signal of the L + M channel, standalone luminous flux values cannot account for perceived quantities such as brightness or visual clarity).

Depending on the type of light source, the loss of CS at the age of 90 ranges between 19% and 34%. Concerning Q_p, the two street lamps high-pressure sodium (HPS) and mercury vapor provide by themselves (i.e., unweighted) very low color quality descriptor values (see the second and third columns of Table 9.2). It is more interesting to consider the two warm white light sources H7 and MLED-WW with high color quality for which a deterioration of 32–36% occurs at the age of 90 in terms of the color quality parameter CQS Q_p.

Finally, S-cone signals (which are relevant to predict visual clarity; see Figure 7.5) decrease rapidly with age. The reason is that the most accentuated deterioration of the spectral transmittance of the human eye lens occurs in the bluish spectral range (see Figure 9.16) in which the S-cones have the highest spectral sensitivity. Thus, it can be expected that visual clarity (which is an important prerequisite for any visual task and the appreciation of color quality as pointed out in Section 7.2) is strongly influenced by the spectral transmittance changes of the human lens with age and this should be compensated for by conscious HCL design that considers the HCL luminaire user's age.

9.5 Conclusions

As was pointed out in this chapter, in HCL, instead of energy efficiency considerations, the human user is in the focus of the attention of lighting design with the multitude of influencing factors summarized in Figures 9.1 and 9.2. Although the ability of influencing human circadian rhythms by artificial illumination is limited, especially in the presence of daylight during the daytime, circadian optimization can be combined with other HCL aspects (depending on the application) including color quality or visual clarity to achieve better work performance or to ensure emotional well-being. The primary aim of circadian optimization should be to support circadian rhythms, especially with the CS = min. criterion in the late afternoon or in the evening. Anyway, to test this HCL concept, further psychophysical experiments are necessary.

In the future, new concepts of (multichannel LED based) HCL luminaires will be economically more and more important according to the multitude of their applications. Modern technologies will be at the lighting engineer's disposal in order to create a high-quality HCL luminaire including optical (e.g., lenses, diffusors, reflectors, daylight combiners, and changing angular light distributions), chemical (e.g., new LED phosphors), electronic (e.g., driving circuits and sensors), photonic (e.g., quantum dot LEDs), and mechanical (e.g., components of thermal management) technologies.

References

1. Khanh, T.Q., Vinh, Q.T., and Bodrogi, P. (2016) A numerical analysis of recent color rendition metrics. *Light. Res. Technol.* doi: 10.1177/1477153516632568
2. Rautkylä, E., Puolakka, M., and Halonen, L. (2012) Alerting effects of daytime light exposure–a proposed link between light exposure and brain mechanisms. *Light. Res. Technol.*, **44**, 238–252.
3. Khanh, T.Q. and Bodrogi, P. (2015) Ganzheitliche Optimierung-Human Centric Lighting und Farbwahrnehmung. *Licht*, **67** (5), 74–80.
4. Bodrogi, P., Brückner, S., Khanh, T.Q., and Winkler, H. (2013) Visual assessment of light source color quality. *Color Res. Appl.*, **38** (1), 4–13.
5. Rea, M.S., Figueiro, M.G., Bierman, A., and Bullough, J.D. (2010) Circadian light. *J. Circadian Rhythms*, **8** (1), 2.
6. Rea, M.S., Figueiro, M.G., Bierman, A., and Bullough, J.D. (2010) Circadian light. *J. Circadian Rhythms*, **8** (2), 1–10.
7. Khanh, T.Q., Bodrogi, P., Vinh, Q.T., and Winkler, H. (2015) *LED Lighting-Technology and Perception*, Wiley-VCH Verlag GmbH & Co. KGaA, Weinheim.
8. Licht.de (ZVEI) (2010) Wirkung des Lichts auf den Menschen, licht.wissen 19.
9. Berson, D.M., Dunn, F.A., and Takao, M. (2002) Phototransduction by retinal ganglion cells that set the circadian clock. *Science*, **295**, 1070–1073.
10. Brainard, G.C., Hanifin, J.P., Greeson, J.M., Byrne, B., Glickman, G., Gerner, E., and Rollag, M.D. (2001) Action spectrum for melatonin regulation in humans: evidence for a novel circadian photoreceptor. *J. Neurosci.*, **21**, 6405–6412.
11. Thapan, K., Arendt, J., and Skene, D.J. (2001) An action spectrum for melatonin suppression: evidence for a novel non-rod, non-cone photoreceptor system in humans. *J. Physiol.*, **535**, 261–267.
12. Rea, M.S., Figueiro, M.G., Bullough, J.D., and Bierman, A. (2005) A model of phototransduction by the human circadian system. *Brain Res. Rev.*, **50**, 213–228.
13. Schanda, J., Morren, L., Rea, M., Rositani-Ronchi, L., and Walraven, P. (2002) Does lighting need more photopic luminous efficiency functions? *Light. Res. Technol.*, **34** (1), 69–78.
14. Bodrogi, P. and Khanh, T.Q. (2013) Review on the spectral sensitivity of the circadian effect and circadian stimulus of artificial and natural light sources. Green Lighting Shanghai 2013, Session P202: Biological and Health Effects of Lighting, 2013.
15. Bodrogi, P. and Khanh, T.Q. (2015) Ganzheitliche Optimierung – Human Centric Lighting und Farbwahrnehmung – Teil 2. *Licht*, **67** (6), 66–73.
16. Dietz, B. (2012) Das Klinikum am Bruderwald in Bamberg – Lichtblicke für die alters- und demenzsensible Krankenhausplanung. *Licht*, **64**, 20–24.
17. Pokorny, J., Smith, V.C., and Lutze, M. (1987) Aging of the human lens. *Appl. Opt.*, **26** (8), 1437–1440.

10

Conclusions: Lessons Learnt for Lighting Engineering

The aim of the present book is to systematize and model all aspects of color rendition (color quality) of light sources and lighting systems with a comprehensive analysis of the results from color science and their application for the use of lighting practice. In order to describe color appearance with its perceptual attributes brightness, lightness, colorfulness, chroma, saturation, and hue, its influencing parameters such as the underlying color-matching functions (2° and 10° CMFs), adaptation luminance, chromatic adaptation, white point, correlated color temperature (CCT), viewing time, and color structure (texture and spatial frequency of the structure of the colored objects) have to be analyzed. In the long history of color science from 1931 to date, the establishment of a uniform color space has turned out to be essential. In this book, the CAM02-UCS color space is generally used as this was considered to be the most uniform practically usable color space.

Uniformity means that equal color distances between two arbitrary color points in this color space correspond to equal perceived color differences. A uniform color space allows the correct design and evaluation of color quality experiments and the establishment of usable color quality metrics to model the aspects color fidelity, color gamut, or chroma enhancement mathematically. With the aid of well-defined visual psychophysical methods with realistic viewing conditions and relevant colored objects, color quality metrics and their combinations can be improved. They serve as color quality criteria for a correct optimization of light source spectra.

Figure 10.1 shows the system of color quality aspects for lighting engineering starting from color space via color metrics and color evaluation until the procedure of light source optimization.

In order to model the color appearance correctly (which is the starting point for color quality modeling), viewing conditions such as the viewing angle, the adaptation luminance level (photopic level, e.g., in a well-illuminated shop or mesopic level in a cinema room), and the white point have to be taken into account. The issue of the white point or predominating white tone in the lit environment was dealt with in Chapter 3 as the starting point of color quality design. It was pointed out that, according to the deviations of the CMFs of the individual observers, differences in white tone perception arise and the degree of interpersonal agreement on the perception of a certain (temporally stabilized and spatially homogeneous) white tone becomes an important prerequisite of spectral design. As it

Color Quality of Semiconductor and Conventional Light Sources, First Edition.
Tran Quoc Khanh, Peter Bodrogi, and Trinh Quang Vinh.
© 2017 Wiley-VCH Verlag GmbH & Co. KGaA. Published 2017 by Wiley-VCH Verlag GmbH & Co. KGaA.

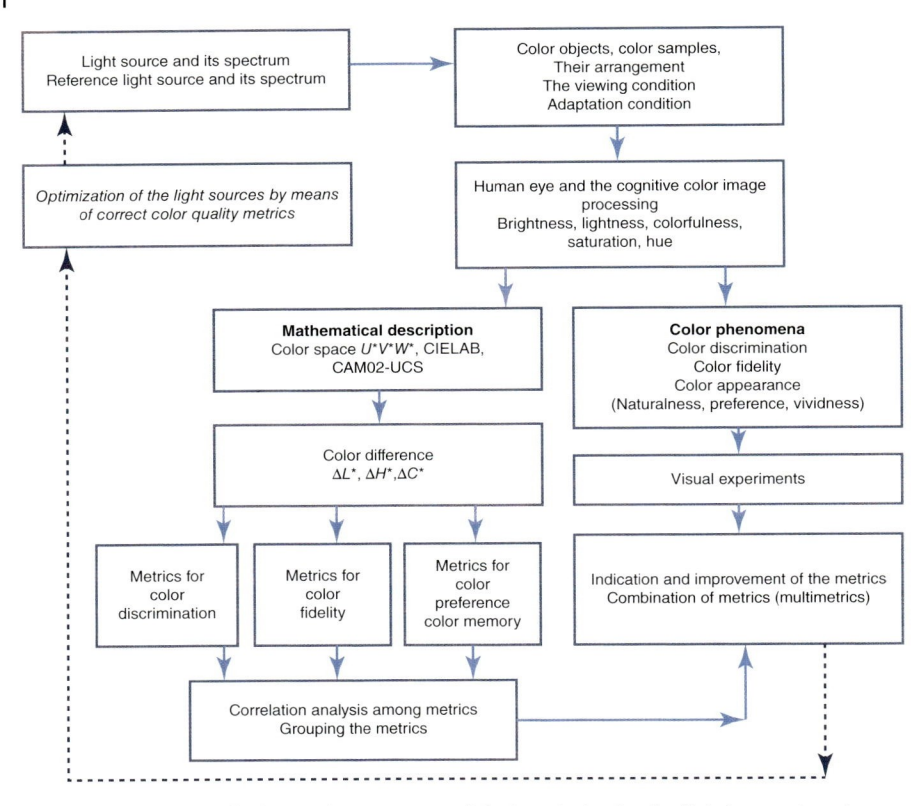

Figure 10.1 System of color quality aspects and their optimization for lighting engineering.

was pointed out, the degree of interpersonal agreement is lower for highly structured spectra with emphasized local minima and maxima (e.g., fluorescent lamps or especially RGB LEDs) and higher for more balanced spectra (e.g., multi-LED with phosphor-converted LEDs or colored phosphor-converted LED channels).

Another important message of Chapter 3 is the distinction between unique white (a white tone that does not contain any chromatic component; it is pure white, that is, neither greenish, nor purplish, bluish, or yellowish) and preferred white. Preferred white tones might contain – depending on CCT and also on the context of lighting application – a (slight) amount of blue (for cold white) and more or less yellow, for warm white color temperatures. For neutral white, around the CCT of 4000 K, the hue content of the preferred white tone tends to vanish; thus it agrees well with unique white (i.e., without any hue content). The other important aspect of white tone preference was described in Sections 7.4 and 7.5: the white tone of the light source is rarely seen separately in a real application. In most of the cases, there are different nearly achromatic objects (walls or white objects of different white tones; Section 7.4) or different colored objects (red, blue, or colorful object combinations; Section 7.5) in the lit scene, which are observed and assessed together with a predominating white tone (which can be different from the one of the light source owing to multiple inter-reflections in the room). In this context, it turned out that white tone preference depends on the culture and gender of the observers and also on the type of colored object combination: warm white is preferred for red objects while neutral or cool white is preferred for bluish or colorful object combinations, according to a cognitive effect of white tone preference.

As mentioned above, in most cases, the color stimulus reaching the human eye is the product of several reflections from the colored objects in the room so that the colorimetric characteristics of the typical colored objects shall be considered depending on the application. After a comprehensive analysis of the test color sample (or color object) collections used by the CIE, appearing in publications of national organizations (e.g., IES) or published by different color scientists, researchers and light source developers shall formulate the requirements to be fulfilled by an optimum set of object colors:

1) The set of colored objects shall be representative for a specific lighting application or many typical lighting applications (e.g., TV and film studios, shop lighting, oil paintings for gallery and museum lighting, food lighting, or the lighting of living rooms).
2) The set of colored objects shall be homogeneously distributed over many typical ranges of lightness L^*, hue H (red, orange, skin tones, yellow, green, blue, or violet), and chroma C^* (strongly saturated or desaturated colors).
3) The spectral reflectance of the collected colors should be uniformly distributed over the visible wavelength range between 400 and 700 nm with their typical band-pass, short-pass, and long-pass characteristics.

The analysis of object colors for flowers, skin tones, art paintings, and the Leeds database in Chapter 4 showed that many colors with moderate and high saturation, positioned at different lightness levels and distributed well around the hue circle are available. These colors have to be grouped in order to define a collection of representative colors to be used to define, in turn, the color fidelity, color gamut, and color saturation metrics appropriately. Regarding the test color sample collections of the current color quality metrics, the 15 CQS (color quality scale) test color samples come close to an ideal color collection for a color gamut index and the description of saturation enhancement. Using the philosophy of the CQS system and expanding it to several lightness levels, lighting engineers will have a reasonable collection for color gamut analysis. For the purpose of color fidelity, the IES color collection of 99 test color samples seems to be nearly optimal if expanding it by some more saturated colors that have to be distributed around the hue circle. For the evaluation of color preference, naturalness, and vividness as multiple color quality metrics (see Chapter 7), a simple combination of the 99 IES and 15 CQS test color samples should be the first step in the right direction before a general final step can be done in order to have about 120 colors fulfilling the requirements mentioned above.

From the point of view of the authors of the present book at the time of writing, the use of the CAM02-UCS color space and the application of the 15 CQS colors for color gamut and saturation analysis and the 99 IES colors for the fidelity issue can be considered as a reasonable solution. However, in the history of color science and color quality research, many trials have been conducted to define color quality (with its color discrimination, color fidelity, and color preference related aspects) using different color spaces and a limited number of color samples by following several different experimental designs and posing different research hypotheses and questions. Before finding new, comprehensive color quality metrics and metric combinations, it is essential to analyze all these modern color quality metrics regarding their correlations. Then, it is also important to group the metrics if some of them exhibit similar behavior for a large set of different light sources. This analysis was carried out in Chapter 6 and the lessons learnt there are summarized as follows:

- The three color fidelity indices $R_{1,14}$, CRI_{2012}, and R_f have high correlation coefficients among each other for the warm white and cold white light sources. The two last mentioned metrics are defined in the uniform color space CAM02-UCS with relatively uniformly distributed test color samples over the hue circle in the $a' - b'$ plane and over the visible wavelength range so that these two color fidelity indices should be preferred for further international color quality studies and discussions.
- There seem to be two subgroups of color gamut indices for the warm white light sources. In one of the subgroups, gamut area index (GAI) and feeling of contrast index (FCI) correlate well with each other while in the other group, Q_g and R_g have a high correlation ($R^2 = 0.88$). However, FCI does not correlate with R_g, neither for the cold white nor for the warm white light sources. For the warm white lamp types, it can be clearly recognized that the correlation between FCI and R_g as well as FCI and Q_g is highly dependent on the light source spectra. From the fact that GAI was established in a nonuniform color space with only eight not saturated surface colors and FCI was defined with only four surface colors, it is reasonable to concentrate on the group of the color gamut metrics R_g and Q_g.
- The R_g metric was defined by using the set of 99 IES colors while Q_g was defined by the 15 saturated CQS colors. An analysis in Chapter 6 for the approximations of memory color rendition index MCRI and Q_p from a color fidelity metric plus R_g or Q_g as a color gamut metric, the conclusion was drawn that the correlation coefficients are slightly better in case of CQS Q_g than in case of R_g.
- The currently most discussed color preference metrics MCRI and CQS Q_p can be approximated by a quadratic combination of color gamut and color fidelity metrics. Q_p can also be built up from linear combinations of color gamut and color fidelity metrics. These approximations yield better correlation coefficients in the case of the cold white light sources than in the case of the warm white light sources. This fact can be explained by the choice of daylight illuminants as reference light sources in the case of the cold white light sources. Additionally, Q_p is defined in the CIELAB $L^*a^*b^*$ color space with the D65 illuminant in order to calculate the reference CCT factor based on the color gamut ratio (so-called CCT factor) and with a chromatic adaptation formula that works better in the case of the light sources with color temperatures higher than 4500–5000 K. Before 4500 K (or 5000 K), the reference light

sources are black body radiators with very small spectral contributions in the wavelength range between 380 and 480 nm.

In Chapter 6, only numeric calculations were conducted to analyze color quality metrics and their mathematical relationships, based on the platform of the specific definition of each color quality metric with its own test color samples. Even in an ideal case in which the international color community would find a color fidelity index and a color gamut index together with a suitable color object collection that optimally describes the attributes *"color fidelity-color realness"* and *"color gamut or (over)saturation for a specific color group,"* the final task would not be solved because these metrics are not able to describe the visually relevant color quality attributes color preference, color naturalness, or color vividness alone. The latter attributes are the most important ones for general interior lighting purposes.

With the development of solid-state lighting products at the end of the last century and the use of white LEDs in all domains of human life (e.g., homes, hospitals, schools, offices, museums), the need for a correct description of color quality has been intensified and this requires a comprehensive and usable color quality metric that is able to describe color quality for the practice of lighting engineering. Therefore, one or several two-metrics solutions shall be defined by conducting a number of dedicated visual color quality experiments in different research institutes with light sources having different chromaticities, color temperatures, white tones, and causing different degrees of (over)saturation. The visual judgment results of reliable test subjects observing different categories of colored objects (e.g., textiles, art objects, cosmetics, foods, and flowers) can be used to establish reasonable two- (or more-) metrics combinations and the best combination with a high correlation over different types of experiments and experimental conditions (lighting contexts) shall be chosen. A series of such experiments and the resulting two- (or more-) metrics combinations were described in Chapter 7. From the considerations in Chapter 7, it was concluded that a comprehensive color quality metric should take the following most important aspects color preference (or, alternatively, naturalness or vividness), color fidelity, and color discrimination into account.

In several visual experiments, the level and range of the illuminance on the colored objects and the CCTs used for different applications and groups of colored objects was also investigated. The results in Section 7.2 imply that an illuminance of higher than 500 lx is reasonable for optimal visual acuity and good color discrimination. In contrast to color discrimination and color fidelity based on the principle of color differences (resulting from lower level mechanisms of the human visual system), the color attributes color preference, color vividness, and color naturalness are strongly related to higher level cognitive color processing and memory effects comparing the illumination scenarios under natural outdoor conditions of high illuminance (30 000–100 000 lx) with artificially lit environments. It was concluded that visual color quality experiments shall be designed at an illuminance level higher than 500 lx. The results in Sections 7.4 and 7.5 showed a range of preferred CCTs between 4000 and 5000 K with a specific characteristic that blue–green objects should be illuminated with lights of higher color temperature and warm white lights should be favored for the illumination of red–orange and skin tone objects in order to achieve the highest degree of user acceptance.

From the visual experiments described in Sections 7.6–7.8 with different setups of colored objects representing different lighting application contexts, general multiple-metrics formulae were derived to predict color quality. The proposed formulae provided high correlation with the mean visual assessments of color preference, naturalness, and vividness. These multiple-metrics formulae represent the idea that these visual attributes can be described as linear combinations of a color fidelity index (IES R_f or CQS Q_f) and the chroma difference ΔC^* between the test and reference appearances of suitable test color samples. The idea behind these formulae was inspired by the concept of the CQS Q_p metric, that is, rewarding the increase of chroma ($\Delta C^* > 0$) in comparison to the color fidelity condition ($\Delta C^* = 0$). However, the authors of the present book believe that the present formulae are more advantageous because

- the coefficients a, b, c for color fidelity and chroma difference were fitted to visual results on three different attributes (color preference, naturalness, vividness);
- color quality (CQ) values, that is, the new index values resulting from the multimetrics combinations *per se* are psychophysically relevant as these scales include a semantic interpretation with criterion values to achieve, for example, the "good" level; and
- by the use of the variable ΔC^*, it turned out to be straightforward to establish a link between color preference and color discrimination ability.

According to the thoughts and findings of Chapter 7, color quality optimization principles of LED-based light sources were described in Chapter 8 including the trade-off of color discrimination. This optimization should include the following three steps:

First step: Color fidelity (e.g., IES R_f) should be optimized. The result is a multi-LED spectrum with a high color-rendering index (CRI) ($R_a > 93$) so that the chroma difference ΔC^* approaches zero.

Second step: According to the color quality index formula $CQ = a + b\ \Delta C^* + c\ R_f$, the spectral power distribution of the multi-LED light engine should be optimized step by step by continuously reducing the color fidelity index and increasing the chroma difference ΔC^* while keeping white tone chromaticity constant. Thus, the optimum color preference rating will be achieved after a few calculation steps.

Third step: Developers should then compare the resulting ΔC^*(VS1VS15) values that provide optimum color preference, naturalness, or vividness with the values ΔC^*(VS1–VS15)$_{crit}$ for acceptable color discrimination. If color discrimination is important in a specific illumination project (e.g., museum lighting), then the values of ΔC^*(VS1–VS15) that provide optimum color preference, naturalness, or vividness should be reduced. Otherwise, the values ΔC^*(VS1–VS15) for optimum color preference, naturalness, or vividness should be adopted.

In the computational approach of this book, the critical value of ΔC^*(VS1–VS15)$_{\text{crit}}$ = 2.35 was specified for the case of the greenish-blue pair of test color samples as a color discrimination criterion. This value corresponds to $R_{\text{a,crit}}$ = 84 and to the visual color preference rating of 77, which is slightly less than "good." The maximum mean color preference assessment (87, i.e., better than "good") corresponds to $R_{\text{a,crit}}$ = 75 at the mean ΔC^*(VS1–VS15) value of 3.3 corresponding to $\Delta E'_{i,0-1}$ = 2.9 with deteriorated color discrimination performance. A mean color preference rating around 65 (which is rather "moderate" than "good") is associated with R_a = 96–98 and with $\Delta E'_{i,0-1}$ = 3.2–3.3, that is, $\Delta C^* < \Delta C^*_{\text{crit}}$ in this case. These trade-off values between color preference and color discrimination in terms of the CRI (CIE R_a), the mean ΔC^*(VS1–VS15) value, and the value of $\Delta E'_{i,0-1}$ are analyzed in detail in Chapter 7 and this knowledge was applied, in turn, in Chapter 8.

The establishment of the formula CQ = $a + b\ \Delta C^* + c\ R_f$ describing color quality according to color preference, naturalness, or vividness calls for the question for which surface colors the chroma difference ΔC^* shall be calculated. In Figure 10.2, the relationship among the chroma differences ΔC^* computed as different average values for different color collections is shown for several LED spectra with the same chromaticity at CCT = 3100 K and with different degrees of chroma differences. These color collections include the 15 CQS test colors, 15 different skin tones, the 14 CIE test color samples of the CIE CRI 1995, the 17 CRI 2012 test colors, the 4 FCI colors, the 99 IES colors, and the 9 MCRI colors. The linear behavior of all relations indicates that the choice of the 15 CQS colors as a representative color object collection for color chroma difference computations is appropriate.

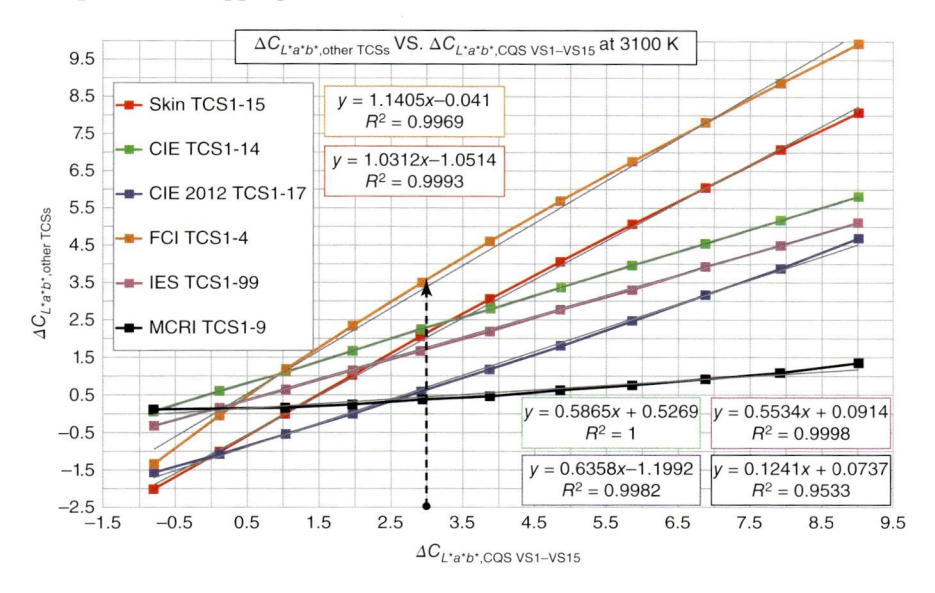

Figure 10.2 Relation between the mean chroma difference (ΔC^*) for the 15 CQS colors (abscissa) and other color collections (ordinate) for several LED spectra with the same chromaticity at CCT = 3100 K and with different degrees of chroma difference. TCS: Test Color Sample

The challenge of the twenty-first century with many social and industrial problems such as the demographic development in the international frame and urbanization with building of large cities in many regions of the world related to the need for improving the life and working conditions and working quality will call for a qualitatively new culture and mentality in illumination design and planning. In this context, the strategy of "Human Centric Lighting" has been established, bringing the individual needs of the human light source users with their age, lighting needs, biological rhythms, life conditions, and their social contacts into the focus of lighting design. This means the implementation of dynamic lighting with different lighting spectra, different illuminance levels, and lighting directions depending on the day time, season, weather, location, and working and living context. Dynamic light with different spectra can only be realized if the color quality of the illumination is taken into account. The integration of the category "color quality" into the context of Human Centric Lighting was discussed in Chapter 9.

Index

a
accent lighting 72
achromatic visual clarity 205–212

b
Berman *et al.* model 44
brightness 26
    Berman *et al.* model 44
    CIE 43–44
    Fairchild and Pirrotta's L^{**} model 45
    feature 43
    Fotios and Levermore model 45
    Helmholtz-Kohlrausch effect 41
    human visual system 41
    impression 41
    luminance channel 41
    WCCf 44
brightness matching experiment 212–218
brightness-luminance discrepancy 41

c
2D & 3D color gamut area 115
CAM02-UCS color difference 55
CAM02-UCS uniform 46–48
chroma difference 33, 116, 363
chroma difference formulae 47
chroma increment factor 60
chroma shifts 243
chromatic adaptation 28–29, 52, 145
chromatic visual clarity 204–212
chromatic visual clarity score (VC_c) 208
chromaticity
    criterion values 273–276
    semantic contours 274, 275
    white tone 72
chromaticity diagram
    $u'-v$ 24
    applications 21
    coordinates 19
chromaticity difference 216
CIE brightness model 43–44
CIE color rendering index (Ra)
    flowchart 50
CIE tristimulus values 52
CIECAM02 color appearance model 41
CIELAB chroma difference 60
CIELAB color difference 45–46, 53
CIELAB color space 36–37
CIELAB values 53
circadian clock 338
circadian stimulus (CS) 338–344
classical approach
    theory of signal processing application 120–121
    visual color model application 121
cognitive color 29–31
cold white light sources 184–189
color appearance 28, 31, 357
    brightness models 41–45
    changing 33
    chromatic adaptation 28
    CIECAM02 model 37–41

color appearance (*contd.*)
   CIELAB color space   36–37
   color attributes, of perceived   26–27
   color differences   45–46
   color stimuli   26
   perceived color differences   29
   semantic interpretation and criterion values   276–277
   similarity of   31
   viewing conditions   28–29
color arrangement   131
color collection   101, 105, 108, 310
   advantages and disadvantages   93
color constancy   28
color difference   91
   CAM02-UCS uniform   46–48
   chroma differences   29
   CIELAB   45–46
   description   131
   semantic interpretation and criterion values   268–277
   spider web diagrams   116
   state-of-the-art   145–154
color discrimination   272
   ability   35
   indices   63
   trade-off values   280
color fidelity   1, 32, 131, 309, 362
   CIE color rendering index   49–52
   CQS   52–53
   CRI201 method   53–56
   deficiencies   57
   index R_f 56
   LED light engines   320–323
   philosophy   155
   RCRI   57
   semantic interpretation   270–272
color gamut   113
color gamut differences   243–246
color gamut index   155–156
   R_g 62–63
   CQS method   62
   CRC   62
   CSA   62
   deficiencies   63
   FCI   62
   GAI   62
color gamut values R_g 133
color grouping
   principles   114–125
   quality   122–123
color harmony   35
color LEDs
   spectral power distribution   302–305
   temperature and current dependence of   299–300
color matching functions   17–19
color memory   30
color naturalness   32
color perception   2, 11, 92
color preference   241
   food products   256–268
   linear fidelity formula   195–196
   makeup products   246–256
   MCRI   190
   quadratic formula   195–196
   real room   234–246
   trade-off values   280
   with constrained linear formula   192–193
   with unconstrained linear formula   194–195
color preference indices   32
   color discrimination indices   63–64
   color gamut indices   61–63
   CQS Q_a 60–61
   GAI   58
   Judd's flattery index   57–58
   MCRI   58–60
   Thornton's   58
color quality   48–64
   prediction potential and correctness   166–171
   acceptable   14
   achromatic visual clarity   204–212
   and quantity   285
   brightness matching experiment   212–218
   chromatic visual clarity   204–212
   color fidelity indices   49
   color preference indices   57–61
   color sources   132–140
   colored object   141–142

colorimetric properties   228
computational methods   49–50
concept   31
correlated color temperature   218–225
correlation   177–189
correlation calculations   161
evaluation   130
impression   31–35
index formula   362
LED wavelengh   311–320
light source   278
methodology   159
metrics   7, 132
metrics and values   110
motivation and aim   201–204
of light sources   100–114
optimization   305–311
R_1-R_2-G-B_1-B_2-W-LED-configuration   327–333
real room, naturalness and vividness   234–246
research   3
RGB-W-LED configuration   323–327
semantic interpretation and criterion values   268–277
state-of-the-art of   154–159
subjective aspects   5
system   358
systematization   315–320
task of   91
type   31
viewing conditions   142–145
visual aspects   48–49
visual assessments   239
color quality optimization
   for human centric lighting   335–338
Color Quality Scale (CQS)   52
color realness   32
color rendering capacity (CRC)   62
color rendering index   135
   numeric scale   52
color rendering index (CRI R_a)   1, 2
color rendering indices   221
color science   17

color space   26, 27
   CAM02-UCS uniform   46
   state-of-the-art   145–154
color temperature preference, colorful object combinations   225–233
color temperatures   132
color vision
   inter-observer variability   20
   mechanisms   12
   visual perception   11
color vividness   32
colored objects
   state-of-the-art   141–142
colorful object combinations   225–233
colorfulness   26
colorimetric data   99, 112
colorimetric purity   24
colorimetry
   u'–v' diagram   24
   applications   21
   basics of   16
   CCT   22
   chromaticity coordinates   19
   color matching functions and tristimulus values   17
   dominant wavelength   24
   inter-observer variability, of color vision   20
   MacAdam ellipses   24
cone
   density   14
   sensitivity curves   16
   spectral sensitivities   15
   types   15
cone surface area (CSA)   62
Constant current dimming (CCD)   301
cornea   12
correlated color temperature (CCT)   1, 21, 218, 219
   colorful object combinations   225
correlated color temperature (T_{cp})   49
   computation method   51
   correlation calculation   166
CQS Q_a 60
CQS Q_p 61
CQS method   62

CRI2012   53
   calculate   55
   components   53
   computation method   54
   flowchart   54
   principles   53
   RMS formula   55
   test colors   55

d
dimming method   300–302
dominant wavelength   23, 24
double Gaussian model   306
driving electronics   288

e
electronics
   control and regulation   289–290
Energy Star Eligibility Criteria   286

f
Fairchild and Pirrotta's L^{**} model   45
feeling of contrast index (FCI)   62
fluorescent lamps   133
food products   256–268
Fotios and Levermore model   45
fovea   14

g
gamut area index (GAI)   58, 62

h
HCL concept
   visual and non-visual aspects   286
Helmholtz-Kohlrausch effect   41
heterochromatic brightness matching experiment   42
horizontal illuminance   33, 34
hue   26
hue circle diagram   118
human centric lighting (HCL)   3, 364
   aspects of   336
   co-optimizing circadian aspects   344–348
   color quality optimization for   335–338
   design concept for   337
   spectral transmittance   348–354
human eye
   structure   13
human vision mechanisms   42
human visual system   11

i
illuminance values (lx)   210, 211
indoor lighting technology   283
industrial electrification   335
inter-observer variability   78–83
intrinsically photoreceptive retinal ganglion cell (ipRGC)   339–341

j
Judd's flattery index   57–58

l
LED   288
LED light engines
   chroma enhancement   320–323
   R_1-R_2-G-B_1-B_2-W-LED-configuration   327–333
   RGB-W-LED configuration   323–327
LED light sources   2
LED luminaries   283
   in real applications   315
LED-luminaire Lunexo   289
lens   12
light source
   electromagnetic radiation   11
   spectral design   33
   white tone   71
light source spectra   3
light sources
   cold white   184
   color quality   100
   colorimetric properties   216, 220
   relative spectral power distributions   149, 150
   relative spectral radiance   205
   state-of-the-art   132
   warm white   178

light sources emitting   339
lighting dynamic properties   284
lighting engineering   357–364
lighting engineers   3
lighting industry   71
lighting practice   277–280
lighting quality optimization   5
lighting quantities   284
lighting systems   129
lightness   26
LMS cone signals   223, 230
long-term memory colors   30, 33
luminaire's luminous flux   291
luminance channel   16, 41
luminous efficiency function
      ($V(\lambda)$)   16

m
MacAdam ellipses   24
makeup products   246–256
MCRI, *see* memory color rendition
      index (MCRI)   59
melatonin   339–340
memory color quality index (MCRI)
      59, 190
memory colors   157
metameric white light sources   212
modern color metrics   123
multi-input multi-output (MIMO)
      system   303
multi-LED spectrum   348
multidimensional scaling (MDS)   164

n
naturalness   241
   food products   256
   makeup products   246–256
   real room   234
   white light chromaticity   84
nCRI   53
non-uniform color space   51

o
object colors   125
   database   98
oil colors   123

optic disk   13
optic nerve   13
optimization   318, 319
optimization of illuminating systems   3

p
Pearson's correlation coefficients   253
Perceived Adequacy of Illumination
      (PAI)   3
perceived color attributes   26
phases of daylight   22
photometric quantities   16
photoreceptors   14
prediction
   of color preference   241
   of naturalness   241
   of vividness   240
preferred colors   157
pulse width modulation (PWM)   301

r
R_g metric   360
radiometric quantities   16
RCRI   57
reference light source   32, 52
retina   12, 13
rod density   14
root mean square (RMS)   53, 55

s
S-cone signals   354
saturation   26
sclera   12
scotopic vision   14
Solid State Lighting (SSL)   286
solid-state-lighting products   361
spectra of Planckian radiators   111
spectral luminous efficiency function
      $V(\lambda)$ 1
spectral power distribution (SPD)   75,
      213, 302
spectral reflectance
   art paintings   97
   curves   221, 237
   flowers   94
   skin tones   96

spectral sensitivities   15
spider web diagrams   118
suprachiasmatic nucleus (SCN)   339

t

test color samples (TCS)   50–52, 103, 114
thermal management   290–291
Thornton's color preference index   58
tri-stimulus values   17, 145
tungsten halogen lamps   133
tungsten incandescent lamps   133
two lighting situations   220

u

unique white
   chromaticity of   76
   L-M and L+M-S signals   77
   location of   74

v

visual clarity   35, 204, 206, 207
   analysis and modeling   208
   variables   209
visual color quality, *see* color quality
visual experiments   160

visual interval scale variables   238
vitreous body   12
vividness   240
   makeup products   246
   real room   234

w

Ware and Cowan Conversion Factor formula (WCCF)   44
warm white light sources   178
warm white pc-LEDs
   color difference   297
   current depenence   297–298
   temperature dependence   295
WCCF, *see* Ware and Cowan Conversion Factor formula (WCCF)
white light chromaticity   84
white tone
   chromaticity   71–73
   inter-observer variability   78
   perceived brightness   85
   preference   83, 358
   unique white   74
white tone chromaticity matching discrepancy   81
white tone quality   35